String Theory: From Gauge Interactions to Cosmology

NATO Science Series

A Series presenting the results of scientific meetings supported under the NATO Science Programme.

The Series is published by IOS Press, Amsterdam, and Springer (formerly Kluwer Academic Publishers) in conjunction with the NATO Public Diplomacy Division

Sub-Series

I. Life and Behavioural Sciences	IOS Press
II. Mathematics, Physics and Chemistry	Springer (formerly Kluwer Academic Publishers)
III. Computer and Systems Science	IOS Press
IV. Earth and Environmental Sciences	Springer (formerly Kluwer Academic Publishers)

The NATO Science Series continues the series of books published formerly as the NATO ASI Series.

The NATO Science Programme offers support for collaboration in civil science between scientists of countries of the Euro-Atlantic Partnership Council. The types of scientific meeting generally supported are "Advanced Study Institutes" and "Advanced Research Workshops", and the NATO Science Series collects together the results of these meetings. The meetings are co-organized by scientists from NATO countries and scientists from NATO's Partner countries –countries of the CIS and Central and Eastern Europe.

Advanced Study Institutes are high-level tutorial courses offering in-depth study of latest advances in a field.
Advanced Research Workshops are expert meetings aimed at critical assessment of a field, and identification of directions for future action.

As a consequence of the restructuring of the NATO Science Programme in 1999, the NATO Science Series was re-organized to the four sub-series noted above. Please consult the following web sites for information on previous volumes published in the Series.

http://www.nato.int/science
http://www.springer.com
http://www.iospress.nl

Series II: Mathematics, Physics and Chemistry – Vol. 208

String Theory: From Gauge Interactions to Cosmology

edited by

Laurent Baulieu
Université Pierre et Marie Curie,
Paris, France

Jan de Boer
University of Amsterdam,
The Netherlands

Boris Pioline
Université Pierre et Marie Curie,
Paris, France

and

Eliezer Rabinovici
Hebrew University,
Jerusalem, Israel

Published in cooperation with NATO Public Diplomacy Division

Proceedings of the NATO Advanced Study Institute on
String Theory: From Gauge Interactions to Cosmology
Cargèse, France
7–19 June 2004

A C.I.P. Catalogue record for this book is available from the Library of Congress.

ISBN-10 1-4020-3732-5 (PB)
ISBN-13 978-1-4020-3732-0 (PB)
ISBN-10 1-4020-3731-7 (HB)
ISBN-13 978-1-4020-3731-3 (HB)
ISBN-10 1-4020-3733-3 (e-book)
ISBN-13 978-1-4020-3733-7 (e-book)

Published by Springer,
P.O. Box 17, 3300 AA Dordrecht, The Netherlands.

www.springer.com

Printed on acid-free paper

Printed in the Netherlands.

Table of Contents

Preface

The present volume is a collection of the lecture and seminar notes delivered at the Summer School " String Theory : from Gauge Interactions to Cosmology", which was held in Cargèse, France, from June 7 to 19, 2004.

The main focus of the school was on the inter-relations between the fields of superstring theory, cosmology and particle physics, which have been the subject of very active research recently. Much of the recent progress on topical problems in the field was covered, including the duality between gravity and gauge theories, strings and cosmology, critical phenomena in statistical mechanics, topological string theory, physics beyond the standard model, and the landscape of vacua of string theory.

The School was an excellent opportunity for the youngest researchers to establish close relationship with the lecturers and among each other, be it during the vibrant Gong Show session, under the shade of the Wisdom Tree or gazing at the beautiful Corsican coastline. We hope that these proceedings will further serve in fixing the acquired knowledge, and hopefully become a valuable reference for anyone working in this fascinating domain of physics.

It is a pleasure to extend our warm thanks to the NATO Division for Scientific Affairs, under the Advanced Study Institute grant PST.ASI 97 85 17, and to the the Human Factor-Mobility & Marie Curie Actions Division of the European Commission, under the High-Level Scientific Conference grant HPCFCT-2001-00298, whose generous funding have allowed this meeting to take place. NATO's announced discontinuation of its support will be sorely felt in forthcoming events. We also wish to thank all the people who have contributed to the organization of this Conference, especially the staff of the Institut d'Etudes Scientifiques de Cargèse its Director, Elisabeth Dubois-Violette. We are also much indebted to Josette Durin, for her invaluable help in preparing this volume.

Finally, we are very grateful to all the participants of the School for a creating a wonderful stimulating atmosphere, and to the contributors of this volume for their dedication in preparing their notes.

Laurent Baulieu
Jan de Boer
Boris Pioline
Eliezer Rabinovici

Part I

Lectures

STRINGS IN A LANDSCAPE

TOM BANKS
Santa Cruz Institute for Particle Physics, Santa Cruz CA 95064
NHETC, Rutgers U., Piscataway, NJ 08854

Abstract. This is a summary of some of the topics covered at the Cargese Summer School on String Theory in June of 2004. Most space is devoted to a discussion of ideas about the String Landscape, which formed the background for some of the talks and much of the discussion at the School.

1. Introduction

This paper is a synthesis of my summary remarks at the Cargese School, and my contribution to the wisdom tree discussion with Arkani-Hamed and Polchinski. The topics of the school were varied. There was a discussion of the interface between cosmology and Planck/String scale physics, in the talks of Brandenberger, Polchinski, Pioline, Elitzur, and Rabinovici. Polchinski and Arkani-Hamed devoted some of their time to discussion of the string landscape and expectations for fine tuning. Arkani-Hamed discussed his beautiful model of spontaneous breakdown of general coordinate invariance (also known as ghost condensation), as well as the split supersymmetry model motivated by the string theory landscape. Polchinski also talked about the emergence of cosmologically stable cosmic strings in one of the regions of moduli space that are being explored in the landscape. Tseytlin's talks were devoted to the string/gauge theory connection, and in particular to the emergence of integrable structures on both sides of the duality between maximally supersymmetric planar Yang-Mills theory and tree level string theory on $AdS_5 \times S^5$. Dijkgraaf and Ooguri discussed the emergent non-perturbative formulation of topological string theory. Morozov's lectures reviewed some of the older connections between exactly soluble string theories, and matrix models. Okun and Pokorski gave us detailed updates on neutrino and collider phenomenology. Finally, D. Bernard reminded us that conformal field theory still has interesting things to say about real problems in condensed matter physics.

I will confine these paragraphs to brief summaries of cosmology, ghost condensation, and a somewhat more detailed discussion of the debate over the landscape. I begin with cosmology.

L. Baulied et al. (eds.), String Theory: From Gauge Interactions to Cosmology, 3–17.

2. Cosmology

Robert Brandberger gave us a review of the calculation of fluctuations in inflationary cosmologies, with emphasis on the possibility of seeing Transplanckian effects in the cosmic microwave background. I will give you my understanding of this issue, which may not coincide with Robert's. The usual argument for the universality and independence of microscopic physics of the inflationary fluctuation spectrum relies on the adiabatic theorem. Given the assumption that the adiabatic theorem applies, effective field theory arguments show that the corrections to the leading order predictions for fluctuations are of order $(H/M_P)^2$, which usually means they are unobservable. Attempts to produce effects of order H/M_P rely on non-adiabatic initial conditions. Brandenberger showed us models of initial states which do produce such effects, as well as models which don't. String theory does not yet have the tools to express results of stringy cosmologies in terms of the language of field theory in which these results are presented. It is not clear which class of models string theory prefers.

What is clear is that models with many e-folds of inflation, which incorporate violations of the adiabatic theorem (and avoid making the UV more singular than field theory), must be very fine tuned in order to make an observational effect. Such models make large modifications to conventional predictions, only over a limited range of scales. It requires fine tuning to make this range of scales coincide with the range we observe in the CMB.

Elitzur, Pioline and Rabinovici talked about models of string cosmology which use the machinery of world-sheet conformal field theory to investigate singular cosmologies. These models fall into two types: those where an infinite universe undergoes a Big Crunch, which may or may not be followed by a Big Bang (with string theory providing the tool to remove and glue together the singularities) and those with closed universes which start with a Big Bang and end with a Big Crunch. Both types of models are generalized orbifold CFTs, *i.e.* they are gauged σ models, where the gauge group may be continuous or discrete.

In the Crunch-Bang models the correlation functions of string vertex operators have the conventional interpretation of a scattering theory. It has been shown pretty definitively that conventional string perturbation theory breaks down near the singularity. Hopes for progress depend on the notion that the production of string winding modes, which are concentrated near the singularity will smooth out the transition without recourse to Planck scale physics. It is hoped that some modified perturbation expansion will give a clear picture of what is going on near the singularity.

In Bang-Crunch models there are no asymptotic regions in which to define scattering amplitudes. Mathematical computations rely on infinite space-like regions with closed time-like curves, which have been dubbed "whiskers" (actually the Crunch-Bang orbifolds have similar whiskers). I do not understand the physical interpretation of the world sheet correlators in this context, or how they are related to physics in the actual cosmological part of the space-time.

3. Ghost condensation

Arkani-Hamed discussed the theory[1] of ghost condensation, which is amusing and beautiful. He and his collaborators have shown that theories which do not have a stable Lorentz

L. Baulied et al. (eds.), String Theory: From Gauge Interactions to Cosmology, 3–17.

invariant solution can nonetheless have stable solutions which have the symmetries of homogeneous isotropic cosmology. Moreover, because these Lagrangians describe a partial Higgsing of gravity, their low energy behavior is almost universal. It is described by a single new massive parameter f, analogous to the pion decay constant. The theory has a new field ϕ, whose low energy excitations have a non-Lorentz invariant dispersion relation. They can modify known theories in ways which are not yet ruled out, but are potentially observable. In particular, they can produce interesting non-Gaussian fluctuations in the Cosmic Microwave Background. The theory can also account for the acceleration of the universe, without a cosmological constant, although it does not solve the cosmological constant problem.

In order to explain the acceleration, f must be taken very small. The new field in the theory *couples only to gravity*, and for small f its effects on ordinary physics are almost negligible.

There are two problems with this model. The first is that we cannot derive it from string theory in any known way. The second is that the assumption that the new physics couples to the old only through gravity is artificial. Standard effective field theory reasoning leads us to expect non-renormalizable couplings of ϕ to standard model fields, scaled by f. For small values of f this is inconsistent with experiment. It appears to be technically natural to omit the direct couplings of the Goldstone field to matter. If we postulate that it couples only to gravity, then the small value of f/M_P ensures that radiative corrections do not induce large direct couplings of the Goldstone field to the standard model. Nonetheless, one would like to have a principle that explained the absence of direct couplings to the standard model, rather than a simple postulate, however self-consistent.

Neither of these criticisms is grounds for abandoning this very interesting new model. In particular, string theory as we know it is built to handle asymptotically flat and AdS backgrounds, and has notorious problems with any kind of real cosmology. Ghost condensation is an intrinsically cosmological model. It does not have any stable, Lorentz invariant background solutions. So, the inability to find it among solutions to the effective field equations of a known string theory, may be more a problem of the limitations of the present formulation of string theory, than an indication that ghost condensation is an incorrect idea.

The wonderful thing about ghost condensation is that it makes predictions for novel phenomena, which are just beyond the current reach of experiment. It should be studied and tested to the best of our ability.

4. Strings in a landscape

Although there were no talks devoted explicitly to the string landscape at Cargese, it pervaded many of the discussions. Arkani-Hamed's split supersymmetry model was motivated by the landscape, as was Polchinski's discussion of cosmic strings in string theory. And our discussion under the wisdom tree was a debate about the landscape.

Part of the confusion in the debate about the landscape is the conflation of two notions of effective action which occur in quantum field theory. The first is the 1PI effective action, which is an exact summary of the entire content of a field theory. Knowledge of it enables us to construct all of the correlation functions of the theory, in any of its vacuum states. In perturbation theory, we can compute the 1PI action around any vacuum state, and get the same result. It is important to realize that there is NO known analog of the 1PI action in string theory, which applies to all solutions of the theory. For example, both $AdS_5 \times S^5$ and ten dimensional Minkowski space, are solutions of the tree level effective action of Type IIB SUGRA, but higher order corrections (not to mention non-perturbative corrections) to the action which generates the S-matrix in Minkowski space will not be the same as those which generate the correlation functions on the boundary of AdS space. There is no single quantum effective action which we can use to get both the S-matrix and the AdS correlation functions by "shifting the field".

The other concept of effective action in field theory is the low energy or Wilsonian action. This is defined, either in a single vacuum, or in a set of quasi-degenerate vacua whose energy density differences (as well as the heights of the barriers between them) are small compared to some cutoff scale. The Wilsonian action only contains degrees of freedom whose fluctuations are significant at these low energy scales. It is important to avoid using it when the conditions of its validity are violated. For example, in minimally SUSic QCD with $N_F \leq N_C - 1$, the low energy degrees of freedom consist of a meson superfield M and one can compute the exact low energy superpotential of M. The low energy Kahler potential is canonical. If one uses this low energy Lagrangian to compute the energy density of states with a given expectation value of M, this Lagrangian gives a divergent answer at the symmetric point of vanishing eigenvalue. This is not the correct physical answer. The true Kahler potential (in the 1PI sense) of M is modified at the origin of moduli space, and the energy density there is really of order the QCD scale. We should stop paying attention to the predictions of the Wilsonian action when they are outside its range of validity.

In string theory, the effective actions we compute are analogous to Wilsonian actions, but their range of validity is even more constrained. In particular, the stringy derivation of the effective action views it as a tool for calculating boundary correlation functions in a *fixed asymptotic space-time background.* We tend to forget this because, particularly in situations with a lot of SUSY, the leading low energy term in the effective action is independent of the background. This fosters the illusion that different backgrounds can be viewed as *vacuum states of the same theory*. In fact, as emphasized in [2], the italicized phrase is borrowed from quantum field theory, and refers to concepts which depend entirely on the separation between IR and UV physics of that formalism. In string theory/quantum gravity, UV and IR physics are much more intimately entangled, and the concept of different vacuum states of the same underlying string theory Hamiltonian is much more circumscribed. When we have a continuous moduli space of super-Poincare invariant S-matrices, we can do experiments at one value of the moduli, which are sensitive to the S-matrix at other values of the moduli[1]. Note however, that the only Hamiltonian form we have for such models is in light cone frame, where different values of the moduli correspond to different Hamiltonians. Similarly, a moduli space of correlation functions on the boundary of Anti-deSitter space, corresponds to a one parameter set of different Hamiltonians, rather than different superselection sectors of the same Hamiltonian.

The recognition that changes in background asymptotics correspond to changes in the Hamiltonian, rather than changes of superselection sectors for a given Hamiltonian

[1]These experiments are much more difficult than they would be in a SUSic quantum field theory, without gravity.

goes back to [3], and has become commonplace with the advent of AdS/CFT. Changes in non-normalizable modes of bulk fields[2] add relevant terms to the Hamiltonian. Changes in the (negative) cosmological constant correspond to changes in the fixed point which defines the boundary CFT, and thus to a completely different set of high energy degrees of freedom.

These facts lead us[3] to be suspicious of attempts to find new string theory models by patching together an effective potential using the degrees of freedom of *e.g.* Type II string theory in flat space-time. Indeed, once we include gravitational effects, the low energy action itself gives us reason to be suspicious of meta-stable de Sitter vacua constructed in this way.

Under the wisdom tree, Arkani-Hamed gave us an example of a situation where we feel pretty confident that we can reliably compute a Wilsonian effective potential with minima that have positive vacuum energy. Simply take the supersymmetric standard model, with cubic (A-term) soft supersymmetry breaking terms. We get a host of minima with energy density differences of order the SUSY breaking scale. There is no doubt that a TeV scale experimenter could verify the existence of these minima by exciting localized coherent excitations of the squark and slepton fields. Let us however assume that we are living in a Poincare invariant state[4], and see what this potential implies about the geometry of space-time when gravitational effects are taken into account. In order to exhibit a meta-stable dS space, we have to excite a region of order the putative Hubble radius into the meta-stable minimum. Old results of Guth and Farhi[4] show that the external observer can never verify the existence of the inflating region. A black hole forms around it. Any observer in the asymptotically flat region, who tries to jump into the black hole to find inflation, first encounters a singularity. So effective field theory tells us that we can find and explore meta-stable positive energy density minima of an effective potential, but not the meta-stable dS spaces that these minima have been thought to imply. That is, the Guth-Farhi results suggest that *the stable asymptotically flat vacuum state does not have excitations which correspond to meta-stable dS vacua, even when it has an effective potential with positive energy meta-stable minima.* Rather, it has excitations in which fields are excited into meta-stable minima only over regions small compared to the Hubble radius at those minima. The attempt to create larger regions succeeds only in creating black holes.

We see that in gravitational theories, the criterion for the validity of non-gravitational effective field theory reasoning depends on more than just the value of the energy density in Planck units. When the Schwarzschild radius of a region exceeds its physical size in the approximation in which gravity is neglected, a black hole forms. Effective field theory remains valid outside the black hole horizon (if it is large enough), but not inside. In the above example, no external observer can probe the putative dS region, without first encountering a singularity.

We also see that the solutions of the same effective equations of motion may not reside in the same quantum theory.

[2]Note that if we want to make finite rather than infinitesimal changes we must restrict our attention to Breitenlohner-Freedman allowed tachyon fields.

[3]well, at least they lead me

[4]In other words, fine tune the c.c. to be zero in one preferred vacuum state.

The results of Coleman and De Lucia[5], on vacuum tunneling in the presence of gravity, give us a sort of converse to this result. Given the same effective Lagrangian we used in the previous paragraph, we can assume the existence of the meta-stable dS space, and ask what it decays into. Here there is a surprise, particularly for those who constantly repeat the mantra "dS space decays into flat space". In fact, the analytic continuation of the CDL instanton is a negatively curved Friedmann Robertson Walker cosmology, which (if the potential has a stable zero cosmological constant minimum), is asymptotically matter dominated. Although it locally resembles flat space on slices of large cosmological time, its global structure is completely different. In particular, an attempt to set up an asymptotically Minkowskian coordinate system, starting from the local Minkowski frame of some late time observer, inevitably penetrates into regions where the energy density is high (the energy density is constant on slices of constant negative spatial curvature, and is of order the energy density of the false vacuum at early FRW times) .

Thus, both analysis of creation and decay of meta-stable dS states, suggests that if a potential has a stable minimum with vanishing cosmological constant, and another with positive energy density, the Minkowski solution and the meta-stable dS solution are simply not part of the same theory. There remains a possibility of the existence of a theory with a stable, matter dominated, FRW cosmological solution with a meta-stable dS excitation. The problem with this is that it is very unlikely that we can make a reliable exploration of this scenario within the realm of low energy effective field theory. The CDL instanton solution is non-singular. The $a = 0$ point of the FRW coordinate system is just a coordinate singularity marking the boundary between the FRW region and a region of the space-time which continues to inflate. However, arbitrary homogeneous perturbations of the CDL solution have curvature singularities. Further, there is a large class of localized perturbations which evolve to Big Crunch singularities, rather than passing smoothly through the $a = 0$ point. For example, consider a localized perturbation on some hyperbolic time slice a finite proper time prior to $a = 0$. Let it be homogeneous in a large enough region that signals from its inhomogeneous tail cannot propagate to the $a = 0$ point. Then we will have a singularity. I will mention a more generic example below.

If we re-examine the Guth-Farhi argument in the FRW context, we see that it continues to hold until we go back in time to a point where the cosmic energy density is of order the barrier to the meta-stable minimum. At high enough energy density, there are FRW solutions in which the field classically evolves into the meta-stable minimum. It will then decay by tunneling, into the FRW continuation of the CDL instanton. Generic FRW solutions (including arbitrary homogeneous perturbations of the CDL instanton) have curvature singularities at a finite cosmic time in the past. To establish their existence as genuine theories of quantum gravity one must go beyond effective field theory, and probably beyond perturbative string theory.

Freivogel and Susskind[6] have suggested a scattering theory in which asymptotic states are associated with incoming and outgoing wave perturbations of the nonsingular, time symmetric Lorentzian continuations of the, CDL instantons for the various meta-stable vacua of string theory. They claim that in this framework, the breakdown of effective field theory is avoided, as long as the effective potential is everywhere smaller than Planck scale. I find this suggestion interesting, but it is not based on reliable calculations. If one considers black holes with radius larger than the dS radius, formed in the remote past of the FRW part of the time symmetric Lorentzian CDL geometry, it is hard to see how these solutions asymptote to the future CDL geometry without encountering a singularity. Near the $a = 0$ point where the FRW coordinates on the CDL solution become singular, most of the space-time is isometric to dS space at its minimal radius. If we have formed a black hole in the past, with radius much larger than this, then the entire space-time must end in a singularity.

Thus, I would claim that unlike asymptotically flat or AdS space-times, we have no reliable effective field theory argument that there are an infinite number of states in space-times that asymptote to the time symmetric Lorentzian CDL instanton. An infinite number of states is a minimal requirement for the existence of a quantum theory that can make precise mathematical predictions, which can be self consistently measured in the theory. We must understand the nature of the singularities in these perturbations of the CDL instanton geometry before we can conclude that the framework makes sense.

I think it is more likely, that meta-stable dS vacua exist only in the context of a Big Bang cosmology. If we consider the problem of accessing a meta-stable dS minimum of an effective potential, at times when the cosmic energy density is of order the barrier height of the potential, then the Guth-Farhi problem does not appear to exist. Starting from a Big Bang singularity, one can find homogeneous solutions where a scalar field wanders over its potential surface at a time when the energy is higher than the barriers between minima, and then settles in to a meta-stable minimum with positive cosmological constant. One must understand the Big Bang to make a reliable theory of such a situation, but apart from that the solution is non-singular. In particular, the problem of large black holes in the initial state, is not present for this situation. The difficulties are all associated with understanding the Big Bang singularity.

To summarize, it is clear from semi-classical calculations alone, that the concept of a vacuum state associated with a point in scalar field space is not a valid one in theories of quantum gravity. A given low energy effective field theory may have different solutions which do not have anything to do with each other in the quantum theory. One solution may be a classical approximation to a well defined quantum theory, while the other is not. It seems likely that the context in which we will have to investigate the existence of non-existence of the landscape is Big Bang cosmology. This is the only situation in which we can reliably construct a universe which gets stuck in a meta-stable dS minimum.

Thus, I claim that if string theory really has a multitude of meta-stable dS states, then exact theory into which they fit is a theory of a Big Bang universe which temporarily gets stuck, with some probability, in each of these states. This is ultimately followed by decay to negatively curved FRW universes. These FRW universes have four infinite dimensions and 6 or 7 large compact dimensions, which are expanding to infinity. It is clear that the probability for finding a particular dS vacuum is partly determined by the density matrix at the Big Bang and not just by counting arguments. In the next subsection I will describe existing proposals for the cosmological distribution of vacua. My main point here is that the nature of the Big Bang will have to be addressed before we can hope to understand the correct statistics of stringy vacua.

4.1. ETERNAL INFLATION

The string landscape seems to fit in well with older ideas which go under the name of *eternal inflation, the self reproducing inflationary universe, etc.*. The simplest model which exhibits this sort of behavior is one with a single scalar field with two minima, one with positive vacuum energy and the other with vanishing energy. One considers the expanding branch of the meta-stable dS universe with the field in the false minimum. In the quantum field theory approximation this seems to produce an ever-expanding region of space and one allows the dS space to decay by CDL bubble formation independently in each horizon volume. In the eternal inflation picture, one tries to interpret the result as a single classical space-time. One obtains a Penrose diagram with a future space-like boundary. The future space-like boundary is fractal, with regions corresponding to singularities[5] as well as FRW asymptotics, interspersed in a causally disconnected way. In the landscape there will be many different singular regions of the boundary, as well as many different FRW regions. Advocates of the landscape/eternal inflation picture then make the analogy to maps of the observable universe, with different causally disconnected regions being the analogs of different planets. Physics it is said, depends on "where you live" and organisms like ourselves can only live in certain regions of the map. A key difference between different minima of the effective potential in eternal inflation, and different planets is that *we cannot, even in principle, communicate with causally disconnected regions of the universe.*

How is one to interpret such a picture in terms of conventional quantum mechanics? I believe that the fundamental clue comes from the principle of Black Hole Complementarity[7]. Black holes also present us with two regions of space-time which are causally disconnected. Hawking showed long ago that this was an artifact of the semi-classical approximation, and that black holes return their energy to the external space-time in which they are embedded. If we assume that the region behind the horizon has independent degrees of freedom, commuting with those in the external space, then we are confronted by the information loss paradox.

String theorists have believed for some years now, that this is not the case. The principle of Black Hole Complementarity is the statement that the observables behind the horizon do not commute with those in the external space-time. For a large black hole, (and for a long but finite time as measured by the infalling observer), these two sets of observables are both individually well described by semiclassical approximations, but the two descriptions are not compatible with each other.

Fischler and I tried to relate this principle to the *Problem of Time*[8]. In the semi-classical quantization of gravity one attempts to solve the Wheeler-DeWitt equation

$$\mathcal{H}\Psi = 0$$

with an ansatz

$$\Psi = e^{iS}\chi(t,\phi),$$

where S is the action of some classical space-time background solution, and χ is the wave functional of a quantum field theory in this space-time

$$i\partial_t\chi = H(t)\chi.$$

[5]The singular regions correspond to decays to parts of the potential where the vacuum energy is negative. They would exist in the landscape context, but not in the simple model we are discussing.

The (generally time dependent) Hamiltonian $H(t)$ depends both on the choice of classical background, and on the particular time slicing chosen for that background. For example, for the Schwarzschild background we could choose H_{Sch}, the Schwarzschild Hamiltonian, or some time dependent $H(t)$ where t is the proper time of a family of in-falling observers. Even ignoring subtle questions of whether these two Hamiltonian evolutions act on the same Hilbert space, it is clear that they are different and that $[H(t), H_{Schw}] \neq 0$ at *any* time t. It is therefore not surprising that the semi-classical observables of different observers do not commute with each other.

Given this description of black holes, it is natural to conjecture that a similar phenomenon occurs for any space-time with horizons. In [8] this was called Cosmological Complementarity for asymptotically dS spaces. E. Verlinde has suggested the name Observer Complementarity for the general case.

If we apply this logic to Eternal Inflation, we obtain a picture quite different from the original description of these space-times. We simply associate a single Hilbert space and many different (generally time dependent) Hamiltonians to the fractal Penrose diagram. Each Hamiltonian is associated with the causal patch of a given observer. Mathematically, the situation can be equally well described by saying we have a collection of different theories of the universe. There is a philosophical cachet, associated with the phrase, "physics depends on where you live in the multiverse", which is absent from this alternative way of describing the physics.

In the formalism described in [6] this is almost precisely what is conjectured. For each meta-stable dS point L in the landscape, which can decay into the Dine-Seiberg region of moduli space, and for each (typically 10 or 11 dimensional) Super-Poincare invariant solution V of string theory into which L can decay there is a different unitary S-matrix $S_{L,V}$[6]. It is claimed that each of these S-matrices contains all the physics of the landscape, because there is a canonical way to compute the unitary equivalence U in the formula $S_{L,V} = U_{L,V,L',V'} S_{L'V'} U^{\dagger}_{L,V,L',V'}$. The statement that there is a theory of eternal inflation in which all of the points L are meta-stable states is really the statement that the theory contains a canonical algorithm for computing the "gauge transformations" U[7]. The authors of [6] claim that all of the meta-stable dS states will show up as resonances in every S-matrix, $S_{V,L}$. This claim is plausible if the S-matrices are indeed related by unitary conjugation. The spectrum of the S-matrix is then gauge invariant. In ordinary scattering theory, time delays, which are related to resonance lifetimes, are related to the spectrum of the S-matrix. In the eternal inflation context, there is no universal notion of time for the different asymptotic states, so more work is necessary to understand these concepts.

If indeed the information about each meta-stable dS vacuum can be extracted from the spectral density of a given S-matrix $S_{V,L}$, and if we can find a reliable framework, for defining and calculating these S-matrices, then the landscape will have a mathematical

[6]We can also consider initial and final states corresponding to different CDL instantons. In[6] these are claimed to be different gauge copies of the same information in the S matrices. I will mention a different interpretation below.

[7]For the moment, no approximate statement of what this algorithm is has been proposed. I would conjecture that, if the formalism makes sense, the transition amplitudes between two different FRW spacetimes, mentioned in the previous footnote, provide the algorithm for calculating the U mappings.

definition. From a practical point of view however, we just say that string theory gives us an algorithm for constructing models of the world (in the landscape context this means choosing particular stable or meta-stable minima) which is not unique and that we are trying to use data to constrain which model we choose. In the next section, I will describe the situation in the language of the approximate Hamiltonian of a given meta-stable dS observer, rather than that of eternal inflation.

I have emphasized the problems with this S-matrix point of view and suggested the alternative notion that the landscape could only make sense in the context of Big Bang Cosmology. One must thus understand how to describe the initial states. There are two possibilities, either there is some principle which picks out a fixed initial state[8] at the Big Bang, or there is a generalized S-matrix in which we relate a particular state at the Big Bang to a particular linear combination of final scattering states in one of the FRW backgrounds defined by a decaying meta-stable dS space. In a manner analogous to Freivogel and Susskind, one would conjecture that the descriptions in terms of different future FRW backgrounds would be unitarily equivalent to each other, by unitary transformations which do not respect locality. Since we are unlikely to have much control over the initial conditions at the Big Bang, one should choose the in-state at the Big Bang to be a high entropy density matrix. Thus, the practical difference between the two proposals is the entropy of the initial state.

I know of two proposals for the initial density matrix at the Big Bang, which might lead to a set of cosmological selection rules for meta-stable points in the landscape. The first, *modular cosmology*[9] postulates an early era in which the universe can be described semi-classically, but the potential on moduli space is smaller than the total energy density. The metric on moduli space then provides a finite volume measure. Furthermore, the motion of the moduli is chaotic. These facts suggest that the probability of finding the universe in a given meta-stable minimum of the potential is the volume of the basin of attraction of that minimum, divided by the volume of moduli space.

Holographic cosmology[10] gives an alternative view of the initial state of the universe, as a "dense black hole fluid" where standard notions of local field theory do not apply. The model contains two phenomenological parameters, which govern the transition between this phase of the universe and a normal phase in which the field theory description is valid. There are indications that the transition occurs at an energy density well below the unification scale of standard model couplings. We might then expect a transition directly into a state with most of the moduli frozen. In order for the model to provide an adequate account of the fluctuations in the CMB, one must have at least one "active" modulus at these low energies, which can provide for a modest number of e-folds of inflation. In such a model, minima of the potential on moduli space with energy higher than the scale at which a field theoretic description of the universe is possible, cannot make any sense. At best a small class of low energy minima could be compatible with holographic cosmology[9]. If holographic cosmology is compatible with landscape ideas, the probability of accessing a particular minimum will be determined by quantum grav-

[8] *e.g.* the Hartle Hawking Wave Function of the Universe.

[9] The potentials calculated in the landscape have no indication of a cut-off at energy scales far below the unification scale. This suggests that the two theoretical frameworks are not compatible, but I am trying to avoid jumping to conclusions.

itational considerations, far removed from effective field theory. In this framework, the dense black hole fluid is stable and is the most probable state of the universe. A normal universe like our own is determined by a somewhat improbable initial condition, but one expects the maximum entropy initial state that does not collapse to a dense black hole fluid. The survival probability of a given normal state depends on both the properties of the black hole fluid and the low energy physics of the normal state, so the determination of the most probable meta-stable minimum would be a complicated quantum gravitational calculation.

In both of these classes of models, simple enumeration of meta-stable minimum is not a good account of the physical probability distributions.

5. Phenomenology of the landscape

For practical purposes , the landscape gives us a large set of alternative effective Lagrangians for describing the physics we have observed or will observe in our universe. These are parametrized by a collection of numbers, which include the dimension of space-time, the name, rank and representation content of the low energy gauge theory, the value of the cosmological constant, and the values of all the coupling constants and masses of fields in the Lagrangian[10]. These numbers can be collected together and viewed as a multidimensional probability space. In the supergravity approximation, we have a way of calculating an *a priori* distribution for these numbers. Proponents of the landscape would claim that this is an approximation to some more exact distribution, though no-one has suggested a procedure for calculating the corrections. In the previous section I have suggested that early universe cosmology may make important modifications of the distribution of metastable minima. If it turned out that the distribution predicted the Lagrangian we observe with high probability, it would be a great triumph for string theory.

The value of the cosmological constant tells us that this is not the case. Weinberg's bound[11],which constrains cosmological parameters by insisting on the existence of galaxies, has the form (in Planck units).

$$\Lambda \leq K\rho_0 Q^3,$$

where ρ_0 is the dark matter density at the beginning of the matter dominated era, Q is the amplitude of primordial density perturbations at horizon crossing, and K is a pure number of order 1.. For any reasonable values of the other parameters, this means Λ is much smaller than the typical value found in the landscape.

It is clear then that we must supply additional data from experiment in order to fix our description of the world. The landscape framework supplies some theoretical guidance - it tells us that there are a finite number of possibilities, of order 10^{10^2} (to order of magnitude accuracy in the logarithm). Various authors [12] have begun to investigate the *a priori* distribution of properties like the gauge group and number of generations, the scale of supersymmetry breaking, the existence of large warp factors which give rise to large hierarchies of energy scales *etc.*, assuming a uniform distribution on the space of

[10]In principle we could also have non-trivial conformal field theories in the low energy world, at least in some approximation.

minima. The hope is that correlations will become evident which will tell us that a small number of inputs is enough to extract the Lagrangian of the world we live in from the ensemble of Lagrangians the landscape presents us with.

The anthropic principle has also been invoked as an input datum to impose on the ensemble of Lagrangians. For its proponents, the attraction of this principle is that the answer to a single yes/no question, "Is there carbon based life?" puts strong constraints on a collection of parameters in the Lagrangian (assuming all others fixed at their real world values). This attraction may be an illusion. In a probability space, the characteristic function of any subset of data points is a single yes/no question. So physicists must ask if there is any special merit to the particular characteristic function chosen by the anthropic principle. This is a hard question to answer, because we do not have much theoretical understanding of life and intelligence, and we have no experimental evidence about other forms of life in the universe we inhabit.

If the typical life form resembles the great red spot on Jupiter rather than us, then this life form would think that the criterion that is most appropriate to apply in our universe is the Redspotthropic principle. To put this in a more positive manner: if considerations of carbon based life lead to explanations of the values of the fundamental parameters, then we are making a prediction about the typical form of life that our descendants will find when they explore the universe. It should look just like us, or at least be sufficiently similar that the criteria for its existence are close to the criteria for ours. If instead, our descendants' explorations show that the typical life form could tolerate much larger variations in the fundamental constants, then our so-called explanations would really be a fine-tuning puzzle. The Red Spot People could calculate and understand that *we* wouldn't be there if the up quark mass were a little bigger, but they might reasonably ask "Who ordered them?".

Of all *soi disant* anthropic arguments, Weinberg's bound on cosmological parameters is the least susceptible to this kind of criticism. If there are no galaxies, there are no planets, no Red Spots, no Black Clouds, perhaps no conceivable form of life. Polchinski and others at the school cited the numerical success of this bound as evidence that anthropic reasoning may have relevance to the real world. It is important to realize that this numerical success depends on keeping all other parameters fixed. Arkani-Hamed reported on unpublished work with Dimopoulos, which showed that if both Λ and M_P (really the ratio of these parameters to particle physics scales, which are held fixed) are allowed to vary subject only to anthropic constraints, then the preferred value of Λ is larger than experimental bounds by many orders of magnitude. Similarly the authors of [31] following [14] argued that if both Λ and Q are allowed to vary then the probability of finding a universe like our own is of order 10^{-4}. A contrary result was reported in [15], but only by assuming an *a priori* probability distribution that favored small values of Q.

In inflationary models of primordial fluctuations, the value of Q depends on details of the inflaton potential at the end of slow roll. We would certainly expect this parameter to vary as we jump around the landscape. Similar remarks apply to ρ_0. Allowing ρ_0 to vary would further reduce the probability that the anthropic distribution favors the real world. Thus, at least with our current knowledge of the landscape, it seems likely that the numerical success of Weinberg's bound in the landscape context is not terribly impressive[11].

[11] I cannot resist remarking that in the context of Cosmological Supersymmetry Breaking[16],

The greatest challenge to all methods of dealing with the landscape is the large number of parameters in the standard model which have to be finely tuned to satisfy experimental constraints. These include the strength of baryon, lepton and flavor violating couplings, θ_{QCD} and the values of many quark and lepton masses. Anthropic reasoning helps with some of these parameters, but not all, and is insufficient to explain the lifetime of the proton, the value of θ_{QCD} and many parameters involving the second and third generation quarks and leptons. From the landscape point of view, the best way to deal with this (in my opinion) is to find classes of vacua in which all of these fine tuning problems are solved, perhaps by symmetries. One can then ask whether there are enough vacua left to solve the cosmological constant problem. This might be a relatively easy task. One could then go on to see whether other features of this class of vacua are in concordance with the real world.

It is clear that at a certain point in this process, if we don't falsify the landscape easily[12], we will run into the problem that current technology does not allow one to calculate the low energy parameters with any degree of precision. Indeed, the error estimates are only guesses because we don't even know in principle how to calculate the next term in the expansion in large fluxes. A more fundamental framework for the discussion of the landscape is a practical necessity as well as a question of principle. I have suggested that if a rigorous framework for the landscape exists, it is probably to be found in the context of a theory of a Big Bang universe with Eternal Inflation, and future FRW asymptotics for any given observer. It is likely that we will be unable to define more precise calculations of the properties of the landscape without finding a rigorous mathematical definition of such a space-time.

The fundamental object in such a space-time would be a scattering matrix[8] relating a complete set of states at the Big Bang to states in Lorentzian CDL bubble space-times corresponding to decays of meta-stable dS landscape states into the Dine-Seiberg region of moduli space. The final states, in addition to particle labels, would carry indices ,(V, L) describing a particular dS minimum L and a particular "asymptotic vacuum", V, into which it decays. Thus, we would have matrix elements

$$S(I|V, L, p_i),$$

where I labels an initial state at the Big Bang and p_i a set of "particle" labels for localized scattering states in a given CDL bubble. An important unanswered question is whether the S-matrix is unitary for each (L, V)[13] or only when all (L, V) sectors are taken into account. The first alternative is analogous to the proposal of [6].

The following argument has a bearing on this question. I have stated above that there was no problem with an infinite number of final states for fixed L. This is not necessarily the case. If I extrapolate scattering data on $\mathcal{I}_+$ backwards, using the classical equations of motion, and assuming a minimal finite energy for each particle, then all but a finite number of states will encounter a space-like singularity before transition to the metastable

only Λ varies, and Weinberg's bound retains its original numerical status.

[12] *e.g.* by showing that the number of vacua left after all the other fine tuning problems are solved is too small to solve the cosmological constant problem.

[13] One would then invoke the existence of unitary mappings taking the different unitary S-matrices into each other.

dS regime. This is the time reverse of the argument about black holes in the initial state of the time symmetric CDL bubble. This suggests that for fixed L, the matrix $S(I|V.L, p_i)$ has finite rank: only a finite subspace of the space of all out states on the CDL geometry labeled by L, V would be allowed. This leads to a modification of the proposal of [6] in which only the S-matrix for fixed V, keeping all possible values of L, is unitary. However, there is no clear reason now to assume that V should be fixed, so perhaps only the full S-matrix is unitary.

A more disturbing conclusion is reached if one combines the claim of [17] that the number of L sectors is finite, with the above argument. One then concludes that the whole S-matrix has finite rank, and that the entire landscape fits into a Hilbert space with a finite number of states. One of the supposed virtues of the landscape picture of metastable dS was that, unlike a stable dS space, the landscape was part of a system with an infinite number of states, which could make infinitely precise quantum measurements on itself. If both Douglas' claim, and that of the last paragraph are true, this is no longer obvious. All but a finite number of the final states in a given CDL instanton geometry, would not connect to a tunneling process from a meta-stable dS vacuum, but instead would evolve directly from a Big Bang. The part of the scattering matrix that involved meta-stable dS resonances would be of finite rank. The whole issue of a rigorous framework for the landscape remains as murky as ever.

Acknowledgements

I would like to thank L. Baulieu for inviting me to the Cargese workshop, and Nima Arkani-Hamed for conversations which contributed to my understanding of the issues discussed here.

This research was supported in part by DOE grant number DE-FG03-92ER40689.

References

1. N. Arkani-Hamed, H-C. Cheng, M.A. Luty, *Ghost Condensation and a Consistent Infrared Modification of Gravity*, JHEP 0405:074,2004, hep-th/0312099.
2. T. Banks, *A Critique of Pure String Theory: Heterodox Opinions of Diverse Dimensions*, hep-th/0306074.
3. N. Seiberg, S. Shenker, *A Note on Background (In)dependence*, Phys. Rev. D45, 4581, (1992), hep-th/9201017.
4. E. Farhi, A. Guth, *An Obstacle to Creating a Universe in the Laboratory*, Phys. Lett. B183, 149, (1987).
5. S. Coleman, F. DeLuccia, *Gravitational Effects On and Of Vacuum Decay*, Phys. Rev. D21, 3305, (1980).
6. B. Freivogel, L. Susskind, *A Framework for the Landscape*, hep-th/0408133.
7. C.R. Stephens, G. 't Hooft, B.F. Whiting, *Black Hole Evaporation Without Information Loss*, Class. Quant. Grav. 11:621, (1994); L. Susskind, L. Thorlacius, J. Uglum, *The Stretched Horizon and Black Hole Complementarity*, Phys. Rev. D48, 3743, (1993).
8. T. Banks, W. Fischler, *M-theory Observables for Cosmological Space-times*, hep-th/0102077.
9. G. Moore, J. Horne, *Chaotic Coupling Constants*, Nucl. Phys. B432, 109, (1994);T. Banks, M. Berkooz, G. Moore, S. Shenker, P.J. Steinhardt, *Modular Cosmology*, Phys. Rev. D52, 3548, (1995).
10. T. Banks, W. Fischler, *An Holographic Cosmology*, hep-th/0111142; *Holographic Cosmology 3.0*, hep-th/0310288; T. Banks, W. Fischler, L. Mannelli, *Microscopic Quantum Mechanics of*

the $p = \rho$*Universe*, hep-th/0408076.
11. S. Weinberg, *Anthropic Bound on the Cosmological Constant*, Phys. Rev. Lett. 59, 2607, (1987).
12. M.R. Douglas, *Statistics of String Vacua*, hep-ph/0401004; F. Denef, M.R. Douglas, *Distributions of Flux Vacua*, JHEP 0405:072, 2004, hep-th/0404116; F. Denef, M.R. Douglas, B. Florea, *Building a Better Racetrack*, JHEP 0406:034, 2004, hep-th/0404257; M.R. Douglas *Statistical Analysis of the Supersymmetry Breaking Scale*, hep-th/0405279; *Basic Results in Vacuum Statistics*, hep-th/0409207; A. Giryvaets, S. Kachru, P.K. Tripathy, *On the Taxonomy of Flux Vacua*, JHEP 0408:002,2004, hep-th/0404243; O. DeWolfe, A. Giryvaets, S. Kachru, W. Taylor, *Enumerating Flux Vacua With Enhanced Symmetries*, hep-th/0411061
13. M.L. Graesser, S.D.H. Hsu, A. Jenkins,M. Wise, *Anthropic Distribution for Cosmological Constant and Primordial Density Perturbations*, Phys. Lett. B600, 15, (2004).
14. M. Rees, M. Tegmark, *Why is the CMB Fluctuation Level* 10^{-5}, Ap. J. 499, 526, (1998).
15. L. Pogosian, M. Tegmark, A. Vilenkin, *Anthropic Predictions for Vacuum Energy and Neutrino Masses*, JCAP 0407:005, 2004, astro-ph/0404497.
16. T. Banks, *Cosmological Breaking of Supersymmetry?*, hep-th/0007146; and Int. J. Mod. Phys. A16, 910, (2001).
17. M.R. Douglas, *The Statistics of String/M-theory Vacua*, JHEP 0305:046,2003, hep-th/0303194.

QUANTIZATION OF HOLOMORPHIC FORMS AND $\mathcal{N} = 1$ SUPERSYMMETRY ON SPECIAL MANIFOLDS

LAURENT BAULIEU
Laboratoire de Physique Théorique et Hautes Energies,
Université de Paris VI and Paris VII, France

Abstract: We study the quantization of a holomorphic two–form coupled to a Yang–Mills field on special manifolds in various dimensions, and we show that it yields twisted supersymmetric theories. For Kähler manifolds in four dimensions, our topological model is related to $\mathcal{N} = 1$ Super Yang–Mills theory. Extended supersymmetries are recovered by considering the coupling with chiral multiplets.

1. Introduction

The idea that Poincaré supersymmetry is a "phase" of a more fundamental symmetry is appealing. For instance, it was shown that Poincaré supersymmetry and topological symmetry are deeply related, and that the field spectrum of dimensionally reduced $\mathcal{N} = 1$ $D = 11$ supergravity can be determined in the context of an 8-dimensional gravitational Topological Quantum Field Theory (TQFT) [1]. One can foresee that many models, which are dimensional reductions and truncations of maximal supergravity, might be possibly related by twist to topological models [2]. The BRST operator that characterizes a topological symmetry is a scalar operator which can be defined in any given curved space, while Poincaré supersymmetry is a delicate concept in curved space. Therefore, topological symmetry could be a more fundamental concept than Poincaré supersymmetry. On the other hand, in order to perform the twist operation that relates Poincaré supersymmetry and topological symmetry, one often needs to use manifolds with special holonomy.

We will discuss the quantization of (holomorphic) two–forms coupled to a Yang-Mills field on special manifolds in various dimensions. These theories are basically ATQFT's (Almost Topological Quantum Field Theories), in the sense that they are defined in terms of a classical action and a set of observables which are invariant under changes of coordinates belonging to restricted classes, for instance, reparametrizations that respect a complex structure. This is to be compared to genuine TQFT that contain observables invariant under all possible changes of metrics. Interesting cases that we will analyse in detail are Kähler manifolds in four dimensions and special manifolds in higher (6,7 and

L. Baulied et al. (eds.), String Theory: From Gauge Interactions to Cosmology, 19–39.

8) dimensions. In particular in seven dimensions we will analyse G_2 manifolds of the kind recently studied by Hitchin [15].

One of the original motivations for this work was to try to understand how (twisted) $N = 1$ supersymmetric theories can be directly constructed as TQFT. As we will see in the next section, this immediately leads to the introduction in the classical action of a "charged" 2-form B, valued in the adjoint representation of a Lie algebra. In these models one can also consider the coupling to chiral multiplets. If these transforms in the adjoint representation, one recover in this way also the extended supersymmetry in a twisted form.

In Sect.2 we introduce the holomorphic BF model and discuss its relationship with $N = 1$ (twisted) supersymmetry. Notice that the quantization of this model requires the use of the Batalin–Vilkoviski formalism. In Sect.3 the four dimensional case is considered, including a detailed discussion on the coupling with a chiral multiplet. In Sect.4 we discuss the six–dimensional case on a Calabi–Yau three–fold and show how two different quantizations yields respectively to theories related to the B model and A model of topological string. In Sect.5 we discuss the eight–dimensional theory on a Calabi–Yau four–fold and its dimensional reduction to $\mathrm{CY}_3 \times S^1$. The eight–dimensional model is discussed also for manifolds with $SU(4)$ structure. This theory can be regarded as a generalization of the four–dimensional self–dual Yang–Mills model [22].

2. $N = 1$ supersymmetry and the holomorphic BF theory

The standard construction of a TQFT leads to models with $N = 2$ supersymmetry. To see this, let us consider the "prototype" case of Topological Yang–Mills theory in four and eight dimensions. The relevant BRST transformations read

$$\begin{array}{rclrcl} \delta A_\mu & = & \Psi_\mu + D_\mu c & \delta \Psi_\mu & = & D_\mu \Phi - [c, \Psi_\mu] \\ \delta c & = & -\Phi - \frac{1}{2}[c, c] & \delta \Phi & = & -[c, \Phi] \end{array} \tag{2.1}$$

These equations stand for the geometrical identity $(\delta + d)(A + c) + \frac{1}{2}[A + c, A + c] = F + \Psi + \Phi$ [23]. There are as many components in the topological ghosts as in the gauge fields, and to gauge fix the topological freedom, one must also introduce as many antighosts as topological ghosts. The antighosts are an anticommuting antiself dual 2-form $\kappa_{\mu\nu}$ and an anticommuting scalar η. For each one of the antighosts, there is an associated Lagrange multiplier field, and their BRST equations are :

$$\begin{array}{rclrcl} \delta \kappa_{\mu\nu} & = & b_{\mu\nu} - [c, \kappa_{\mu\nu}] & \delta b_{\mu\nu} & = & -[c, b_{\mu\nu}] \\ \delta \bar{\Phi} & = & \eta - [c, \bar{\Phi}] & \delta \eta & = & [c, \eta] \end{array} \tag{2.2}$$

The twist operation is a mapping from these ghost and antighost fermionic degrees of freedom on a pair of spinors, which leads one to reconstruct the spinor spectrum of $N = 2$ supersymmetry, both in 4 and 8 dimensions. The scalar BRST operator δ can then be identified as a Lorentz scalar combination of the $N = 2$ Poincaré supersymmetry generators. However, this "twist" operation has different geometrical interpretation in 4 and 8 dimensions. In the former case, it is a redefinition of the Euclidean Lorentz group contained in the global $SU_L(2) \times SU_R(2) \times SU(2)$ invariance of the supersymmetric theory. In the latter case, it uses the triality of 8-dimensional space. In the previous works

[4, 5, 6], a constant covariant spinor has been used, which implies that one uses $Spin(7)$ invariant manifolds; one can also use a manifold with $SU(4)$ holonomy.

Using self-duality equations as gauge functions, one can build a δ-exact action that provides twisted supersymmetric theories with a δ-exact energy momentum tensor. The cohomology of the δ symmetry determines therefore a ring of topological observables, which is a subsector of the familiar set of observables for the gauge particles. The latter is selected from the cohomology of the ordinary gauge invariance.

From (2.1) and (2.2) one concludes that in these TQFT one has twice as many fermionic degrees of freedom than bosonic ones. This makes seemingly impossible to determine $N = 1$ models as a twist of TQFTs. We may, however, look for models with a "milder" topological symmetry, such as BF models or Chern–Simons type model, characterized by metric independent classical actions. Such actions are not boundary terms, and thus their topological symmetry cannot be as large as that displayed in (2.1) and (2.2). This might lead us to models that are twisted $N = 1$ supersymmetric theories. In this paper, we will consider the following holomorphic BF action

$$I_{n-BF} = \int_{M_{2n}} \mathrm{Tr}(B_{n,n-2} \wedge F_{0,2}) \tag{2.3}$$

which is defined on any complex manifold M of complex dimension n. Some aspects of the classical action (2.3) were studied in [24]. It would be interesting to study its quantization and possible relation with supersymmetry. For the moment we only consider some particular models which can be obtained from (2.3) by choosing a particular form for the field $B_{n,n-2}$. Notice that the equations of motion for this field coming from (2.3) implies that $B_{n,n-2}$ is a holomorphic $(n, n-2)$ form.

One might try to define a theory that is classically invariant under the following (almost) topological symmetry, which is localised in the holomorphic sector [1]:

$$\begin{array}{ll} QA_m = \Psi_m + D_m c & Q\Psi_m = -[c, \Psi_m] \\ Qc = -\frac{1}{2}[c,c] & QA_{\bar{m}} = D_{\bar{m}} c \end{array} \tag{2.4}$$

This "heterotic" symmetry was already used [26, ?, ?, ?] in four dimensions and in [4, ?] in higher dimensions. Here we recover it as a symmetry associated to the classical action (2.3). If we count the ghost degrees of freedom, we have 2 components for Ψ_m and one for c. Notice that Ψ_m cannot have a ghost of ghost symmetry with a ghost of ghost Φ, since $Q\Psi_m = D_m\Phi - [c, \Psi_m]$ and $Qc = \Phi - \frac{1}{2}[c, c]$ would imply that $Q^2 A_{\bar{m}} \neq 0$. Modulo gauge transformations, only one degree of freedom for the field A is left free by the symmetry in (2.4). Moreover, if we succeed in writing a BRST gauge–fixed action for the classical symmetry in (2.4), this will depend on the ghosts Ψ_m and as many antighost components as there are in Ψ_m (four components). Then the number of fermion degrees of freedom will fit with those of a single Majorana spinor and we have a chance to eventually reach $N = 1$ supersymmetry, as we will explain in detail in the next section. Notice that in these models one can also recover the coupling to a chiral multiplet in the adjoint representation and the corresponding extended supersymmetry in the twisted form.

[1] We use the standard notation where the complex indices are denoted with latin letters m, n and $\bar{m}, \bar{n}$, and the complex coordinates are given by z^m and $\bar{z}^{\bar{m}}$.

3. Four dimensions: Kähler manifold

3.1. THE CLASSICAL ACTION FOR A BF SYSTEM ON A KÄHLER MANIFOLD

On a Kähler manifold one can define a complex structure

$$J^{mn} = 0\,, \qquad J^{\bar{m}\bar{n}} = 0\,, \qquad J^{m\bar{n}} = ig^{m\bar{n}} \tag{3.1}$$

which allows one to introduce complex coordinates z^m and $z^{\bar{m}}$ and $1 \leq m, \bar{m} \leq N$ in 2N dimensions by

$$J^m_n z^n = iz^m\,, \qquad J^{\bar{m}}_{\bar{n}} z^{\bar{n}} = -iz^{\bar{m}}\,. \tag{3.2}$$

In four dimensions, the action (2.3) reads

$$I_{cl}(A,B) = \int_{M_4} \mathrm{Tr} B_{2,0} \wedge F_{0,2} = \int_{M_4} d^4x \sqrt{g}\, \mathrm{Tr}\Big(\epsilon^{mn\bar{m}\bar{n}} B_{mn} F_{\bar{m}\bar{n}}\Big) \tag{3.3}$$

where $F = dA + A \wedge A$ is the curvature of the Yang–Mills field A. The equations of motion are

$$F_{\bar{m}\bar{n}} = 0 \qquad \epsilon^{mn\bar{m}\bar{n}} D_{\bar{n}} B_{mn} = 0 \tag{3.4}$$

Classically, A_m is undetermined, $A_{\bar{m}}$ is a pure gauge and B_{mn} is holomorphic. Notice that $B_{2,0}$ has no vector gauge invariance. It counts for one propagating degree of freedom. Altogether, there are two gauge invariant degrees of freedom that are not specified classically. Modulo gauge invariance, there is a mixed propagation between A and B. The symmetries of the action (3.3) are

$$\begin{array}{ll} QA_m = \Psi_m + D_m c & Q\Psi_m = -[c, \Psi_m] \\ Qc = -\frac{1}{2}[c,c] & QA_{\bar{m}} = D_{\bar{m}} c \\ QB_{mn} = -[c, B_{mn}] & \end{array} \tag{3.5}$$

In the first two lines of the above equation we can recognize the symmetry (2.4). The geometrical interpretation of complete charged 2-forms is a non-trivial issue. However, in the language where form degree and positive and negative ghost number are unified within a bigrading, the charged 2-form can be understood as a sort of Hodge dual to the Yang–Mills field [30]. Here we only consider the $(2,0)$ component of such an object, and we hopefully avoid the ambiguities for defining its theory. We now explain the BRST quantization of the action (3.3), for the sake of inserting it in a path integral.

3.2. QUANTIZATION OF THE BF SYSTEM ON A KÄHLER MANIFOLD

In order to define a quantum theory, that is, a path integral, we need to gauge fix the (almost) topological symmetry of the BF system, in a way that respect the BRST symmetry associated to this symmetry. As it is well-known, the anti–self–duality condition in 4 dimensions can be expressed in complex coordinates as:

$$F_{mn} = 0\,, \qquad F_{\bar{m}\bar{n}} = 0\,, \qquad J_{m\bar{n}} F^{m\bar{n}} = 0 \tag{3.6}$$

and one has the identity

$$\mathrm{Tr}\left(F_{\bar{m}\bar{n}}F^{\bar{m}\bar{n}} + \tfrac{1}{2}|J_{m\bar{n}}F^{m\bar{n}}|^2\right) = \frac{1}{4}\mathrm{Tr}(F_{\mu\nu}F^{\mu\nu} + F_{\mu\nu}\tilde{F}^{\mu\nu}) \tag{3.7}$$

Modulo ordinary gauge invariance, we have two topological freedoms, corresponding to the two components in Ψ_m. In order to perform a suitable gauge fixing for the two–form $B_{2,0}$ and for $A_{1,0}$, which is the part of the gauge connection absent from the classical action (3.3), we introduce two anticommuting antighosts κ^{mn} and κ, and two Lagrange multipliers b^{mn} and b:

$$\begin{array}{rclrcl} Q\kappa^{mn} & = & b^{mn} & Qb^{mn} & = & 0 \\ Q\kappa & = & b & Qb & = & 0 \end{array} \tag{3.8}$$

Since $Q^2 = 0$ from the beginning, we have a first order Batalin-Vilkoviski (BV) system. However, since the treatment of the chiral multiplet in the next Section will produce a non-trivial second order BV system, we find convenient to introduce right now BV anti-fields for A and B and their ghosts, antighosts and Lagrangian multipliers. The upper left notation * labels antifields[2]. Let us recall that the antifield $^*\phi$ of a field with ghost number g has ghost number $-g-1$ and opposite statistics. For a Q-invariant BV action S one has $Q\phi = \frac{\partial_l S}{\partial^*\phi}$ and $Q^*\phi = -\frac{\partial_l S}{\partial\phi}$. The property $Q^2 = 0$ is equivalent to the master equation

$$\frac{\partial_r S}{\partial\phi}\frac{\partial_l S}{\partial^*\phi} = 0 \tag{3.9}$$

where (∂_r, ∂_l) indicate respectively the derivatives from the left and from the right.

The following BV action encodes at once the classical action (3.3) and the definition of its BRST symmetry:

$$\begin{aligned} S \;=\; & \int_{M_4} d^4x\sqrt{g}\,\mathrm{Tr}\Big(\frac{1}{4}\epsilon^{mn\bar{m}\bar{n}}B_{mn}F_{\bar{m}\bar{n}} \\ & +{}^*A^m(\Psi_m + D_m c) +{}^*A^{\bar{m}}(D_{\bar{m}}c) \\ & -{}^*B^{mn}[c, B_{mn}] -{}^*\Psi^m[c,\Psi_m] - \tfrac{1}{2}{}^*c[c,c] \\ & +{}^*\kappa_{mn}b^{mn} +{}^*\kappa b\Big) \end{aligned} \tag{3.10}$$

The BV master equation (3.9) is satisfied, which implies the gauge invariance of the classical action as well as the nilpotency $Q^2 = 0$ on all the fields. It is actually important to note that the invariance of the action (3.10) implies that

$$Q^*B^{mn} = \epsilon^{mn\bar{m}\bar{n}}F_{\bar{m}\bar{n}} - [c,{}^*B^{mn}]. \tag{3.11}$$

This equation will shortly play a key role for defining the coupling to scalar fields.

The topological gauge fixing corresponds to the elimination of antifields by a suitable choice of a gauge function Z. The antifields are to be replaced in the path integral by the BV formula:

$$^*\phi = \frac{\delta Z}{\delta\phi} \tag{3.12}$$

[2]As stressed in [30], the antifields of A appear in the ghost expansion of B and vice-versa.

The Q-invariant observables are formally independent on the choice of Z. In particular, their mean values are expected to be independent on small changes of the metric that one must introduce to define Z.

In order to concentrate path integral around the anti-self-duality condition (3.6) we choose:

$$Z = \kappa^{mn}(B_{mn} - \epsilon_{mn\bar{m}\bar{n}}F^{\bar{m}\bar{n}}) + \kappa(\tfrac{1}{2}b + iJ_{m\bar{n}}F^{m\bar{n}}) \tag{3.13}$$

The BV equation (3.12) implies that $B_{2,0}$ is eliminated in the path integral with $B_{mn} = \epsilon_{mn\bar{m}\bar{n}}F^{\bar{m}\bar{n}}$. After Gaussian integration on b, the gauge-fixed action reads

$$\begin{aligned} S^{g.f.} &= \int_{M_4} d^4x\sqrt{g}\,\mathrm{Tr}\Big(F_{\bar{m}\bar{n}}F^{\bar{m}\bar{n}} + \tfrac{1}{2}|J_{m\bar{n}}F^{m\bar{n}}|^2 \\ &\quad -2\epsilon^{\bar{m}\bar{n}pq}\kappa_{\bar{m}\bar{n}}D_p\Psi_q + i\kappa J^{\bar{m}l}D_{\bar{m}}\Psi^l\Big) \end{aligned} \tag{3.14}$$

Here and in the following discussions we omit the c–dependent terms in the action. In fact these terms express the covariance of the gauge–fixing conditions (3.6) with respect to the gauge symmetry, and vanish when these conditions are enforced. Moreover, we leave aside the standard gauge fixing of the ordinary gauge invariance $\partial^\mu A_\mu = 0$.

The action (3.14) can be compared with that of $N = 1$ SYM on a Kähler manifold. It is known [31] that on a complex spin manifold the complex spinors can be identified with forms $S_\pm \otimes \mathbb{C} \sim \Omega^{0,{even \atop odd}}$, so that we can identify our topological ghost Ψ_m as a left–handed Weyl spinor λ_α and the topological anti–ghosts $(\kappa_{\bar{m}\bar{n}}, \kappa)$ as a right–handed Weyl spinor $\bar{\lambda}^{\dot{\alpha}}$. More explicitly, the holonomy group of a four–dimensional Kähler manifold is locally given by $U(2) \sim SU(2)_L \times U(1)_R \subset SU(2)_L \times SU(2)_R$, so that one can naturally identify the forms $\sigma_{\mu\ \alpha 1}dx^\mu$ and $\sigma_{\mu\ \alpha\dot{2}}dx^\mu$ as $(1,0)$ and $(0,1)$ forms respectively [26, 27]. Then the twist reads[3]

$$\begin{aligned} \Psi_m &= \lambda^\alpha \sigma_{\mu\ \alpha 1} e^\mu_m\,, \\ \kappa_{\bar{m}\bar{n}} &= \bar{\lambda}_{\dot{\alpha}}\,\bar{\sigma}_{\mu\nu}{}^{\dot{\alpha}}{}_{\dot{2}}\, e^\mu_{\bar{m}} e^\nu_{\bar{n}}\,, \\ \kappa &= \delta_{\dot{2}}^{\dot{\alpha}}\bar{\lambda}_{\dot{\alpha}} \end{aligned} \tag{3.15}$$

On a Hyperkähler manifold, the twist formula can be reinterpreted by making explicit a constant spinor dependence in (3.15). With this change of variables, it is immediate to recognize that the action (3.14) is the $N = 1, D = 4$ Yang–Mills action

$$S_{SYM} = \int_{M_4} d^4x\sqrt{g}\,\frac{1}{4}\mathrm{Tr}\Big(F_{\mu\nu}F^{\mu\nu} + F_{\mu\nu}\tilde{F}^{\mu\nu} + \bar{\lambda}\gamma^\mu D_\mu\lambda\Big) \tag{3.16}$$

As compared to [26], we started from a classical BF system, which, eventually, gives the $N = 1, D = 4$ Yang–Mills theory as a microscopic theory in a twisted form. Let us notice that it should be possible to cast the topological BRST symmetry in the form of conditions on curvatures yielding descent equations with asymmetric holomorphic decompositions and eventually solve the cocycle equations for Q, similarly to what has been done in [23] for the Topological Yang–Mills theory.

[3] We define the euclidean σ matrices as $\sigma_\mu = (i\tau^c, \mathbf{1})$, τ^c, $c = 1, 2, 3$ being the Pauli matrices.

3.3. COUPLING OF THE BF TO A CHIRAL MULTIPLET

Since the $N = 2$ SYM theory is an ordinary TQFT, and since its Poincaré supersymmetric version can be obtained by coupling the $N = 1$ Yang–Mills multiplet to a chiral multiplet in the adjoint representation of the gauge group, one expects to have an expression of the $N = 1$ scalar theory as an (almost)TQFT on a Kähler manifold. As we shall see, this is slightly more complicated than the $N = 1$ Super Yang-Mills theory, since it will involve the vector gauge symmetry of a (0,2)-charged form, and thus a 2nd rank BV system arises.

In order to introduce the chiral multiplet, we extend the set of classical fields of the previous section as

$$B_{2,0} \rightarrow (B_{2,0}, B_{0,2}), \tag{3.17}$$

However, we keep the same classical action as in (3.3):

$$I_{cl}(A, B_{0,2}, B_{0,2}) = \int_{M_4} \text{Tr}(B_{2,0} \wedge F_{0,2}) = \int_{M_4} \text{Tr}(\epsilon^{mn\bar{m}\bar{n}} B_{mn} F_{\bar{m}\bar{n}}) \tag{3.18}$$

Having a classical action that is independent of $B_{0,2}$ is equivalent of having the following symmetry for $B_{0,2}$

$$\begin{aligned} QB_{\bar{m}\bar{n}} &= D_{[\bar{m}}\Psi_{\bar{n}]} - [c, B_{\bar{m}\bar{n}}] - \frac{1}{4}\epsilon_{\bar{m}\bar{n}mn}[{}^{*}B^{mn}, \Phi] \\ Q\Psi_{\bar{m}} &= D_{\bar{m}}\Phi - [c, \Psi_{\bar{m}}] \\ Q\Phi &= -[c, \Phi]. \end{aligned} \tag{3.19}$$

In fact, the unique degree of freedom carried by $B_{\bar{m}\bar{n}}$ is canceled by the two degrees of freedom of the topological ghost $\Psi_{\bar{n}}$, defined modulo the ghost of ghost symmetry generated by Φ. The presence of the antifield ${}^{*}B^{mn}$ in (3.19) is necessary in order that $Q^2 = 0$, as one can verify by using (3.19), and (3.11), together with the usual BRST variation of the ghost c, $Qc = -c^2$. We thus have a second order BV system, since the BRST variations of the fields depend linearly on the antifield. This non trivial property justifies, a posteriori, that the classical action depend on the charged $(2,0)$-form $B_{2,0}$ in an "almost topological way", as in (3.3), with the ordinary gauge symmetry $QA_{\bar{m}} = D_{\bar{m}}c$ and $QB_{2,0} = -[c, B_{2,0}]$. This determines the relevant Q-transformation of the antifield of the 2-form, which is eventually necessary to obtain a closed symmetry.

The fate of the $(0,2)$-form $B_{0,2}$ is to be gauge-fixed and eliminated from the action, as it was the case for $B_{2,0}$, but wit a different gauge function. For this purpose we choose a topological antighosts that is a (0,2)-form $\kappa^{\bar{m}\bar{n}}$, with bosonic Lagrange multiplier $b^{\bar{m}\bar{n}}$. The ghost of ghost symmetry of $\Psi_{\bar{m}}$ must be gauge fixed, and we introduce a bosonic antighost $\bar{\Phi}$ with fermionic Lagrange multiplier $\bar{\eta}$. $\kappa^{\bar{m}\bar{n}}$ and $\bar{\eta}$ will be eventually untwisted and provide half of a Majorana spinor for $N = 1$ supersymmetry

One has in fact an ordinary pyramidal structure for a 2-form gauge field [4], which shows that $B_{\bar{m}\bar{n}}$ truly carries zero degrees of freedom, and can be consistently gauge-

[4] In this table, we indicate explicitly the ghost number of the fields by a superscript. The BRST symmetry acts on the South–West direction, as indicated by the arrows.

fixed to zero

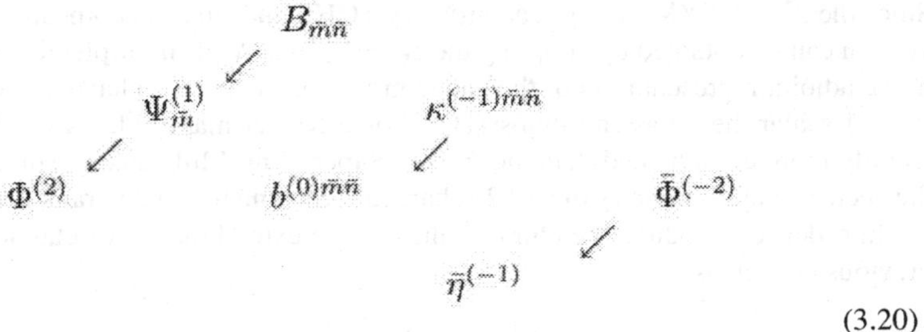

$$\begin{array}{ccccccc} & & B_{\bar m\bar n} & & & & \\ & \swarrow & & & & & \\ & \Psi^{(1)}_{\bar m} & & \kappa^{(-1)\bar m\bar n} & & & \\ \swarrow & & & \swarrow & & & \\ \Phi^{(2)} & & b^{(0)\bar m\bar n} & & & \bar\Phi^{(-2)} & \\ & & & & \swarrow & & \\ & & & \bar\eta^{(-1)} & & & \end{array} \tag{3.20}$$

The vector ghost symmetry of the charged $(0, 2)$-form with ghost $\Psi^{(1)}_{\bar m}$ plays the role of a topological symmetry. In the untwisted theory, Φ and $\bar\Phi$ will be identified as the complex scalar field for the $N = 1$ chiral multiplet in 4 dimensions. The BV action for the fields in the Table (3.20) and their antifields is

$$\begin{aligned} S_{matter} &= \int_{M_4} d^4x\sqrt{g}\,\mathrm{Tr}\Big({}^*B^{\bar m\bar n}(D_{[\bar m}\Psi_{\bar n]} - [c, B_{\bar m\bar n}] - \frac{1}{4}\epsilon_{\bar m\bar n mn}[{}^*B^{mn}, \Phi]) \\ &\quad + {}^*\Psi^{\bar m}(D_{\bar m}\Phi - [c, \Psi_{\bar m}]) - {}^*\Phi[c, \Phi] \\ &\quad + {}^*\kappa_{\bar m\bar n}b^{\bar m\bar n} + {}^*\bar\Phi\bar\eta\Big) \end{aligned} \tag{3.21}$$

For consistency, we have to add to the above action the action (3.10) in order to properly define the variation of ${}^*B^{mn}$ as in (3.11).

In order to gauge–fix S_{matter}, we choose the following BV gauge function:

$$Z' = \kappa^{\bar m\bar n}B_{\bar m\bar n} + \bar\Phi D^{\bar m}\Psi_{\bar m} \tag{3.22}$$

Using the BV equation (3.12) and integrating on the Lagrangian multiplier $b^{\bar m\bar n}$ one gets $B_{\bar m\bar n} = 0$, and finds

$$S^{g.f.}_{matter} = \int_{M_4} d^4x\sqrt{g}\,\mathrm{Tr}\Big(\kappa^{\bar m\bar n}D_{[\bar m}\Psi_{\bar n]} + \bar\Phi D^{\bar m}D_{\bar m}\Phi + \bar\eta D^{\bar m}\Psi_{\bar m}\Big) \tag{3.23}$$

As for the Yang–Mills supermultiplet, we can perform a mapping of the topological ghost $\Psi_{\bar m}$ and of the topological anti–ghosts $(\kappa^{\bar m\bar n}, \bar\eta)$ on left and right handed spinors $(\psi^\alpha, \bar\psi_{\dot\alpha})$ respectively. With this change of variables, we now recognize that the action (3.23) is the $N = 1, D = 4$ chiral multiplet action

$$S_{SYM} = \int_{M_4} d^4x\sqrt{g}\,\mathrm{Tr}\Big(\bar\Phi D^\mu D_\mu\Phi + \bar\psi\gamma^\mu D_\mu\psi\Big) \tag{3.24}$$

The sum of both actions (3.10) and (3.21), when the suitable gauge fixing conditions (3.13) and (3.22) are enforced, corresponds to the twisted $N = 2$ Super Yang–Mills action, with notations that are adapted to a Kähler manifold [5].

[5] In order to recover the Yukawa couplings and the quartic term in the potential $[\bar\Phi, \Phi]^2$ typical of the $N = 2$ SYM one should slightly modify the gauge–fixing fermion Z' in (3.22), but this doesn't change the results on the topological observables.

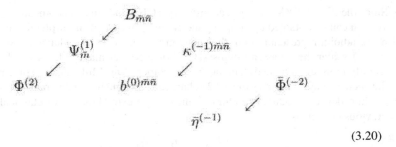

However, the BRST algebra we discussed so far is mapped only to a $N=1$ subsector of the $N=2$ supersymmetry. The complete $N=2$ superalgebra on a Kähler manifold has been discussed in [25, ?, ?]. Let us briefly display how these results can be recovered in our model. In our construction, we can interchange the role of holomorphic and antiholomorphic coordinates. We can thus consider another operator $\bar{\Delta}$

$$\begin{array}{rclrcl} \Delta A_m &=& \Psi_m & \bar{\Delta} A_m &=& 0 \\ \Delta A_{\bar{m}} &=& 0 & \bar{\Delta} A_{\bar{m}} &=& \Psi_{\bar{m}} \\ \Delta \Psi_m &=& 0 & \bar{\Delta}\Psi_m &=& D_m\Phi \\ \Delta \Psi_{\bar{m}} &=& D_{\bar{m}}\Phi & \bar{\Delta}\Psi_{\bar{m}} &=& 0 \\ \Delta\Phi &=& 0 & \bar{\Delta}\Phi &=& 0 \end{array} \tag{3.25}$$

where with $\Delta, \bar{\Delta}$ we indicate the equivariant BRST operators, with the ghost field c associated to the gauge symmetry set to zero. One has:

$$\begin{array}{rcl} (\Delta+\bar{\Delta})A_m &=& \Psi_m \\ (\Delta+\bar{\Delta})A_{\bar{m}} &=& \Psi_{\bar{m}} \\ (\Delta+\bar{\Delta})\Psi_m &=& D_m\Phi \\ (\Delta+\bar{\Delta})\Psi_{\bar{m}} &=& D_{\bar{m}}\Phi \\ (\Delta+\bar{\Delta})\Phi &=& 0 \end{array} \tag{3.26}$$

Thus, $\Delta^2 = \bar{\Delta}^2 = 0$, and $(\Delta+\bar{\Delta})^2 = \{\Delta, \bar{\Delta}\} = \delta_\Phi$, where δ_Φ is a gauge transformation with parameter Φ. The operator $(\Delta+\bar{\Delta})$ is the topological BRST symmetry operator (for $c=0$), corresponding to the twisted $N=2$ supersymmetry. The classical action which is invariant under the symmetry (3.26) is the (real) BF action plus a "cosmological term" $\mathrm{Tr}(B\wedge B)$, with B the complete real two–form. This field transform as $(\Delta+\bar{\Delta})B = D\Psi$. In a space where one cannot consistently separate holomorphic and antiholomorphic components of forms, the only admissible operation is $(\Delta+\bar{\Delta})$, which is Lorentz invariant. Then, to close the BRST symmetry, and get $\delta^2 = (Q+\bar{Q})^2 = 0$, one must redefine the BRST transformation of the Faddeev–Popov ghost c, as follows:

$$Qc = -\tfrac{1}{2}[c,c] \rightarrow Qc = -\Phi - \tfrac{1}{2}[c,c] \tag{3.27}$$

In this way one recover the complete symmetry of the Topological Yang–Mills action (2.1).

4. Six dimensions: Calabi–Yau three–fold

On a Calabi–Yau three–fold CY_3 we can use the holomorphic closed $(3,0)$–form $\Omega_{3,0}$ and define $B_{3,1} = \Omega_{3,0}\wedge B_{0,1}$. The classical action (2.3) then becomes

$$I_{cl}(A, B_{0,1}) = \int_{M_6} \Omega_{3,0}\wedge \mathrm{Tr}(B_{0,1}\wedge F) \tag{4.1}$$

The BRST symmetry corresponding to the action (4.1) is

$$\begin{aligned} QA_m &= \Psi_m + D_m c \\ QA_{\bar m} &= D_{\bar m} c\,, \\ QB_{\bar m} &= D_{\bar m}\chi - [c, B_{\bar m}]\,, \\ Q\chi &= -[c,\chi]\,, \\ Qc &= -\tfrac{1}{2}[c,c]\,, \end{aligned} \tag{4.2}$$

The invariance of the action (4.1) under the transformation of the B field is guaranteed by part of the Bianchi identity

$$D_{[\bar m}F_{\bar n \bar l]} = 0 \tag{4.3}$$

and the fact that Ω is closed. Notice also that the action (4.1) is invariant under the *complexified* gauge group $GL(N,\mathbb{C})$. The BRST symmetry (4.2) follows from the following Batalin–Vilkoviski action

$$\begin{aligned} S &= \int_{M_6} \sqrt{g}\, d^6x\, \mathrm{Tr}\Big(\epsilon^{\bar m \bar p \bar q} B_{\bar m} F_{\bar p \bar q} +^{*} B^{\bar m}(D_{\bar m}\chi - [c, B_{\bar m}]) \\ &\quad +^{*} A^m(\Psi_m + D_{\bar m} c) +^{*} A^{\bar m} D_{\bar m} c -^{*} \Psi^m [c, \Psi_m] \\ &\quad +^{*} c(-\tfrac{1}{2}[c,c])\Big)\,, \end{aligned} \tag{4.4}$$

where we have normalized the $(3,0)$–form Ω such that $\Omega \wedge \bar\Omega$ is the volume form.

Let us now proceed to the quantization of the model: this can be performed in different ways, which lead us to the study of different sets of observables. If one chooses to quantize the theory around the perturbative vacuum corresponding to holomorphic flat connections, the corresponding observables will depend on the *complex* structure Ω of the manifold, as usually happens in type B topological string theories. In fact, in this case the holomorphic BF model has a deep relationship with the holomorphic Chern–Simons theory, which can be regarded as an effective action for D5 branes in type B topological string [32]. This relationship should be a generalization of that between BF and Chern Simons theories in three (real) dimensions [33], and it deserves further investigations.

If instead one quantize the theory around a non-perturbative vacuum corresponding to a stable holomorphic vector bundle, one can show that the BF model correspond to the twisted version of a supersymmetric Yang–Mills theory! In this case the observables are dependent on the *Kähler* data of the manifold, as happens in type A topological string. In fact a direct relationship between a twisted $U(1)$ maximally supersymmetric action and the topological vertex has been shown in [7].

The study of the holomorphic BF model in the abelian case can be then useful to clarify the issue of S–duality in topological strings pointed out in [8, 9, 10], and the relationship between Gromov–Witten and Donaldson–Thomas invariants discussed in [18]. Let us now show the details of the two quantizations.

4.1. PERTURBATIVE QUANTIZATION AND B MODEL

In this case, the symmetries are treated as ordinary gauge symmetries and fixed with transversality conditions on the $A_{\bar{m}}$ and $B_{\bar{m}}$ fields:

$$\begin{aligned} D^{\bar{m}} A_{\bar{m}} &= 0 \\ D^{\bar{m}} B_{\bar{m}} &= 0 \end{aligned} \tag{4.5}$$

The BV fermion corresponding to these conditions is

$$Z = \bar{\chi} D^{\bar{m}} B_{\bar{m}} + \bar{c} D^{\bar{m}} A_{\bar{m}} \tag{4.6}$$

Once the gauge fixing conditions (4.5) are enforced, one has a well defined mixed propagator between the A and B field, and can use it to evaluate the path integral in a perturbative expansion. The partition function of this model in the semiclassical limit should be related to the Ray–Singer holomorphic torsion [34] similarly to what happens for the holomorphic Chern–Simons theory analysed in [35, 36]. The higher order terms in the perturbative expansion should be related to other manifold invariants. It would be interesting to study these invariants along the lines of the perturbative analysis of three–dimensional Chern–Simons theory [37].

4.2. NON–PERTURBATIVE QUANTIZATION AND A MODEL

The shift symmetry on the $(0, 1)$ part of the connection gives rise to three degrees of freedom, while the symmetry on the B field to one. These are collected into the ghost fields (Ψ_m, χ) respectively. In the non–perturbative case, the gauge fixing conditions are chosen as follows

$$\begin{aligned} F_{mn} &= -\frac{4}{3}\epsilon_{mnp} B^p \\ J^{\bar{m}n} F_{\bar{m}n} &= 0 \end{aligned} \tag{4.7}$$

and amount to three conditions for the first line and one for the second. The reason for the particular choice of the coefficient in the first equation of (4.7) will be evident shortly. Notice that the second equation in (4.7) reduces the complex gauge group $GL(N, \mathbb{C})$ to the unitary group $U(N)$, and as such can be considered as a partial gauge–fixing for the complex gauge symmetry of the classical action (4.1). This has to be completed with a further gauge–fixing for the unitary group, as for example the ordinary Landau gauge $\partial^\mu A_\mu = 0$. We will discuss this issue in more detail in Sect.5.1. The BV fermion corresponding to (4.7) is given by

$$Z = \bar{\chi}^{\bar{m}\bar{n}} (F_{\bar{m}\bar{n}} + \frac{4}{3}\epsilon_{mnp} B^p) + \bar{\eta}(2i J^{\bar{m}n} F_{\bar{m}n} - h) \ , \tag{4.8}$$

where $(\bar{\chi}^{\bar{m}\bar{n}}, \bar{\eta})$ are the antighosts associated to the gauge–fixing conditions (4.7), whose BV action is given by

$$S_{aux} = \int_{M_6} \sqrt{g}\, d^6x \, \mathrm{Tr}({}^*\bar{\chi}_{\bar{m}\bar{n}} h^{\bar{m}\bar{n}} +^* \bar{\eta} h +^* \bar{c} b) \tag{4.9}$$

Eliminating the anti–fields by means of (3.12) and implementing the gauge–fixing conditions (4.7) by integration on the Lagrangian multipliers, we get from (4.4) and (4.9)

$$
\begin{aligned}
S^{g.f.} &= \int_{M_6} d^6x\sqrt{g}\,\mathrm{Tr}\Big(-\frac{3}{2}F^{\bar{m}\bar{n}}F_{\bar{m}\bar{n}} + |J^{m\bar{n}}F_{m\bar{n}}|^2) \\
&\quad +\bar{\chi}^{\bar{m}\bar{n}}D_{[\bar{m}}\Psi_{\bar{n}]} + 2i\bar{\eta}J^{\bar{m}n}D_{\bar{m}}\Psi_n + \frac{4}{3}\epsilon_{\bar{m}\bar{n}\bar{p}}\bar{\chi}^{\bar{m}\bar{n}}D^{\bar{p}}\chi\Big) \; . \qquad (4.10)
\end{aligned}
$$

By using the identity [38]

$$
\begin{aligned}
&-\frac{1}{4}\mathrm{Tr}(F\wedge *F) + J\wedge\mathrm{Tr}(F\wedge F) = \\
&\mathrm{Tr}\Big(-\frac{3}{2}F^{\bar{m}\bar{n}}F_{\bar{m}\bar{n}} + |J^{\bar{m}n}F_{\bar{m}n}|^2\Big) \qquad (4.11)
\end{aligned}
$$

we can recognize in the first line of (4.10) the bosonic part of the N=1 D=6 SYM action, modulo the topological density $J\wedge\mathrm{Tr}(F\wedge F)$, where J is the Kähler two–form. Concerning the fermionic part, we can make use of the mapping between chiral fermions and complex forms $S_\pm\otimes\mathbb{C}\sim\Omega^{odd}_{even}$ to map the topological ghosts (Ψ_m,χ) into the right–handed spinor $\bar{\lambda}$ and the topological antighosts into the left–handed spinor λ. More explicitly, we can use the covariantly constant spinor ζ of the Calabi–Yau three–fold to perform the mapping

$$
\begin{aligned}
\Psi_m &\rightarrow \bar{\lambda}\Gamma_m\zeta \\
\chi &\rightarrow \bar{\lambda}\zeta \\
\bar{\chi}^{\bar{m}\bar{n}} &\rightarrow \zeta\Gamma^{\bar{m}\bar{n}}\lambda \\
\bar{\eta} &\rightarrow \epsilon_{\bar{m}\bar{n}\bar{p}}\zeta\Gamma^{\bar{m}}\Gamma^{\bar{n}}\Gamma^{\bar{p}}\lambda \qquad (4.12)
\end{aligned}
$$

In this way, one can recognize in (4.10) the twisted version of the $N=1$ $D=6$ Super Yang–Mills action. In order to reproduce the $U(1)$ twisted maximally supersymmetric action discussed in [7], one has to add to the classical action (4.1) the higher Chern class $F\wedge F\wedge F$ and couple this theory to an hypermultiplet, with a procedure similar to that discussed in the four dimensional case. As in Sect.3.3, one has to consider the quantization of a $(0,2)$–form $B_{0,2}$. The corresponding BRST complex is the same as in Table (3.20), but now with six–dimensional fields $(\bar{m},\bar{n}=\bar{1},\bar{2},\bar{3})$. It is straightforward to realize that the fields appearing in the Table (3.20) together with the multiplet discussed in this subsection give rise exactly to the spectrum of the twisted maximally supersymmetric Yang–Mills discussed in [7]. An alternative and more economical way would be to proceed from the dimensional reduction of the eight-dimensional model that we are going to discuss in the next section. Notice that, as discussed in Sect.3.3, the coupling to the hypermultiplet does not change the classical action (4.1). Moreover, the higher Chern class $(F)^3$ is only a boundary term which does not affect the propagation of the A and B fields. Thus the relationship with the perturbatively quantized model of the previous subsection still holds.

5. Eight dimensions

5.1. CALABI–YAU FOUR–FOLD

On a Calabi–Yau four–fold we can write the following generalization of the action (3.3)

$$I_{cl}(A, B_{0,2}) = \int_{M_8} \Omega_{4,0} \wedge \mathrm{Tr}(B_{0,2} \wedge F_{0,2}) \tag{5.1}$$

Here $\Omega_{4,0}$ is the holomorphic covariantly *closed* (4,0)-form. This, together with part of the Bianchi identity, ensures the invariance of the classical action (5.1) analogously to the CY_3 case of the previous section. Also, as in the previous section, we normalize Ω such that $\Omega \wedge \bar{\Omega}$ is the volume element on M_8. The action (5.1) displays the symmetry

$$\begin{aligned}
QA_M &= \Psi_M + D_M c \\
QA_{\bar{M}} &= D_{\bar{M}} c\,, \\
QB_{\bar{M}\bar{N}} &= D_{[\bar{M}}\chi_{\bar{N}]} - [c, B_{\bar{M}\bar{N}}] - \frac{1}{4}\epsilon_{\bar{M}\bar{N}\bar{P}\bar{Q}}[{}^*B^{\bar{P}\bar{Q}}, \phi]\,, \\
Q\chi_{\bar{N}} &= D_{\bar{N}}\phi - [c, \chi_{\bar{N}}]\,, \\
Qc &= -\tfrac{1}{2}[c, c]\,, \\
Q\phi &= -[c, \phi]
\end{aligned} \tag{5.2}$$

Notice that c is the complexified Faddeev–Popov ghost. The BV action corresponding to (5.1) is given by

$$\begin{aligned}
S = \int_{M_8} \sqrt{g}\, d^8x\, \mathrm{Tr}\Big(&\epsilon^{\bar{M}\bar{N}\bar{P}\bar{Q}} B_{\bar{M}\bar{N}} F_{\bar{P}\bar{Q}} \\
&+{}^* B^{\bar{M}\bar{N}}(D_{[\bar{M}}\chi_{\bar{N}]} - [c, B_{\bar{M}\bar{N}}] - \frac{1}{4}\epsilon_{\bar{M}\bar{N}\bar{P}\bar{Q}}[{}^*B^{\bar{P}\bar{Q}}, \phi]) \\
&+{}^* A^M(\Psi_M + D_{\bar{M}} c) +{}^* A^{\bar{M}} D_{\bar{M}} c +{}^* \chi^{\bar{N}}(D_{\bar{N}}\phi - [c, \chi_{\bar{N}}]) \\
&+{}^* \bar{\chi}_{\bar{M}\bar{N}} h^{\bar{M}\bar{N}} +{}^* \bar{\chi} h +{}^* \bar{c} b +{}^* \bar{\phi}\eta \\
&+{}^* c(-\tfrac{1}{2}[c, c]) -{}^* \phi[c, \phi]\Big)
\end{aligned} \tag{5.3}$$

The action (5.1) only define the propagation of part of the gauge field, as in the case studied in section 2. It can be gauge–fixed in by imposing six complex conditions for $B_{\bar{M}\bar{N}}$

$$\begin{aligned}
B^{+}_{\bar{M}\bar{N}} &= 0\,, \\
B^{-}_{\bar{M}\bar{N}} &= F^{-}_{\bar{M}\bar{N}}\,,
\end{aligned} \tag{5.4}$$

and a gauge–fixing for $\chi_{\bar{M}}$

$$D^{\bar{M}}\chi_{\bar{M}} = 0. \tag{5.5}$$

The projection on self-dual or anti-self-dual part $B^{\pm}_{0,2}$ of the $(0, 2)$–forms can be done by using the anti-holomorphic $(0, 4)$ form. The conditions (5.4) can be enforced by using the BRST doublets of complex antighosts and Lagrangian multipliers $(\bar{\chi}^{\bar{M}\bar{N}}, h^{\bar{M}\bar{N}})$ and

$(\bar{\phi}, \bar{\chi})$ respectively. Then, as a generalization of [4], we complete the above six complex conditions for $B_{\bar{M}\bar{N}}$ by the following complex condition :

$$D^{\bar{M}\ \mathrm{c}}A_{\bar{M}} = 0 \tag{5.6}$$

The real part of (5.6) is the ordinary Landau gauge condition. The imaginary part gives instead a condition analogous to the second line of (4.7):

$$\begin{aligned} ImD^{\bar{M}}A_{\bar{M}} &= 0 \Rightarrow J^{M\bar{N}}F_{M\bar{N}} = 0 \\ ReD^{\bar{M}}A_{\bar{M}} &= 0 \Rightarrow \partial^{\mu}A_{\mu} = 0 \end{aligned} \tag{5.7}$$

The gauge–fixing fermion corresponding to the gauge conditions (5.4), (5.6) and (5.7) is

$$\begin{aligned} Z &= \bar{\chi}^{\bar{M}\bar{N}^+}B_{\bar{M}\bar{N}} + \bar{\chi}^{\bar{M}\bar{N}^-}(B_{\bar{M}\bar{N}} - 2F_{\bar{M}\bar{N}}) + \bar{\phi}D^{\bar{M}}\chi_{\bar{M}} \\ &\quad + \bar{\chi}(iJ^{M\bar{N}}F_{M\bar{N}} + \tfrac{1}{2}h) + \bar{c}(\partial^{\mu}A_{\mu} + \tfrac{1}{2}b) \end{aligned} \tag{5.8}$$

By using the BV equation (3.12) and enforcing the gauge conditions (5.4), (5.6) and (5.7) by integration on the Lagrangian multipliers, we get the wanted action, as a twisted form of the $D = 8$ supersymmetric Yang–Mills action. Its gauge invariant part is :

$$\begin{aligned} S^{g.f.} &= \int_{M_8} d^8x\sqrt{g}\,\mathrm{Tr}\Big(2F^{\bar{M}\bar{N}-}F^{-}_{\bar{M}\bar{N}} + \tfrac{1}{2}|J^{M\bar{N}}F_{M\bar{N}}|^2) + \bar{\phi}D^{\bar{M}}D_{\bar{M}}\phi \\ &\quad + \bar{\chi}^{\bar{M}\bar{N}-}D_{[\bar{M}}\chi_{\bar{N}]} + \bar{\chi}^{\bar{M}\bar{N}+}D_{[\bar{M}}\Psi_{\bar{N}]} + \bar{\chi}D^{M}\Psi_{M} + \bar{\eta}D^{\bar{M}}\chi_{\bar{M}}\Big) \end{aligned} \tag{5.9}$$

As for the previous cases, we do not display in (5.9) the gauge dependent part of the action. By using the identity

$$\int_{M_8} d^8x\sqrt{g}\,\mathrm{Tr}(2F^{\bar{M}\bar{N}-}F^{-}_{\bar{M}\bar{N}} + \tfrac{1}{2}|J^{M\bar{N}}F_{M\bar{N}}|^2) + S_0 = \frac{1}{4}\int_{M_8} d^8x\sqrt{g}\,\mathrm{Tr}(F^{\mu\nu}F_{\mu\nu}) \tag{5.10}$$

where $S_0 = \int_{M_8}\Omega \wedge \mathrm{Tr}(F_{0,2} \wedge F_{0,2})$ is a surface term [4], we can recognize the first line of the action (5.9) as the bosonic part of the $D = 8$ SYM action. Concerning the fermionic part, one can use the identification of fermions with forms $S_{\pm} \sim \Omega^{{even \atop odd},0}$, which for $D = 8$ reads

$$S_- \sim \Omega^{1,0} \oplus \Omega^{3,0}\,, \tag{5.11}$$

$$S_+ \sim \Omega^{0,0} \oplus \Omega^{2,0} \oplus \Omega^{4,0} \tag{5.12}$$

to identify the topological ghosts $(\Psi_M, \chi_{\bar{M}})$ with the right–handed projection of the Majorana spinor $\bar{\lambda}^{\dot{a}}$, $\dot{a} = 1, \ldots, 8$, and the topological anti–ghosts $(\bar{\chi}, \bar{\chi}^{\bar{M}\bar{N}}, \bar{\eta})$ with the left–handed projection λ_a. Notice in fact that on a Calabi–Yau four–fold one can use the holomorphic four–form Ω to identify the $(3,0)$–form appearing on the r.h.s. of (5.11) with the the field $\chi_{\bar{M}}$ [31]. Analogously, one can identify the scalar $\bar{\eta}$ with the $(4,0)$–form appearing in the r.h.s. of (2.15). More explicitly, one can use the two left–handed covariantly constant spinors (ζ_1, ζ_2) of the Calabi–Yau four–fold to identify

$$\begin{aligned} \chi^{M} &= \epsilon^{MNPQ}\zeta_1\Gamma_N\Gamma_P\Gamma_Q\bar{\lambda} \\ \bar{\chi}_{\bar{M}\bar{N}} &= \zeta_1\Gamma_{\bar{M}\bar{N}}\lambda \\ \bar{\chi} &= \zeta_1\lambda \end{aligned} \tag{5.13}$$

and

$$\begin{aligned}\Psi_M &= \zeta_2 \Gamma_M \bar{\lambda} \\ \bar{\chi}^+_{MN} &= \zeta_2 \Gamma^+_{MN} \lambda \\ \bar{\eta} &= \epsilon^{MPNQ} \zeta_2 \Gamma_M \Gamma_N \Gamma_P \Gamma_Q \lambda \end{aligned} \tag{5.14}$$

moduli space and $Spin(7)$ **theory**: the moduli space probed by the above TQFT is a holomorphic $(0,2)$–form $\bar{D}B_{02} = 0$ and

$$F^+_{\bar{M}\bar{N}} = 0 \tag{5.15}$$

$$D^{\bar{M}\mathbf{c}} A_{\bar{M}} = 0 \tag{5.16}$$

Notice that the classical action (5.1) is invariant under the group of *complex* gauge transformations $GL(N, \mathbb{C})$. The moduli space described by (5.15) and (5.16) with complex gauge group $GL(N, \mathbb{C})$ should be equivalent to that described by (5.15) and (5.7) with the unitary group $U(N)$. This last moduli space is directly related with that explored by a $Spin(7)$ invariant topological action [35]. In fact, by using (5.7) one realizes that the imaginary part of (5.16) together with (5.15) amount to seven real conditions, which fit in the **7** part of the $Spin(7)$ decomposition of the (real) two–forms $\mathbf{28} = \mathbf{7} \oplus \mathbf{21}$. The real part of (5.16) is the ordinary transversality condition for the unitary gauge group. Then the theory defined by the action (5.1) should be equivalent to that defined by the $Spin(7)$–invariant action

$$I_{\Psi-BF} = \int_{M_8} \Psi \wedge \mathrm{Tr}(B \wedge F) \tag{5.17}$$

where Ψ is the real $Spin(7)$–invariant Cayley four–form and B, F are real two–forms. The mapping can be done by identifying the fundamental representation of the $SU(4)$ group with the real spinor representation of $Spin(7)$.

5.2. SEVEN DIMENSIONS: FROM CY$_4$ TO CY$_3$ × S^1

In [4], the case of writing a BRSTQFT for a G_2 manifold was directly done by starting from the topological action

$$\int_{M_7} d^7x c^{ijk} D_i \varphi F_{jk} \tag{5.18}$$

where c^{ijk} stand for the G_2-invariant tensor made of octonionic structure coefficients, and φ is a Higgs field. The BRST quantization of this topological term yields a twisted version of the dimensional reduction to seven dimensions of the D=8 super Yang–Mills action on a manifold with $Spin(7)$ or $SU(4)$ holonomy. Basically, the topological gauge functions are the generalization of Bogolmony equations in 7 dimensions, as shown in [4].

G_2 manifolds of the kind $\Sigma_6 \times S_1$, where Σ_6 is a Calabi–Yau 3-fold, are of special interest both for mathematical [15] and physical [12] applications. A topological theory for such manifolds can be obtained by considering the dimensional reduction of the model discussed in the previous section, for Calabi–Yau four–fold, although we will shortly give the classical 7-dimensional action that one can directly quantize on such manifolds.

Starting from the ATQFT for CY_4, one can set the fourth component of the gauge connection to $A_{\bar{4}} = A_7 - iL$, where 7 is the direction along the circle S_1 and L a real scalar field. (L will shortly have a special interpretation in seven dimensions.) Moreover, the dimensional reduction impose to set $i(\partial_4 - \partial_{\bar{4}})A_{\bar{M}} = \partial_8 A_{\bar{M}} = 0$ for any $\bar{M} = \bar{1}, \ldots, \bar{4}$. Then one gets from the classical action (5.1) the following action

$$S = \int_{\Sigma_6 \times S^1} \sqrt{g}\, d^7x \,\mathrm{Tr}\Big[2\epsilon^{\bar{m}\bar{n}\bar{p}}(B_{\bar{m}\bar{n}}F_{\bar{p}\bar{4}} + B_{\bar{m}}F_{\bar{n}\bar{p}})\Big] \tag{5.19}$$

with $F_{\bar{p}\bar{4}} = F_{\bar{p}7} - iD_{\bar{p}}L$ and $B_{\bar{m}} = B_{\bar{m}\bar{4}}$. In this section we define the complex three–form $\Omega_{0,3}$ on Σ_6 starting from the (normalized) complex four–form in eight dimensions as $\epsilon_{\bar{m}\bar{n}\bar{p}} = \epsilon_{\bar{m}\bar{n}\bar{p}\bar{4}}$. We thus consider the following classical action:

$$I_{cl}(A, B_{0,2}) = \int_{M_7} \Omega_{3,0} \wedge \mathrm{Tr}(B \wedge F) \tag{5.20}$$

Here the coordinates on the manifold are $z^m, z^{\bar{m}}$ for Σ_6 and the periodic real coordinate x^7 for the circle. The only components of the 2-form that have a relevant propagation are $B_{\bar{m}\bar{n}}, B_{\bar{m}7}$.

The covariant quantization of $B_{\bar{m}\bar{n}}, B_{\bar{m}}(= B_{\bar{m}7})$ and A requires seven topological antighosts $(\kappa^{\bar{m}\bar{n}}, \kappa^{\bar{m}}, \kappa)$. Modulo the ordinary gauge symmetry, we have indeed seven freedoms for the classical action (5.19) associated to the topological ghosts $(\chi_{\bar{m}}, \chi)$ for the fields $(B_{\bar{m}\bar{n}}, B_{\bar{m}})$ and the ghost Ψ_m for the field A_m. The relevant invariance is $B_{\bar{m}\bar{n}} \sim B_{\bar{m}\bar{n}} + D_{[\bar{m}}\chi_{n]}$, $B_{\bar{m}} \sim B_{\bar{m}} + D_m\chi$ $A_m \sim A_m + \Psi_m$, $A_{\bar{m}} \sim A_{\bar{m}}$, $A_7 \sim A_7$, modulo ordinary gauge transformations. Moreover, in the BRST complex for the field B it will appear also three commuting scalar ghost of ghosts (Φ, L, $\bar{\Phi}$), for the "gauge symmetries" of the topological ghosts and antighosts of B_2. All these fields are conveniently displayed as elements of the following pyramidal diagram:

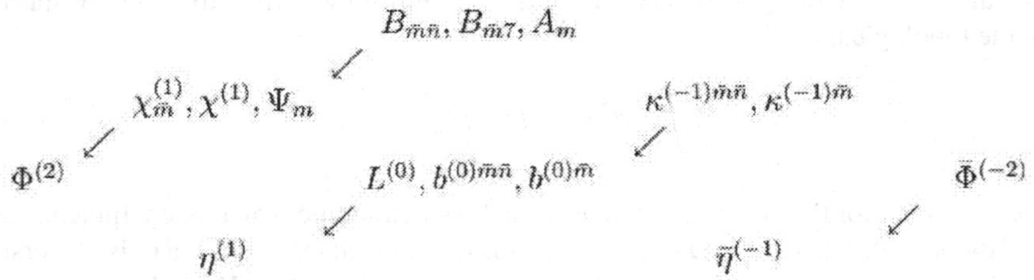

$$\begin{array}{ccccccc}
 & & & B_{\bar{m}\bar{n}}, B_{\bar{m}7}, A_m & & & \\
 & & \swarrow & & & & \\
 & \chi^{(1)}_{\bar{m}}, \chi^{(1)}, \Psi_m & & & & \kappa^{(-1)\bar{m}\bar{n}}, \kappa^{(-1)\bar{m}} & \\
 \swarrow & & & & \swarrow & & \\
\Phi^{(2)} & & & L^{(0)}, b^{(0)\bar{m}\bar{n}}, b^{(0)\bar{m}} & & & \bar{\Phi}^{(-2)} \\
 & & \swarrow & & & \swarrow & \\
 & \eta^{(1)} & & & & \bar{\eta}^{(-1)} &
\end{array}$$

From the point of view of the dimensional reduction of the CY$_4$ theory, it is interesting to observe that the medium ghost of ghost L of the B field can be identified with the component A_8 of the eight dimensional gauge field. The BRST transformations of the fields can be read from the following Batalin–Vilkoviski action :

$$\begin{aligned} S &= \int_{\Sigma_6\times S^1} \sqrt{g}\, d^7x \,\mathrm{Tr}\Big[2\epsilon^{\bar m\bar n\bar p}(B_{\bar m\bar n}F_{\bar p\bar 4} + B_{\bar m}F_{\bar n\bar p}) \\ &\quad +{}^*B^{\bar m\bar n}(D_{[\bar m}\chi_{\bar n]} - [c, B_{\bar m\bar n}] + \tfrac{1}{2}\epsilon_{\bar m\bar n\bar p}[{}^*B^{\bar p}, \Phi]) \\ &\quad +{}^*B^{\bar m}(2D_{[\bar m}\chi_{7]} - 2[c, B_{\bar m}] + \tfrac{1}{2}\epsilon_{\bar m\bar n\bar p}[{}^*B^{\bar n\bar p}, \Phi]) \\ &\quad +{}^*\chi^{\bar m}(D_{\bar m}\Phi - [c,\chi_{\bar m}]) +{}^*\chi(D_{\bar 4}\Phi - [c,\chi]) \\ &\quad +{}^*A^m(\Psi_m + D_{\bar m}c) +{}^*A^{\bar m}D_{\bar m}c +{}^*A^7D_7c -{}^*\Psi^m[c,\Psi_m] \\ &\quad +{}^*c(-\tfrac{1}{2}[c,c]) +{}^*\bar\Phi\bar\eta +{}^*L\eta \\ &\quad +{}^*\kappa_{\bar m\bar n}b^{\bar m\bar n} + 2{}^*\kappa_{\bar m}b^{\bar m} +{}^*\kappa b\Big] \end{aligned} \tag{5.21}$$

This action is actually the dimensional reduction of action (5.3). However, the interpretation of the ghost of ghost system is quite different. Notice that the vectorial part of the BRST transformation for the field $B_{\bar m}$ is

$$\delta B_{\bar m} = D_{\bar m}\chi - D_{\bar 4}\chi_{\bar m} = D_{\bar m}\chi - D_7\chi_{\bar m} \tag{5.22}$$

The variation of the action due to the first term in (5.22) simply vanishes after integration by parts because of the identity (4.3). In the context of the dimensional reduction, this identity can be read as the $\bar M = \bar 4$ component of the eight–dimensional one $\epsilon^{\bar M\bar N\bar P\bar Q}D_{\bar M}F_{\bar N\bar Q} = 0$. The variation associated to the last term in (5.22), together with that coming from the variation of the field $B_{\bar m\bar n}$, $QB_{\bar m\bar n} = D_{[\bar m}\chi_{\bar n]}$, gives after integration by parts the other three components $\bar M = \bar 1, \bar 2, \bar 3$ of the above eight–dimensional identity. The topological freedom of the classical action (5.19) can be fixed by choosing seven independent gauge functions. The first six can be directly obtained from the dimensional reduction of (5.4)

$$\begin{aligned} B_{\bar m\bar n} &= \tfrac{1}{2}(F_{\bar m\bar n} - \epsilon_{\bar m\bar n\bar p}F^{\bar p\bar 4}) \\ B_{\bar m} &= \tfrac{1}{2}(F_{\bar n\bar 4} - \tfrac{1}{2}\epsilon_{\bar n\bar p\bar q}F^{\bar p\bar q}) \\ &= \tfrac{1}{2}(F_{\bar n7} - iD_{\bar n}L - \tfrac{1}{2}\epsilon_{\bar n\bar p\bar q}F^{\bar p\bar q}) \end{aligned} \tag{5.23}$$

while the seventh one corresponds to the imaginary part of the complex gauge–fixing condition (5.6)

$$\tfrac{1}{2}J^{m\bar n}F_{m\bar n} + D_7L = 0 \tag{5.24}$$

Notice in fact that the reduction on a manifold of the kind $\Sigma_6 \times S^1$ breaks the complex group of gauge invariance of the eight–dimensional action (5.1) to the unitary group. The residual gauge invariance under this group can be fixed by the ordinary transversality condition $\partial^{\bar m}A_{\bar m} + \partial^m A_m + \partial^7A_7 = 0$. The gauge–fixing fermion corresponding to the conditions (5.23) and (5.24) is

$$\begin{aligned} Z &= \kappa^{\bar m\bar n}\left[B_{\bar m\bar n} - \tfrac{1}{2}\left(F_{\bar m\bar n} - \epsilon_{\bar m\bar n\bar p}(F^{\bar p7} + iD^{\bar p}L)\right)\right] \\ &\quad +\kappa^{\bar m}\left(B_{\bar m} - \frac{3}{2}(F_{\bar n7} - iD_{\bar n}L - \tfrac{1}{2}\epsilon_{\bar n\bar p\bar q}F^{\bar p\bar q})\right) \\ &\quad +\kappa\Big(2(J^{\bar m n}F_{\bar m n} + D_7L) - b\Big) \\ &\quad +\bar\Phi(D^{\bar m}\chi_{\bar m} + D^7\chi) \end{aligned} \tag{5.25}$$

The bosonic part of the gauge–fixed action, which can be obtained as usual by using the BV equation (3.12) and enforcing the gauge conditions (5.23) and (5.24) by integration on the Lagrangian multipliers, reads

$$\begin{aligned} S^{g.f.} &= \int_{M_7} d^7x\sqrt{g}\,\mathrm{Tr}\Big(-\frac{3}{2}F^{\bar{m}\bar{n}}F_{\bar{m}\bar{n}} + |J^{m\bar{n}}F_{m\bar{n}}|^2 - \tfrac{1}{2}F^{\bar{m}7}F_{\bar{m}7} \\ &\quad +\bar{\Phi}(D^{\bar{m}}D_{\bar{m}} + D^7D^7)\Phi + L(D^{\bar{m}}D_{\bar{m}} + D^7D^7)L\Big) \end{aligned} \tag{5.26}$$

and can be identified with the bosonic part of the seven dimensional Super Yang–Mills on $M_7 = \Sigma_6 \times S^1$.

Concerning the fermionic sector, we have 8 topological ghosts, $\Psi_{\bar{m}}, \chi_{\bar{m}}, \chi$ and η, and 8 topological antighosts $\kappa^{\bar{m}\bar{n}}, \kappa^{\bar{m}}, \kappa$ and $\bar{\eta}$. The mapping with spinors can be obtained from the dimensional reduction of the eight–dimensional mapping (5.13) and (5.14). After the dimensional reduction to $\Sigma_6 \times S^1$, the two covariantly–constant eight–dimensional chiral spinors (ζ_1, ζ_2) are identified with the unique covariantly constant Majorana spinor ξ of the Calabi–Yau three–fold Σ_6. On the other side, the eight dimensional spinor λ yields two seven dimensional Majorana spinors (λ_1, λ_2). This results in the mapping

$$\begin{aligned} &\Psi_m \to \xi\Gamma_m\lambda_1 \ , \ \eta \to \xi\Gamma_7\lambda_1 \\ &\chi^{\bar{m}} \to \epsilon^{mnp}\xi\Gamma_{np}\lambda_1 \ , \ \chi \to \epsilon^{mnp}\xi\Gamma_{mnp}\lambda_1 \end{aligned} \tag{5.27}$$

for the topological ghosts and

$$\begin{aligned} &\kappa_{mn} \to \xi\Gamma_{mn}\lambda_2 \ , \ \kappa_m \to \xi\Gamma_{m7}\lambda_2 \\ &\kappa \to \xi\lambda_2 \ , \ \bar{\eta} \to \epsilon^{mnp}\Gamma_{mnp7}\lambda_2 \end{aligned} \tag{5.28}$$

for the topological antighosts. By this mapping we can identify the topological action (5.19) gauge fixed with the conditions (5.23) and (5.24) as the twisted version of $N = 2$ seven dimensional Super Yang–Mills[6]. The observables of the topological model can be identified with the dimensional reduction of the eight-dimensional cocycles.

5.3. MANIFOLDS WITH $SU(4)$ GROUP STRUCTURE AND SELF–DUAL YANG–MILLS IN EIGHT–DIMENSIONS

On a Kähler manifold with a $SU(4)$ group structure we can choose $B_{4,2} = J \wedge J \wedge B^+_{2,0}$, where J is as usual the Kähler $(1,1)$ form. Here the 2-form $B^+_{2,0}$ is self dual in the indices $[mn]$. Self duality is defined from a $SU(4)$ invariant 4-form $\Omega_{4,0}$, which is globally well defined, but not necessarily closed. It counts for 3 degrees of freedom, according to the $SU(4)$ independent decomposition of a 2-form in 8 dimensions:

$$28 = 6 \oplus \bar{6} \oplus 15 \oplus 1 \tag{5.29}$$

and a further decomposition 6 as $6 = 3 \oplus 3$, using the ϵ_{mnpq} tensor.

[6]As discussed for the four dimensional case, in order to recover the Yukawa couplings and the quartic term in the potential $[\bar{\Phi}, \Phi]^2$ typical of the $N = 2$ SYM one should slightly modify the gauge–fixing fermion (5.25).

The novelty of this case is that neither Kähler $(1,1)$-form nor $\Omega_{4,0}$ are necessarily closed. Both the forms can also be rewritten in terms of spinors, which correspondingly are not covariantly constant with respect to the usual spin connection, but only with respect to a modified connection including torsion terms. The corresponding classical action is a generalisation of that in (3.3):

$$I_{k-BF}(A, B^+_{2,0}, B^+_{0,2}) = \int_{M_8} J \wedge J \wedge \mathrm{Tr}(B^+_{2,0} \wedge F_{0,2}) \tag{5.30}$$

The symmetries of this action are

$$\begin{aligned}
&QA_m = \Psi_m \qquad QA_{\bar m} = D_{\bar m} c \\
&QB^+_{mn} = -[c, B^+_{mn}]\,, \qquad QB_{\bar m\bar n}{}^+ = \left(D_{[\bar m}\Psi_{\bar n]}\right)^+ - [c, B^+_{\bar m\bar n}] + [{}^*B^{mn\,+}, \Phi] \\
&Q\Psi_m = 0\,, \qquad Q\Psi_{\bar m} = D_{\bar n}\Phi - [c, \Psi_{\bar n}] \\
&Q\Phi = -[c, \Phi]\,, \qquad Qc = -\tfrac{1}{2}[c, c]
\end{aligned} \tag{5.31}$$

The quantization of this action can be worked out using the BV formalism, and is very similar to that already discussed in [4].

We have 8 freedoms for gauge-fixing the system $(B^+_{2,0}, B^+_{0,2}, A_{1,0}, A_{0,1})$. Indeed, Ψ_m, $\Psi_{\bar m}$ and c have respectively 4,4 and 1 components, but $\Psi_{\bar m}$ has a ghost of ghost symmetry with ghost of ghost Φ, so it only counts for 3=4-1 freedoms. We can choose the following 7 gauge–fixing conditions in the gauge covariant sector:

$$\begin{aligned}
B^+_{mn} &= \epsilon_{mnlp\bar m\bar n\bar l\bar p} F^{\bar m\bar n +} J^{l\bar l} J^{p\bar p} \\
B_{\bar m\bar n^+} &= 0 \\
J^{m\bar n} F_{m\bar n} &= 0
\end{aligned} \tag{5.32}$$

plus the ordinary transversality condition for the gauge field, $(\partial^m A_m + \partial^{\bar m} A_{\bar m} = 0)$. The transformation law of B^+_{mn} implies that a gauge fixing for $\Psi_{\bar m}$ must be also done, with a gauge function :

$$D^{\bar m}\Psi_{\bar m} = 0 \tag{5.33}$$

The Q-invariant gauge fixing of the action with these functions is standard, and reproduces the action as in [4], that is the twisted form of the eight–dimensional Super Yang–Mills theory. The classical action (5.30) can be considered as an eight–dimensional generalization of the self–dual Yang–Mills in four dimensions, in particular of its realisation studied in [22]. We should notice that we are at the extreme point of the definition of an ATQFT. The action is the sum of a d-closed term and a Q-exact term. Thus, there is a ring of observables defined from the cohomology of the BRST operator. However, the classical action is actually completely dependent of the metrics of the manifold, since it depends on both the Kähler form and the complex form at the same time.

Acknowledgements

This work was done in collaboration with A. Tanzini. L.B. is grateful to I.M. Singer for interesting discussions.

References

1. L. Baulieu, *Gravitational Topological Quantum Field Theory Versus N = 2, D = 8 Supergravity and its lift to N = 1 D = 11 Supergravity*, JHEP 0404 (2004) 044, hep-th/0304221 ; L. Baulieu, M. Bellon and A. Tanzini, *"Supergravity and the knitting of the Kalb-Ramond two-form in eight-dimensional topological gravity,"*, Phys. Lett. B **565** (2003) 211 [arXiv:hep-th/0303165]; *"Eight-dimensional topological gravity and its correspondence with supergravity,"* Phys. Lett. B **543** (2002) 291 [arXiv:hep-th/0207020].
2. D. Anselmi and P. Fre, *"Twisted N=2 supergravity as topological gravity in four-dimensions,"* Nucl. Phys. B **392** (1993) 401 [arXiv:hep-th/9208029]; L. Baulieu and A. Tanzini, *Topological gravity versus supergravity on manifolds with special holonomy,* JHEP **0203** (2002) 015, [hep-th/0201109]; P. de Medeiros and B. Spence, *"Four-dimensional topological Einstein-Maxwell gravity,"* Class. Quant. Grav. **20** (2003) 2075 [arXiv:hep-th/0209115].
3. L. Baulieu, *"Going down from a 3-form in 16 dimensions"*, Phys.Lett. B544 (2002) 367-373, hep-th/0207184.
4. L. Baulieu, H. Kanno and I.M. Singer, *Special Quantum Field Theories In Eight And Other Dimensions,* Commun.Math.Phys.**194** (1998) 149, [hep-th/9704167] and [hep-th/9705127]; L. Baulieu and C. Laroche, *On Generalized Self-Duality Equations Toward Supersymmetric Quantum Field Theories of Forms,* Mod.Phys Lett. **A13** (1998) 1115, [hep-th/9801014].
5. B. S. Acharya, J. M. Figueroa-O'Farrill, B. Spence and M. O'Loughlin, *"Euclidean D-branes and higher-dimensional gauge theory,"* Nucl. Phys. B **514** (1998) 583 [arXiv:hep-th/9707118]; *"Higher-dimensional analogues of Donaldson-Witten theory,"* Nucl. Phys. B **503** (1997) 657 [arXiv:hep-th/9705138].
6. M. Blau and G. Thompson, *"Euclidean SYM theories by time reduction and special holonomy manifolds,"* Phys. Lett. B **415** (1997) 242 [arXiv:hep-th/9706225].
7. A. Iqbal, N. Nekrasov, A. Okounkov and C. Vafa, *"Quantum foam and topological strings,"*, hep-th/0312022.
8. A. Neitzke and C. Vafa, *"N = 2 strings and the twistorial Calabi-Yau,"* hep-th/0402128.
9. N. Nekrasov, H. Ooguri and C. Vafa, *"S-duality and topological strings,"* JHEP **0410** (2004) 009, hep-th/0403167.
10. A. Kapustin, *"Gauge theory, topological strings, and S-duality,"* JHEP **0409** (2004) 034 hep-th/0404041.
11. H. Ooguri, A. Strominger and C. Vafa, *"Black hole attractors and the topological string,"* hep-th/0405146.
12. R. Dijkgraaf, S. Gukov, A. Neitzke and C. Vafa, *Topological M-theory as Unification of Form Theories of Gravity* , hep-th/0411073.
13. A. Gerasimov and S. L. Shatashvili, *Towards Integrability of Topological Strings I: Three-forms on Calabi-Yau manifolds*, hep-th/0409238.
14. P. A. Grassi and P. Vanhove, *Topological M theory from pure spinor formalism,* hep-th/0411167.
15. N. Hitchin, *Stable forms and special metrics*, "Proceedings of the Congress in memory of Alfred Gray", (eds M. Fernandez and J. Wolf), AMS Contemporary Mathematics Series, math.DG/0107101; *The geometry of three-forms in six and seven dimensions*, math.DG/0010054.
16. A. S. Cattaneo, P. Cotta-Ramusino, F. Fucito, M. Martellini, M. Rinaldi, A. Tanzini and M. Zeni, *"Four-dimensional Yang-Mills theory as a deformation of topological BF theory,"* Commun. Math. Phys. **197** (1998) 571 [arXiv:hep-th/9705123].
17. L. Baulieu and S. L. Shatashvili, *"Duality from topological symmetry,"* JHEP **9903** (1999) 011 [arXiv:hep-th/9811198].
18. D. Maulik, N.Nekrasov, A. Okounkov and R. Pandharipande, math.AG/0312059, math.AG/0406092.
19. M. J. Plebanski, *"On the separation of Einsteinian substructures,"* J. Math. Phys. 18 (1977), 2511; G. T. Horowitz, *"Exactly Soluble Diffeomorphism Invariant Theories,"* Commun. Math. Phys. **125** (1989) 417; C. G. Torre, *"Perturbations Of Gravitational Instantons,"* Phys. Rev. D **41** (1990) 3620; M. Abe, A. Nakamichi and T. Ueno, *"Gravitational instantons and moduli*

spaces in topological two form gravity," Phys. Rev. D **50**, 7323 (1994), hep-th/9408178.
20. work in progress.
21. E. Witten, *"(2+1)-Dimensional Gravity As An Exactly Soluble System,"* Nucl. Phys. B **311** (1988) 46.
22. W. Siegel, *"N=2, N=4 string theory is selfdual N=4 Yang-Mills theory,"* Phys. Rev. D **46** (1992) 3235.
23. L. Baulieu, I. M. Singer, Nucl.Phys.Proc.Suppl. **B5** (1988) 12.
24. A. D. Popov, *"Holomorphic analogs of topological gauge theories,"* Phys. Lett. B **473** (2000) 65 [arXiv:hep-th/9909135]; T. A. Ivanova and A. D. Popov, *"Dressing symmetries of holomorphic BF theories,"* J. Math. Phys. **41** (2000) 2604 [arXiv:hep-th/0002120].
25. J. S. Park, *N=2 topological Yang-Mills theory on compact Kahler surfaces,* Commun. Math. Phys. **163** (1994) 113 [arXiv:hep-th/9304060]; J. S. Park, *Holomorphic Yang-Mills theory on compact Kahler manifolds,* Nucl. Phys. B **423** (1994) 559 [arXiv:hep-th/9305095].
26. A. Johansen, *Twisting of N=1 SUSY Gauge Theories and Heterotic Topological Theories* , Int.J.Mod.Phys. A10 (1995) 4325-4358, hep-th/9403017;
27. E. Witten, *Supersymmetric Yang-Mills Theory On A Four-Manifold,* J.Math.Phys. 35 (1994) 5101-5135, hep-th/9403195.
28. R. Dijkgraaf, J. S. Park and B. J. Schroers, *N = 4 supersymmetric Yang-Mills theory on a Kaehler surface,* arXiv:hep-th/9801066.
29. C. Hofman and J. S. Park, *"Cohomological Yang-Mills theories on Kaehler 3-folds,"* Nucl. Phys. B **600** (2001) 133 [arXiv:hep-th/0010103].
30. L. Baulieu, *Algebraic Construction Of Gauge Invariant Theories,* Particles and Fields, Cargese, Published in Cargese Summer Inst. 1983, 1; *On Forms With Non-Abelian Charges And TheirDualities,* Phys.Lett. **B441** (1998) 250 , hep-th/9808055; L. Baulieu, E. Rabinovici, *Selfduality And New TQFTs For Forms,* JHEP 9806 (1998) 0068.
31. H.B. Lawson and M.-L. Michelsohn, *Spin geometry*, Princeton University Press, Princeton, New Jersey, 1989.
32. E. Witten, *"Chern-Simons gauge theory as a string theory,"* Prog. Math. **133** (1995) 637 [arXiv:hep-th/9207094].
33. A. S. Cattaneo, P. Cotta-Ramusino, J. Frohlich and M. Martellini, *"Topological BF theories in three-dimensions and four-dimensions,"* J. Math. Phys. **36** (1995) 6137 [arXiv:hep-th/9505027].
34. D. Ray and A. Singer, *"Analytic torsion for complex manifdlds,"* Ann. Math. **98** (1973) 154–177.
35. S. K. Donaldson and R. P. Thomas, *"Gauge theory in higher dimensions,", Prepared for Conference on Geometric Issues in Foundations of Science in honor of Sir Roger Penrose's 65th Birthday, Oxford, England, 25-29 Jun 1996*; R. P. Thomas, PhD thesis, Oxford, 1997.
36. I. Frenkel, B. Khesin and A. Todorov, *Complex counterpart of Chern–Simons–Witten theory and holomorphic linking*, preprint.
37. S. Axelrod and I. M. Singer, *"Chern-Simons perturbation theory,"*, hep-th/9110056; J. Diff. Geom. **39** (1994) 173 [arXiv:hep-th/9304087].
38. P. de Bartolomeis and G. Tian, J. Diff. Geom. **43** (1996) 231.
39. V.G. Turaev and O. Yu. Viro, *"Sum Invariants of 3–Manifolds and Quantum 6j–symbols"*, Topology **31** (1992) 865.
40. N. Reshetikhin and V.G. Turaev, *"Invariants of 3–Manifolds via Link Polynomials and Quantum Groups"*, Invent. Math. **103** (1991) 547.

LOEWNER CHAINS

MICHEL BAUER AND DENIS BERNARD
Service de Physique Théorique de Saclay CEA-Saclay,
91191 Gif-sur-Yvette, France

Abstract. These lecture notes on 2D growth processes are divided in two parts. The first part is a non-technical introduction to stochastic Loewner evolutions (SLEs). Their relationship with 2D critical interfaces is illustrated using numerical simulations. Schramm's argument mapping conformally invariant interfaces to SLEs is explained. The second part is a more detailed introduction to the mathematically challenging problems of 2D growth processes such as Laplacian growth, diffusion limited aggregation (DLA), etc. Their description in terms of dynamical conformal maps, with discrete or continuous time evolution, is recalled. We end with a conjecture based on possible dendritic anomalies which, if true, would imply that the Hele-Shaw problem and DLA are in different universality classes.

Growth phenomena are ubiquitous in the physical world at many scales, from crystals to plants to dunes and larger. They can be studied in many frameworks, deterministic of probabilistic, in discrete or continuous space and time. Understanding growth is usually a very difficult task. This is true even in two dimensions, the case we concentrate on in these notes.

Yet two dimensions is a highly favorable situation because it allows to make use of the power of complex analysis in one variable. In many interesting cases, the growing object in two dimensions can be seen as a domain, i.e. a contractible open subset of the Riemann sphere (the complex plane with a point at infinity added). A deep theorem of Riemann asserts that such a domain, whatever complicated and fancy, is conformally equivalent to a simple reference domain, which is usually taken as the upper-half plane or the unit disk. This conformal equivalence is unique once an appropriate normalization, which may depend on the growth problem at hand, has been chosen. Cauchy's theorem allows to write down an integral representation for the conformal map as an integral along the boundary of the reference domain, involving a density. This density is time dependent. Then the time derivative of the conformal map has an analogous representation and a nice way to specify the growth rule is often directly on this density. This leads to the concept of Loewner chains, which is the central theme of these notes. We shall illustrate Loewner chains in several situations.

L. Baulied et al. (eds.), String Theory: From Gauge Interactions to Cosmology, 41–77.

Our aim is to give a pedagogical introduction to a beautiful subject. We wanted to show that it leads to many basic mathematical structures whose appearance in the growth context is not so easy to foresee, like Brownian motion, integrable systems and anomalies to mention just a few. We have also tried to stress that some growth processes have rules which are easy to simulate on the computer. A few minutes of CPU are enough to get an idea of the shape of the growing patterns, to be convinced that something interesting and non trivial is going on, and even sometimes to get an idea of fractal dimensions. This is of course not to be compared with serious large scale simulations, but it is a good illustration of the big contrast between simple rules, complex patterns and involved mathematical structures. However, other growth models, and among those some have been conjectured to be equivalent to simple ones, have resisted until recently to precise numerical calculations due to instabilities.

To avoid any confusion, let us stress that being able to describe a growth process using tools from complex analysis and conformal geometry does not mean that the growth process itself is conformally invariant at all. Conformal invariance of the growth process itself puts rather drastic conditions on the density that appears in the Loewner chain.

This is illustrated by the first part of these notes, which deals with conformally invariant interfaces and their relation to stochastic Loewner evolutions. This part is an elaboration of the the main points developed during the lectures.

The second part is an introduction to a larger class of processes describing the growth of possibly random fractal planar domains and a review of some of their basic properties. Due to lack of time, this part was not presented during the lectures.

The study of the continuum limit of non-intersecting curves on the lattice has been a subject of lasting interest both in mathematics and in physics. A famous example is given by self-avoiding random walks. The motivation comes from combinatorics, but also from statistical mechanics. Two dimensions is most interesting because a non-intersecting curve is the boundary between two domains and can very often be interpreted as an interface separating two coexisting phases.

At a critical point and for short range interactions, such interfaces are expected to be conformally invariant. The argument for that was given two decades ago in the seminal paper on conformal field theory [1]. The rough idea is the following. At a critical point, a system becomes scale invariant. If the interactions on the lattice are short range, the model is described in the continuum limit by a local field theory and scale invariance implies that the stress tensor is traceless. In two dimensions this is enough to ensure that the theory transforms simply –no dynamics is involved, only pure kinematics– when the domain where it is defined is changed by a conformal transformation.

The local fields are classified by representations of the infinite dimensional Virasoro algebra and this dictates the way correlation functions transform. This has led to a tremendous accumulation of exact results using conformal field theory (CFT). A situation that is well under control is that of unitary minimal models. The Hilbert space of the system splits as a finite sum of representations of the Virasoro algebra, each associated to a (local) primary field, and the corresponding correlation functions can be described rather explicitly. The study of non-local objects like interfaces at criticality has not seen such a systematic development and only isolated though highly artful results [2, 3, 12] have been discovered using conformal field theory techniques.

Non local objects like interfaces are not classified by representations of the Virasoro algebra but the reasoning that led O. Schramm to the crucial breakthrough [4], i.e. the definition of stochastic Loewner evolutions (SLEs), rests on a fairly obvious but cleverly exploited statement of what conformal invariance means for an interface. Surprisingly it allows to turn this problem into growth problem, something which looks natural only a posteriori. For some years, probabilistic techniques have been applied to interfaces, leading to a systematic understanding that was lacking on the CFT side. A sample, surely biased by our ignorance, can be found in refs.[5, 6, 7, 8, 9]. There is now a satisfactory understanding of interfaces in the continuum limit. However, from a mathematical viewpoint, giving proofs that a discrete interface on the lattice has a conformally invariant limit remains a hard challenge and only a handful of cases has been settled up to now.

There is now a good explanation of the –initially mysterious– relationship between SLE and CFT. This was one of the main topics of the lectures. Though this aspect is mostly due to the work of the present authors ([10], but see also [11]), we have decided to be very sketchy. The interested reader is referred to the literature.

Stochastic Loewner evolution is a simple but particularly interesting example of growth process for which the growth is local and continuous so that the resulting set is a curve without branching. Of course other examples have been studied in connection with 2d physical systems. The motivations are sometimes very practical. For instance, is it efficient to put a pump in the center of oil film at the surface of the ocean to fight against pollution? The answer has to do with the Laplacian growth or Hele-Shaw problem [14]. The names diffusion limited aggregation [15], see Figure (1), and dielectric breakdown [16] speak for themselves. Various models have been invented, sometimes with less physical motivation, but in order to find more manageable growth processes. These include various models of iterated conformal maps [17], etc. See refs.[18] for recent reviews. As mentioned before, in most cases the shape of the growing domains is encoded in a uniformizing conformal map whose evolution describes the evolution of the domain. The dynamics can be either discrete or continuous in time, it can be either deterministic or stochastic. But the growth process is always described by a simple generalization of the Loewner equation called a Loewner chain.

The general understanding of these Loewner chains is still embryonic, in strong contrast with the case of Loewner evolutions, which deal with local growth. For many problems the most basic question are unanswered. For instance, as we shall explain below, Laplacian growth is at the same time a completely integrable system and an ill-posed problem because it develops singularities in finite time. Hydrodynamics gives a natural regularization via the introduction of surface tension. It has been argued for some time that in the limit of vanishing surface tension one retrieves a model which is in the same universality class as diffusion limited aggregation. But the experimental and numerical evidence is inconclusive and there is no consensus. In fact, at the end of these notes we shall give an argument suggesting that they belong to different universality classes. This is very conjectural, but as should be clear from this introduction, non-local growth processes are certainly a source of interesting and challenging problems.

Figure 1. An example of DLA cluster obtained by iterating conformal maps.

1. Critical interfaces and stochastic processes

The following pedagogical introduction to conformally invariant random curves breaks in two part. The first is a list of examples of critical interfaces on the lattice and some of their properties. The second is a presentation of O. Schramm's derivation of SLE.

In this part the upper-half plane is used as a reference geometry.

1.1. THREE EXAMPLES AND A GENERALITY

Let us start with three examples. Our aim is to explain their definitions, to show a few samples and, in the first two cases, to give a numerical estimate of the fractal dimension.

The first, loop-erased random walks, belongs to the realm of "pure combinatorics".

The second, percolation, is at the frontier between pure combinatorics and statistical mechanics because it arises very naturally as the domain boundary of a statistical mechanics model, but the Boltzmann weights are trivial. It is however a limiting case of a family of models with non trivial Boltzmann weights and, from that point of view, quantities relevant for percolation are obtained by exploring an infinitesimal neighborhood around the trivial Boltzmann weights.

The third example, the Ising model, is deeply rooted in statistical mechanics.

We shall end this section by abstracting a crucial property of interfaces which allows to make an efficient use of conformal invariance and derive the Markov property for SLEs.

1.1.1. *Loop-erased random walks*

This example is purely of combinatorial nature. A loop-erased random walk is a random walk with loops erased along as they appear. More formally, if $X_0, X_1, \cdots, X_n$ is a finite sequence of abstract objects, we define the associated loop-erased sequence by the following recursive algorithm.

Until all terms in the sequence are distinct,

Step 1 Find the couple (l, m) with $0 \leq l < m$ such that the terms with indexes from 0 to $m-1$ are all distinct but the terms with indexes m and l coincide.

Step 2 Remove the terms with indexes from $l+1$ to m, and shift the indexes larger than m by $l-m$ to get a new sequence.

Let us look at two examples.

For the "month" sequence $j, f, m, a, m, j, j, a, s, o, n, d$, the first loop is m, a, m, whose removal leads to $j, f, m, j, j, a, s, o, n, d$, then j, f, m, j, leading to j, j, a, s, o, n, d, then j, j leading to j, a, s, o, n, d where all terms are distinct.

For the "reverse month" sequence $d, n, o, s, a, j, j, m, a, m, f, j$, the first loop is j, j, leading to $d, n, o, s, a, j, m, a, m, f, j$, then a, j, m, a leading to d, n, o, s, a, m, f, j.

This shows that the procedure is not "time-reversal" invariant. Moreover, terms that are within a loop can survive: in the second example m, f, which stands in the j, m, a, m, f, j loop, survives because the first j is inside the loop a, j, m, a which is removed first.

A loop-erased random walk is when this procedure is applied to a (two dimensional for our main interest) random walk. This is very easy to simulate. Fig.2 represents a loop-erased walk of 200 steps. The thin lines build the shadow of the random walk (where shadow means that we do not keep track of the order and multiplicity of the visits) and the thick line is the corresponding loop-erased walk. The time asymmetry is clearly visible and allows to assert with little uncertainty that the walk starts on the left.

Figure 2. A loop-erased random walk with its shadow.

To fit with the general SLE framework, let us restrict to loop-erased random walks in the upper-half plane. There are a few options for the choice of boundary conditions.

A first choice is to consider reflecting boundary conditions on the real axis for the random walk.

Another choice is annihilating boundary conditions: if the random walk hits the real axis, one forgets everything and starts anew at the origin. Why this is a natural boundary condition has to wait until section 1.1.4.

Due to the fact that on a two-dimensional lattice a random walk is recurrent (with probability one it visits any site infinitely many times), massive rearrangement occur with probability one. This means that if one looks at the loop-erased random walk associated to a given random walk, it does not have a limit in any sense when the size of the random walk goes to infinity. Let us illustrate this point. The samples in fig.3 were obtained with reflecting boundary conditions. It takes 12697 random walk steps to build a loop-erased walk of length 633, but step 12698 of the random walk closes a long loop, and then the first occurrence of a loop-erased walk of length 634 is after 34066 random walk steps. Observe that in the mean time most of the initial steps of the loop-erased walk have been reorganized.

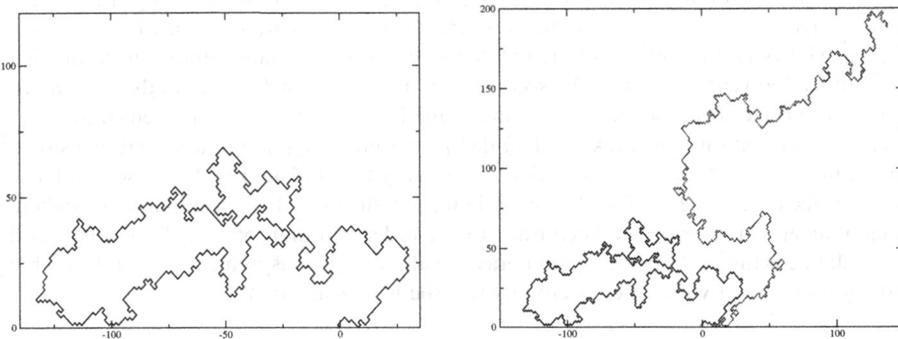

Figure 3. On the left: a large loop is about to be created. On the right: the massive rearrangement to go from 633 to 634 steps.

However, simulations are possible because when the length of the random walk tends to infinity, so does the maximal length of the corresponding loop-erased walk with probability one.

Though annihilating boundary conditions lead to remove even more parts of the random walk than the reflecting ones, the corresponding process can be arranged (conditioned in probabilistic jargon) to solve the problem of convergence as follows.

Instead of stopping the process when the loop-erased walk has reached a given length, one can stop it when it reaches a certain altitude, say n, along the y-axis. Whatever the corresponding random walk has been, the only thing that matters is the last part of it, connecting the origin to altitude n without returning to altitude 0. Moreover, the first time the loop-erased walk reaches altitude n is exactly the first time the random walk reaches altitude n. Now a small miracle happens: if a 1d symmetric random walk is conditioned to reach altitude n before it hits the origin again, the resulting walk still has the Markov property. It is a discrete equivalent to the 3d Bessel process (a Bessel process describes the norm of a Brownian motion, however no knowledge of Bessel processes is needed here, we just borrow the name). When at site m, $0 < m < n$, the probability to go to $m \pm 1$ is $(1 \pm 1/m)/2$, independently of all previous steps. Observe that there is no n dependence so that we can forget about n, i.e. let it go to infinity. The discrete 3d Bessel process is not recurrent and tends to infinity with probability one: for any altitude l there is with probability one a time after which the discrete 3d Bessel process remains above l

for ever. Henceforth, we choose to simulate a symmetric simple random walk along the x axis and the discrete 3d Bessel process along the y-axis and we erase the loops of this new process. This leads to the convergence of the loop-erased walk and numerically to a more economical simulation.

Fig.4 is a simulation of about 10^5 steps, both for reflecting and annihilating boundary conditions. At first glance, one observes in both cases similar simple (no multiple points) but irregular curves with possibly fractal behavior.

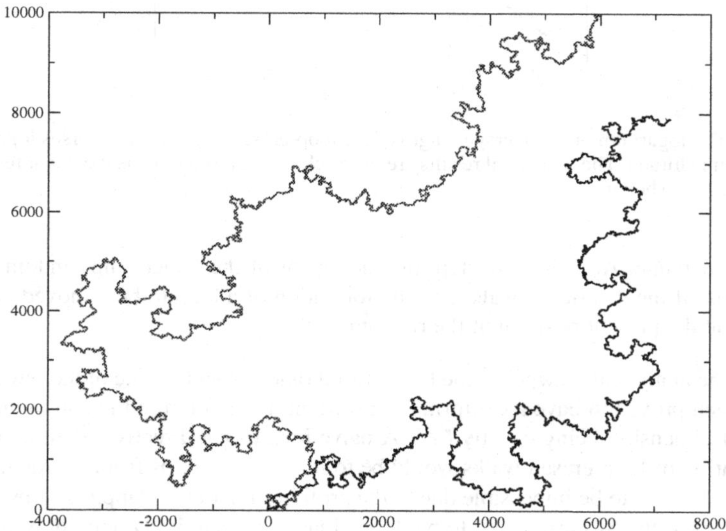

Figure 4. A sample of the loop-erased random walk for the two boundary conditions.

To estimate the Hausdorff dimensions in both cases, we have generated samples of random walks, erased the loops and made the statistics of the number of steps S of the resulting walks compared to a typical length L (end-to-end distance for reflecting boundary conditions, maximal altitude for annihilating boundary conditions). In both cases, one observes that $S \propto L^\delta$ and a modest numerical effort (a few hours of CPU) leads to $\delta = 1.25 \pm .01$. This is an indication that the boundary conditions do not change the universality class.

To get an idea of how small the finite size corrections are, observe fig.5. The altitude was sampled from 2^4 to 2^{13}. The best fit gives a slope 1.2496 and the first two points already give 1.2403.

As recalled in the introduction, it is believed on the basis of intuitive arguments that in two dimensions scale invariance implies conformal invariance, providing there are no long range interactions. What does this absence of long range interactions mean for loop-erased random walks? Clearly along the loop-erased walk there are long range correlations, if only because a loop-erased random walk cannot cross itself. However, the relevant feature for the intuitive argument is that, in the underlying 2d physical space, interactions are

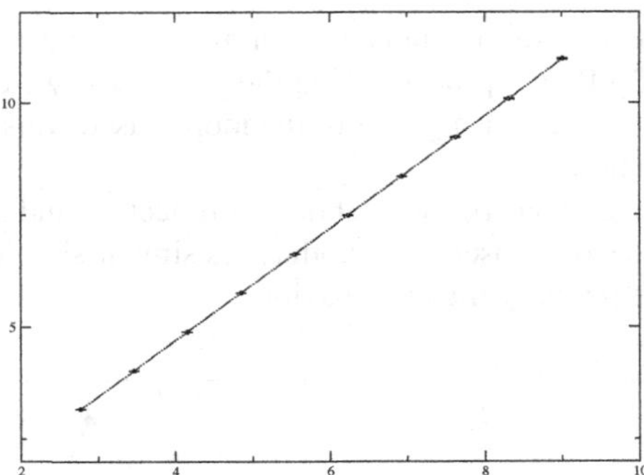

Figure 5. The logarithm of the average length of the loop-erased random walk versus logarithm of the maximum altitude. The numerical results are the circles, the straight line is the linear regression, the error bars are shown.

indeed short range. At each time step, the increment of the underlying random walk is independent of the rest of the walk, and the formation of a loop to be removed is known from data at the present position of the random walk.

From the analytical viewpoint, the loop-erased random walk is one of the few systems that has been proved to have a conformally invariant distribution in the continuum limit, the fractal dimension being exactly $5/4$. A naive idea to get directly a continuum limit representation of loop-erased walks would be to remove the loops from a Brownian motion. This turns out to be impossible due to the proliferation of overlapping loops of small scale. However, the SLE_2 process, to be defined later, gives a direct definition. In fact, it is the consideration of loop-erased random walks that led Schramm [4] to propose SLE as a description of interfaces.

1.1.2. *Percolation*

To define a random interface for percolation, one possibility is to pave the upper-half plane with regular hexagons. Each one is, independently of the others, occupied or empty with probability $1/2$, with the exception of the ones along the real axis, which are all empty on the positive real axis and occupied on the negative real axis, see fig.6. Then a continuous interface, starting at the origin and separating occupied and unoccupied hexagons, is uniquely defined. As before, this defines a simple curve on the lattice.

The percolation interface has an obvious but very remarkable property that singles it out: *locality*. Locality means that the percolation interface does not depend on the distribution of occupied and empty sites away from itself. Equivalently, if $\mathbb{D}$ is a domain (i.e. in this discrete setting a connected and simply connected family of hexagons in the upper half plane) containing the origin, the law of the percolation interface in $\mathbb{D}$ before it hits the boundary of $\mathbb{D}$ for the first time is independent of the distribution of occupied/empty sites outside $\mathbb{D}$. It is the same law as for what one would define as the percolation interface in $\mathbb{D}$ without ever mentioning the world outside $\mathbb{D}$.

This observation makes the percolation interface easy to simulate. Indeed, the con-

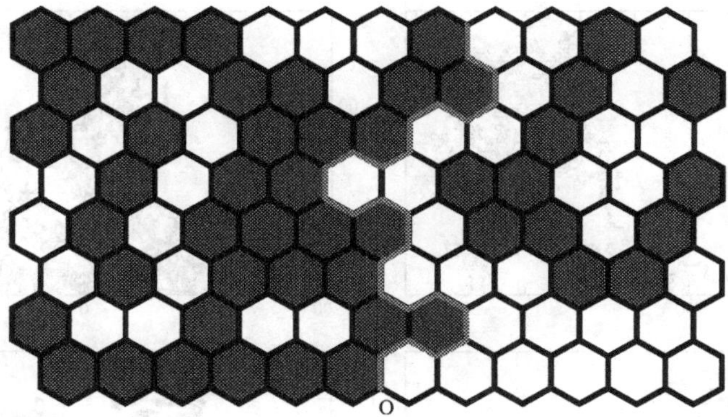

Figure 6. The definition of the percolation interface.

struction of a percolation interface proceeds inductively, as shown on the fig.7.

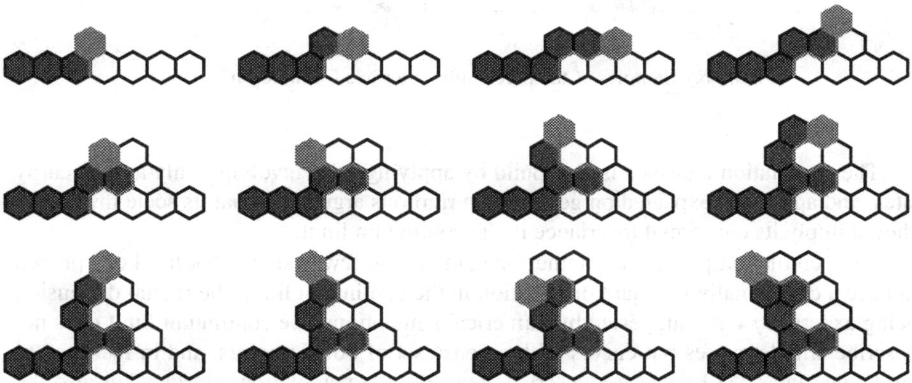

Figure 7. The percolation interface as a growth process.

Fig.8 shows a few samples of increasing size.

The contrast with the previous example is rather striking. Even for small samples, the percolation interface makes many twists and turns. With the resolution of the figure, the percolation interface for large samples does not look like a simple curve at all! The intuitive explanation why this does not occur for loop-erased walks is that if it comes close to itself, then with large probability a few more steps of the random walk will close a loop.

Getting an estimate of the fractal dimension of the percolation interface proceeds along lines similar to the first example. If the number of steps S of the percolation interface is compared to a typical length L one observes that $S \propto L^{\delta}$ and a modest numerical effort (a few hours of CPU) leads to $\delta = 1.75 \pm .01$.

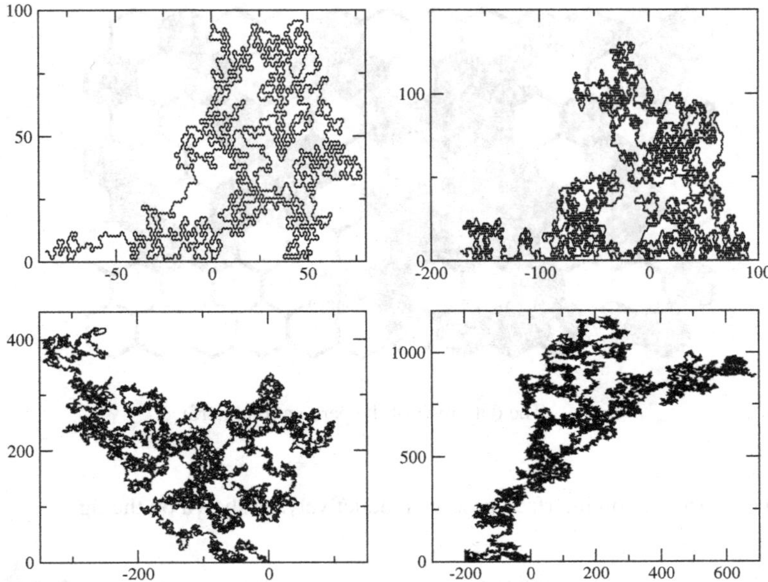

Figure 8. Samples of the percolation interface for increasing sizes.

The percolation interface is also build by applying rules involving only a few nearby sites, and again it is expected on general non rigorous arguments that its scale invariance should imply its conformal invariance in the continuum limit.

The percolation process is another one among the few systems that has been proved to have a conformally invariant distribution in the continuum limit, the fractal dimension being exactly $7/4$. As suggested by numerical simulations, the continuum limit does not describe simple curves but curves with a dense set of double points, and in fact the –to be defined later– SLE_6 process describes not only the percolation interface but also the percolation hull [2], which is the complement of the set of points that can be joined to infinity by a continuous path that does not intersect the percolation interface.

1.1.3. *The Ising model*

Our next example makes also use of the same pictorial representation: it is the Ising model on the triangular lattice in the low temperature expansion. The spins are fixed to be *up* on the left and down on the right of the origin along the real axis. The energy of a configuration is proportional to the length of the curves separating up and down islands. The proportionality constant has to be adjusted carefully to lead to a critical system with long range correlations. This time, making accurate simulations is much more demanding. On the square lattice, the definition of the interface suffers from ambiguities, but these become less relevant for larger sample sizes.

Although there is no question that the fractal dimension of the Ising interface with the above boundary conditions is $11/8$ and is described by –to be defined later– SLE_3,

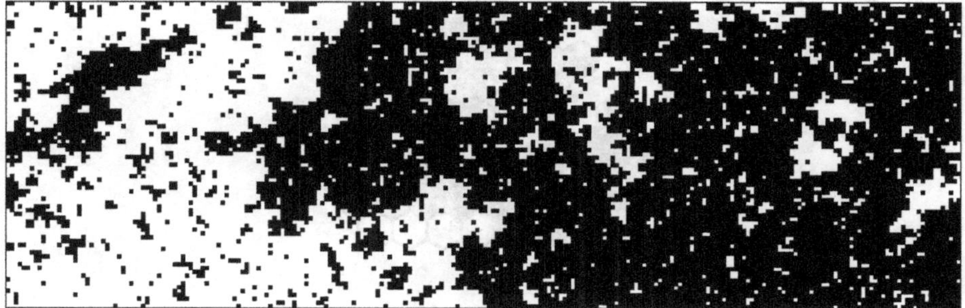

Figure 9. A sample for the critical Ising model. The bottom line, where the spins are frozen –black on the right, white on the left– is not represented. Courtesy of J. Houdayer.

a mathematical proof that a continuum limit distribution for the interface exists and is conformally invariant is still out of reach.

We conclude this list of examples with two remarks.

First, the pictorial representation by a gas of non-intersecting curves on the hexagonal lattice that we used in the second and the third example applies to another family of models, the $O(n)$ models, to which similar interface considerations would apply.

Second, non branching interfaces, described on the lattice by simple curves, are not the generic situation. For instance the $Q = 3$ states Potts model, three different phases coexist and the physical interfaces have branch points.

1.1.4. *A generality*

Up to now, we have discussed interfaces in –lattice approximations of– the upper half plane. Let us note that they make sense for more general domains.

Start with the case of the percolation or of the Ising model. In the plane, take any connected and simply connected collection of hexagons, split it into an interior and a connected boundary, and split the boundary in two connected pieces in such a way that exactly two pairs of adjacent boundary hexagons, marked a and b, carry different colors, see fig.10. Observe that we allow a cut which would correspond to the beginning of the interface. Hexagons separated by the cut are not to be counted as neighbors.

Then any coloring of the hexagons in the interior will lead to a well-defined interface. It will also lead to a well defined energy, and then the question of the distribution of the interface in such a domain is meaningful. For percolation the energy is independent of the configuration. For the Ising model, it is proportional to the length of the curves separating up and down sites. Observe that –as the possible cut separates fixed spins– counting interactions across the cut or not in the energy just adds a constant to it, so that it has no influence on probabilities.

For the loop-erased random case, the idea is similar. One takes a simply connected piece of the square lattice with two points a and b on the boundary, again allowing some cuts, see fig.11. Consider all walks from a to b that do not touch the boundary except at a before the first step and at b after the last step and give each such walk of length

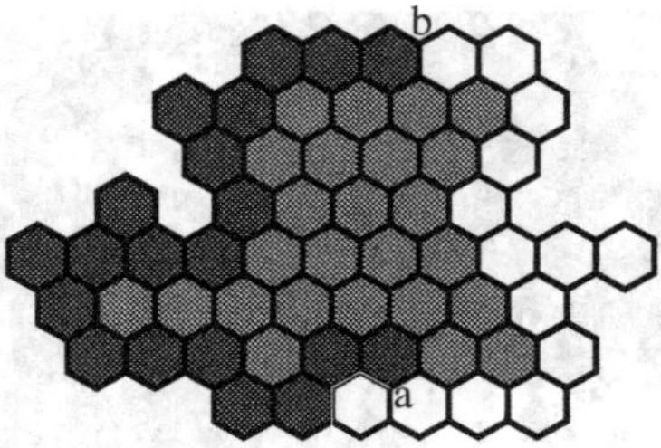

Figure 10. A domain on which percolation and the Ising model can be defined.

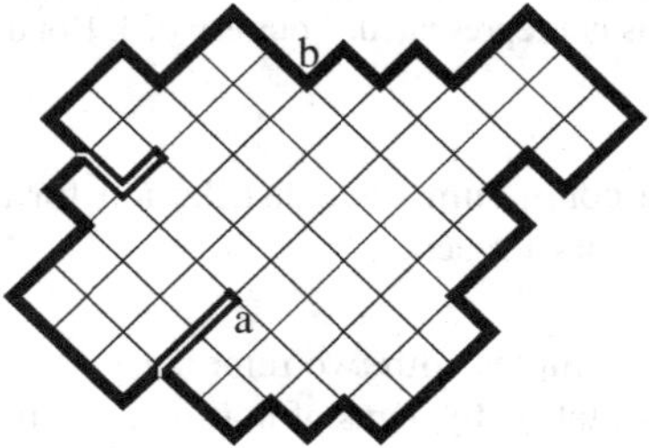

Figure 11. A domain on which loop-erased random walks can be defined.

l a weight 4^{-l}. Observe that this choice is exactly the annihilating boundary condition: in the half plane geometry, a was the origin and b the point at infinity and, due to the infinite extension of the boundary, we had to go through a limiting procedure. Then erase the loops to get a probability distribution for loop-erased random walks from a to b in the domain. The probability for the simple symmetric random walk to hit the boundary for the first time at b starting from a can be interpreted as the partition function for loop-erased walks.

We can now go to the point we want to make, valid for all the above examples. For any of these, we use $P_{(\mathbb{D},a,b)}$ to denote the probability distribution for the interface $\gamma_{[ab]}$ from a to b in $\mathbb{D}$.

Suppose that we fix the beginning $\gamma_{[ac]}$ of a possible interface in domain $\mathbb{D}$, up to a certain point c. Then 1) we can consider the conditional distribution for the rest of the interface and 2) we can remove the beginning of the interface from the domain to create a new domain and consider the distribution of the interface in this new domain. This is illustrated on fig.12.

We claim that the distributions defined in 1) and 2) coincide. For reasons to be explained in a moment, we call this property "locality at the interface". In equations

$$P_{(\mathbb{D},a,b)}(\,\cdot\,|\gamma_{[ac]}) = P_{(\mathbb{D}\backslash\gamma_{[ac[},c,b)}(\,\cdot\,).$$

It is obvious that these two probabilities are supported on the same set, namely simple

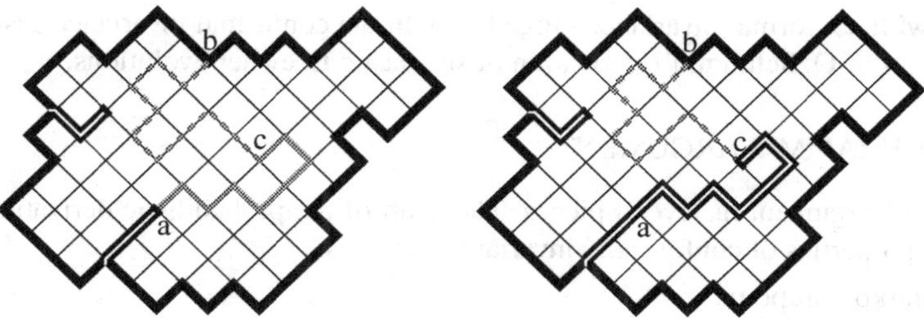

Figure 12. An illustration of situations 1) and 2) for the case of loop-erased walks. What is the distribution of the dotted curve in both situations ?

curves along the edges of the lattice, going from c to b in $\mathbb{D}\backslash\gamma_{[ac[}$. Let us however note that for loop-erased random walks, annihilating boundary conditions are crucial. Reflecting boundary conditions clearly do not work, if only because the supports do not coincide in that case.

For the case of loop-erased random walks, the argument for the equivalence of 1) and 2) goes as follows. Take any random walk (possibly with loops) $W_0 = a, W_1, \cdots, W_l = b$ that contributes to an interface $\gamma_{[ab]}$ which is $\gamma_{[ac]}$ followed by some $\gamma_{[cb]}$. Let m be the largest index for which the walk visits c. Because the interface has to start with $\gamma_{[ac]}$, the walk $W_m = c, \cdots, W_l = b$ cannot cross $\gamma_{[ac[}$ again, so it is in fact a walk in $\mathbb{D}\backslash\gamma_{[ac[}$ from c to b leading to the interface $\gamma_{[cb]}$. The weight for the walk $W_0 = a, W_1, \cdots, W_l = b$ is 4^{-l}, i.e. simply the product of weights for the walks $W_0 = a, W_1, \cdots, W_m = c$ and $W_m = c = a, \cdots, W_l = b$. Then a simple manipulation of weights leads directly to the announced result.

For the case of percolation and the Ising model, in fact more is true: we can view $P_{(\mathbb{D},a,b)}$ not only as a probability distribution for the interface, but as the full probability distribution for the colors of the hexagons and still check the identity of 1) and 2). Again, the supports are the same for 1) and 2), namely any configuration of the colors, except that the colors on both sides of $\gamma_{[ac]}$ are fixed. For the case of percolation, the colors are independent of each other so the identity of 1) and 2) is clear. For the Ising model, the difference is that the conditional probabilities in 1) take into account the interactions between the colors along the interface, whereas the probability in 2) does not take into account the interactions between the spins along the cut left by the removal of the interface. However, as already mentioned above, the corresponding colors are fixed anyway, so the Boltzmann weights for the configurations that are in the support of 1) or 2) differ by a multiplicative constant, which disappears when probabilities are computed.

This argument extends immediately to systems with only nearest neighbor interactions. They can be defined on any graph. If any subset of edges is chosen and the configuration at both end of each edge is frozen, it makes no difference for probabilities to consider the model on a new graph in which the frozen edges have been deleted.

Instead of looking for further generalizations, we argue more heuristically that the continuum limit for a system with short range interactions should satisfy locality at the interface. The use of this locality property –which, as should be amply evident, has nothing to do with conformal invariance– together with the conformal invariance assumption is at the heart of O. Schramm's derivation of stochastic Loewner evolutions.

1.2. O. SCHRAMM'S ARGUMENT

We break the argument in several pieces. The heart of the probabilistic derivation establishes two properties of conformally invariant interfaces:

- the Markov property,
- the stationarity of increments,

see sections 1.2.3, 1.2.2 and 1.2.5. The crucial results needed for the probabilistic part, in particular the basics of Loewner evolutions, are collected in 1.2.1 and 1.2.4. The last two subsections are not directly related to the argument. They collect some basic facts on the fractal properties of SLE interfaces and on the connection with CFT.

1.2.1. *Riemann's theorem and hulls*

In this section we collect a few indispensable results used in the rest of the course.

A domain is a non empty connected and simply connected open set strictly included in the complex plane $\mathbb{C}$. Simple connectedness is a notion of purely topological nature which in two dimensions asserts essentially that a domain has no holes and is contractible. But it is a deep theorem of Riemann that two domains are always conformally equivalent, i.e. there is an invertible holomorphic map between them. For instance, the upper-half plane $\mathbb{H}$ is a domain. It is well known that it has a three dimensional Lie group of conformal automorphisms, $PSL_2(\mathbb{R})$, that also acts on the boundary of $\mathbb{H}$. There is a unique automorphism, possible followed by a transposition, that maps any triple of boundary points to any other triple of boundary points. By Riemann's theorem, this is also true for any other domain, at least if the boundary is not too wild.

Riemann's theorem is used repeatedly in the rest of this course and is the starting point of many approaches to growth phenomena in two dimensions.

For later use, we note that one can be a bit more explicit when the domain $\mathbb{D}$ differs only locally from the upper half plane $\mathbb{H}$, that is if $\mathbb{K} = \mathbb{H} \setminus \mathbb{D}$ is bounded. Such a set $\mathbb{K}$ is called a hull. The real points in the closure of $\mathbb{K}$ in $\mathbb{C}$ form a compact set which we call $\mathbb{K}_{\mathbb{R}}$. Let $f : \mathbb{H} \mapsto \mathbb{D}$ be a conformal bijection. As the boundary of $\mathbb{H}$ is smooth, f has a continuous extension to $\overline{\mathbb{R}} \equiv \mathbb{R} \cup \infty$, and $f^{-1}(\overline{\mathbb{R}} \setminus \mathbb{K}_{\mathbb{R}})$ is a non-empty open set in $\overline{\mathbb{R}}$ with compact complement. We call the complement the cut of f. By the Schwarz symmetry principle, defining $f(z) = \overline{f(\bar{z})}$ for $\Im\mathrm{m}\, z \leq 0$ gives an analytic extension of f to the whole Riemann sphere minus the cut. Across the cut, f has a purely imaginary nonnegative discontinuity which we write as a Radon-Nikodym derivative $d\mu_f/dx$.

One can use the $PSL_2(\mathbb{R})$ automorphism group of $\mathbb{H}$ to ensure that f is holomorphic at ∞ and $f(w) - w = O(1/w)$ there. This is called the hydrodynamic normalization. It involves three conditions, so there is no further freedom left. We shall denote this special representative by $f_{\mathbb{K}}$, which is uniquely determined by $\mathbb{K}$: any property of $f_{\mathbb{K}}$ is an intrinsic property of $\mathbb{K}$.

Cauchy's theorem yields

$$f_{\mathbb{K}}(w) = w + \frac{1}{2\pi} \int_{\mathbb{R}} \frac{d\mu_{f_{\mathbb{K}}}(x)}{x - w}, \tag{1.1}$$

A quantity that plays an important role in the sequel is

$$C_{\mathbb{K}} \equiv \frac{1}{2\pi} \int_{\mathbb{R}} d\mu_{f_{\mathbb{K}}}(x),$$

a positive (unless $\mathbb{K} = \emptyset$) number called the capacity of $\mathbb{K}$, which is such that $f_{\mathbb{K}}(w) = w - C_{\mathbb{K}}/w + O(1/w^2)$ at infinity. The usefulness of capacity stems from its good behavior under compositions: if $\mathbb{K}$ and $\mathbb{K}'$ are two hulls, $\mathbb{K} \cup f_{\mathbb{K}}(\mathbb{K}')$ is a hull and

$$C_{\mathbb{K} \cup f_{\mathbb{K}}(\mathbb{K}')} = C_{\mathbb{K}} + C_{\mathbb{K}'}, \tag{1.2}$$

as seen by straightforward expansion at infinity of $f_{\mathbb{K}} \circ f_{\mathbb{K}'}$, the map associated to $\mathbb{K} \cup f_{\mathbb{K}}(\mathbb{K}')$. In particular capacity is a continuous increasing function on hulls.

Anticipating a little bit, let us note immediately that giving a dynamical rule for the evolution of the finite positive measure $d\mu_{f_{\mathbb{K}}}(x)$ is a good way to define growth processes.

1.2.2. *Conformally invariant interfaces*

Consider a domain $\mathbb{D}$, with two distinct points on its boundary, which we call a and b. A simple curve, denoted by $\gamma_{[ab]}$, from a to b in $\mathbb{D}$ is the image of a continuous one-to-one map γ from the interval $[0, +\infty]$ to $\mathbb{D} \cup \{a, b\}$ such that $\gamma(0) = a$, $\gamma(\infty) = b$ and $\gamma_{]ab[} \equiv \gamma(]0, \infty[) \subset \mathbb{D}$. Alternatively, a simple curve from a to b is an equivalence class of such maps under increasing reparametrizations. A point on it has no preferred coordinate but is has a past and a future. If $c \in \mathbb{D}$ is an interior point, we use a similar definition for a simple curve $\gamma_{[ac]}$ from a to c in $\mathbb{D}$.

Note that, apart from the fact that on the lattice we could use lattice length as a parameter along the interface, we have just rephrased in the continuum –but with the same notations– what we did before in a discrete setting.

Our aim is to study conformally invariant probability measures on the set of simple curves from a to b in $\mathbb{D}$. There is a purely kinematical step, which demands that if h is any conformal map that sends $\mathbb{D}$ to another domain $h(\mathbb{D})$, the measure for $(h(\mathbb{D}), h(a), h(b))$ should be the image by h of the measure for $(\mathbb{D}, a, b)$:

$$P_{(\mathbb{D},a,b)}(\gamma_{[ab]} \subset U) = P_{(h(\mathbb{D}),h(a),h(b))}(h(\gamma_{[ab]}) \subset h(U)),$$

where $P_{(\mathbb{D},a,b)}(\gamma_{[ab]} \subset U)$ denotes the probability for the curve $\gamma_{[ab]}$ to remain in a subset U of $\mathbb{D}$. See fig.13.

This condition is natural and it is the one that conformal field theory suggests immediately. Let us note however that a totally different definition of conformal invariance is understood in the familiar statement "two dimensional Brownian motion is conformally invariant".

Observe that we could take any measure for $(\mathbb{D}, a, b)$ –well, with the invariance under the one parameter group of automorphisms that fixes $(\mathbb{D}, a, b)$– and declare that the measure in $h(\mathbb{D})$ is obtained by definition by the rule above. To make progress, we need to combine conformal invariance with locality at the interface.

1.2.3. *Markov property and stationarity of increments*

This short section establishes the most crucial properties of conformally invariant interfaces.

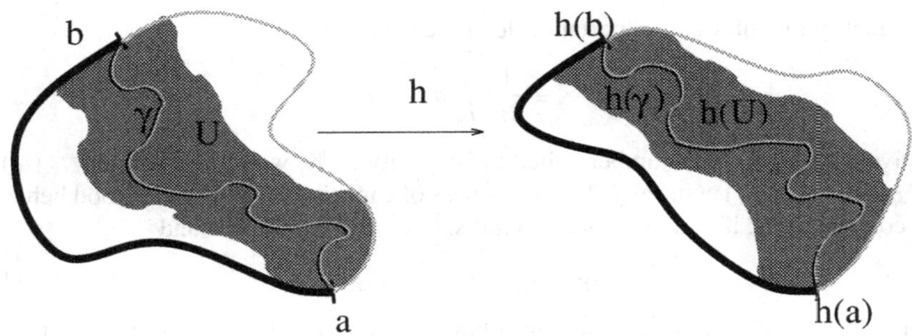

Figure 13. Conformal invariance for change of domain.

Take $c \in \mathbb{D}$ and let $\gamma_{[ac]}$ be a simple curve from a to c in $\mathbb{D}$. Observe that $\mathbb{D} \setminus \gamma_{]ac]}$ is a domain. To answer the question "if the beginning of the interface is fixed to be $\gamma_{[ac]}$, what is the distribution of the rest $\gamma'_{[cb]}$ of the interface ?" we apply locality at the interface to argue that this is exactly the distribution of the interface in $\mathbb{D} \setminus \gamma_{]ac]}$. We map this domain conformally to $\mathbb{D}$ via a map h_γ sending b to b and c to a, see fig.14.

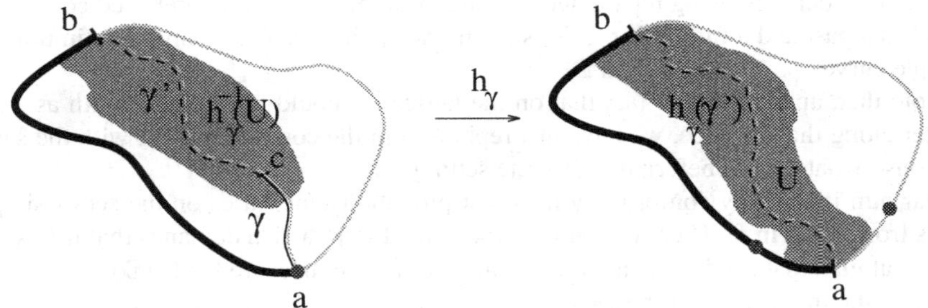

Figure 14. Conformal invariance for conditional probabilities.

Conformal invariance implies that the image measure is the original measure in $P_{(\mathbb{D},a,b)}$, where $h_\gamma(\gamma'_{[cb]})$ is an interface that has forgotten $\gamma_{[ac]}$. To summarize :

$h_\gamma(\gamma'_{[cb]})$ is independent of $\gamma_{[ac]}$ (the Markov property) and has the same distribution as the original interface itself (stationarity of increments).

1.2.4. *Local growth and Loewner evolutions*

The Markov and stationarity of increments property make it plain that to understand the distribution of the full interface, it is enough to understand the distribution of a small, or even infinitesimal, initial segment, and then glue segments via conformal maps.

This calls for a description by differential equations.

For this purpose, it is very convenient –even if by no means mandatory– to have a natural parametrization of interfaces. Using conformal invariance, we can restrict ourselves to the situation when $(\mathbb{D}, a, b) = (\mathbb{H}, 0, \infty)$. If $\gamma_{[0\infty]}$ is a simple curve from 0 to ∞ in $\mathbb{H}$, and c a point on it, we know that $\mathbb{H} \setminus \gamma_{]0c]}$ is a domain, and $\gamma_{]0c]}$ itself is a hull. We use capacity as a parametrization and define a time parameter by $2t(c) \equiv C_{\gamma_{]0c]}}$. The factor of 2 is just historical. The map t is a continuous increasing function of c varying from 0 to ∞ in $\bar{\mathbb{R}}$ when c moves from 0 to ∞ along $\gamma_{[0\infty]}$. The inverse map $t \mapsto c(t)$ is well-defined and gives a parametrization of the curve. So we introduce an increasing family of hulls $\mathbb{K}_t, t \in [0, \infty[$, by $\mathbb{K}_t = \gamma_{]0c(t)]}$ which has capacity $2t$. Let $f_t \equiv f_{\mathbb{K}_t}$ be the conformal homeomorphism from $\mathbb{H}$ to $\mathbb{H} \setminus \mathbb{K}_t$ normalized to satisfy $f_t(w) = w - 2t/w + O(1/w^2)$ at infinity. Define $g_t : \mathbb{H} \setminus \mathbb{K}_t \mapsto \mathbb{H}$ to be the inverse of f_t. Then $g_t(z) = z + 2t/z + O(1/z^2)$ at infinity.

To study the evolution of the family of hulls $\mathbb{K}_t$, fix $\varepsilon \geq 0$ and consider the hull $\mathbb{K}_{\varepsilon,t} \equiv g_t(\mathbb{K}_{t+\varepsilon} \setminus \mathbb{K}_t)$, which has capacity 2ε by eq.(1.2). Define $f_{\varepsilon,t} \equiv f_{\mathbb{K}_{\varepsilon,t}}$. Then $g_t = f_{\varepsilon,t} \circ g_{t+\varepsilon}$ on $\mathbb{H} \setminus \mathbb{K}_{t+\varepsilon}$. Using the representation of $f_{\mathbb{K}_{\varepsilon,t}}$ in terms of its discontinuity eq.(1.1), we obtain

$$g_{t+\varepsilon} - f_{\varepsilon,t} \circ g_{t+\varepsilon} = g_{t+\varepsilon} - g_t = \frac{1}{2\pi} \int_{\mathbb{R}} \frac{d\mu_{f_{\varepsilon,t}}(x)}{g_{t+\varepsilon} - x}$$

We introduce now the notion of local growth which is crucial for interfaces. When ε is small, $\mathbb{K}_{\varepsilon,t}$ is a tiny piece of curve and the support of $d\mu_{f_{\varepsilon,t}}$ is small and becomes a point when ε goes to 0. Measures supported at a point are δ functions, so there is a point ξ_t such that, as a measure, $d\mu_{f_{\varepsilon,t}}/dx \sim 2\varepsilon\delta(x - \xi_t)$ as $\varepsilon \to 0^+$. If $\mathbb{K}_t$ is a more general increasing family of hulls of capacity $2t$, we say that the condition of local growth is satisfied if the above small ε behavior holds. At first sight, it might seem that local growth is only true for curves, but this is not true. We shall give an example below.

Letting $\varepsilon \to 0^+$, from the local growth condition, we infer the existence of a real function ξ_t such that

$$\frac{dg_t}{dt}(z) = \frac{2}{g_t(z) - \xi_t}. \tag{1.3}$$

It is useful to look at this equation from a slightly different point of view, taking the function ξ_t as the primary data. The solutions of this equation for a given function ξ_t with initial condition $g_0(z) = z$ is called a Loewner evolution. The image of ξ_t by g_t^{-1} is the tip of the curve at time t.

The cases when the local growth condition is not satisfied are called Loewner chains, see below. Had we used another parametrization of the curve, the 2 in the numerator would be replaced by a positive function of the parameter along the curve.

Informally, if $\mathbb{K}_t$ is a growing curve, we expect that $g_{t+\varepsilon}(z) - g_t(z)$ describes an infinitesimal cut. This is confirmed by the explicit solution of eq.(1.3) for the trivial case $\xi_t \equiv 0$, which yields $g_t(z)^2 = z^2 + 4t$, the branch to be chosen being such that at large z, $g_t(z) \sim z$. This describes a growing segment along the imaginary axis. So intuitively, the simple pole in eq.(1.3) accounts for the existence of a cut and different functions ξ_t account for the different shapes of curves.

One can also solve the case when $\mathbb{K}_t$ is an arc of circle starting from the origin. It can be obtained from the trivial solution above by applying appropriate time dependent $PSL_2(\mathbb{R})$ transformations both to g_t and z and then by a time change to recover the capacity parametrization. This is an illuminating exercise that we leave to the reader.

Take an arc going from 0 to $2R$ along a circle of radius R. It turns out that when the arc approaches the real axis to close a half disk, the function ξ_t has a square root singularity $\xi_t \propto \sqrt{R^2-2t}$. The capacity remains finite and goes to R^2 and the map itself has a limit $g_{R^2/2}(z) = z + R^2/(z+R)$ which has swallowed the half disk without violating the local growth condition. One can start the growth process again. Making strings of such maps with various values of the radii is a simple way to construct growing families of hulls that are not curves and that nevertheless grow locally. Note that a square root singularity for ξ_t is the marginal behavior: if ξ_t is Hölder of exponent $> 1/2$, Loewner evolution yields a simple curve.

1.2.5. *Stochastic (or Schramm) Loewner evolutions*

If we sample locally growing hulls with a certain distribution, we get an associated random process ξ_t. In the case of a conformally invariant distribution, we have established two crucial properties: Markov property and stationarity of increments. To finish Schramm's argument leading to SLE, what remains is to see the implications of these properties on the distribution of ξ_t.

The argument and expressions for the meaning of Markov property and stationarity of increments involved a map h that mapped the tip of the piece of interface to the initial marked point a and the final marked point b to itself. The map $h_t(z) = g_t(z) - \xi_t$ has the required property when the domain is the upper-half plane with 0 and ∞ as marked points. It behaves like $h_t(z) = z - \xi_t + 2t/z + O(1/z^2)$ at infinity. We infer that for $s > t$, $h_t(\mathbb{K}_s \setminus \mathbb{K}_t)$ is independent of $\mathbb{K}_{t'}$, $t' \leq t$ (Markov property) and is distributed like a hull of capacity $s - t = C_{h_t(\mathbb{K}_s \setminus \mathbb{K}_t)}$ (stationarity of increments).

The hull determines the corresponding map h, so this can be rephrased as: $h_s \circ h_t^{-1}$ (which uniformizes $h_t(\mathbb{K}_s \setminus \mathbb{K}_t)$) is independent of $h_{t'}$, $t' \leq t$, and distributed like an h_{s-t}. As $h_s \circ h_t^{-1} = z - (\xi_s - \xi_t) + (s-t)/z + O(1/z^2)$ at infinity, the driving parameter for the process $h_s \circ h_t^{-1}$ is $\xi_s - \xi_t$. To summarize:

> the Markov property and stationarity of increments for the interface lead to the familiar statement for the process ξ_t: for $s > t$, $\xi_s - \xi_t$ is independent of $\xi_{t'}$, $t' \leq t$, (Markov property) and distributed like a ξ_{s-t} (stationarity of increments).

To conclude, a last physical input is needed: one demands that the interface does not branch, which means that at two nearby times the growth is at nearby points. This implies that ξ_t is a continuous process, in the sense that it has continuous trajectories.

One is now in position to apply a mathematical theorem: a 1d Markov process with continuous trajectories and stationary increments is proportional to a Brownian motion. We conclude that there is a real positive number κ such that $\xi_t = \sqrt{\kappa} B_t$ for some normalized Brownian motion B_t with covariance $\mathbb{E}[B_s B_t] = \min(s,t)$. The same argument without imposing that the time parametrization is given by the capacity of the hull would lead to the conclusion that the driving parameter is a continuous martingale, which is nothing but a Brownian motion after a possibly random time change.

A solution of

$$\frac{dg_t}{dt}(z) = \frac{2}{g_t(z) - \sqrt{\kappa} B_t} \tag{1.4}$$

is called a chordal Schramm-Loewner evolution of parameter κ, in short a chordal SLE$_\kappa$. The connection of this equation with interfaces relies mainly on conformal invariance.

But local growth, absence of branches and to a lower level locality at the interface also play a crucial role.

1.2.6. *Miscellanea on SLE*

Up to now we have only discussed the situation when the interface goes from 0 to ∞ in the upper half plane, or more generally from one point on the boundary of a domain to another one. This is called chordal SLE. The argument can be extended easily to two other situations. The first is for an interface starting at a fixed point on the boundary of a domain and ending at a fixed point in the bulk. This is called radial SLE. For the second situation, three points are chosen on the boundary and the interface starts from the first point and ends at a random point between the two others. This is called dipolar SLE. Just as the upper-half plane is a convenient geometry for chordal SLE, a semi infinite cylinder is nice for radial SLE and an infinite strip is nice for dipolar SLE.

For radial SLE on a cylinder of circumference π, the equation reads

$$\frac{dg_t}{dt}(z) = \frac{2}{\tan(g_t(z) - \sqrt{\kappa}B_t)} \tag{1.5}$$

and for dipolar SLE on a strip of width $\pi/2$, one has

$$\frac{dg_t}{dt}(z) = \frac{2}{\tanh(g_t(z) - \sqrt{\kappa}B_t)} \tag{1.6}$$

The generalization of SLE to Riemann surfaces with moduli is easy for the annulus, but raises non-trivial problems in general, currently subject of active research.

The set of exact results obtained for SLE forms an impressive body of knowledge. We shall not mention the ones dealing with the explicit computation of certain crossing probabilities but we list just a few "pictorial" properties with some comments. They (the properties and the comments) should be understood with the standard proviso "almost surely" or "with probability 1".

We start with a surprisingly difficult result[4, 5, 6, 7].

- Whatever the value of κ, the pre-image of the driving parameter $\lim_{w \to \sqrt{\kappa}B_t} g_t^{-1}(w)$ is a continuous curve γ_t, called the SLE trace. The trace never crosses itself. This property is crucial if the trace is to be interpreted as a curve separating two phases.
- For $\kappa \in [0, 4]$ the SLE trace is a simple curve. For $\kappa \in]4, 8[$, it has double points. For $\kappa \in [8, \infty[$, it is space filling.
- The fractal dimension d_κ of the trace is $1 + \kappa/8$ for $\kappa \leq 8$ and 2 for $\kappa \geq 8$.

Using the formula for the dimension of the trace and confronting with the numerical simulations, it is plausible (actually, these are among the few cases for which a mathematical proof exists) that loop-erased random walks correspond to $\kappa = 2, d = 5/4$ and percolation to $\kappa = 6, d = 7/4$. This is also compatible with the general shape of the numerical samples, which indicate that loop-erased random walks indeed lead to simple curves and that percolation doesn't.

The hull $\mathbb{K}_t$ is by definition $\mathbb{H} \setminus g_t^{-1}(\mathbb{H})$. It has the following properties

- The hull $\mathbb{K}_t$ is the complement of the connected component of ∞ in $\mathbb{H} \setminus \gamma_{]0,t]}$.
- For $\kappa \in [0, 4]$, the SLE hull is a simple curve coinciding with the trace. For $\kappa \in]4, \infty[$, the SLE hull has an nonempty connected and relatively dense interior.

This may seem surprising at first sight. It is the sign that for $\kappa > 4$, the drift $\sqrt{\kappa}B_t$ goes fast enough for the swallowing procedure to take place, as described in the closing arc example, but on all scales.

This is summarized by fig.15.

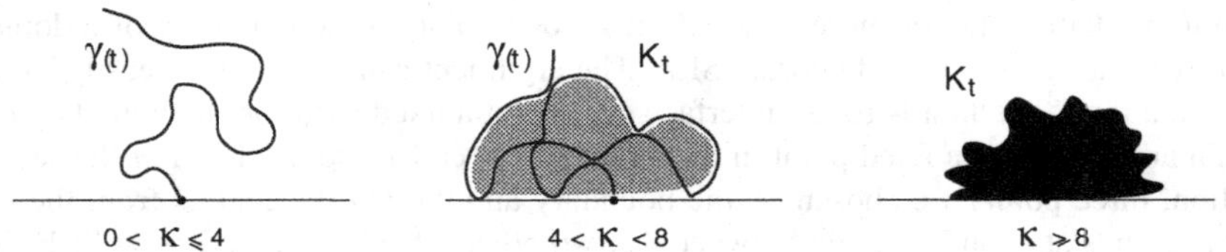

Figure 15. The phases of SLE.

For the many other properties known about SLE, the reader is invited to read the vast literature [4, 5, 6, 7, 8, 9].

1.3. SHORT REMARKS ON SLE AND CFT

As already mentioned, it is by no means an easy task to prove that a discrete random curve converges to an SLE. For geometric models like percolation, loop-erased random walks, self avoiding walks, it is often easier to compute the appropriate κ using known special properties of the discrete model that are expected to survive in the continuum limit.

For instance, one can formulate locality in the continuum. Showing that SLE_κ has the locality property only for $\kappa = 6$ is a standard computation in stochastic Ito calculus, see e.g. [9].

For critical statistical mechanics models with non trivial Boltzmann weights, things are more complicated. The system contains local dynamical degrees of freedom independently from the interface and, in the continuum limit, these local degrees of freedom are expected to be described by a conformal field theory (CFT). The most basic parameter of a CFT is its central charge c. Computing it (even non rigorously) from the lattice model can be a challenge comparable to the one of computing κ.

On the other hand, the relation between c and κ can be worked out in general. The idea is elementary. From a technical viewpoint, it rests on assumptions similar to the ones needed in O. Schramm's argument.

In the discrete setting, take O to be any observable. If one computes $\langle O\rangle_{\mathbb{D}\setminus\gamma_{]ac]}}$ in $\mathbb{D}\setminus\gamma_{]ac]}$, i.e. with part of the interface fixed, and then averages over $\gamma_{]ac]}$ one retrieves $\langle O\rangle_{\mathbb{D}}$:

$$\mathbb{E}[\langle O\rangle_{\mathbb{D}\setminus\gamma_{]ac]}}] = \langle O\rangle_{\mathbb{D}}$$

where the expectation is the average over γ, see the last reference in [10]. This is a straightforward application of the usual rules of statistical mechanics

This basic property, which we call the martingale property because in probabilistic jargon it would be translated as "$\langle O\rangle_{\mathbb{D}\setminus\gamma_{]ac]}}$ is a closed martingale for any observable O" is expected to survive in the continuum limit. In this limit, one has two powerful tools inherited from conformal invariance at hand, CFT to compute correlators and SLE to average over the piece of interface. Doing this for arbitrary values of κ and c means mixing the degrees of freedom from two different models and there is a priori no reason for the martingale property to hold. An explicit computation shows that it holds only if

$$2\kappa c = (6-\kappa)(3\kappa-8).$$

The martingale property also gives information on the operator content of the CFT. As the discrete statistical mechanics examples show, there is a change in boundary conditions at the tip of the interface. The martingale property allows to identify this boundary changing condition operator as a primary field of weight $h = (6-\kappa)/(2\kappa)$ degenerate at level two.

This approach also exhibits a family of martingales which are at the heart of many probabilistic computations. For instance, it gives a systematic way to interpret probabilities for SLE events as correlation functions of a CFT, and shows how the changes of behavior of the trace at $\kappa = 4, 8$ are related to operator product expansions. The interested reader is refereed to [10].

2. Loewner chains

This section deals with more general 2D growth processes. Although, they do not fulfill the local growth and conformal invariance properties of SLEs, they are nevertheless described by dynamical conformal maps. We first present systems whose conformal maps have a time continuous evolution and give examples. We then go on by presenting a discrete version thereof in terms of iterated conformal maps. We explain integrability of Laplacian growth. The last part is a discussion concerning the limit of small ultraviolet cutoff and the consequences of possible dendritic anomalies.

2.1. CONTINUOUS LOEWNER CHAINS

In this part, the exterior of the unit disk is used as the reference geometry.

2.1.1. *Radial Loewner chains*

Let $\mathbb{K}_t$ be a family of growing closed planar domains with the topology of a disk. Let $\mathbb{O}_t \equiv \mathbb{C} \setminus \mathbb{K}_t$ be their complements in the complex plane. See Figure (16). To fix part of translation invariance we assume that the origin belongs to $\mathbb{K}_t$ and the point at infinity to $\mathbb{O}_t$.

Loewner chains describe the evolution of family of conformal maps f_t uniformizing $\mathbb{D} = \{w \in \mathbb{C};\ |w| > 1\}$ onto $\mathbb{O}_t$. It thus describes the evolution of the physical domains $\mathbb{O}_t$. We normalize the maps $f_t : \mathbb{D} \to \mathbb{O}_t$ by demanding that they fix the point at infinity, $f_t(\infty) = \infty$ and that $f_t'(\infty) > 0$. With t parameterizing time, Loewner equation reads:

$$\frac{\partial}{\partial t} f_t(w) = w f_t'(w) \oint \frac{du}{2i\pi u} \left(\frac{w+u}{w-u} \right) \rho_t(u) \tag{2.1}$$

The integration is over the unit circle $\{u \in \mathbb{C}, |u| = 1\}$. The Loewner density $\rho_t(u)$ codes for the time evolution. It may depends on the map f_t in which case the growth process in non-linear. For the inverse maps $g_t \equiv f_t^{-1} : \mathbb{O}_t \to \mathbb{D}$, Loewner equation reads:

$$\frac{\partial}{\partial t} g_t(z) = -g_t(z) \oint \frac{du}{2i\pi u} \left(\frac{g_t(z)+u}{g_t(z)-u} \right) \rho_t(u) \tag{2.2}$$

The behavior of f_t at infinity fixes a scale since at infinity, $f_t(w) \simeq R_t w + O(1)$ where $R_t > 0$, with the dimension of a `[length]`, is called the conformal radius of $\mathbb{K}_t$ viewed from infinity. R_t may be used to analyze scaling behaviors, since Kobe 1/4-theorem (see e.g. [13]) ensures that R_t scales as the size of the domain. In particular, the (fractal) dimension D of the domains $\mathbb{K}_t$ may be estimated by comparing their area $\mathcal{A}_t$ with their linear size measured by R_t: $\mathcal{A}_t \asymp R_t^D$ for large t – the proportionality factor contains a cutoff dependence which restores naive dimensional analysis.

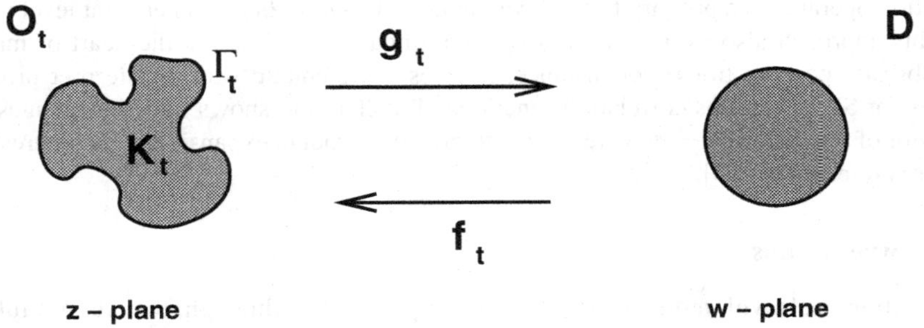

Figure 16. Uniformizing maps intertwining the 'physical' z-plane and the 'mathematical' w-plane.

2.1.2. *The boundary curves*

The boundary curve $\Gamma_t \equiv \partial\mathbb{O}_t$ is the image of the unit circle by f_t. We may parameterize the boundary points by $\gamma_{t;\alpha} = f_t(u)$ with $u = e^{i\alpha}$. The Loewner equation codes for the evolution of the shape of $\mathbb{O}_t$ and thus for the normal velocity of the boundary points. Only the normal velocity is relevant as the tangent velocity is parameterization dependent. The tangent to the curve is $\tau = iuf_t'(u)/|f_t'(u)|$ and the outward normal is $n = -i\tau$ so that the normal velocity at γ_t is $v_n = \Re\mathrm{e}[\bar{n}\,\partial_t f_t(u)]$, or

$$v_n = |f_t'(u)|\,\Re\mathrm{e}[\partial_t f_t(u)/uf_t'(u)].$$

The r.h.s. is determined by the Loewner equation (2.1) because this equation may be viewed as providing the solution of a boundary value problem. Indeed, recall that for $\hat{h}(u)$ a real function on the unit circle, $h(w) = \oint \frac{du}{2i\pi u}\left(\frac{w+u}{w-u}\right)\hat{h}(u)$ is the unique function analytic outside the unit disk whose real part on the unit circle is $\hat{h}$, i.e. $\Re\mathrm{e}[\,h(u)\,] = \hat{h}(u)$. Thus, since $\partial_t f_t(w)/wf_t'(w)$ is analytic in $\mathbb{D}$, the Loewner equation (2.1) is equivalent to:

$$v_n = |f_t'(u)|\,\rho_t(u)$$

or more explicitly [1]:

$$(\partial_\alpha\gamma_{t;\alpha})\,(\partial_t\overline{\gamma_{t;\alpha}}) - (\partial_\alpha\overline{\gamma_{t;\alpha}})\,(\partial_t\gamma_{t;\alpha}) = 2i\,|f_t'(u)|^2\,\rho_t(u) \tag{2.3}$$

[1]For SLE this equation has to be modified according to Itô calculus

Hence, the evolution of the domain may be encoded either in the evolution law for its uniformizing conformal map as in eq.(2.1) or in the boundary normal velocity as in eq.(2.3). The two equations are equivalent.

2.2. EXAMPLES

2.2.1. *Stochastic Loewner evolution (SLE)*

From this point of view, SLE is singular as it corresponds to a Dirac measure $\delta_{U_t}(u)$ centered at the position of a Brownian motion on the unit circle. The locality of this measure reflects the fact that SLE describes growing curves. Continuity of the Brownian motion reflects the absence of branching in the SLE growths. To make contact with the previous sections, the reader should check that eq.(1.5) becomes of the form (2.2) after the cylinder has been mapped to the outside of the unit disk.

2.2.2. *Laplacian growth (LG)*

This is a process in which the growth of the domain is governed by the solution of Laplace equation, i.e. by an harmonic function, in the exterior of the domain with appropriate boundary conditions. It originates from the hydrodynamical Hele-Shaw problem to be described below.

To be precise, let P be the real solution of Laplace equation, $\nabla^2 P = 0$, in $\mathbb{O}_t$ with the boundary behavior $P = -\log|z| + \cdots$ at infinity and $P = 0$ on the boundary curve $\Gamma_t = \partial\mathbb{O}_t$. The time evolution of the domain is then defined by demanding that the normal velocity of points on the boundary curve be equal to minus the gradient of P: $v_n = -(\nabla P)_n$.

This problem may be written as a Loewner chain since, as is well known, Laplace equation is solved via complex analysis by writing P as the real part of an analytic function. One first solves Laplace equation in the complement of the unit disk with the appropriate boundary conditions and then transports it back to the physical domain $\mathbb{O}_t$ using the map f_t. This gives:

$$P = -\Re\text{e}\,\Phi_t \quad \text{with} \quad \Phi_t(z) = \log g_t(z)$$

The evolution equation for the map f_t is derived using that the boundary normal velocity is $v_n = -(\nabla P)_n$. The above expression for P gives:

$$v_n = -(\nabla P)_n = |f_t'(u)|^{-1}$$

at point $\gamma_t = f_t(u)$ on the boundary curve. As explained in the previous section, this is enough to determine $\partial_t f_t(w)$ for any $|w| > 1$ since this data specifies the real part on the unit circle of the analytic function $\partial_t f_t(w)/w f_t'(w)$ on the complement of the unit disk. The result is:

$$\partial_t f_t(w) = w f_t'(w) \oint_{|u|=1} \frac{du}{2i\pi u\,|f_t'(u)|^2} \Big(\frac{w+u}{w-u}\Big) \tag{2.4}$$

It is a Loewner chain with $\rho_t(u) = |f_t'(u)|^{-2}$.

As we shall see below, Laplacian growth is an integrable system, which may be solved exactly, but it is ill-posed as the domain develops singularities (cusps $y^2 \simeq x^3$) in finite time. It thus needs to be regularized. There exist different ways of regularizing it.

One may also formulate Laplacian growth using a language borrowed from electrostatics by imagining that the inner domain is a perfect conductor. Then $V = \Re\mathrm{e}\,\Phi_t$ is the electric potential which vanishes on the conductor but with a charge at infinity. The electric field $\vec{E} = \vec{\nabla} V$ is $\bar{E} \equiv E_x - iE_y = \partial_z \Phi_t$. Its normal component $E_n = |f_t'(u)|^{-1}$ is proportional to the surface charge density.

2.2.3. *The Hele-Shaw problem (HS)*

This provides a hydrodynamic regularization of Laplacian growth. The differences with Laplacian growth are in the boundary conditions which now involve a term proportional to the surface tension. It may be formulated as follows [14].

One imagines that the domain $\mathbb{K}_t$ is filled with a non viscous fluid, say air, and the domain $\mathbb{O}_t$ with a viscous one, say oil. Air is supposed to be injected at the origin and there is an oil drain at infinity. The pressure in the air domain $\mathbb{K}_t$ is constant and set to zero by convention. In $\mathbb{O}_t$ the pressure satisfies the Laplace equation $\nabla^2 P = 0$ with boundary behavior $P = -\phi_\infty \log|z| + \cdots$ at infinity reflecting the presence of the oil drain. The boundary conditions on the boundary curve are now $P = -\sigma\kappa_t$ with σ the surface tension and κ_t the curvature of the boundary curve [2]. The fluid velocity in the oil domain $\mathbb{O}_t$ is $\vec{v} = -\vec{\nabla} P$. Laplace equation for P is just a consequence of incompressibility. The evolution of the shape of the domain is specified by imposing that this relation holds on the boundary so that the boundary normal velocity is $v_n = -(\nabla P)_n$ as in Laplacian growth.

Compared to Laplacian growth, the only modification is the boundary condition on the boundary curve. This term prevents the formation of cusps with infinite curvature singularities. The parameter ϕ_∞ sets the scale of the velocity at infinity. In the following we set $\phi_\infty = 1$. By dimensional analysis this implies that $[\mathtt{time}]$ scales as $[\mathtt{length}^2]$ and the surface tension σ has dimension of a $[\mathtt{length}]$. It plays the role of an ultraviolet cut-off.

A standard procedure [14] to solve the equations for the Hele-Shaw problem is by first determining the pressure using complex analysis and then computing the boundary normal velocity. By Laplace equation, the pressure is the real part of an analytic function, $P = -\Re\mathrm{e}\Phi_t$. The complex velocity $v = v_x + iv_y$ is $\bar{v} = \partial_z \Phi_t$. At infinity $\Phi_t(z) \simeq \log z + \cdots$ and $\bar{v} \simeq 1/z + \cdots$. The boundary conditions on P demand that

$$(\Phi_t \circ f_t)(w) = \log w + \sigma\vartheta_t(w)$$

where $\vartheta_t(w)$ is analytic in $\mathbb{D}$ with boundary value $\Re\mathrm{e}[\vartheta_t(u)] = \kappa_t(f_t(u))$ with κ_t the curvature. Explicitly

$$\vartheta_t(w) = \oint \frac{du}{2i\pi u}\left(\frac{w+u}{w-u}\right)\kappa_t(f_t(u))$$

The evolution of f_t is then found by evaluating the boundary normal velocity $v_n = \Re\mathrm{e}(\nabla\Phi)_n$ at point $\gamma_t = f_t(u)$:

$$v_n = \Re\mathrm{e}[\,n\partial_z\Phi_t\,] = |f_t'(u)|^{-1}\,\Re\mathrm{e}[1 + \sigma u\partial_u\theta_t(u)]$$

[2]The curvature is defined by $\kappa \equiv -\vec{n}.\partial_s\vec{\tau}/\vec{\tau}^2 = \Im\mathrm{m}[\bar{\tau}\partial_s\tau/|\tau|^3]$ with $\vec{\tau}$ the tangent and $\vec{n}$ the normal vectors. An alternative formula is: $\kappa = |f_t'(u)|^{-1}\Re\mathrm{e}[1 + \frac{uf_t''(u)}{f_t'(u)}]$. For a disk of radius R, the curvature is $+1/R$.

As above, this determines uniquely $\partial_t f_t(u)$ and it leads to a Loewner chain (2.1) with density:

$$\rho_t(u) = |f_t'(u)|^{-2}\left(1 + \sigma\epsilon_t(u)\right) \quad , \quad \epsilon_t(u) = \Re\mathrm{e}[u\partial_u \vartheta_t(u)] \tag{2.5}$$

The difference with Laplacian growth is in the extra term proportional to σ. It is highly non-linear and non-local. This problem is believed to be well defined at all times for σ positive.

2.2.4. *Other regularized Laplacian growth (rLG)*

These regularizations amount to introduce an UV cutoff δ in the physical space by evaluating $|f_t'|$ at a finite distance away from $\partial\mathbb{O}_t$. A possible choice [20] is $\rho_t(u)^{1/2} = \delta^{-1}\inf\{\varepsilon : \mathrm{dist}[f_t(u+\varepsilon u);\partial\mathbb{O}_t] = \delta\}$. An estimation gives $\rho_t(u) \asymp |f_t'(u+\hat{\varepsilon}_u u)|^{-2}$ where $\hat{\varepsilon}_u$ goes to 0 with δ, so that it naively approaches $|f_t'(u)|^{-2}$ as $\delta \to 0$.

Another possible, but less physical, regularization consists in introducing an UV cutoff ν in the mathematical space so that $\rho_t(u) = |f_t'(u+\nu u)|^{-2}$.

2.2.5. *Dielectric breakdown and generalizations*

A larger class of problems generalizing Laplacian growth have been introduced. Their Loewner measures are as in Laplacian growth but with a different exponent:

$$\rho_t(u) = |f_t'(u)|^{-\alpha} \quad , \quad 0 \leq \alpha \leq 2.$$

Using an electrostatic interpretation of the harmonic potential, one usually refers to the case $\alpha = 1$ as a model of dielectric breakdown because the measure is then proportional to the local electric field $E_n = |f_t'(u)|^{-1}$. This is a phenomenological description. Just as the naive Laplacian growth these models are certainly ill-posed. They also require ultraviolet regularizations, one of which is described below using iterated conformal maps.

2.3. CUSP SINGULARITIES IN LG

The naive LG problem, without regularization, corresponds to the Loewner density $\rho_t(u) = |f_t'(u)|^{-2}$. The occurrence of singularities may be grasped by looking for the evolution of domains with a Z_n symmetry uniformized by the maps

$$f_t(w) = R_t w\left(1 + \frac{\beta_t}{n-1} w^{-n}\right)$$

for some $n > 2$ and with $|\beta_t| \leq 1$. This form of conformal maps is preserved by the dynamics. The conformal radius R_t and the coefficients β_t evolve with time according to $\partial_t R_t^2 = 2/(1-\beta_t^2)$ and $\beta_t = (R_t/R_c)^{n-2}$ with R_c some integration constant. The singularity appears when β_t touches the unit circle which arises at a finite time t_c. At that time the conformal radius is R_c.

At t_c the boundary curve Γ_{t_c} has cusp singularities of the generic local form

$$\ell_c\,(\delta y)^2 \simeq (\delta x)^3$$

with ℓ_c a characteristic local length scale. In the present simple case $\ell_c \simeq R_c$. At time $t \nearrow t_c$, the dynamics is regular in the dimensionless parameter $\ell_c^{-1}\sqrt{t_c - t}$. The maximum curvature of the boundary curve scales as $\kappa_{\max} \simeq \ell_c/(t_c - t)$ near t_c and it is localized at a distance $\sqrt{t_c - t}$ away from the would be cusp tip. See Figure (17).

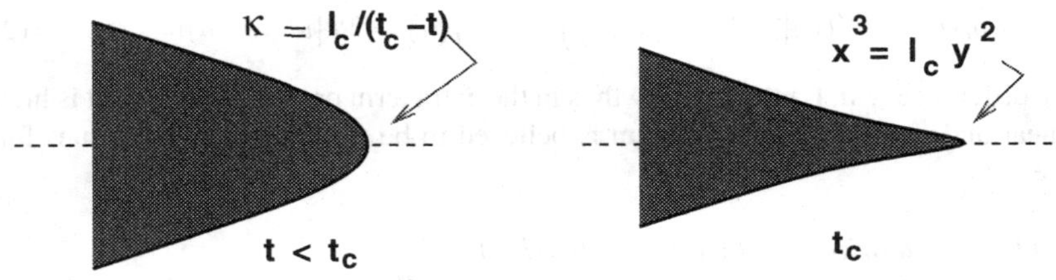

Figure 17. Cups formation in Laplacian growth.

This behavior is quite generic. As we shall see, conformal maps $f_t(w)$ such that their derivatives are polynomials in w^{-1} are stable by the Laplacian growth dynamics. By construction, their zeroes are localized inside the unit disk. A singularity in the boundary curve occurs if one of these zeroes converges to the unit circle. The singularity is then a cusp $\ell_c y^2 \simeq x^3$ as can be seen by expanding locally the conformal map around its singular point.

Once regularized with an explicit ultraviolet cut-off, the processes are believed to be well defined for all time. The effect of the regularization procedure on the domain properties is presently unclear. The domain structures may a priori depend on how the problem has been regularized. In the hydrodynamic regularization –the Hele-Shaw problem– the cusp production is expected to be replaced by unlimited ramifications leading to dendritic growth.

In the regularized model, the curvature of Γ_t is expected to remain finite at all time. Using scaling theory, a crude estimate of its maximum around the would be singularities may be obtained by interchanging the short distance scale $\sqrt{(t_c - t)}$ near the singularity in the unregularized theory with the UV cutoff of the regularized theory. In the hydrodynamic regularization this gives $\kappa_{\max} \simeq \ell_c/\sigma^2$ as $\sigma \to 0$.

2.4. DISCRETE LOEWNER CHAINS

2.4.1. *DLA*

DLA stands for diffusion limited aggregation [15]. It refers to processes in which the domains grow by aggregating diffusing particles. Namely, one imagines building up a domain by clustering particles one by one. These particles are released from the point at infinity, or uniformly from a large circle around infinity, and diffuse as random walkers. They will eventually hit the domain and the first time this happens they stick to it. By convention, time is incremented by unity each time a particle is added to the domain. Thus at each time step the area of the domain is increased by the physical size of the particle. The position at which the particle is added depends on the probability for a random walker to visit the boundary for the first time at this position.

In a discrete approach one may imagine that the particles are tiny squares whose centers move on a square lattice whose edge lengths equal that of the particles, so that particles fill the lattice when they are glued together. The center of a particle moves as a random walker on the square lattice. The probability $Q(x)$ that a particle visits a given site x of the lattice satisfies the lattice version of the Laplace equation $\nabla^2 Q = 0$. It vanishes on the boundary of the domain, i.e. $Q = 0$ on the boundary, because the probability for a particle to visit a point of the lattice already occupied, i.e. a point of the growing cluster, is zero. The local speed at which the domain is growing is proportional to the probability for a site next to the interface but on the outer domain to be visited. This probability is proportional to the discrete normal gradient of Q, since the visiting probability vanishes on the interface. So the local speed is $v_n = (\nabla Q)_n$. It is not so easy to make an unbiased simulation of DLA on the lattice. One of the reasons is that on the lattice there is no such simple boundary as a circle, for which the hitting distribution from infinity is uniform. The hitting distribution on the boundary of a square is not such a simple function. Another reason is that despite the fact that the symmetric random walk is recurrent is 2d, each walk takes many steps to glue to the growing domain. The typical time to generate a single sample of reasonable size with an acceptable bias is comparable to the time it takes to make enough statistics on loop-erased random walks or percolation to get the scaling exponent with two significant digits. Still this is a modest time, but it is enough to reveal the intricacy of the patterns that are formed. Fig.18 is such a sample. The similarity with the sample in fig.1, obtained by iteration of conformal maps, is striking. But a quantitative comparison of the two models is well out of analytic control and belongs to the realm of extensive simulations.

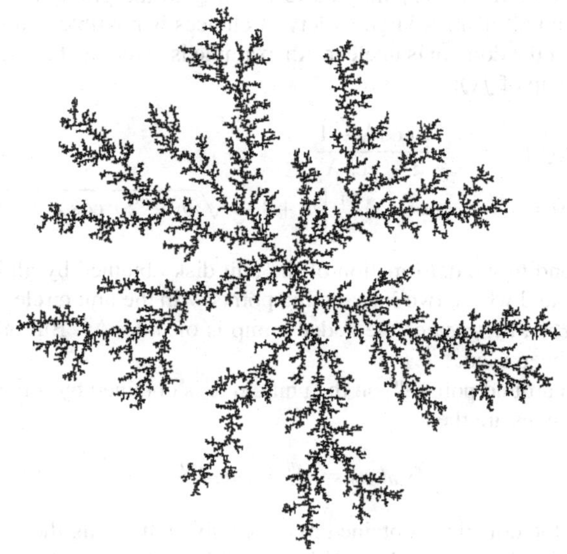

Figure 18. A DLA sample.

DLA provides a discrete analogue of Laplacian growth. The particle size plays the role of an ultraviolet cutoff. Since DLA only assumes that the growth is governed by the diffusion of elementary patterns, DLA has been applied to a large variety of aggregation or deposition phenomena, see eg.[18].

During this process the clustering domain gets ramified and develops branches and fjords of various scales. The probability for a particle to stick on the cluster is much higher on the tip of the branches than deep inside the fjords. This property, relevant at all scales, is responsible for the fractal structure of the DLA clusters.

Since its original presentation [15], DLA has been studied numerically quite extensively. There is now a consensus that the fractal dimension of 2d DLA clusters is $D_{\rm dla} \simeq 1.71$. There is actually a debate on whether this dimension is geometry dependent but a recent study [19] seems to indicate that DLA clusters in a radial geometry and a channel geometry have identical fractal dimension. To add a new particle to the growing domain, a random walk has to wander around and the position at which it finally sticks is influenced by the whole domain. To rephrase this, for each new particle one has to solve the outer Laplace equation, a non-local problem, to know the sticking probability distribution. This is a typical example when scale invariance is not expected to imply conformal invariance.

2.4.2. *Iterated conformal maps*

As proposed in [17], an alternative way to mimic the gluing of elementary particles consists in composing elementary conformal maps, each of which corresponds to adding an elementary particle to the domain.

One starts with an elementary map corresponding to the gluing of a tiny bump, of linear size λ, to the unit disk. A large variety of choices is possible, whose influence on the final structure of the domain is unclear. An example is given by the following formulæ (g_λ is the inverse map of f_λ):

$$\begin{aligned} g_\lambda(z) &= z\,\frac{z\cos\lambda - 1}{z - \cos\lambda} \\ f_\lambda(w) = &= (2\cos\lambda)^{-1}\left[w + 1 + \sqrt{w^2 - 2w\cos 2\lambda + 1}\right] \end{aligned}$$

where f_λ correspond to the deformation of the unit disk obtained by gluing a semi-disk centered at point 1 and whose two intersecting points with the unit circle define a cone of angle 2λ. For $\lambda \ll 1$, the area of the added bump is of order λ^2. But other choices are possible and have been used.

Gluing a bump around point $e^{i\theta}$ on the unit circle is obtained by rotating these maps. The uniformizing maps are then

$$f_{\lambda;\theta}(w) = e^{i\theta}\, f_\lambda(we^{-i\theta})$$

The growth of the domain is obtained by successively iterating the maps $f_{\lambda_n;\theta_n}$ with various values for the size λ_n and the position θ_n of the bumps. See Figure (19). Namely, if after n iterations the complement of the unit disk is uniformized into the complement of the domain by the map $F_{(n)}(w)$, then at the next $(n+1)^{\rm th}$ iteration the uniformizing map is given by:

$$F_{(n+1)}(w) = F_{(n)}(\,f_{\lambda_{n+1};\theta_{n+1}}(w)\,) \tag{2.6}$$

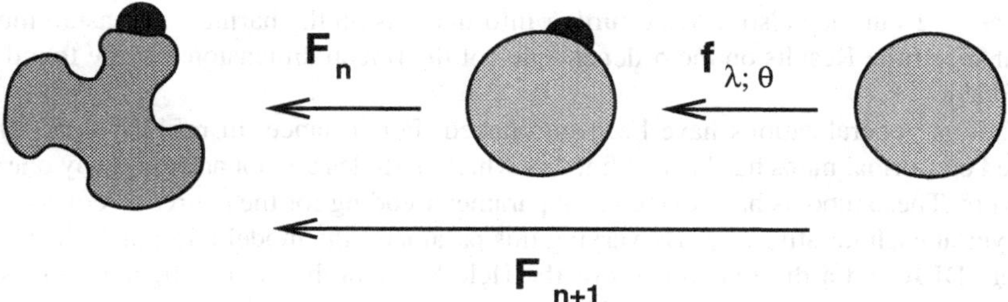

Figure 19. Iteration of conformal maps.

For the inverse maps, this becomes $G_{(n+1)} = g_{\lambda_{n+1};\theta_{n+1}} \circ G_{(n)}$.

To fully define the model one has to specify the choice of the parameter λ_n and θ_n at each iteration. Since λ_n codes for the linear size of the added bump and since locally conformal maps act as dilatations, the usual choice is to rescale λ_{n+1} by a power of $|F'_{(n)}(e^{i\theta_n})|$ as:

$$\lambda_{n+1} = \lambda_0\, |F'_{(n)}(e^{i\theta_n})|^{-\alpha/2}, \quad 0 \leq \alpha \leq 2$$

The case $\alpha = 2$ corresponds to DLA as the physical area of the added bump are approximatively constant and equal to λ_0 at each iterations. In the other case, the area of the added bump scales as $|F'_{(n)}(e^{i\theta_n})|^{2-\alpha}$.

The positions of the added bump are usually taken uniformly distributed on the mathematical unit circle with a measure $d\theta/2\pi$.

It is clear that this discrete model with $\alpha = 2$ provides a regularization of Laplacian growth with λ_0 playing the role of an ultraviolet cutoff. This may also be seen by looking at the *naive* limit of a small cutoff. Indeed, a naive expansion as $\lambda_n \ll 1$ gives that $F_{(n+1)} = F_{(n)} + \delta F_{(n)}$ with

$$\delta F_{(n)}(w) \simeq \lambda_n\, w F'_{(n)}(w) \left(\frac{w + e^{i\theta_n}}{w - e^{i\theta_n}}\right)$$

where we used the expression of f_λ for $\lambda \ll 1$. Using the recursive expression for λ_n and averaging over θ with a uniform distribution yields:

$$\langle \delta F(w) \rangle = \lambda_0\, w F'(w) \oint \frac{d\theta}{2\pi} |F'(e^{i\theta})|^{-\alpha}\, \frac{w + e^{i\theta}}{w - e^{i\theta}}$$

For $\alpha = 2$ this reproduces the Loewner chain for Laplacian growth. But this computation is too naive as the small cutoff limit is not smooth, a fact which is at the origin of the non trivial fractal dimensions of the growing domains.

There are only very few mathematical results on these discrete models. The most recent one [27] deals with the simplest (yet interesting but not very physical) model with $\alpha = 0$. It proves the convergence of the iteration to well-defined random maps uniformizing domains of Hausdorff dimension 1. However, these models have been studied numerically extensively. There exists a huge literature on this subject but see ref.[24] for instance. These studies confirm that the fractal dimension of DLA clusters with $\alpha = 2$ is

$D_{\text{dla}} \simeq 1.71$ but they also provide further informations on the harmonic measure multi-fractal spectrum. Results on the α dependence of the fractal dimension may be found eg. in ref.[25].

Various generalizations have been introduced. For instance, in ref.[26] a model of iterated conformal maps has been defined in which particles are not added one by one but by layers. These models have one control parameter coding for the degree of coverage of the layer at each iterative step. By varying this parameter the model interpolates between discrete DLA and a discrete version of the Hele-Shaw problem. The fractal dimension of the resulting clusters varies with this parameter and seems to indicate that the fractal dimension of this discrete analogue of the Hele-Shaw problem is 2, a point that we shall discuss further in the last section.

2.5. INTEGRABILITY OF LAPLACIAN GROWTH

Laplacian growth is an integrable system, at least up to the cusp formation. Let us recall that it corresponds to a Loewner chain with a density $\rho_t(u) = |f_t(u)|^{-2}$, or equivalently to the quadratic equation

$$(\partial_\alpha \gamma_{t;\alpha})\,(\partial_t \overline{\gamma_{t;\alpha}}) - (\partial_\alpha \overline{\gamma_{t;\alpha}})\,(\partial_t \gamma_{t;\alpha}) = 2i \tag{2.7}$$

for the dynamics of the boundary points $\gamma_{t;\alpha} = f_t(u)$, $u = e^{i\alpha}$. What makes the model integrable is the fact that the r.h.s of eq.(2.7) is constant. Eq.(2.7) is then similar to a quadratic Hirota equation. Hints on the integrable structure were found in [21] and much further developed in [22, 23].

2.5.1. *Conserved quantities*

We now define an infinite set of quantities which are conserved in the naive unregularized LG problem. They reflect its integrability. We follow ref.[22, 23]. These quantities may be defined via a Riemann-Hilbert problem on Γ_t specified by,

$$S_+(\gamma) - S_-(\gamma) = \bar{\gamma} \quad , \quad \gamma \in \Gamma_t \tag{2.8}$$

for functions S_- and S_+ respectively analytic in the outer domain $\mathbb{O}_t$ and in the inner domain $\mathbb{K}_t$. We fix normalization by demanding $S_-(\infty) = 0$. We assume Γ_t regular enough for this Riemann-Hilbert problem to be well defined. As usual, $S_\pm$ may be presented as contour integrals:

$$S_\pm(z) = -\oint_{\Gamma_t} \frac{d\gamma}{2i\pi} \frac{\bar{\gamma}}{z-\gamma}.$$

The conserved quantities are going to be expressed in terms of $S_\pm$. We thus need their time evolution. Differentiation of eq.(2.8) with respect to time and use of the evolution equation (2.7) gives:

$$\partial_t S_+(\gamma) - \partial_t S_-(\gamma) = 2g_t'(\gamma)/g_t(\gamma)$$

Notice now that $g_t'(\gamma)/g_t(\gamma)$ is the boundary value of $(\log g_t)'$ which by construction is analytic in $\mathbb{O}_t$. We may thus rewrite this equation as a trivial Riemann-Hilbert problem, $\partial_t S_+(\gamma) - (\partial_t S_- + 2(\log g_t)')(\gamma) = 0$, so that both terms vanish:

$$\partial_t S_+(z) = 0 \quad \text{and} \quad (\partial_t S_- + 2(\log g_t)')(z) = 0 \tag{2.9}$$

Since S_+ is analytic around the origin, we may expand it in power of z. Equation $\partial_t S_+(z) = 0$ then tells us that $S_+(z)$ is a generating function of conserved quantities: $S_+(z) = \sum_{k\geq 0} z^k I_k$ with

$$I_k = \oint_{\Gamma_t} \frac{d\gamma}{2i\pi} \bar{\gamma}\gamma^{-k-1} \quad , \quad \partial_t I_k = 0. \tag{2.10}$$

This provides an infinite set of conserved quantities.

Since S_- is analytic around infinity, it may be expanded in power of $1/z$: $S_-(z) = -\mathcal{A}_t/\pi z + \cdots$ with $\mathcal{A}_t = -\frac{i}{2}\oint_{\Gamma_t} d\gamma\bar{\gamma}$ the area of $\mathbb{K}_t$. The second equation $(\partial_t S_- + 2(\log g_t)')(z) = 0$ with $g_t(z) = R_t^{-1} z + O(1)$ then implies $\partial_t \mathcal{A}_t = 2\pi$. The area of the domain grows linearly with time, up to the time at which the first cusp singularity appears.

2.5.2. *Simple solutions*

A particularly simple class of conformal maps, solutions of the Laplacian growth equation, are those such that their derivatives are polynomials in w^{-1}. They may be expanded as:

$$f_t(w) = \sum_{n=0}^{N} f_n w^{1-n}, \quad f_0 = R_t > 0 \tag{2.11}$$

with N finite but arbitrary. The dynamical variables are the $N+1$ coefficient $f_0, \cdots, f_N$. They are all complex except f_0 which is real. It will be convenient to define the function $\bar{f}_t$ by $\bar{f}_t(w) = \overline{f_t(\overline{w})}$.

The fact that this class is stable under the dynamics follows from the Loewner equation (2.4). The trick consists in using the fact that the integration contour is on the unit circle so that $|f_t'(u)|^2 = f_t'(u)\bar{f}_t'(1/u)$. The contour integral then involves a meromorphic function of u so that it can be evaluated by deforming the contour to pick the residues. This is enough to prove that $\partial_t f_t(w)$ possesses the same structure as $f_t(w)$ itself so that the class of functions (2.11) is stable under the dynamics.

Alternatively one may expand the quadratic equation (2.7) to get a hierarchy of equations:

$$\sum_{n\geq 0}(1-n)[f_n \dot{\bar{f}}_{j+n} + \bar{f}_n \dot{f}_{-j+n}] = 2\delta_{j;0}$$

For $j = 0$, this equation tells us again that the area of the domain grows linearly with time. Besides this relation there are only N independent complex equations for $j = 1, \cdots, N$ which actually code for the conserved quantities.

To really have an integrable system we need to have as many independent integrals of motion as dynamical variables. Thus we need to have N conserved quantities. These are given by the I_k's defined above which may rewritten as

$$I_k = \oint_{|u|=1} \frac{du}{2i\pi} \frac{f_t'(u)\bar{f}_t(1/u)}{f_t(u)^{k+1}}$$

Only the first N quantities, $I_0, \cdots, I_{N-1} = R^{1-N}\bar{f}_N$ are non-vanishing. They are independent. They can be used to express algebraically all f_n's, $n \geq 0$, in terms of the real parameter $f_0 = R_t$. The area law,

$$\mathcal{A}_t = \pi[R_t^2 + \sum_{n\geq 1}(1-n)|f_n|^2] = 2\pi t,$$

with the f_n's expressed in terms of R_t, then reintroduces the time variable by giving its relation with the conformal radius.

2.5.3. *Algebraic curves*
As was pointed out in [23], solutions of Laplacian growth and their cusp formations have an elegant geometrical interpretation involving Riemann surfaces.

Recall that given a sufficiently smooth real curve Γ_t drawn on the complex plane one may define a function $S(z)$, called the Schwarz function, analytic in a ribbon enveloping the curve such that

$$S(\gamma) = \overline{\gamma}, \quad \gamma \in \Gamma_t$$

By construction, the Schwarz function may be expressed in terms of uniformizing maps of the domain bounded by the curves as $S(z) = \bar{f}_t(1/g_t(z))$.

The Riemann-Hilbert problem (2.8) defining the conserved charges then possesses a very simple interpretation: $S_\pm$ are the polar part of the Schwarz function respectively analytic inside or outside Γ_t, i.e. $S(z) = S_+(z) - S_-(z)$. Thus the polar part S_+, analytic in the inner domain, is conserved. The polar part S_-, analytic in the outer domain, evolves according to eqs.(2.9). Since $\log g_t(z)$ is analytic in the outer domain, these equations are equivalent to the single equation:

$$\partial_t S(z) = -2(\log g_t(z))' \tag{2.12}$$

Now the physical curve Γ_t may be viewed as a real slice of a complex curve, alias a Riemann surface. The latter is constructed using the Schwarz function as follows. Recall that $s = S(z)$ is implicitly defined by the relations $z = f_t(w)$, $s = \bar{f}_t(1/w)$. In the case of polynomial uniformizing maps we get the pair of equations

$$\begin{aligned} z &= f_0 w + f_1 + f_2 w^{-1} + \cdots + f_N w^{1-N} \\ s &= \bar{f}_0 w^{-1} + \bar{f}_1 + \bar{f}_2 w + \cdots + \bar{f}_N w^{N-1} \end{aligned}$$

Eliminating w yields an algebraic equation for z and s only:

$$\mathbf{R}: \quad R(z,s) = 0 \tag{2.13}$$

with R a polynomial of degree N in both variables, $R(z,s) = \bar{f}_N z^N + f_N s^N + \cdots$. Eq.(2.13) defines an algebraic curve $\mathbf{R}$. It is of genus zero since by construction it is uniformized by points w of the complex sphere. It has many singularities which have to be resolved to recover a smooth complex manifold.

The Riemann surface $\mathbf{R}$ may be viewed as a N-sheeted covering of the complex z plane: each sheet corresponds to a determination of s above point z. At infinity, the physical sheet corresponds to $z \simeq f_0 w$ with $w \to \infty$ so that $s \simeq (z/f_0)^{N-1}\,\bar{f}_N$, the other $N-1$ sheets are ramified and correspond to $z \simeq f_N/w^{N-1}$ and $s \simeq \bar{f}_0/w$ with $w \to 0$ so that $z \simeq (s/\bar{f}_0)^{N-1}\,f_N$. Hence infinity is a branch point of order $N-1$.

By the Riemann-Hurwitz formula the genus g is $2g - 2 = -2N + \nu$ with ν the branching index of the covering. Since the point at infinity counts for $\nu_\infty = N - 2$, there should be N other branch points generically of order two. By definition they are determined by solving the equations $R(z,s) = 0$ and $\partial_s R(z,s) = 0$. Since the curve is uniformized by $w \in \mathbb{C}$, these two equations imply that $z'(w)\partial_z R(z(w), s(w)) = 0$. Hence either $z'(w) = 0$, $\partial_z R \neq 0$, and the point is a branch point, or $z'(w) \neq 0$, $\partial_z R = 0 = \partial_s R$, and the point is actually a singular point which needs to be desingularized. So the N branch points at finite distance are the critical points of the uniformizing map $z = f_t(w)$.

The curve $\mathbf{R}$ possesses an involution $(z,s) \to (\bar{s}, \bar{z})$ since $R(\bar{s}, \bar{z}) = \overline{R(z,s)}$ by construction. The set of points fixed by this involution has two components: (i) a continuous one parametrized by points $w = u$, $|u| = 1$ –this is the real curve Γ_t that we started with– and (ii) a set of N isolated points which are actually singular points.

The cusp singularity of the real curve Γ_t arises when a isolated real point merges with the continuous real slice Γ_t. Locally the behavior is as for the curve $u^2 = \varepsilon\, v^2 + v^3$ with $\varepsilon \to 0$.

The simplest example is for $N = 3$ with $\mathbb{Z}_3$ symmetry so that $f_t(w) = w + b/w^2$ and

$$w^2\, z = w^3 + b \quad , \quad w\, s = 1 + bw^3$$

We set $f_0 = 1$ and $f_3 = b$. Without lost of generality we assume b real. The algebraic curve is then

$$R(z,s) \equiv bz^3 + bs^3 - b^2 s^2 z^2 + (b^2 - 1)(2b^2 + 1)sz - (b^2 - 1)^3 = 0$$

Infinity is a branch point of order two. The three other branch points are at $z = 3\omega\,(b/4)^{1/3}$, $s = \omega^2\,(2b^2+1)(2b)^{-1/3}$ corresponding to $w = \omega(2b)^{1/3}$ with ω a third root of unity. They are critical points of $z(w)$. There are three singular points at $z = \omega\,(1 - b^2)/b$, $s = \omega^2\,(1 - b^2)/b$ corresponding to $w = \omega(1 \pm \sqrt{1 - 4b^2})/2b$. The physical regime is for $b < 1/2$ in which case the real slice $\Gamma_t = \{z(u),\ |u| = 1\}$ is a simple curve. The singular points are then in the outer domain and the branch points in the inner domain. The cusp singularities arise for $b = 1/2$. For $b > 1/2$ there are no isolated singular points, they are all localized on the real slice so that Γ_t possesses double points. See Figure (20).

2.6. DENDRITIC ANOMALIES

We now would like to discuss a few points concerning the regularized models and their small cut-off limits –which are actually what we are interested in. We shall point out that the hydrodynamic regularization used in the Hele-Shaw problem possesses essential differences with that used in other regularized models, say DLA. This opens the possibility for the Hele-Shaw problem and DLA not to be in the same universality class.

This observation is based on the conjectural existence of anomalies in the Hele-Shaw problem –a word which refers to quantities which, although naively vanishing in the small cut-off limit, are actually non zero but finite in this limit due to compensating effects.

An analogy with Burgers turbulence may be useful. The 1d Burgers equation is $\partial_t u + u\partial_x u - \nu\partial_x^2 u = 0$ for a velocity field $u(t,x)$. The viscosity ν plays the role of ultraviolet cut-off. At $\nu = 0$ this equation is simply Euler equation which may be easily solved.

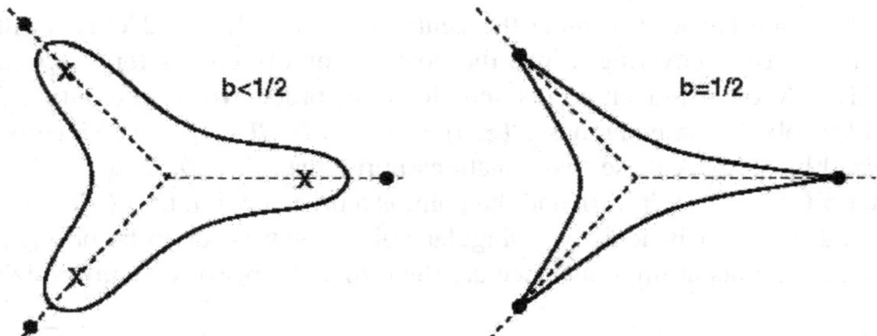

Figure 20. Subcritical and critical algebraic curves. Black circles are singular points. Crosses are branch points.

At $\nu = 0$, any smooth initial data produces shocks in finite time with a discontinuity of u. In the regularized equation with $\nu \neq 0$, there is no shock production but the field $\hat{\epsilon} \equiv \nu(\partial_x u)^2$, which naively vanishes when $\nu \longrightarrow 0$, has actually a non trivial limit. It does not vanish because there is a compensation between small viscosity ν and large velocity gradient $(\partial_x u)$. The field $\hat{\epsilon}$ codes for the amount of energy dissipated inside the shocks.

The analogy that we would like to put forward is that the cusp formation in Laplacian growth is analogous to shock formation in Burgers and that the surface tension σ in Laplacian growth plays the role of the viscosity in Burgers. There are potential compensating effects between small surface tension and large curvature in the Hele-Shaw problem.

2.6.1. *Almost conserved quantities*

We now describe the effect of the regularization procedure on the conserved quantities of Laplacian growth. Recall that they were defined starting from the Riemann-Hilbert problem:

$$h_+(\gamma) - h_-(\gamma) = \bar{\gamma} \quad , \quad \gamma \in \Gamma_t \tag{2.14}$$

for functions h_- and h_+ respectively analytic in $\mathbb{O}_t$ and $\mathbb{K}_t$. We use the notation $h_\pm$ instead of $S_\pm$ in order to avoid confusion with the unregularized Laplacian growth. We again fix normalization by demanding $h_-(\infty) = 0$ so that $h_\pm(z) = -\oint_{\Gamma_t} \frac{d\gamma}{2i\pi} \frac{\bar{\gamma}}{z-\gamma}$. Time differentiating eq.(2.14) using the evolution equation (2.3) now gives:

$$\partial_t h_+(\gamma) - \partial_t h_-(\gamma) = 2 \frac{g_t'(\gamma)}{g_t(\gamma)} \frac{v_n(\gamma)}{|g_t'(\gamma)|} \tag{2.15}$$

with v_n the boundary normal velocity –in LG, we had $v_n = |g_t'(\gamma)|$.

In the case of the Hele-Shaw (HS) problem, the hydrodynamic regularization of Laplacian growth, the normal velocity possesses a particular form (2.5) involving the derivative of ϑ. Eq.(2.15) may then be rewritten in terms of the potential, $\Phi_t(z) = \log g_t(z) + \sigma\vartheta_t(g_t(z))$, and of the curvature as:

$$\partial_t h_+(\gamma) - \partial_t h_-(\gamma) = 2\Phi_t'(\gamma) - 2\sigma\kappa_t'(\gamma)$$

Since Φ_t is analytic in $\mathbb{O}_t$, the solution of this equation is given by that of the Riemann-Hilbert problem for κ_t. So, let κ_- and κ_+ be respectively analytic in $\mathbb{O}_t$ and $\mathbb{K}_t$ with boundary values $\kappa_+ - \kappa_- = \kappa_t$ on Γ_t. Then:

$$\begin{aligned} \partial_t h_+(z) &= -2\sigma\partial_z\,\kappa_+(z) \\ \partial_t h_-(z) &= -2\partial_z\Phi_t(z) - 2\sigma\partial_z\,\kappa_-(z) \end{aligned} \tag{2.16}$$

These equations codes for the behavior of $h_\pm$ in the limit of vanishing surface tension.

A noticeable point is that the above r.h.s. are total derivatives. In particular, contribution to the $1/z$ term in $\partial_t h_-$ only comes from the logarithmic factor in the potential. This implies that at $\sigma \neq 0$ the area of the domains also grows linearly with time, $\mathcal{A}_t \simeq 2\pi t$ at large time [3], and it is independent of the UV cutoff σ. This is contrast with other regularizations for which the area growth may be cutoff dependent.

2.6.2. *A conjecture*

The supplementary terms proportional to $\sigma\kappa'_\pm$ in the quasi conservation laws (2.16) naively vanish when the UV cutoff is removed. *Dendritic anomalies* refer to the possibilities for these terms not to vanish as the cutoff goes to zero, say $\sigma \longrightarrow 0$ in the Hele-Shaw (HS) problem. Such possibility arises due to a balance between small surface tension and large curvature.

Recall the previous HS estimation of the maximum curvature $\kappa_{\max} \simeq \ell_c/\sigma^2$ with ℓ_c the local scale associated to the cusp. These maxima are localized around points γ_c which would turn into cusp singularities in absence of surface tension. So, we may expect that in the HS problem the product $\sigma\kappa(\gamma)$ is a function picked around its maximum values and is of the form $\frac{\ell_c}{\sigma}\varphi(\frac{\gamma-\gamma_c}{\sigma})$ when the point γ moves away from γ_c with φ finite at the origin and decreasing rapidly away from it. For isolated singular points γ_c, this would converge in the limit $\sigma \longrightarrow 0$ to a sum of Dirac point measures.

This leads us to conjecture that the anomalous term $\sigma\kappa(\gamma)$ goes to a non trivial but finite distribution as $\sigma \longrightarrow 0$. This distribution then contributes to eq.(2.16) to break the 'classical' conservation laws valid in the unregularized theory, hence the name *dendritic anomaly*.

The Loewner equation (2.1) –or the evolution equation (2.3)– are then expected to have a finite limit as $\sigma \longrightarrow 0$ with a non trivial right hand side. The solution would be well-defined for all times and would describe a non trivial familly $\hat{\mathbb{K}}_t$ such that, for fixed t, $\hat{\mathbb{K}}_t$ is the limit of the domains $\mathbb{K}_t$ as $\sigma \longrightarrow 0$. This is consistent with the observation that the area of $\mathbb{K}_t$ is independant of σ and grows linearly with time for any σ, ie. $\mathcal{A}_t \simeq 2\pi t$, so that the existence of a limiting set $\hat{\mathbb{K}}_t$ is not ruled out.

It is implicit in the previous conjecture that the uniformizing maps f_t have a finite limit, say $\hat{f}_t$, as $\sigma \longrightarrow 0$ such that the image of the unit circle by $\hat{f}_t$ is the boundary of $\mathbb{C} \setminus \hat{\mathbb{K}}_t$. This leads us to conjecture that the limiting domain $\hat{\mathbb{O}}_t$, which is the component of $\mathbb{C} \setminus \hat{\mathbb{K}}_t$ containing infinity, is uniformized onto $\mathbb{D}$ by $\hat{g}_t \equiv \hat{f}_t^{-1}$.

Let R_t be the conformal radius computed at finite σ using the maps f_t. Recall that if $D_{\rm hs}$ is the fractal dimension we have $R_t^{D_{\rm hs}} \asymp \mathcal{A}_t$ for large t. By dimensional analysis

[3]This result concerning the area is simply a consequence of the fluid incompressibility. But the general method allows to deal with the complete hierarchy of quasi conserved charges.

$R_t \simeq \sigma(\mathcal{A}_t/\sigma^2)^{1/D_{\rm hs}}$ since σ has dimension of a `[length]`. Because in the HS problem the area grows linearly with time without any dependence on σ, this also reads:

$$R_t^{D_{\rm hs}} \simeq \sigma^{D_{\rm hs}-2}\, t$$

Since we expect the limiting domains and the limiting conformal maps to exit, R_t should have a finite limit as $\sigma \to 0$ which coincides with the conformal radius $\hat{R}_t$ computed using $\hat{f}_t$. As consequence, we expect:

$$D_{\rm hs} = 2$$

That is: the fractal dimension of Laplacian growth clusters, within the hydrodynamic regularization, is 2. If these conjectures are true, they imply that the Hele-Shaw problem is not in the same universality class as DLA.

These conjectures are compatible with that of ref.[26] based on numerical studies of generalized iterated conformal maps. They are not expected to apply to arbitrary regularizations because in these cases the area may depend on the UV cutoff and the previous dimensional analysis does not apply. It would then be natural to wonder whether we may define renormalized uniformizing maps in the limit of a vanishing cutoff.

Acknowledgements

It is a pleasure thank Vincent Pasquier and Paul Wiegmann for discussions on this problem and the organizers of the Cargèse 2004 ASI for providing us the opportunity to write these notes.

Work supported in part by EC contract number HPRN-CT-2002-00325 of the EUCLID research training network.

References

1. A. Belavin, A. Polyakov, A. Zamolodchikov, *Infinite conformal symmetry in two-dimensional quantum field theory*, Nucl. Phys. **B241**, 333-380, (1984).
2. J. Cardy, J. Phys. **A25** (1992) L201–206.
   The rigourous proof is in
   S. Smirnov, C.R. Acad. Sci. Paris **333** (2001) 239–244.
3. G. Watts, J. Phys **A29** (1996) L363–368.
   The rigourous proof is in
   J. Dubédat, *Excursion Decompositions for SLE and Watts' crossing formula*, preprint, arXiv: math.PR/0405074
4. O. Schramm, Israel J. Math., **118**, 221–288, (2000);
5. G. Lawler, O. Schramm and W. Werner, (I): Acta Mathematica **187** (2001) 237–273; arXiv:math.PR/9911084
   G. Lawler, O. Schramm and W. Werner, (II): Acta Mathematica **187** (2001) 275–308; arXiv:math.PR/0003156
   G. Lawler, O. Schramm and W. Werner, (III): Ann. Henri Poincaré **38** (2002) 109–123. arXiv:math.PR/0005294.

6. S. Rohde and O. Schramm, arXiv:math.PR/0106036; and references therein.
7. V. Beffara, *The dimension of the SLE curves*, arXiv: math.PR/0211322.
8. J. Dubédat, *$SLE(\kappa, \rho)$ martingales and duality*, to appear in Ann. Probab. arXiv:math.PR/0303128
9. G. Lawler, introductory texts, including the draft of a book, may be found at http://www.math.cornell.edu/~lawler
   W. Werner, *Lectures notes of the 2002 Saint Flour summer school*
10. M. Bauer and D. Bernard, Commun. Math. Phys. **239** (2003) 493–521, arXiv:hep-th/0210015, and Phys. Lett. **B543** (2002) 135–138;
    M. Bauer and D. Bernard, Phys. Lett. **B557** (2003) 309–316, arXiv-hep-th/0301064;
    M. Bauer and D. Bernard, Ann. Henri Poincaré **5** (2004) 289–326, arXiv:math-ph/0305061.
    M. Bauer and D. Bernard, *SLE, CFT and zig-zag probabilities*, arXiv:math-ph/0401019.
    M. Bauer, D. Bernard and J. Houdayer, *Dipolar SLEs*, arXiv:math-ph/0411038
11. R. Friedrich and W. Werner, *Conformal restriction, highest weight representations and SLE*, arXiv:math-ph/0301018.
12. , B. Duplantier *Conformal Fractal Geometry and Boundary Quantum Gravity*, arXiv:math-ph/0303034.
13. John B. Conway, *Functions of One Complex Variable II*, Springer-Verlag, New York, New York, 1995
14. D. Bensimon, L. Kadanoff, S. Liang, B. Shraiman, C. Tang, Rev. Mod. Phys. **58** (1986) 977, and references therein.
15. T. Witten and L. Sander, Phys. Rev. Lett. **47** (1981) 1400.
16. L. Niemeyer, L. Pietronero and H. Wiesmann, Phys. Rev. Lett. **52** (1984) 1033.
17. M. Hasting and S. Levitov, Physica **D116** (1998) 244.
18. T. Hasley, Physics Today, **53**, Nov. 2000, 36-41;
    M. Bazant, D. Crowdy, *Conformal methods for interfacial dynamics*, arXiv:cond-mat/0409439, and references therein.
19. E. Somfai, R. Ball, J. deVita, L. Sander, ArXiv:cond-mat/0304458.
20. L. Carleson and N. Makarov, Commun. Math. Phys. **216** (2001) 583.
21. B. Shraiman, D. Bensimon, Phys. Rev A **30** (1984) 2840.
22. S. Richardson, Euro. J. of Appl. Math. **12** (2001) 571.
23. I. Krichever, M. Mineev-Weinstein, P. Wiegmann and A. Zabrodin, arXiv:nlin.SI/0311005, and refs. therein.
24. B. Davidovitch, H. Hentschel, Z. Olami, I. Procaccia, L. Sander, E. Somfai, Phys. Rev. **E59** (1999) 1368.
    M. Jensen, A. Levermann, J. Mathiesen, I. Procaccia, Phys. Rev. **E65** (2002) 046109.
25. M. Hastings, arXiv:cond-mat/0103312.
26. H. Hentschel, A. Levermann, I. Procaccia, arXiv:cond-mat/0111567.
27. S. Rohde, M. Zinsmeister, *Some remarks on Laplacian growth*, preprint April 2004.

THEORY OF COSMOLOGICAL PERTURBATIONS AND APPLICATIONS TO SUPERSTRING COSMOLOGY

ROBERT H. BRANDENBERGER
Physics Department, McGill University 3600 rue Université
Montreal, QC, H3A 2T8, CANADA
and
Brown University Physics Department Providence, RI 02912, USA

Abstract.
The theory of cosmological perturbations is the main tool which connects theories of the early Universe (based on new fundamental physics such as string theory) with cosmological observations. In these lectures, I will provide an introduction to this theory, beginning with an overview of the Newtonian theory of fluctuations, moving on to the analysis of fluctuations in the realm of classical general relativity, and culminating with a discussion of the quantum theory of cosmological perturbations. I will illustrate the formalism with applications to inflationary cosmology. I will review the basics of inflationary cosmology and discuss why - through the evolution of fluctuations - inflation may provide a way of observationally testing Planck-scale physics.

1. Introduction

Recent years have provided a wealth of observational data about the cosmos. We have high resolution maps of the anisotropies in the temperature of the cosmic microwave background (CMB) [1], surveys of the large-scale structure (LSS) - the distribution of galaxies in three-dimensional space - are increasing in size and in accuracy (see e.g. [2] and [3]), and new techniques which will allow us to measure the distribution of the dark matter are being pioneered. All of this data involves small deviations of the cosmos from homogeneity and isotropy. The cosmological observations reveal that the Universe has non-random fluctuations on all scales smaller than the present Hubble radius.

Parallel to this spectacular progress in observational cosmology, new cosmological scenarios have emerged within which it is possible to explain the origin of non-random inhomogeneities by means of causal physics. The scenario which has attracted most attention is inflationary cosmology [4, 5], according to which there was a period in the early Universe in which space was expanding at an accelerated rate. However, there are also alternative proposals [7, 8] in which our current stage of cosmological expansion

L. Baulied et al. (eds.), String Theory: From Gauge Interactions to Cosmology, 79–116.

is preceded by a phase of contraction. These scenarios have in common the fact that for scales of cosmological interest today, although their physical wavelength is larger than the Hubble length during most of the history of the universe, it is smaller than the Hubble radius at very early times, thus in principle allowing for a causal origin of the cosmological fluctuations.

In order to connect theories of fundamental physics providing an origin of perturbations with the data on the late time universe, one must be able to evolve cosmological fluctuations from earliest times to today. Since on large scales (scales larger than about 10 Mpc - 1 Mpc being roughly three million light years) the relative density fluctuations are smaller than one today, and since these relative fluctuations grow in time as a consequence of gravitational instability, they were smaller than one throughout their history - at least in a universe which is always expanding. Thus, it is reasonable to expect that a linearized analysis of the fluctuations will give reliable results.

In most current models of the very early universe it is assumed that the perturbations originate as quantum vacuum fluctuations. Thus, quantum mechanics is important. On the other hand, since for most of the history of the universe the wavelengths corresponding to scales of cosmological interest today were larger than the Hubble radius, it is crucial to consider the general relativistic theory of fluctuations. Hence, a quantum theory of general relativistic fluctuations is required. At the level of the linearized theory of cosmological perturbations, a unified quantum theory of the formation and evolution of fluctuations exists - and this will be the main topic of these lectures. Note that all of the conceptual problems of merging quantum mechanics and general relativity have been thrown out by hand by restricting attention to the linearized analysis. The question of "linearization stability" of the system, namely whether the solutions of the linearized equations of motion in fact correspond to the linearization of solutions of the full equations, is a deep one and will not be addressed here (see e.g. [6] for discussions of this issue).

Inflationary cosmology is at the present time the most successful framework of connecting physics of the very early universe with the present structure (although alternatives such as the Pre-Big-Bang [7] and Ekpyrotic [8] scenarios have been proposed and may turn out to be successful as well). I will thus begin these lectures with a review of inflationary cosmology, and of scalar field-driven models for inflation. Next, I will provide a detailed discussion of the theory of cosmological perturbations, beginning with the Newtonian theory, moving on to the classical general relativistic analysis, and ending with the quantum theory of cosmological perturbations [1] In the final sections of these notes, I will return to inflationary cosmology, and focus on some important conceptual problems which are not addressed in current realizations of inflation in which the accelerated expansion of space is driven by a scalar matter field. Addressing these conceptual problems is a challenge and great opportunity for string theory. Since cosmological inflation leads to a quasi-exponential increase in the wavelength of inhomogeneities, it provides a microscope with which string-scale physics can in principle be probed in current observations. I will conclude these lectures with a discussion of this "window of opportunity" for string theory which inflation provides [9].

[1] These sections are an updated version of the lecture notes [10].

2. Overview of Inflationary Cosmology

To establish our notation and framework, we will be taking the background space-time to be homogeneous and isotropic, with a metric given by

$$ds^2 = dt^2 - a(t)^2 d\mathbf{x}^2 \,, \tag{2.1}$$

where t is physical time, $d\mathbf{x}^2$ is the Euclidean metric of the spatial hypersurfaces (here taken for simplicity to be spatially flat), and $a(t)$ is the scale factor. The scale factor determines the *Hubble expansion rate* via

$$H(t) = \frac{\dot{a}}{a}(t) \,. \tag{2.2}$$

The coordinates $\mathbf{x}$ used above are *comoving* coordinates, coordinates painted onto the expanding spatial hypersurfaces.

In standard big bang cosmology, the universe is decelerating, i.e. $\ddot{a} < 0$. As a consequence, the *Hubble radius*

$$l_H(t) = H^{-1}(t) \tag{2.3}$$

is increasing in comoving coordinates. As will be explained later mathematically, the Hubble radius is the maximal distance that microphysics can act coherently over a Hubble expansion time - in particular it is the maximal distance on which any causal process could create fluctuations. If the universe were decelerating forever, then scales of cosmological interest today would have had a wavelength larger than the Hubble radius at all early times. This gives rise to the *fluctuation problem* for Standard Big Bang (SBB) cosmology, namely the problem that there cannot be any causal process which at early time creates perturbations on scales which are being probed in current LSS and CMB observations [2].

The idea of inflationary cosmology is to assume that there was a period in the very early Universe during which the scale factor was accelerating, i.e. $\ddot{a} > 0$. This implies that the Hubble radius was shrinking in comoving coordinates, or, equivalently, that fixed comoving scales were "exiting" the Hubble radius. In the simplest models of inflation, the scale factor increases nearly exponentially. As illustrated in Figure (1), the basic geometry of inflationary cosmology provides a solution of the fluctuation problem. As long as the phase of inflation is sufficiently long, all length scales within our present Hubble radius today originate at the beginning of inflation with a wavelength smaller than the Hubble radius at that time. Thus, it is possible to create perturbations locally using physics obeying the laws of special relativity (in particular causality). As will be discussed later, it is quantum vacuum fluctuations of matter fields and their associated curvature perturbations which are responsible for the structure we observe today.

Postulating a phase of inflation in the very early universe also solves the *horizon problem* of the SBB, namely it explains why the causal horizon at the time t_{rec} when photons last scatter is larger than the radius of the past light cone at t_{rec}, the part of the last scattering surface which is visible today in CMB experiments. Inflation also explains the near flatness of the universe: in a decelerating universe spatial flatness is an unstable fixed

[2]The reader is encouraged to find the hole in this argument, and is referred to [11, 7, 13] for the answer.

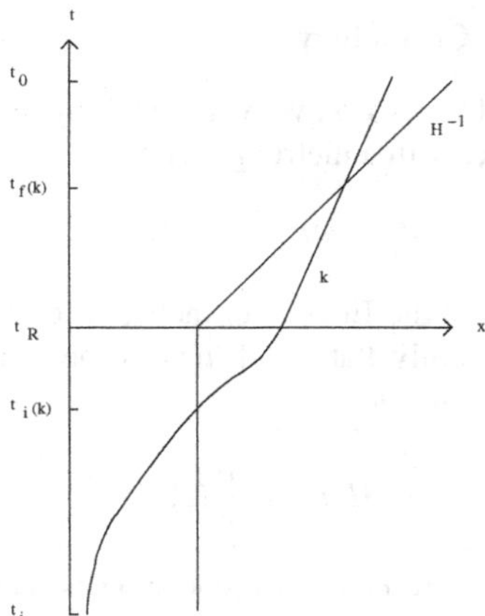

Figure 1. Space-time diagram (sketch) showing the evolution of scales in inflationary cosmology. The vertical axis is time, and the period of inflation lasts between t_i and t_R, and is followed by the radiation-dominated phase of standard big bang cosmology. During exponential inflation, the Hubble radius H^{-1} is constant in physical spatial coordinates (the horizontal axis), whereas it increases linearly in time after t_R. The physical length corresponding to a fixed comoving length scale labelled by its wavenumber k increases exponentially during inflation but increases less fast than the Hubble radius (namely as $t^{1/2}$), after inflation.

point of the dynamics, whereas in an accelerating universe it becomes an attractor. Another important aspect of the inflationary solution of the *flatness problem* is that inflation exponentially increases the volume of space. Without inflation, it is not possible that a Planck scale universe at the Planck time evolves into a sufficiently large universe today.

Let us now consider how it is possible to obtain a phase of cosmological inflation. We will assume that space-time is to be described using the equations of General Relativity[3]. In this case, the dynamics of the scale factor $a(t)$ is determined by the Friedmann-Robertson-Walker (FRW) equations

$$(\frac{\dot{a}}{a})^2 = 8\pi G\rho \tag{2.4}$$

and

$$\frac{\ddot{a}}{a} = -4\pi G(\rho + 3p) \tag{2.5}$$

where for simplicity we have omitted the contributions of spatial curvature (since spatial curvature is diluted during inflation) and of the cosmological constant (since any small cosmological constant which might be present today has no effect in the early Universe since the associated energy density does not increase when going into the past). In the above, ρ and p denote the energy density and pressure, respectively. From (2.5) it is clear

[3]Note, however, that the first model of exponential expansion of space [14] made use of a higher derivative gravitational action.

that in order to obtain an accelerating universe, matter with sufficiently negative pressure

$$p < -\frac{1}{3}\rho \tag{2.6}$$

is required. Exponential inflation is obtained for $p = -\rho$.

Conventional perfect fluids have positive semi-definite pressure and thus cannot yield inflation. In addition, we know that a description of matter in terms of classical perfect fluids must break down at early times. An improved description of matter will be given in terms of quantum fields. Scalar matter fields are special in that they allow at the level of a renormalizable action the presence of a potential energy term. The energy density and pressure of a scalar field φ with canonically normalized action [4]

$$\mathcal{L} = \sqrt{-g}\big[\frac{1}{2}\partial_\mu\varphi\partial^\mu\varphi - V(\varphi)\big] \tag{2.7}$$

(where Greek indices are space-time indices, $V(\varphi)$ is the potential energy density, and g is the determinant of the metric) are given by

$$\begin{aligned} \rho &= \frac{1}{2}(\dot{\varphi})^2 + \frac{1}{2}a^{-2}(\nabla\varphi)^2 + V(\varphi) \\ p &= \frac{1}{2}(\dot{\varphi})^2 - \frac{1}{6}a^{-2}(\nabla\varphi)^2 - V(\varphi). \end{aligned} \tag{2.8}$$

Thus, it is possible to obtain an almost exponentially expanding universe provided the scalar field configuration [5] satisfies

$$\frac{1}{2}(\nabla_p\varphi)^2 \ll V(\varphi), \tag{2.9}$$

$$\frac{1}{2}(\dot{\varphi})^2 \ll V(\varphi). \tag{2.10}$$

In the above, $\nabla_p \equiv a^{-1}\nabla$ is the gradient with respect to physical as opposed to comoving coordinates. Since spatial gradients redshift as the universe expands, the first condition will (for single scalar field models) always be satisfied if it is satisfied at the initial time [6]. It is the second condition which is harder to satisfy. In particular, this condition is in general not preserved in time even it is initially satisfied.

It is sufficient to obtain a period of cosmological inflation that the *slow-roll conditions* for φ are satisfied. Recall that the equation of motion for a homogeneous scalar field in a cosmological space-time is (as follows from (2.7)) is

$$\ddot{\varphi} + 3H\dot{\varphi} = -V'(\varphi), \tag{2.11}$$

where a prime indicates the derivative with respect to φ. In order that the scalar field roll slowly, it is necessary that

$$\ddot{\varphi} \ll 3H\dot{\varphi} \tag{2.12}$$

[4] See [15] for a discussion of fields with non-canonical kinetic terms.

[5] The scalar field yielding inflation is called the *inflaton*.

[6] In fact, careful studies [16] show that since the gradients decrease even in a non-inflationary backgrounds, they can become subdominant even if they are not initially subdominant.

such that the first term in the scalar field equation of motion (2.11) is negligible. In this case, the condition (2.10) becomes

$$(\frac{V'}{V})^2 \ll 48\pi G \tag{2.13}$$

and (3) becomes

$$\frac{V''}{V} \ll 24\pi G\,. \tag{2.14}$$

In the initial model of inflation using scalar fields ("old inflation" [4]), it was assumed that φ was initially in a false vacuum with large potential energy. Hence, the conditions for inflation are trivially satisfied. To end inflation, a quantum tunneling event from the false vacuum to the true vacuum [75] was invoked (see e.g. [18] for a pedagogical review). This model, however, has a graceful exit problem since the tunneling leads to an initially microscopical bubble of the true vacuum which cannot grow to encompass our presently observed universe - the flatness problem of SBB cosmology in a new form. Hence, attention shifted to models in which the scalar field φ is slowly rolling during inflation.

There are many models of scalar field-driven inflation. Many of them can be divided into three groups [19]: small-field inflation, large-field inflation and hybrid inflation. *Small-field inflationary models* are based on ideas from spontaneous symmetry breaking in particle physics. We take the scalar field to have a potential of the form

$$V(\varphi) = \frac{1}{4}\lambda(\varphi^2 - \sigma^2)^2\,, \tag{2.15}$$

where σ can be interpreted as a symmetry breaking scale, and λ is a dimensionless coupling constant. The hope of initial small-field models ("new inflation" [20]) was that the scalar field would begin rolling close to its symmetric point $\varphi = 0$, where thermal equilibrium initial conditions would localize it in the early universe. At sufficiently high temperatures, $\varphi = 0$ is a stable ground state of the one-loop finite temperature effective potential $V_T(\varphi)$ (see e.g. [18] for a review). Once the temperature drops to a value smaller than the critical temperature T_c, $\varphi = 0$ turns into an unstable local maximum of $V_T(\varphi)$, and φ is free to roll towards a ground state of the zero temperature potential (2.15). The direction of the initial rolling is triggered by quantum fluctuations. The reader can easily check that for the potential (2.15) the slow-roll conditions cannot be satisfied if $\sigma \ll m_{pl}$, where m_{pl} is the Planck mass which is related to G. If the potential is modified to a Coleman-Weinberg [21] form

$$V(\varphi) = \frac{\lambda}{4}\varphi^4 + \frac{\lambda^2}{44\pi^2}\varphi^4[\ln(\frac{\varphi^2}{M^2}) - \frac{25}{6}] \tag{2.16}$$

(where M denotes some renormalization scale) then the slow-roll conditions can be satisfied. However, this corresponds to a severe fine-tuning of the shape of the potential. A further problem for most small-field models of inflation (see e.g. [22] for a review) is that the slow-roll trajectory is not an attractor in phase space. In order to end up close to the slow-roll trajectory, the initial field velocity must be constrained to be very small. This *initial condition problem* of small-field models of inflation effects a number of recently proposed brane inflation scenarios, see e.g. [23] for a discussion.

There is another reason for abandoning small-field inflation models: in order to obtain a sufficiently small amplitude of density fluctuations, the interaction coefficients of φ must be very small (this problem is discussed at length towards the end of these lectures). In particular, this makes it inconsistent to assume that φ started out in thermal equilibrium. In the absence of thermal equilibrium, the phase space of initial conditions is much larger for large values of φ.

This brings us to the discussion of large-field inflation models, initially proposed in [24] under the name "chaotic inflation". The simplest example is provided by a massive scalar field with potential

$$V(\varphi) = \frac{1}{2}m^2\varphi^2\,, \tag{2.17}$$

where m is the mass. It is assumed that the scalar field rolls towards the origin from large values of $|\varphi|$. It is a simple exercise for the reader to verify that the slow-roll conditions (2.13) and (2.14) are satisfied provided

$$|\varphi| > \frac{1}{\sqrt{12\pi}}m_{pl}\,. \tag{2.18}$$

Values of $|\varphi|$ comparable or greater than m_{pl} are also required in other realizations of large-field inflation. Hence, one may worry whether such a toy model can consistently be embedded in a realistic particle physics model, e.g. supergravity. In many such models $V(\varphi)$ receives supergravity-induced correction terms which destroy the flatness of the potential for $|\varphi| > m_{pl}$. However, as recently discussed in [25], if the flatness of the potential is protected by some symmetry, then it can survive inclusion of the correction terms. As will be discussed later, a value of $m \sim 10^{13}$GeV is required in order to obtain the observed amplitude of density fluctuations. Hence, the configuration space of field values with $|\varphi| > m_{pl}$ but $V(\varphi) < m_{pl}^4$ is huge. It can also be verified that the slow-roll trajectory is a local attractor in field initial condition space [16], even including metric fluctuations at the perturbative level [26].

With two scalar fields it is possible to construct a class of models which combine some of the nice features of large-field inflation (large phase space of initial conditions yielding inflation) and of small-field inflation (better contact with conventional particle physics). These are models of hybrid inflation [27]. To give a prototypical example, consider two scalar fields φ and χ with a potential

$$V(\varphi,\chi) = \frac{1}{4}\lambda_\chi(\chi^2-\sigma^2)^2 + \frac{1}{2}m^2\varphi^2 - \frac{1}{2}g^2\varphi^2\chi^2\,. \tag{2.19}$$

In the absence of thermal equilibrium, it is natural to assume that $|\varphi|$ begins at large values, values for which the effective mass of χ is positive and hence χ begins at $\chi = 0$. The parameters in the potential (2.19) are now chosen such that φ is slowly rolling for values of $|\varphi|$ somewhat smaller than m_{pl}, but that the potential energy for these field values is dominated by the first term on the right-hand side of (2.19). The reader can easily verify that for this model it is no longer required to have values of $|\varphi|$ greater than m_{pl} in order to obtain slow-rolling [7] The field φ is slowly rolling whereas the potential

[7]Note that the slow-roll conditions (2.13) and (2.14) were derived assuming that H is given by the contribution of φ to V which is not the case here.

energy is determined by the contribution from χ. Once $|\varphi|$ drops to the value

$$|\varphi_c| = \frac{\sqrt{\lambda_\chi}}{g}\sigma \tag{2.20}$$

the configuration $\chi = 0$ becomes unstable and decays to its ground state $|\chi| = \sigma$, yielding a graceful exit from inflation. Since in this example the ground state of χ is not unique, there is the possibility of the formation of topological defects at the end of inflation (see [11, 7, 13] for reviews of topological defects in cosmology, and the lectures by Polchinski at this school [28] for a discussion of how this scenario arises in brane inflation models).

After the slow-roll conditions break down, the period of inflation ends, and the inflaton begins to oscillate around its ground state. Due to couplings of φ to other matter fields, the energy of the universe, which at the end of the period of inflation is stored completely in φ, gets transferred to the matter fields of the particle physics Standard Model. Initially, the energy transfer was described perturbatively [29, 30]. Later, it was realized [31, 32, 33, 34] that through a parametric resonance instability, particles are very rapidly produced, leading to a fast energy transfer ("preheating"). The quanta later thermalize, and thereafter the universe evolves as described by SBB cosmology.

After this review of inflationary cosmology (see e.g. [35] for a more complete recent review), we turn to the discussion of the main success of inflationary cosmology, namely the fact that it provides a causal mechanism for generating small inhomogeneities. The reader is referred to [5] for a comprehensive analysis of this theory of cosmological perturbations.

3. Newtonian Theory of Cosmological Perturbations

3.1. INTRODUCTION

The growth of density fluctuations is a consequence of the purely attractive nature of the gravitational force. Imagine (first in a non-expanding background) a density excess $\delta\rho$ localized about some point $\mathbf{x}$ in space. This fluctuation produces an attractive force which pulls the surrounding matter towards $\mathbf{x}$. The magnitude of this force is proportional to $\delta\rho$. Hence, by Newton's second law

$$\ddot{\delta\rho} \sim G\delta\rho\,, \tag{3.1}$$

where G is Newton's gravitational constant. Hence, there is an exponential instability of flat space-time to the development of fluctuations.

Obviously, in General Relativity it is inconsistent to consider density fluctuations in a non-expanding background. If we consider density fluctuations in an expanding background, then the expansion of space leads to a friction term in (3.1). Hence, instead of an exponential instability to the development of fluctuations, the growth rate of fluctuations in an expanding Universe will be as a power of time. It is crucial to determine what this power is and how it depends both on the background cosmological expansion rate and on the length scale of the fluctuations.

Note that in the following two subsections $\mathbf{x}$ will denote the physical coordinates, and $\mathbf{q}$ the comoving ones. The materials covered in this section are discussed in several excellent textbooks on cosmology, e.g. in [37, 38, 39, 40].

3.2. PERTURBATIONS ABOUT MINKOWSKI SPACE-TIME

To develop some physical intuition, we first consider the evolution of hydrodynamical matter fluctuations in a fixed non-expanding background.

In this context, matter is described by a perfect fluid, and gravity by the Newtonian gravitational potential φ. The fluid variables are the energy density ρ, the pressure p, the fluid velocity $\mathbf{v}$, and the entropy density S. The basic hydrodynamical equations are

$$\begin{aligned} \dot{\rho} + \nabla \cdot (\rho \mathbf{v}) &= 0 \\ \dot{\mathbf{v}} + (\mathbf{v} \cdot \nabla)\mathbf{v} + \frac{1}{\rho}\nabla p + \nabla \varphi &= 0 \\ \nabla^2 \varphi &= 4\pi G \rho \\ \dot{S} + (\mathbf{v} \cdot \nabla) S &= 0 \\ p &= p(\rho, S)\,. \end{aligned} \tag{3.2}$$

The first equation is the continuity equation, the second is the Euler (force) equation, the third is the Poisson equation of Newtonian gravity, the fourth expresses entropy conservation, and the last describes the equation of state of matter. The derivative with respect to time is denoted by an over-dot.

The background is given by the background energy density ρ_o, the background pressure p_0, vanishing velocity, constant gravitational potential φ_0 and constant entropy density S_0. As mentioned above, it does not satisfy the background Poisson equation.

The equations for cosmological perturbations are obtained by perturbing the fluid variables about the background,

$$\begin{aligned} \rho &= \rho_0 + \delta\rho \\ \mathbf{v} &= \delta \mathbf{v} \\ p &= p_0 + \delta p \\ \varphi &= \varphi_0 + \delta\varphi \\ S &= S_0 + \delta S\,, \end{aligned} \tag{3.3}$$

where the fluctuating fields $\delta\rho, \delta\mathbf{v}, \delta p, \delta\varphi$ and δS are functions of space and time, by inserting these expressions into the basic hydrodynamical equations (3.2), by linearizing, and by combining the resulting equations which are of first order in time. We get the following differential equations for the energy density fluctuation $\delta\rho$ and the entropy perturbation δS

$$\begin{aligned} \ddot{\delta\rho} - c_s^2 \nabla^2 \delta\rho - 4\pi G \rho_0 \delta\rho &= \sigma \nabla^2 \delta S \\ \dot{\delta S} &= 0\,, \end{aligned} \tag{3.4}$$

where the variables c_s^2 and σ describe the equation of state

$$\delta p = c_s^2 \delta\rho + \sigma \delta S \tag{3.5}$$

with

$$c_s^2 = \left(\frac{\delta p}{\delta \rho}\right)_{|S} \tag{3.6}$$

denoting the square of the speed of sound.

Sincethe equations are linear, we can work in Fourier space. Each Fourier component $\delta\rho_k(t)$ of the fluctuation field $\delta\rho(\mathbf{x}, t)$

$$\delta\rho(\mathbf{x}, t) = \int e^{i\mathbf{k}\cdot\mathbf{x}} \delta\rho_k(t) \quad (3.7)$$

evolves independently.

The fluctuations can be classified as follows: If δS vanishes, we have **adiabatic** fluctuations. If the δS is non-vanishing but $\dot{\delta\rho} = 0$, we speak of an **entropy** fluctuation.

The first conclusions we can draw from the basic perturbation equations (3.4) are that
1) entropy fluctuations do not grow,
2) adiabatic fluctuations are time-dependent, and
3) entropy fluctuations seed an adiabatic mode.

Taking a closer look at the equation of motion for $\delta\rho$, we see that the third term on the left hand side represents the force due to gravity, a purely attractive force yielding an instability of flat space-time to the development of density fluctuations (as discussed earlier, see (3.1)). The second term on the left hand side of (3.4) represents a force due to the fluid pressure which tends to set up pressure waves. In the absence of entropy fluctuations, the evolution of $\delta\rho$ is governed by the combined action of both pressure and gravitational forces.

Restricting our attention to adiabatic fluctuations, we see from (3.4) that there is a critical wavelength, the Jeans length, whose wavenumber k_J is given by

$$k_J = \left(\frac{4\pi G\rho_0}{c_s^2}\right)^{1/2}. \quad (3.8)$$

Fluctuations with wavelength longer than the Jeans length ($k \ll k_J$) grow exponentially

$$\delta\rho_k(t) \sim e^{\omega_k t} \text{ with } \omega_k \sim 4(\pi G\rho_0)^{1/2} \quad (3.9)$$

whereas short wavelength modes ($k \gg k_J$) oscillate with frequency $\omega_k \sim c_s k$. Note that the value of the Jeans length depends on the equation of state of the background. For a background dominated by relativistic radiation, the Jeans length is large (of the order of the Hubble radius $H^{-1}(t)$), whereas for pressure-less matter it goes to zero.

3.3. PERTURBATIONS ABOUT AN EXPANDING BACKGROUND

Let us now improve on the previous analysis and study Newtonian cosmological fluctuations about an expanding background. In this case, the background equations are consistent (the non-vanishing average energy density leads to cosmological expansion). However, we are still neglecting general relativistic effects (the fluctuations of the metric). Such effects turn out to be dominant on length scales larger than the Hubble radius $H^{-1}(t)$, and thus the analysis of this section is applicable only to smaller scales.

The background cosmological model is given by the energy density $\rho_0(t)$, the pressure $p_0(t)$, and the recessional velocity $\mathbf{v}_0 = H(t)\mathbf{x}$, where $\mathbf{x}$ is the Euclidean spatial coordinate vector ("physical coordinates"). The space- and time-dependent fluctuating fields are defined as in the previous section:

$$\begin{aligned}\rho(t,\mathbf{x}) &= \rho_0(t)\big(1+\delta_\epsilon(t,\mathbf{x})\big)\\ \mathbf{v}(t,\mathbf{x}) &= \mathbf{v}_0(t,\mathbf{x})+\delta\mathbf{v}(t,\mathbf{x})\\ p(t,\mathbf{x}) &= p_0(t)+\delta p(t,\mathbf{x}),\end{aligned} \tag{3.10}$$

where δ_ϵ is the fractional energy density perturbation (we are interested in the fractional rather than in the absolute energy density fluctuation!), and the pressure perturbation δp is defined as in (3.5). In addition, there is the possibility of a non-vanishing entropy perturbation defined as in (3.3).

We now insert this ansatz into the basic hydrodynamical equations (3.2), linearize in the perturbation variables, and combine the first order differential equations for δ_ϵ and δp into a single second order differential equation for $\delta\rho_\epsilon$. The result simplifies if we work in "comoving coordinates" $\mathbf{q}$ which are the coordinates painted onto the expanding background, i.e.

$$\mathbf{x}(t) = a(t)\mathbf{q}(t). \tag{3.11}$$

After some algebra, we obtain the following equation which describes the time evolution of density fluctuations:

$$\ddot{\delta_\epsilon} + 2H\dot{\delta_\epsilon} - \frac{c_s^2}{a^2}\nabla_q^2\delta_\epsilon - 4\pi G\rho_0\delta_\epsilon = \frac{\sigma}{\rho_0 a^2}\delta S, \tag{3.12}$$

where the subscript q on the ∇ operator indicates that derivatives with respect to comoving coordinates are used. In addition, we have the equation of entropy conservation

$$\dot{\delta S} = 0. \tag{3.13}$$

Comparing with the equations (3.4) obtained in the absence of an expanding background, we see that the only difference is the presence of a Hubble damping term in the equation for δ_ϵ. This term will moderate the exponential instability of the background to long wavelength density fluctuations. In addition, it will lead to a damping of the oscillating solutions on short wavelengths. More specifically, for physical wavenumbers $k_p \ll k_J$ (where k_J is again given by (3.8)), and in a matter-dominated background cosmology, the general solution of (3.12) in the absence of any entropy fluctuations is given by

$$\delta_k(t) = c_1 t^{2/3} + c_2 t^{-1}, \tag{3.14}$$

where c_1 and c_2 are two constants determined by the initial conditions, and we have dropped the subscript ϵ in expressions involving δ_ϵ. There are two fundamental solutions, the first is a growing mode with $\delta_k(t) \sim a(t)$, the second a decaying mode with $\delta_k(t) \sim t^{-1}$. On short wavelength, one obtains damped oscillatory motion:

$$\delta_k(t) \sim a^{-1/2}(t) exp\big(\pm i c_s k \int dt' a^{-1}(t')\big). \tag{3.15}$$

As a simple application of the Newtonian equations for cosmological perturbations derived above, let us compare the predicted cosmic microwave background (CMB) anisotropies in a spatially flat Universe with only baryonic matter - Model A - to the corresponding anisotropies in a flat Universe with mostly cold dark matter (pressure-less non-baryonic dark matter) - Model B. We start with the observationally known amplitude of the relative density fluctuations today (time t_0), and we use the fact that the amplitude of the CMB anisotropies on the angular scale $\theta(k)$ corresponding to the comoving wavenumber k is set by the primordial value of the gravitational potential ϕ - introduced in the following section - which in turn is related to the primordial value of the density fluctuations at Hubble radius crossing (and **not** to its value of the time t_{rec}). See e.g. Chapter 17 of [5]).

In Model A, the dominant component of the pressure-less matter is coupled to radiation between t_{eq} and t_{rec}, the time of last scattering. Thus, the Jeans length is comparable to the Hubble radius. Therefore, for comoving galactic scales, $k \gg k_J$ in this time interval, and thus the fractional density contrast decreases as $a(t)^{-1/2}$. In contrast, in Model B, the dominant component of pressure-less matter couples only weakly to radiation, and hence the Jeans length is negligibly small. Thus, in Model B, the relative density contrast grows as $a(t)$ between t_{eq} and t_{rec}. In the time interval $t_{rec} < t < t_0$, the fluctuations scale identically in Models A and B. Summarizing, we conclude, working backwards in time from a fixed amplitude of δ_k today, that the amplitudes of $\delta_k(t_{eq})$ in Models A and B (and thus their primordial values) are related by

$$\delta_k(t_{eq})|_A \simeq (\frac{a(t_{rec})}{a(t_{eq})})\delta_k(t_{eq})|_B \,. \tag{3.16}$$

Hence, in Model A (without non-baryonic dark matter) the CMB anisotropies are predicted to be a factor of about 30 larger [76] than in Model B, way in excess of the recent observational results. This is one of the strongest arguments for the existence of non-baryonic dark matter. Note that the precise value of the enhancement factor depends on the value of the cosmological constant Λ - the above value holds for $\Lambda = 0$.

3.4. CHARACTERIZING PERTURBATIONS

Let us consider perturbations on a fixed comoving length scale given by a comoving wavenumber k. The corresponding physical length increases as $a(t)$. This is to be compared to the Hubble radius $H^{-1}(t)$ which scales as t provided $a(t)$ grows as a power of t. In the late time Universe, $a(t) \sim t^{1/2}$ in the radiation-dominated phase (i.e. for $t < t_{eq}$, and $a(t) \sim t^{2/3}$ in the matter-dominated period ($t_{eq} < t < t_0$). Thus, we see that at sufficiently early times, all comoving scales had a physical length larger than the Hubble radius. If we consider large cosmological scales (e.g. those corresponding to the observed CMB anisotropies or to galaxy clusters), the time $t_H(k)$ of "Hubble radius crossing" (when the physical length was equal to the Hubble radius) was in fact later than t_{eq}. As we will see in later sections, the time of Hubble radius crossing plays an important role in the evolution of cosmological perturbations.

Cosmological fluctuations can be described either in position space or in momentum space. In position space, we compute the root mean square mass fluctuation $\delta M/M(k,t)$ in a sphere of radius $l = 2\pi/k$ at time t. A scale-invariant spectrum of fluctuations is defined by the relation

$$\frac{\delta M}{M}(k, t_H(k)) = \text{const.}\,. \tag{3.17}$$

Such a spectrum was first suggested by Harrison [42] and Zeldovich [43] as a reasonable choice for the spectrum of cosmological fluctuations. We can introduce the "spectral index" n of cosmological fluctuations by the relation

$$(\frac{\delta M}{M})^2(k, t_H(k)) \sim k^{n-1}, \tag{3.18}$$

and thus a scale-invariant spectrum corresponds to $n = 1$.

To make the transition to the (more frequently used) momentum space representation, we Fourier decompose the fractional spatial density contrast

$$\delta_\epsilon(\mathbf{x}, t) = \int d^3k \tilde{\delta}_\epsilon(\mathbf{k}, t) e^{i\mathbf{k}\cdot\mathbf{x}}. \tag{3.19}$$

The **power spectrum** P_δ of density fluctuations is defined by

$$P_\delta(k) = k^3 |\tilde{\delta}_\epsilon(k)|^2, \tag{3.20}$$

where k is the magnitude of $\mathbf{k}$, and we have assumed for simplicity a Gaussian distribution of fluctuations in which the amplitude of the fluctuations only depends on k.

We can also define the power spectrum of the gravitational potential φ:

$$P_\varphi(k) = k^3 |\tilde{\delta\varphi}(k)|^2. \tag{3.21}$$

These two power spectra are related by the Poisson equation (3.2)

$$P_\varphi(k) \sim k^{-4} P_\delta(k). \tag{3.22}$$

In general, the condition of scale-invariance is expressed in momentum space in terms of the power spectrum evaluated at a fixed time. To obtain this condition, we first use the time dependence of the fractional density fluctuation from (3.14) to determine the mass fluctuations at a fixed time $t > t_H(k) > t_{eq}$ (the last inequality is a condition on the scales considered)

$$(\frac{\delta M}{M})^2(k, t) = (\frac{t}{t_H(k)})^{4/3} (\frac{\delta M}{M})^2(k, t_H(k)). \tag{3.23}$$

The time of Hubble radius crossing is given by

$$a(t_H(k))k^{-1} = 2t_H(k), \tag{3.24}$$

and thus

$$t_H(k)^{1/2} \sim k^{-1}. \tag{3.25}$$

Inserting this result into (3.23) making use of (3.18) we find

$$(\frac{\delta M}{M})^2(k, t) \sim k^{n+3}. \tag{3.26}$$

Since, for reasonable values of the index of the power spectrum, $\delta M/M(k, t)$ is dominated by the Fourier modes with wavenumber k, we find that (3.26) implies

$$|\tilde{\delta}_\epsilon|^2 \sim k^n, \tag{3.27}$$

or, equivalently,

$$P_\varphi(k) \sim k^{n-1}. \tag{3.28}$$

4. Relativistic Theory of Cosmological Fluctuations

4.1. INTRODUCTION

The Newtonian theory of cosmological fluctuations discussed in the previous section breaks down on scales larger than the Hubble radius because it neglects perturbations of the metric, and because on large scales the metric fluctuations dominate the dynamics.

Let us begin with a heuristic argument to show why metric fluctuations are important on scales larger than the Hubble radius. For such inhomogeneities, one should be able to approximately describe the evolution of the space-time by applying the first FRW equation (2.4) of homogeneous and isotropic cosmology to the local Universe (this approximation is made more rigorous in [44]). Based on this equation, a large-scale fluctuation of the energy density will lead to a fluctuation ("δa") of the scale factor a which grows in time. This is due to the fact that self gravity amplifies fluctuations even on length scales λ greater than the Hubble radius.

This argument is made rigorous in the following analysis of cosmological fluctuations in the context of general relativity, where both metric and matter inhomogeneities are taken into account. We will consider fluctuations about a homogeneous and isotropic background cosmology, given by the metric (2.1), which can be written in conformal time η (defined by $dt = a(t)d\eta$) as

$$ds^2 = a(\eta)^2(d\eta^2 - d\mathbf{x}^2). \tag{4.1}$$

The theory of cosmological perturbations is based on expanding the Einstein equations to linear order about the background metric. The theory was initially developed in pioneering works by Lifshitz [45]. Significant progress in the understanding of the physics of cosmological fluctuations was achieved by Bardeen [46] who realized the importance of subtracting gauge artifacts (see below) from the analysis (see also [47]). The following discussion is based on Part I of the comprehensive review article [5]. Other reviews - in some cases emphasizing different approaches - are [48, 49, 50, 51].

4.2. CLASSIFYING FLUCTUATIONS

The first step in the analysis of metric fluctuations is to classify them according to their transformation properties under spatial rotations. There are scalar, vector and second rank tensor fluctuations. In linear theory, there is no coupling between the different fluctuation modes, and hence they evolve independently (for some subtleties in this classification, see [52]).

We begin by expanding the metric about the FRW background metric $g^{(0)}_{\mu\nu}$ given by (4.1):

$$g_{\mu\nu} = g^{(0)}_{\mu\nu} + \delta g_{\mu\nu}. \tag{4.2}$$

The background metric depends only on time, whereas the metric fluctuations $\delta g_{\mu\nu}$ depend on both space and time. Since the metric is a symmetric tensor, there are at first sight 10 fluctuating degrees of freedom in $\delta g_{\mu\nu}$.

There are four degrees of freedom which correspond to scalar metric fluctuations (the only four ways of constructing a metric from scalar functions):

$$\delta g_{\mu\nu} = a^2 \begin{pmatrix} 2\phi & -B_{,i} \\ -B_{,i} & 2(\psi\delta_{ij} - E_{,ij}) \end{pmatrix}, \tag{4.3}$$

where the four fluctuating degrees of freedom are denoted (following the notation of [5]) ϕ, B, E, and ψ, a comma denotes the ordinary partial derivative (if we had included spatial curvature of the background metric, it would have been the covariant derivative with respect to the spatial metric), and δ_{ij} is the Kronecker symbol.

There are also four vector degrees of freedom of metric fluctuations, consisting of the four ways of constructing metric fluctuations from three vectors:

$$\delta g_{\mu\nu} = a^2 \begin{pmatrix} 0 & -S_i \\ -S_i & F_{i,j} + F_{j,i} \end{pmatrix}, \tag{4.4}$$

where S_i and F_i are two divergence-less vectors (for a vector with non-vanishing divergence, the divergence contributes to the scalar gravitational fluctuation modes).

Finally, there are two tensor modes which correspond to the two polarization states of gravitational waves:

$$\delta g_{\mu\nu} = -a^2 \begin{pmatrix} 0 & 0 \\ 0 & h_{ij} \end{pmatrix}, \tag{4.5}$$

where h_{ij} is trace-free and divergence-less

$$h_i^i = h_{ij}^j = 0\,. \tag{4.6}$$

Gravitational waves do not couple at linear order to the matter fluctuations. Vector fluctuations decay in an expanding background cosmology and hence are not usually cosmologically important. The most important fluctuations, at least in inflationary cosmology, are the scalar metric fluctuations, the fluctuations which couple to matter inhomogeneities and which are the relativistic generalization of the Newtonian perturbations considered in the previous section.

4.3. GAUGE TRANSFORMATION

The theory of cosmological perturbations is at first sight complicated by the issue of gauge invariance (at the final stage, however, we will see that we can make use of the gauge freedom to substantially simplify the theory). The coordinates $t, \mathbf{x}$ of space-time carry no independent physical meaning. They are just labels to designate points in the space-time manifold. By performing a small-amplitude transformation of the space-time coordinates (called "gauge transformation" in the following), we can easily introduce "fictitious" fluctuations in a homogeneous and isotropic Universe. These modes are "gauge artifacts".

We will in the following take an "active" view of gauge transformation. Let us consider two space-time manifolds, one of them a homogeneous and isotropic Universe $\mathcal{M}_0$, the other a physical Universe $\mathcal{M}$ with inhomogeneities. A choice of coordinates can be considered to be a mapping $\mathcal{D}$ between the manifolds $\mathcal{M}_0$ and $\mathcal{M}$. Let us consider a second mapping $\tilde{\mathcal{D}}$ which will map the same point (e.g. the origin of a fixed coordinate system) in $\mathcal{M}_0$ into different points in $\mathcal{M}$. Using the inverse of these maps $\mathcal{D}$ and $\tilde{\mathcal{D}}$, we can assign two different sets of coordinates to points in $\mathcal{M}$.

Consider now a physical quantity Q (e.g. the Ricci scalar) on $\mathcal{M}$, and the corresponding physical quantity $Q^{(0)}$ on $\mathcal{M}_0$ Then, in the first coordinate system given by the mapping $\mathcal{D}$, the perturbation δQ of Q at the point $p \in \mathcal{M}$ is defined by

$$\delta Q(p) = Q(p) - Q^{(0)}\left(\mathcal{D}^{-1}(p)\right). \tag{4.7}$$

Analogously, in the second coordinate system given by $\tilde{\mathcal{D}}$, the perturbation is defined by

$$\tilde{\delta Q}(p) = Q(p) - Q^{(0)}\left(\tilde{\mathcal{D}}^{-1}(p)\right). \tag{4.8}$$

The difference

$$\Delta Q(p) = \tilde{\delta Q}(p) - \delta Q(p) \tag{4.9}$$

is obviously a gauge artifact and carries no physical significance.

Some of the metric perturbation degrees of freedom introduced in the first subsection are gauge artifacts. To isolate these, we must study how coordinate transformations act on the metric. There are four independent gauge degrees of freedom corresponding to the coordinate transformation

$$x^\mu \rightarrow \tilde{x}^\mu = x^\mu + \xi^\mu. \tag{4.10}$$

The zero (time) component ξ^0 of ξ^μ leads to a scalar metric fluctuation. The spatial three vector ξ^i can be decomposed as

$$\xi^i = \xi^i_{tr} + \gamma^{ij}\xi_{,j} \tag{4.11}$$

(where γ^{ij} is the spatial background metric) into a transverse piece ξ^i_{tr} which has two degrees of freedom which yield vector perturbations, and the second term (given by the gradient of a scalar ξ) which gives a scalar fluctuation. To summarize this paragraph, there are two scalar gauge modes given by ξ^0 and ξ, and two vector modes given by the transverse three vector ξ^i_{tr}. Thus, there remain two physical scalar and two vector fluctuation modes. The gravitational waves are gauge-invariant.

Let us now focus on how the scalar gauge transformations (i.e. the transformations given by ξ^0 and ξ) act on the scalar metric fluctuation variables ϕ, B, E, and ψ. An immediate calculation yields:

$$\begin{aligned}
\tilde{\phi} &= \phi - \frac{a'}{a}\xi^0 - (\xi^0)' \\
\tilde{B} &= B + \xi^0 - \xi' \\
\tilde{E} &= E - \xi \\
\tilde{\psi} &= \psi + \frac{a'}{a}\xi^0,
\end{aligned} \tag{4.12}$$

where a prime indicates the derivative with respect to conformal time η.

There are two approaches to deal with the gauge ambiguities. The first is to fix a gauge, i.e. to pick conditions on the coordinates which completely eliminate the gauge freedom, the second is to work with a basis of gauge-invariant variables.

If one wants to adopt the gauge-fixed approach, there are many different gauge choices. Note that the often used synchronous gauge determined by $\delta g^{0\mu} = 0$ does not totally fix the gauge. A convenient system which completely fixes the coordinates is the so-called **longitudinal** or **conformal Newtonian gauge** defined by $B = E = 0$.

If one prefers a gauge-invariant approach, there are many choices of gauge-invariant variables. A convenient basis first introduced by [46] is the basis Φ, Ψ given by

$$\Phi = \phi + \frac{1}{a}\left[(B - E')a\right]' \tag{4.13}$$

$$\Psi = \psi - \frac{a'}{a}(B - E') . \tag{4.14}$$

It is obvious from the above equations that the gauge-invariant variables Φ and Ψ coincide with the corresponding diagonal metric perturbations ϕ and ψ in longitudinal gauge.

Note that the variables defined above are gauge-invariant only under linear space-time coordinate transformations. Beyond linear order, the structure of perturbation theory becomes much more involved. In fact, one can show [53] that the only fluctuation variables which are invariant under all coordinate transformations are perturbations of variables which are constant in the background space-time.

4.4. EQUATIONS OF MOTION

We begin with the Einstein equations

$$G_{\mu\nu} = 8\pi G T_{\mu\nu} , \tag{4.15}$$

where $G_{\mu\nu}$ is the Einstein tensor associated with the space-time metric $g_{\mu\nu}$, and $T_{\mu\nu}$ is the energy-momentum tensor of matter, insert the ansatz for metric and matter perturbed about a FRW background $\left(g^{(0)}_{\mu\nu}(\eta),\, \varphi^{(0)}(\eta)\right)$:

$$g_{\mu\nu}(\mathbf{x},\eta) = g^{(0)}_{\mu\nu}(\eta) + \delta g_{\mu\nu}(\mathbf{x},\eta) \tag{4.16}$$

$$\varphi(\mathbf{x},\eta) = \varphi_0(\eta) + \delta\varphi(\mathbf{x},\eta) , \tag{4.17}$$

(where we have for simplicity replaced general matter by a scalar matter field φ) and expand to linear order in the fluctuating fields, obtaining the following equations:

$$\delta G_{\mu\nu} = 8\pi G \delta T_{\mu\nu} . \tag{4.18}$$

In the above, $\delta g_{\mu\nu}$ is the perturbation in the metric and $\delta\varphi$ is the fluctuation of the matter field φ.

Note that the components δG^{μ}_{ν} and δT^{μ}_{ν} are not gauge invariant. If we want to use the gauge-invariant approach, we note [5] that it is possible to construct a gauge-invariant tensor $\delta G^{(gi)\,\mu}_{\nu}$ via

$$\begin{aligned}
\delta G^{(gi)\,0}_{0} &\equiv \delta G^{0}_{0} + ({}^{(0)}G^{'0}_{0})(B - E') \\
\delta G^{(gi)\,0}_{i} &\equiv \delta G^{0}_{i} + ({}^{(0)}G^{0}_{i} - \frac{1}{3}{}^{(0)}G^{k}_{k})(B - E')_{,i} \\
\delta G^{(gi)\,i}_{j} &\equiv \delta G^{i}_{j} + ({}^{(0)}G^{'i}_{j})(B - E') ,
\end{aligned} \tag{4.19}$$

where ${}^{(0)}G^{\mu}_{\nu}$ denote the background values of the Einstein tensor. Analogously, a gauge-invariant linearized stress-energy tensor $\delta T_{\nu}^{(gi)\,\mu}$ can be defined. In terms of these tensors, the gauge-invariant form of the equations of motion for linear fluctuations reads

$$\delta G^{(gi)}_{\mu\nu} \, = \, 8\pi G \delta T^{(gi)}_{\mu\nu} \, . \tag{4.20}$$

If we insert into this equation the ansatz for the general metric and matter fluctuations (which depend on the gauge), only gauge-invariant combinations of the fluctuation variables will appear.

In a gauge-fixed approach, one can start with the metric in longitudinal gauge

$$ds^2 \, = \, a^2\big[(1+2\phi)d\eta^2 - (1-2\psi)\gamma_{ij}dx^i dx^j\big] \tag{4.21}$$

and insert this ansatz into the general perturbation equations (4.18). The shortcut of inserting a restricted ansatz for the metric into the action and deriving the full set of variational equations is justified in this case.

Both approaches yield the following set of equations of motion:

$$\begin{aligned} -3\mathcal{H}\big(\mathcal{H}\Phi + \Psi^{'}\big) + \nabla^2\Psi \, &= \, 4\pi G a^2 \delta T_0^{(gi)\,0} \\ \big(\mathcal{H}\Phi + \Psi^{'}\big)_{,i} \, &= \, 4\pi G a^2 \delta T_i^{(gi)\,0} \\ \big[(2\mathcal{H}^{'} + \mathcal{H}^2)\Phi + \mathcal{H}\Phi^{'} + \Psi^{''} + 2\mathcal{H}\Psi^{'}\big]\delta^i_j & \\ +\frac{1}{2}\nabla^2 D\delta^i_j - \frac{1}{2}\gamma^{ik}D_{,kj} \, &= \, -4\pi G a^2 \delta T_j^{(gi)\,i} \, , \end{aligned} \tag{4.22}$$

where $D \equiv \Phi - \Psi$ and $\mathcal{H} = a'/a$. If we work in longitudinal gauge, then $\delta T_j^{(gi)\,i} = \delta T^i_j$, $\Phi = \phi$ and $\Psi = \psi$.

The first conclusion we can draw is that if no anisotropic stress is present in the matter at linear order in fluctuating fields, i.e. $\delta T^i_j = 0$ for $i \neq j$, then the two metric fluctuation variables coincide:

$$\Phi \, = \, \Psi \, . \tag{4.23}$$

This will be the case in most simple cosmological models, e.g. in theories with matter described by a set of scalar fields with canonical form of the action, and in the case of a perfect fluid with no anisotropic stress.

Let us now restrict our attention to the case of matter described in terms of a single scalar field φ which can be expanded as

$$\varphi(\mathbf{x},\eta) \, = \, \varphi_0(\eta) + \delta\varphi(\mathbf{x},\eta) \tag{4.24}$$

in terms of background matter φ_0 and matter fluctuation $\delta\varphi(\mathbf{x},\eta)$. Then, in longitudinal gauge, (4.22) reduce to the following set of equations of motion (making use of (4.23))

$$\begin{aligned} \nabla^2\phi - 3\mathcal{H}\phi^{'} - (\mathcal{H}^{'} + 2\mathcal{H}^2)\phi \, &= \, 4\pi G(\varphi_0^{'}\delta\varphi^{'} + V^{'}a^2\delta\varphi) \\ \phi^{'} + \mathcal{H}\phi \, &= \, 4\pi G\varphi_0^{'}\delta\varphi \\ \phi^{''} + 3\mathcal{H}\phi^{'} + (\mathcal{H}^{'} + 2\mathcal{H}^2)\phi \, &= \, 4\pi G(\varphi_0^{'}\delta\varphi^{'} - V^{'}a^2\delta\varphi) \, , \end{aligned} \tag{4.25}$$

where $V^{'}$ denotes the derivative of V with respect to φ. These equations can be combined to give the following second order differential equation for the relativistic potential ϕ:

$$\phi^{''} + 2\left(\mathcal{H} - \frac{\varphi_0^{''}}{\varphi_0^{'}}\right)\phi^{'} - \nabla^2\phi + 2\left(\mathcal{H}^{'} - \mathcal{H}\frac{\varphi_0^{''}}{\varphi_0^{'}}\right)\phi = 0\,. \tag{4.26}$$

This is the final result for the classical evolution of cosmological fluctuations. First of all, we note the similarities with the equation (3.12) obtained in the Newtonian theory. The final term in (4.26) is the force due to gravity leading to the instability, the second to last term is the pressure force leading to oscillations (relativistic since we are considering matter to be a relativistic field), and the second term is the Hubble friction term. For each wavenumber there are two fundamental solutions. On small scales ($k > H$), the solutions correspond to damped oscillations, on large scales ($k < H$) the oscillations freeze out and the dynamics is governed by the gravitational force competing with the Hubble friction term. Note, in particular, how the Hubble radius naturally emerges as the scale where the nature of the fluctuating modes changes from oscillatory to frozen.

Considering the equation in a bit more detail, observe that if the equation of state of the background is independent of time (which will be the case if $\mathcal{H}^{'} = \varphi_0^{''} = 0$), then in an expanding background, the dominant mode of (4.26) is constant, and the sub-dominant mode decays. If the equation of state is not constant, then the dominant mode is not constant in time. Specifically, at the end of inflation $\mathcal{H}^{'} < 0$, and this leads to a growth of ϕ (see the following subsection).

To study the quantitative implications of the equation of motion (4.26), it is convenient to introduce [54, 55] the variable ζ (which, up to correction term of the order $\nabla^2\phi$ which is unimportant for large-scale fluctuations, is equal to the curvature perturbation $\mathcal{R}$ in comoving gauge [56]) by

$$\zeta \equiv \phi + \frac{2}{3}\frac{\left(H^{-1}\dot{\phi} + \phi\right)}{1+w}\,, \tag{4.27}$$

where

$$w = \frac{p}{\rho} \tag{4.28}$$

characterizes the equation of state of matter. In terms of ζ, the equation of motion (4.26) takes on the form

$$\frac{3}{2}\dot{\zeta}H(1+w) = \mathcal{O}(\nabla^2\phi)\,. \tag{4.29}$$

On large scales, the right hand side of the equation is negligible, which leads to the conclusion that large-scale cosmological fluctuations satisfy

$$\dot{\zeta}(1+w) = 0. \tag{4.30}$$

This implies that ζ is constant except possibly if $1 + w = 0$ at some point in time during the cosmological evolution (which occurs during reheating in inflationary cosmology if the inflaton field undergoes oscillations - see [57] and [74, 59] for discussions of the consequences in single and double field inflationary models, respectively). In single matter field models it is indeed possible to show that $\dot{\zeta} = 0$ on super-Hubble scales independent of assumptions on the equation of state [66, 67]. This "conservation law" makes it easy to relate initial fluctuations to final fluctuations in inflationary cosmology, as will be illustrated in the following subsection.

4.5. APPLICATION TO INFLATIONARY COSMOLOGY

Let us now return to the space-time sketch of the evolution of fluctuations in inflationary cosmology - see Figure (1) - and use the conservation law (4.30) - in the form $\zeta = \text{const}$ on large scales - to relate the amplitude of ϕ at initial Hubble radius crossing during the inflationary phase (at $t = t_i(k)$) with the amplitude at final Hubble radius crossing at late times (at $t = t_f(k)$). Since both at early times and at late times $\dot{\phi} = 0$ on super-Hubble scales as the equation of state is not changing, (4.30) and (4.27) lead to

$$\phi(t_f(k)) \simeq \frac{(1+w)(t_f(k))}{(1+w)(t_i(k))}\phi(t_i(k)) \,. \tag{4.31}$$

This equation will allow us to evaluate the amplitude of the cosmological perturbations when they re-enter the Hubble radius at time $t_f(k)$, under the assumption (discussed in detail in the following section) that the origin of the primordial fluctuations is quantum vacuum oscillations.

The time-time perturbed Einstein equation (the first equation of (4.22)) relates the value of ϕ at initial Hubble radius crossing to the amplitude of the fractional energy density fluctuations. This, together with the fact that the amplitude of the scalar matter field quantum vacuum fluctuations is of the order H, yields

$$\phi(t_i(k)) \sim H\frac{V'}{V}(t_i(k)) \,. \tag{4.32}$$

In the late time radiation dominated phase, $w = 1/3$, whereas during slow-roll inflation

$$1 + w(t_i(k)) \simeq \frac{\dot{\varphi}_0^2}{V}(t_i(k)) \,. \tag{4.33}$$

Making, in addition, use of the slow roll conditions satisfied during the inflationary period

$$\begin{aligned} H\dot{\varphi}_0 &\simeq -V' \\ H^2 &\simeq \frac{8\pi G}{3}V \,, \end{aligned} \tag{4.34}$$

we arrive at the final result

$$\phi(t_f(k)) \sim \frac{V^{3/2}}{V' m_{pl}^3}(t_i(k)) \,, \tag{4.35}$$

which gives the position space amplitude of cosmological fluctuations on a scale labelled by the comoving wavenumber k at the time when the scale re-enters the Hubble radius at late times, a result first obtained in the case of the Starobinsky model [14] of inflation in [68], and later in the context of scalar field-driven inflation in [69, 70, 71, 54].

In the case of slow roll inflation, the right hand side of (4.35) is, to a first approximation, independent of k, and hence the resulting spectrum of fluctuations is nearly scale-invariant.

5. Quantum Theory of Cosmological Fluctuations

5.1. OVERVIEW

As already mentioned in the previous section, in many models of the very early Universe, in particular in inflationary cosmology, but also in the Pre-Big-Bang and in the Ekpyrotic scenarios, primordial inhomogeneities emerge from quantum vacuum fluctuations on microscopic scales (wavelengths smaller than the Hubble radius). The wavelength is then stretched relative to the Hubble radius, becomes larger than the Hubble radius at some time and then propagates on super-Hubble scales until re-entering at late cosmological times. In the context of a Universe with a de Sitter phase, the quantum origin of cosmological fluctuations was first discussed in [68] - see [72] for a more general discussion of the quantum origin of fluctuations in cosmology, and also [73, 74] for earlier ideas. In particular, Mukhanov [68] and Press [73] realized that in an exponentially expanding background, the curvature fluctuations would be scale-invariant, and Mukhanov provided a quantitative calculation which also yielded the logarithmic deviation from exact scale-invariance.

To understand the role of the Hubble radius, consider the equation of a free scalar matter field φ on an unperturbed expanding background:

$$\ddot{\varphi} + 3H\dot{\varphi} - \frac{\nabla^2}{a^2}\varphi = 0\,. \tag{5.1}$$

The second term on the left hand side of this equation leads to damping of φ with a characteristic decay rate given by H. As a consequence, in the absence of the spatial gradient term, $\dot{\varphi}$ would be of the order of magnitude $H\varphi$. Thus, comparing the second and the third term on the left hand side, we immediately see that the microscopic (spatial gradient) term dominates on length scales smaller than the Hubble radius, leading to oscillatory motion, whereas this term is negligible on scales larger than the Hubble radius, and the evolution of φ is determined primarily by gravity. Note that in general cosmological models the Hubble radius is much smaller than the horizon (the forward light cone calculated from the initial time). In an inflationary universe, the horizon is larger by a factor of at least $\exp(N)$, where N is the number of e-foldings of inflation, and the lower bound is taken on if the Hubble radius and horizon coincide until inflation begins. It is very important to realize this difference, a difference which is obscured in most articles on cosmology in which the term "horizon" is used when "Hubble radius" is meant. Note, in particular, that the homogeneous inflaton field contains causal information on super-Hubble but sub-horizon scales. Hence, it is completely consistent with causality [57] to have a microphysical process related to the background scalar matter field lead to exponential amplification of the amplitude of fluctuations during reheating on such scales, as it does in models in which entropy perturbations are present and not suppressed during inflation [74, 59].

There are, however, general relativistic conservation laws [60] which imply that adiabatic fluctuations produced locally must be Poisson-statistic suppressed on scales larger than the Hubble radius. For example, fluctuations produced by the formation of topological defects at a phase transition in the early universe are initially isocurvature (entropy) in nature (see e.g. [61] for a discussion). Via the source term in the equation of motion

(3.4), a growing adiabatic mode is induced, but at any fixed time the spectrum of the curvature fluctuation on scales larger than the Hubble radius has index $n = 4$ (Poisson). A similar conclusion applies to recently discussed models [62, 63] of modulated reheating as a new source of density perturbations (see [64] for a nice discussion), and to models in which moduli fields obtain masses after some symmetry breaking, their quantum fluctuations then inducing cosmological fluctuations. A prototypical example is given by axion fluctuations in an inflationary universe (see e.g. [65] and references therein).

To understand the generation and evolution of fluctuations in current models of the very early Universe, we need both Quantum Mechanics and General Relativity, i.e. quantum gravity. At first sight, we are thus faced with an intractable problem, since the theory of quantum gravity is not yet established. We are saved by the fact that today on large cosmological scales the fractional amplitude of the fluctuations is smaller than 1. Since gravity is a purely attractive force, the fluctuations had to have been - at least in the context of an eternally expanding background cosmology - very small in the early Universe. Thus, a linearized analysis of the fluctuations (about a classical cosmological background) is self-consistent.

From the classical theory of cosmological perturbations discussed in the previous section, we know that the analysis of scalar metric inhomogeneities can be reduced - after extracting gauge artifacts - to the study of the evolution of a single fluctuating variable. Thus, we conclude that the quantum theory of cosmological perturbations must be reducible to the quantum theory of a single free scalar field which we will denote by v. Since the background in which this scalar field evolves is time-dependent, the mass of v will be time-dependent. The time-dependence of the mass will lead to quantum particle production over time if we start the evolution in the vacuum state for v. As we will see, this quantum particle production corresponds to the development and growth of the cosmological fluctuations. Thus, the quantum theory of cosmological fluctuations provides a consistent framework to study both the generation and the evolution of metric perturbations. The following analysis is based on Part II of [5].

5.2. OUTLINE OF THE ANALYSIS

In order to obtain the action for linearized cosmological perturbations, we expand the action to quadratic order in the fluctuating degrees of freedom. The linear terms cancel because the background is taken to satisfy the background equations of motion.

We begin with the Einstein-Hilbert action for gravity and the action of a scalar matter field (for the more complicated case of general hydrodynamical fluctuations the reader is referred to [5])

$$S = \int d^4x \sqrt{-g} \big[-\frac{1}{16\pi G} R + \frac{1}{2} \partial_\mu \varphi \partial^\mu \varphi - V(\varphi) \big] , \tag{5.2}$$

where R is the Ricci curvature scalar.

The simplest way to proceed is to work in longitudinal gauge, in which the metric and matter take the form

$$\begin{aligned} ds^2 &= a^2(\eta) \big[(1 + 2\phi(\eta, \mathbf{x})) d\eta^2 - (1 - 2\psi(t, \mathbf{x})) d\mathbf{x}^2 \big] \qquad (5.3) \\ \varphi(\eta, \mathbf{x}) &= \varphi_0(\eta) + \delta\varphi(\eta, \mathbf{x}) . \end{aligned}$$

The next step is to reduce the number of degrees of freedom. First, as already mentioned in the previous section, the off-diagonal spatial Einstein equations force $\psi = \phi$ since $\delta T^i_j = 0$ for scalar field matter (no anisotropic stresses to linear order). The two remaining fluctuating variables ϕ and φ must be linked by the Einstein constraint equations since there cannot be matter fluctuations without induced metric fluctuations.

The two nontrivial tasks of the lengthy [5] computation of the quadratic piece of the action is to find out what combination of φ and ϕ gives the variable v in terms of which the action has canonical kinetic term, and what the form of the time-dependent mass is. This calculation involves inserting the ansatz (4.7) into the action (2.1), expanding the result to second order in the fluctuating fields, making use of the background and of the constraint equations, and dropping total derivative terms from the action. In the context of scalar field matter, the quantum theory of cosmological fluctuations was developed by Mukhanov [75, 76] (see also [77]). The result is the following contribution $S^{(2)}$ to the action quadratic in the perturbations:

$$S^{(2)} = \frac{1}{2}\int d^4x \left[v'^2 - v_{,i}v_{,i} + \frac{z''}{z}v^2\right], \tag{5.4}$$

where the canonical variable v (the "Mukhanov variable" introduced in [76] - see also [72]) is given by

$$v = a\left[\delta\varphi + \frac{\varphi_0'}{\mathcal{H}}\phi\right], \tag{5.5}$$

with $\mathcal{H} = a'/a$, and where

$$z = \frac{a\varphi_0'}{\mathcal{H}}. \tag{5.6}$$

In both the cases of power law inflation and slow roll inflation, $\mathcal{H}$ and φ_0' are proportional and hence (as long as the equation of state does not change over time)

$$z(\eta) \sim a(\eta). \tag{5.7}$$

Note that the variable v is related to the curvature perturbation $\mathcal{R}$ in comoving coordinates introduced in [56] and closely related to the variable ζ used in [54, 55]:

$$v = z\mathcal{R}. \tag{5.8}$$

The equation of motion which follows from the action (5.4) is

$$v'' - \nabla^2 v - \frac{z''}{z}v = 0, \tag{5.9}$$

or, in momentum space,

$$v_k'' + k^2 v_k - \frac{z''}{z}v_k = 0, \tag{5.10}$$

where v_k is the k'th Fourier mode of v. As a consequence of (5.7), the mass term in the above equation is given by the Hubble scale

$$k_H^2 \equiv \frac{z''}{z} \simeq H^2. \tag{5.11}$$

Thus, it immediately follows from (5.10) that on small length scales, i.e. for $k > k_H$, the solutions for v_k are constant amplitude oscillations . These oscillations freeze out at Hubble radius crossing, i.e. when $k = k_H$. On longer scales ($k \ll k_H$), the solutions for v_k increase as z:

$$v_k \sim z \, , \quad k \ll k_H \, . \tag{5.12}$$

Given the action (5.4), the quantization of the cosmological perturbations can be performed by canonical quantization (in the same way that a scalar matter field on a fixed cosmological background is quantized [1]).

The final step in the quantum theory of cosmological perturbations is to specify an initial state. Since in inflationary cosmology all pre-existing classical fluctuations are red-shifted by the accelerated expansion of space, one usually assumes (we will return to a criticism of this point when discussing the trans-Planckian problem of inflationary cosmology) that the field v starts out at the initial time t_i mode by mode in its vacuum state. Two questions immediately emerge: what is the initial time t_i, and which of the many possible vacuum states should be chosen. It is usually assumed that since the fluctuations only oscillate on sub-Hubble scales, the choice of the initial time is not important, as long as it is earlier than the time when scales of cosmological interest today cross the Hubble radius during the inflationary phase. The state is usually taken to be the Bunch-Davies vacuum (see e.g. [1]), since this state is empty of particles at t_i in the coordinate frame determined by the FRW coordinates (see e.g. [79] for a discussion of this point), and since the Bunch-Davies state is a local attractor in the space of initial states in an expanding background (see e.g. [80]). Thus, we choose the initial conditions

$$\begin{aligned} v_k(\eta_i) &= \frac{1}{\sqrt{2\omega_k}} \\ v_k'(\eta_i) &= \frac{\sqrt{\omega_k}}{\sqrt{2}} \end{aligned} \tag{5.13}$$

where here $\omega_k = k$, and η_i is the conformal time corresponding to the physical time t_i.

Let us briefly summarize the quantum theory of cosmological perturbations. In the linearized theory, fluctuations are set up at some initial time t_i mode by mode in their vacuum state. While the wavelength is smaller than the Hubble radius, the state undergoes quantum vacuum fluctuations. The accelerated expansion of the background red-shifts the length scale beyond the Hubble radius. The fluctuations freeze out when the length scale is equal to the Hubble radius. On larger scales, the amplitude of v_k increases as the scale factor. This corresponds to the squeezing of the quantum state present at Hubble radius crossing (in terms of classical general relativity, it is self-gravity which leads to this growth of fluctuations). As discussed e.g. in [81], the squeezing of the quantum vacuum state sets up the classical correlations in the wave function of the fluctuations which are an essential ingredient in the classicalization of the perturbations.

5.3. APPLICATION TO INFLATIONARY COSMOLOGY

We will now use the quantum theory of cosmological perturbations developed above to calculate the spectrum of curvature fluctuations in inflationary cosmology.

We need to compute the power spectrum $\mathcal{P}_{\mathcal{R}}(k)$ of the curvature fluctuation $\mathcal{R}$ defined in (5.8), namely

$$\mathcal{R} = z^{-1}v = \phi + \delta\varphi\frac{\mathcal{H}}{\varphi_0'} \tag{5.14}$$

The idea in calculating the power spectrum at a late time t is to first relate the power spectrum via the growth rate (5.12) of v on super-Hubble scales to the power spectrum at the time $t_H(k)$ of Hubble radius crossing, and to then use the constancy of the amplitude of v on sub-Hubble scales to relate it to the initial conditions (5.13). Thus

$$\begin{aligned}\mathcal{P}_{\mathcal{R}}(k,t) \equiv k^3\mathcal{R}_k^2(t) &= k^3z^{-2}(t)|v_k(t)|^2 \\ &= k^3z^{-2}(t)\Big(\frac{z(t)}{z(t_H(k))}\Big)^2|v_k(t_H(k))|^2 \\ &= k^3z^{-2}(t_H(k))|v_k(t_H(k))|^2 \\ &\sim k^3a^{-2}(t_H(k))|v_k(t_i)|^2\,,\end{aligned} \tag{5.15}$$

where in the final step we have used (5.7) and the constancy of the amplitude of v on sub-Hubble scales. Making use of the condition

$$a^{-1}(t_H(k))k = H \tag{5.16}$$

for Hubble radius crossing, and of the initial conditions (5.13), we immediately see that

$$\mathcal{P}_{\mathcal{R}}(k,t) \sim k^3k^{-2}k^{-1}H^2\,, \tag{5.17}$$

and that thus a scale invariant power spectrum with amplitude proportional to H^2 results, in agreement with what was argued on heuristic grounds in Section (4.5).

5.4. QUANTUM THEORY OF GRAVITATIONAL WAVES

The quantization of gravitational waves parallels the quantization of scalar metric fluctuations, but is more simple because there are no gauge ambiguities. Note that at the level of linear fluctuations, scalar metric fluctuations and gravitational waves are independent. Both can be quantized on the same cosmological background determined by the background scale factor and the background matter. However, in contrast to the case of scalar metric fluctuations, the tensor modes are also present in pure gravity (i.e. in the absence of matter).

Starting point is the action (2.1). Into this action we insert the metric which corresponds to a classical cosmological background plus tensor metric fluctuations:

$$ds^2 = a^2(\eta)\big[d\eta^2 - (\delta_{ij} + h_{ij})dx^i dx^j\big]\,, \tag{5.18}$$

where the second rank tensor $h_{ij}(\eta, \mathbf{x})$ represents the gravitational waves, and in turn can be decomposed as

$$h_{ij}(\eta, \mathbf{x}) = h_+(\eta, \mathbf{x})e^+_{ij} + h_x(\eta, \mathbf{x})e^x_{ij} \tag{5.19}$$

into the two polarization states. Here, e^+_{ij} and e^x_{ij} are two fixed polarization tensors, and h_+ and h_x are the two coefficient functions.

To quadratic order in the fluctuating fields, the action consists of separate terms involving h_+ and h_x. Each term is of the form

$$S^{(2)} = \int d^4x \frac{a^2}{2} \left[h'^2 - (\nabla h)^2 \right], \tag{5.20}$$

leading to the equation of motion

$$h_k^{''} + 2\frac{a^{'}}{a} h_k^{'} + k^2 h_k = 0. \tag{5.21}$$

The variable in terms of which the action (5.20) has canonical kinetic term is

$$\mu_k \equiv a h_k , \tag{5.22}$$

and its equation of motion is

$$\mu_k^{''} + (k^2 - \frac{a^{''}}{a}) \mu_k = 0. \tag{5.23}$$

This equation is very similar to the corresponding equation (5.10) for scalar gravitational inhomogeneities, except that in the mass term the scale factor $a(\eta)$ replaces $z(\eta)$, which leads to a very different evolution of scalar and tensor modes during the reheating phase in inflationary cosmology during which the equation of state of the background matter changes dramatically.

Based on the above discussion we have the following theory for the generation and evolution of gravitational waves in an accelerating Universe (first developed by Grishchuk [82]): waves exist as quantum vacuum fluctuations at the initial time on all scales. They oscillate until the length scale crosses the Hubble radius. At that point, the oscillations freeze out and the quantum state of gravitational waves begins to be squeezed in the sense that

$$\mu_k(\eta) \sim a(\eta) , \tag{5.24}$$

which, from (5.22) corresponds to constant amplitude of h_k. The squeezing of the vacuum state leads to the emergence of classical properties of this state, as in the case of scalar metric fluctuations.

6. Conceptual Problems of Inflationary Cosmology

After this detailed survey of the theory of cosmological perturbations applied to inflationary cosmology we can now turn to some conceptual problems of cosmological inflation, and ways in which string theory may help address these issues. The first two problems relate to the cosmological perturbations we have just discussed.

The first problem concerns the amplitude of the cosmological fluctuations. Considering the simplest large-field potential

$$V(\varphi) = \frac{1}{2} m^2 \varphi^2 , \tag{6.1}$$

the result (4.35) for the amplitude of the gravitational potential ϕ at late times and large scales (which modulo a factor of order unity gives the amplitude of the CMB fluctuations on large angular scales and hence should be of the order 10^{-5}) yields (making use of the fact that the result from (4.35) must be evaluated for field values $\varphi \sim m_{pl}$ when the relevant scales exit the Hubble radius)

$$\phi(t_f(k)) \sim \frac{m}{m_{pl}}. \tag{6.2}$$

Hence, the value of m must be chosen to be about 10^{13}GeV, introducing a new hierarchy problem into particle physics model building. In a model with quartic potential

$$V(\varphi) = \frac{1}{4}\lambda\varphi^4 \tag{6.3}$$

we obtain a severe constraint on the value of the self coupling constant ($\lambda \ll 10^{-10}$ modulo factors of 2π), a constraint which implies that the inflaton cannot be in thermal equilibrium before inflation. To reach this conclusion we have used "naturalness" considerations on coupling constants which state that the lower bound on the self coupling constant λ implies lower bounds on the coupling constants describing interactions of φ with other matter fields, since such interactions generate at higher loop order contributions to the renormalized value of λ. As shown in [83], this hierarchy problem is quite general.

As has recently been shown [59], this problem is worse in many inflationary models with entropy fluctuations. If the entropy perturbations are not suppressed during inflation, they can be parametrically amplified during reheating [84, 57, 74, 59]. This results in fluctuations which are nonlinear after inflation independent of the values of the coupling constants, a result derived including the back-reaction of these fluctuations in the Hartree approximation [85]. Such models are thus phenomenologically ruled out.

The second problem is more important and will be discussed at length in the next section. Basically, since in most scalar field-driven inflationary models the period of inflation lasts much longer than the minimal number of e-foldings required in order that scales of current cosmological interest start out inside the Hubble radius at the beginning of inflation, in such models these scales thus originate with a wavelength much smaller than the Planck length, and hence the justification for using the formalism of the previous section to compute the evolution of fluctuations is doubtful. This is the *trans-Planckian problem* for inflationary cosmology [9] which becomes the *trans-Planckian window of opportunity* for string theory.

Scalar field-driven inflationary models have been shown to be geodesically incomplete in the past [86]. Hence, we know that this model cannot describe the very early Universe. A major challenge for string cosmology is to provide this description.

Most importantly, scalar field-driven inflation uses the time-independent part of $V(\varphi)$ to generate inflation. However, it is observationally known that (but theoretically not understood why) the time independent quantum vacuum contribution to the energy density of any quantum field does not gravitate. This is the famous cosmological constant problem. It may turn out that the solution of the cosmological constant problem will remove not only quantum vacuum energy but also the part of $V(\varphi)$ which generates inflation. I view this issue as the Achilles heel of scalar field-driven inflationary cosmology.

Finally, standard model particle physics does not provide a candidate for the inflaton. Models of particle physics beyond the standard model open the window for providing realizations of inflation.

Why can string theory help? First of all, string theory contains many fields which are massless in the early Universe, namely the moduli fields. Thus, it provides many candidates for an inflaton. In addition, it is quite possible that the hierarchy of scales required to give the right magnitude of density fluctuations emerges from a hierarchy of symmetry breaking scales in string theory. Secondly, string theory is supposed to describe the physics on all scales. Thus, in string cosmology the equations for evolving the fluctuations will be unambiguous (even if they are not known today), and the trans-Planckian problem will be solved. One of the main goals of string theory is to provide a nonsingular cosmology (and a concrete realization of this goal was provided in [87]). Thus, string theory should be able to provide a consistent theory of the very early Universe. This theory might connect with late-time cosmology through a period of inflation, or through scenarios more similar to [7, 8].

7. The Trans-Planckian Window for Superstring Cosmology

The same background dynamics which yields the causal generation mechanism for cosmological fluctuations, the most spectacular success of inflationary cosmology, bears in it the nucleus of the "trans-Planckian problem". This can be seen from Fig. (2). If inflation lasts only slightly longer than the minimal time it needs to last in order to solve the horizon problem and to provide a causal generation mechanism for CMB fluctuations, then the corresponding physical wavelength of these fluctuations is smaller than the Planck length at the beginning of the period of inflation. The theory of cosmological perturbations is based on classical general relativity coupled to a weakly coupled scalar field description of matter. Both the theories of gravity and of matter will break down on trans-Planckian scales, and this immediately leads to the trans-Planckian problem: are the predictions of standard inflationary cosmology robust against effects of trans-Planckian physics [9]?

This question has recently been addressed using a variety of techniques. The simplest method is to replace the usual dispersion relation for cosmological perturbations by an ad-hoc modified relation, as was done in [88, 89] in the context of studying the dependence of the thermal spectrum of black hole radiation on trans-Planckian physics. We will discuss the application of this method to cosmology [90, 91, 92] below. Other methods include considerations of modifications of the evolution of cosmological fluctuations coming from string-motivated space-space uncertainty relations [93, 94, 95, 96, 97], string-motivated space-time uncertainty relations [98] (reviewed below), from minimal length considerations [99], from an effective action analysis [100], from a minimal trans-Planckian physics viewpoint (starting each mode out in the vacuum state of the usual action for cosmological perturbations at the time when the physical wavelength is equal to the new fundamental length) [101, 102], and from the point of view of boundary renormalization group analysis [103].

The simplest way of modeling the possible effects of trans-Planckian physics [90, 91, 92] on the evolution of cosmological perturbations, while keeping the mathematical analysis simple, is to replace the linear dispersion relation $\omega_{\rm phys} = k_{\rm phys}$ of the usual equation for cosmological perturbations by a non standard dispersion relation $\omega_{\rm phys} =$

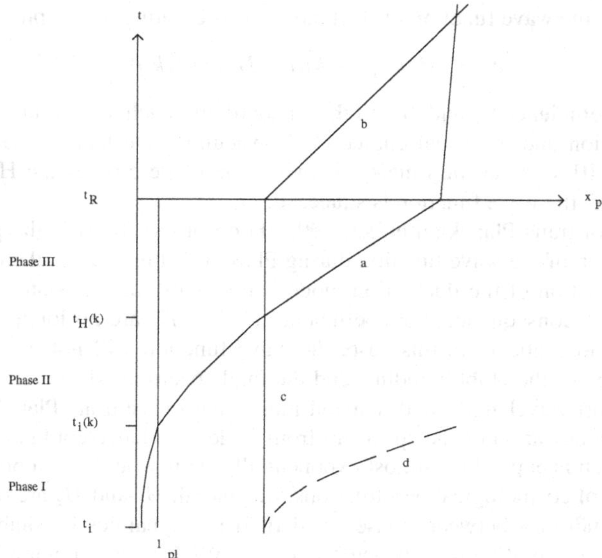

Figure 2. Space-time diagram (physical distance vs. time) showing the origin of the trans-Planckian problem of inflationary cosmology: at very early times, the wavelength is smaller than the Planck scale ℓ_{Pl} (Phase I), at intermediate times it is larger than ℓ_{Pl} but smaller than the Hubble radius H^{-1} (Phase II), and at late times during inflation it is larger than the Hubble radius (Phase III). The line labeled a) is the physical wavelength associated with a fixed comoving scale k. The line b) is the Hubble radius or horizon in SBB cosmology. Curve c) shows the Hubble radius during inflation. The horizon in inflationary cosmology is shown in curve d).

$\omega_{\rm phys}(k_{\rm phys})$ which differs from the standard one only for physical wavenumbers larger than the Planck scale. This amounts to replacing k^2 appearing in (5.10) with $k^2_{\rm eff}(n,\eta)$ defined by

$$k^2 \rightarrow k^2_{\rm eff}(k,\eta) \equiv a^2(\eta)\omega^2_{\rm phys}\left(\frac{k}{a(\eta)}\right). \tag{7.1}$$

For a fixed comoving mode, this implies that the dispersion relation becomes time-dependent. Therefore, the equation of motion of the quantity $v_k(\eta)$ takes the form (with $z(\eta) \propto a(\eta)$)

$$v_k'' + \left[k^2_{\rm eff}(k,\eta) - \frac{a''}{a}\right]v_k = 0\,. \tag{7.2}$$

A more rigorous derivation of this equation, based on a variational principle, has been provided [104] (see also Ref. [105]).

The evolution of modes thus must be considered separately in three phases, see Fig. (2). In Phase I the wavelength is smaller than the Planck scale, and trans-Planckian physics can play an important role. In Phase II, the wavelength is larger than the Planck scale but smaller than the Hubble radius. In this phase, trans-Planckian physics will have a negligible effect (this statement can be quantified [106]). Hence, by the analysis of the previous section, the wave function of fluctuations is oscillating in this phase,

$$v_k = B_1 \exp(-ik\eta) + B_2 \exp(ik\eta) \tag{7.3}$$

with constant coefficients B_1 and B_2. In the standard approach, the initial conditions are fixed in this region and the usual choice of the vacuum state leads to $B_1 = 1/\sqrt{2k}$, $B_2 = 0$. Phase III starts at the time $t_H(k)$ when the mode crosses the Hubble radius. During this phase, the wave function is squeezed.

One source of trans-Planckian effects [90, 91] on observations is the possible non-adiabatic evolution of the wave function during Phase I. If this occurs, then it is possible that the wave function of the fluctuation mode is not in its vacuum state when it enters Phase II and, as a consequence, the coefficients B_1 and B_2 are no longer given by the standard expressions above. In this case, the wave function will not be in its vacuum state when it crosses the Hubble radius, and the final spectrum will be different. In particular, since short wavelength modes spend more time in the trans-Planckian phase, it is possible that a deviation of the spectrum from scale-invariance could be induced in a background which is expanding almost exponentially. This would be an order one effect on the spectrum of cosmological perturbations. In general, B_1 and B_2 are determined by the matching conditions between Phase I and II. If the dynamics is adiabatic throughout (in particular if the a''/a term is negligible), the WKB approximation holds and the solution is always given by

$$v_k(\eta) \simeq \frac{1}{\sqrt{2k_{\text{eff}}(k,\eta)}} \exp\left(-i \int_{\eta_i}^{\eta} k_{\text{eff}} d\tau\right), \tag{7.4}$$

where η_i is some initial time. Therefore, if we start with a positive frequency solution only and use this solution, we find that no negative frequency solution appears. Deep in Region II where $k_{\text{eff}} \simeq k$ the solution becomes

$$v_k(\eta) \simeq \frac{1}{\sqrt{2k}} \exp(-i\phi - ik\eta), \tag{7.5}$$

i.e. the standard vacuum solution times a phase which will disappear when we calculate the modulus. To obtain a modification of the inflationary spectrum, it is necessary to find a dispersion relation such that the WKB approximation breaks down in Phase I.

A concrete class of dispersion relations for which the WKB approximation breaks down is [89]

$$k_{\text{eff}}^2(k,\eta) = k^2 - k^2 |b_m| \left[\frac{\ell_{\text{pl}}}{\lambda(\eta)}\right]^{2m}, \tag{7.6}$$

where $\lambda(\eta) = 2\pi a(\eta)/k$ is the wavelength of a mode. If we follow the evolution of the modes in Phases I, II and III, matching the mode functions and their derivatives at the junction times, the calculation [90, 91, 107] demonstrates that the final spectral index is modified and that superimposed oscillations appear. It has recently been shown [108] that in the case of this class of dispersion relations, the spectrum of black hole Hawking radiation is also affected.

However, the above example suffers from several problems. First, in inflationary models with a long period of inflationary expansion, the dispersion relation (7.6) leads to complex frequencies at the beginning of inflation for scales which are of current interest in

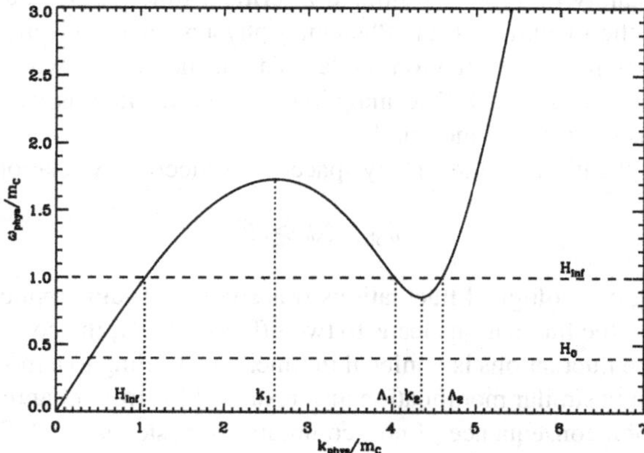

Figure 3. Sketch of the dispersion relation of [104]. The adiabaticity condition on the evolution of fluctuations is broken when the physical frequency (vertical axis) is smaller than the Hubble expansion rate. During the phase of inflation, this is the case when the physical wavenumber (horizontal axis) is in the interval between Λ_1 and Λ_2. Ultraviolet scales are $k > k_0$, and the value $k = k_1$ is the value when the dispersion relation turns over.

cosmology. Furthermore, the initial conditions for the Fourier modes of the fluctuation field have to be set in a region where the evolution is non-adiabatic and the use of the usual vacuum prescription can be questioned. These problems can be avoided in a toy model in which we follow the evolution of fluctuations in a bouncing cosmological background which is asymptotically flat in the past and in the future. The analysis of [109] shows that even in this case the final spectrum of fluctuations depends on the specific dispersion relation used. An example (see Fig. (3)) of a dispersion relation which breaks the WKB approximation in the trans-Planckian regime but does not lead to the problems mentioned in the previous paragraph was investigated in [104]. It is a dispersion relation which is linear for both small and large wavenumbers, but has an intermediate interval during which the frequency decreases as the wavenumber increases, much like what happens in (7.6). The violation of the WKB condition occurs for wavenumbers near the local minimum of the $\omega(k)$ curve. In this model, modes can be set up in the far ultraviolet in their Bunch-Davies vacuum. During the time interval when the physical wavenumber k_p passes through the interval $\Lambda_1 < k_p < \Lambda_2$ when the physical frequency is smaller than the value of the Hubble constant H_{inf} in the inflationary phase, the mode is squeezed and thus is no longer in the local vacuum state when $k_p < \Lambda_1$.

A justified criticism against the method summarized in the previous analysis is that the non-standard dispersion relations used are completely ad hoc, without a clear basis in trans-Planckian physics. Other recently explored approaches are motivated by string theory. For example, there has been a lot of recent work [93, 94, 95, 96, 97] on the implication of space-space uncertainty relations [110, 111] for the evolution of fluctuations. The application of the uncertainty relations to the fluctuations lead to two effects: Firstly, the equation of motion of the fluctuations is modified. Secondly, for fixed comoving length scale k, the uncertainty relation is saturated at critical time $t_i(k)$. Thus, in addition to a modification of the evolution, trans-Planckian physics leads to a modification of the boundary conditions for the fluctuation modes. The upshot of this work is that the spectrum of fluctuations is modified. The magnitude of the deviations can be of the order H/m_{pl}, and is thus in principle measurable.

In [98], the implications of the stringy space-time uncertainty relation [112, 113]

$$\Delta x_{\rm phys} \Delta t \geq l_s^2 \tag{7.7}$$

on the spectrum of cosmological fluctuations was studied. Again, application of this uncertainty relation to the fluctuations leads to two effects. Firstly, the coupling between the background and the fluctuations is nonlocal in time, thus leading to a modified dynamical equation of motion (a similar modification also results [114] from quantum deformations, another example of a consequence of non-commutative basic physics). Secondly, the uncertainty relation is saturated at the time $t_i(k)$ when the physical wavelength equals the string scale l_s. Before that time it does not make sense to talk about fluctuations on that scale. By continuity, it makes sense to assume that fluctuations on scale k are created at time $t_i(k)$ in the local vacuum state (the instantaneous WKB vacuum state).

Let us for the moment neglect the nonlocal coupling between background and fluctuation, and thus consider the usual equation of motion for fluctuations in an accelerating background cosmology. We assume that $a(t)$ scales as a power of time, i.e. we consider power-law inflation, and we distinguish two ranges of scales. Ultraviolet modes are generated at late times when the Hubble radius is larger than l_s. On these scales, the spectrum of fluctuations does not differ from what is predicted by the standard theory, since at the time of Hubble radius crossing the fluctuation modes will be in their vacuum state. However, the evolution of the infrared modes which are created when the Hubble radius is smaller than l_s is different. The fluctuations undergo *less* squeezing than they do in the absence of the uncertainty relation, and hence the final amplitude of fluctuations is lower. From the equation (5.15) for the power spectrum of fluctuations, and making use of the condition

$$a(t_i(k)) = k l_s \tag{7.8}$$

for the time $t_i(k)$ when the mode is generated, it follows immediately that the power spectrum is scale-invariant

$$\mathcal{P}_{\mathcal{R}}(k) \sim k^0 \, . \tag{7.9}$$

In the standard scenario of power-law inflation the spectrum is red ($\mathcal{P}_{\mathcal{R}}(k) \sim k^{n-1}$ with $n < 1$). Taking into account the effects of the nonlocal coupling between background and fluctuation mode leads [98] to a modification of this result: the spectrum of fluctuations in a power-law inflationary background is in fact blue ($n > 1$).

Note that, if we neglect the nonlocal coupling between background and fluctuation mode, the result of (7.9) also holds in a cosmological background which is NOT accelerating. Thus, we have a method of obtaining a scale-invariant spectrum of fluctuations without inflation. This result was also subsequently obtained in [115], however without a micro-physical basis for the prescription for the initial conditions.

An approach to the trans-Planckian issue pioneered by Danielsson [101] which has recently received a lot of attention is to avoid the issue of the unknown trans-Planckian physics and to start the evolution of the fluctuation modes at the mode-dependent time when the wavelength equals the limiting scale. Obviously, the resulting spectrum will depend sensitively on which state is taken to be the initial state. The vacuum state is not unambiguous, and the choice of a state minimizing the energy density depends on the space-time splitting [102]. The signatures of this prescription are typically oscillations superimposed on the usual spectrum. The amplitude of this effect depends sensitively on the prescription of the initial state, and for a fixed prescription also on the background cosmology. For a discussion of these issues and a list of references on this approach the reader is referred to [116]. Note, in particular, that for a fixed background cosmology and for a fixed initial condition prescription, the amplitude of the correction terms in the spectrum may be different for scalar cosmological perturbations on one hand and gravitational waves or test scalar matter fields on the other hand.

If the ultraviolet modes are not in their vacuum state for wavelengths between the Planck or string scale and the Hubble radius, then the question of their back-reaction on the background geometry [117, 118, 119] arises. The naive expectation is that such ultraviolet modes have the equation of state of radiation, and that the mode occupation number hence must be very small (of order $(H_{inf}/m_{pl})^2$) in order for the total energy density carried by these modes not to prevent inflation. This issue can be analyzed in the model of [104] in a setting in which there are no ultraviolet problems (in the naive approach one needs to assume that modes are continuously generated). In this approach, all modes start in their vacuum state, they are squeezed when $\Lambda_1 < k_p < \Lambda_2$, and they oscillate at an excited level thereafter. In an exponentially expanding background it is clear by time-translation invariance that the energy density in the ultraviolet modes is constant. The stress-energy density of a test scalar field with this dispersion relation has recently been calculated [120], yielding the result that up to correction terms suppressed by $(H_{inf}/m_{pl})^2$, the equation of state is that of a cosmological constant. Thus, the back-reaction of the excited states does not prevent inflation but rather leads to a renormalization of the cosmological constant. The back-reaction does effect the speed of rolling of the scalar field, and this yields constraints on the mode occupation numbers in the ultraviolet, constraints which, however, are much weaker than the ones conjectured in [117, 118, 119].

Another constraint on trans-Planckian physics arises from the observational limits on the flux of ultra-high-energy cosmic rays. Such cosmic rays would be produced [121] in the present Universe if Trans-Planckian effects would lead to non-adiabatic mode evolution in the ultraviolet today. In the model of [104] this does not happen if the present Hubble rate H_0 is smaller than the local minimum of the dispersion relation, as in the situation depicted in Fig. (3).

In summary, due to the exponential red-shifting of wavelengths, present cosmological scales originate at wavelengths smaller than the Planck length early on during the period of inflation. Thus, Planck-scale physics may well encode information in these modes which can now be observed in the spectrum of microwave anisotropies. Two examples have been shown to demonstrate the existence of this "window of opportunity" to probe trans-Planckian physics in cosmological observations. The first method makes use of modified dispersion relations to probe the robustness of the predictions of inflationary cosmology, the second applies the stringy space-time uncertainty relation on the fluctuation modes. Both methods yield the result that trans-Planckian physics may lead to measurable effects in cosmological observables.

8. Conclusions

These lectures have focused on the quantum theory of the generation and evolution of cosmological fluctuations. This is the theory which connects the fundamental physics of the very early Universe with current cosmological observations. The general theory was illustrated in the context of the current paradigm of early universe cosmology, namely the inflationary universe scenario. In the context of the current approach to early universe cosmology which involves the coupling of matter described by quantum field theory to classical general relativity, inflationary cosmology suffers from important conceptual problems. Resolving these problems is a major goal for superstring cosmology. Thus, I would argue that inflationary cosmology requires string theory.

Whether string theory requires inflation is a different issue. String theory provides many candidates for an inflaton, and thus it is possible (maybe even likely) that inflation can be implemented in string theory. The study of this issue has recently attracted a lot of attention. However, one should keep in mind that current observations do not prove the correctness of inflation. They show that the spectrum of fluctuations is nearly scale-invariant and nearly adiabatic. However, there are other ways to obtain such a spectrum, although the existing alternatives such as the ad hoc model mentioned in the previous section of these lecture notes, or the Pre-Big-Bang [7] and Ekpyrotic [8] scenarios are either not based on fundamental physics or not (yet) as developed as inflationary cosmology. Thus, it may be possible that string theory connects with observations using a cosmological scenario different than inflation. Even in this case, however, the theory of cosmological perturbations developed here is applicable.

A stringy early Universe may involve the dynamics of higher dimensions in an important way. This is the case for the Ekpyrotic scenario (see e.g. [122, 123] for some initial work). In this case, the formalism of cosmological perturbations needs to be extended (see e.g. [124] for a formalism which is a direct generalization of what was discussed here). There may be many interesting effects of bulk fluctuation modes which cannot be seen in a four-dimensional effective field theory approach. This is a rich field which merits much more work.

Acknowledgements

The author wishes to thank the organizers of this Cargèse school for the invitation to lecture and for their hospitality. He also wishes to thank Pei-Ming Ho and Jérôme Martin for collaborations on the issue of the trans-Planckian problem of inflationary cosmology. He is grateful to the Perimeter Institute for hospitality and financial support during the time when these lectures were prepared. He is supported in part by an NSERC Discovery grant (at McGill) and by the US Department of Energy under Contract DE-FG02-91ER40688, TASK A.

References

1. C. L. Bennett *et al.*, Astrophys. J. Suppl. **148**, 1 (2003) [arXiv:astro-ph/0302207].

2. M. Colless *et al.*, arXiv:astro-ph/0306581.
3. K. Abazajian *et al.* [SDSS Collaboration], Astron. J. **128**, 502 (2004) [arXiv:astro-ph/0403325].
4. A. H. Guth, Phys. Rev. D **23**, 347 (1981).
5. A. Linde: *Particle Physics and Inflationary Cosmology*, (Harwood, Chur, 1990).
6. V. Moncrief, J. Math. Phys. **17**, 1893 (1976);
V. Moncrief, *Prepared for Directions in General Relativity: An International Symposium in Honor of the 60th Birthdays of Dieter Brill and Charles Misner, College Park, MD, 27-29 May 1993*
7. M. Gasperini and G. Veneziano, M. Gasperini and G. Veneziano, Astropart. Phys. **1**, 317 (1993) [arXiv:hep-th/9211021].
8. J. Khoury, B. A. Ovrut, P. J. Steinhardt and N. Turok, Phys. Rev. D **64**, 123522 (2001) [arXiv:hep-th/0103239].
9. R. H. Brandenberger: Inflationary cosmology: Progress and problems. In: *1st Iranian International School on Cosmology: Large Scale Structure Formation, Kish Island, Iran, 22 Jan - 4 Feb 1999* Kluwer, Dordrecht, 2000. (Astrophys. Space Sci. Libr. ; 247), ed. by R. Mansouri and R. Brandenberger (Kluwer, Dordrecht, 2000). [arXiv:hep-ph/9910410].
10. R. H. Brandenberger, Lect. Notes Phys. **646**, 127 (2004) [arXiv:hep-th/0306071].
11. A. Vilenkin and E.P.S. Shellard; *Cosmic Strings and Other Topological Defects*, (Cambridge Univ. Press, Cambridge, 1994).
12. M. B. Hindmarsh and T. W. Kibble, Rept. Prog. Phys. **58**, 477 (1995) [arXiv:hep-ph/9411342].
13. R. H. Brandenberger, Int. J. Mod. Phys. A **9**, 2117 (1994) [arXiv:astro-ph/9310041].
14. A. A. Starobinsky, Phys. Lett. B **91**, 99 (1980).
15. C. Armendariz-Picon, T. Damour and V. Mukhanov, Phys. Lett. B **458**, 209 (1999) [arXiv:hep-th/9904075].
16. R. H. Brandenberger and J. H. Kung, Phys. Rev. D **42**, 1008 (1990).
17. S. R. Coleman, Phys. Rev. D **15**, 2929 (1977) [Erratum-ibid. D **16**, 1248 (1977)];
C. G. . Callan and S. R. Coleman, Phys. Rev. D **16**, 1762 (1977).
18. R. H. Brandenberger, Rev. Mod. Phys. **57**, 1 (1985).
19. S. Dodelson, W. H. Kinney and E. W. Kolb, Phys. Rev. D **56**, 3207 (1997) [arXiv:astro-ph/9702166].
20. A. D. Linde, Phys. Lett. B **108**, 389 (1982);
A. Albrecht and P. J. Steinhardt, Phys. Rev. Lett. **48**, 1220 (1982).
21. S. R. Coleman and E. Weinberg, Phys. Rev. D **7**, 1888 (1973).
22. D. S. Goldwirth and T. Piran, Phys. Rept. **214**, 223 (1992).
23. R. Brandenberger, G. Geshnizjani and S. Watson, Phys. Rev. D **67**, 123510 (2003) [arXiv:hep-th/0302222].
24. A. D. Linde, Phys. Lett. B **129**, 177 (1983).
25. A. Linde, arXiv:hep-th/0402051.
26. H. A. Feldman and R. H. Brandenberger, Phys. Lett. B **227**, 359 (1989);
R. H. Brandenberger, H. Feldman and J. Kung, Phys. Scripta **T36**, 64 (1991).
27. A. D. Linde, Phys. Rev. D **49**, 748 (1994) [arXiv:astro-ph/9307002].
28. J. Polchinski, arXiv:hep-th/0412244.
29. A. D. Dolgov and A. D. Linde, Phys. Lett. B **116**, 329 (1982).
30. L. F. Abbott, E. Farhi and M. B. Wise, Phys. Lett. B **117**, 29 (1982).
31. J. H. Traschen and R. H. Brandenberger, Phys. Rev. D **42**, 2491 (1990).
32. L. Kofman, A. D. Linde and A. A. Starobinsky, Phys. Rev. Lett. **73**, 3195 (1994) [arXiv:hep-th/9405187].
33. Y. Shtanov, J. H. Traschen and R. H. Brandenberger, Phys. Rev. D **51**, 5438 (1995) [arXiv:hep-ph/9407247].
34. L. Kofman, A. D. Linde and A. A. Starobinsky, Phys. Rev. D **56**, 3258 (1997) [arXiv:hep-ph/9704452].
35. A. Liddle and D. Lyth, *Cosmological Inflation and Large-Scale Structure*, (Cambridge Univ. Press, Cambridge, 2000).
36. V. F. Mukhanov, H. A. Feldman and R. H. Brandenberger, Phys. Rept. **215**, 203 (1992).

37. S. Weinberg: *Gravitation and Cosmology*, (Wiley, New York, 1972).
38. P.J.E. Peebles: *The Large-Scale Structure of the Universe*, (Princeton Univ. Press, Princeton, 1980).
39. T. Padmanabhan: *Structure Formation in the Universe*, (Cambridge Univ. Press, Cambridge, 1993).
40. J. Peacock: *Cosmological Physics*, (Cambridge Univ. Press, Cambridge, 1999).
41. R. K. Sachs and A. M. Wolfe, Astrophys. J. **147**, 73 (1967).
42. E. R. Harrison, Phys. Rev. D **1**, 2726 (1970).
43. Y. B. Zeldovich, Mon. Not. Roy. Astron. Soc. **160**, 1 (1972).
44. N. Afshordi and R. H. Brandenberger, Phys. Rev. D **63**, 123505 (2001) [arXiv:gr-qc/0011075].
45. E. Lifshitz, J. Phys. (USSR) **10**, 116 (1946);
E. M. Lifshitz and I. M. Khalatnikov, Adv. Phys. **12**, 185 (1963).
46. J. M. Bardeen, Phys. Rev. D **22**, 1882 (1980).
47. W. Press and E. Vishniac, Astrophys. J. **239**, 1 (1980).
48. H. Kodama and M. Sasaki, Prog. Theor. Phys. Suppl. **78**, 1 (1984).
49. M. Bruni, G. F. Ellis and P. K. Dunsby, Class. Quant. Grav. **9**, 921 (1992).
50. J. c. Hwang, Astrophys. J. **415**, 486 (1993).
51. R. Durrer, Helv. Phys. Acta **69**, 417 (1996).
52. J. Stewart, Class. Quant. Grav. **7**, 1169 (1990).
53. J. Stewart and M. Walker, Proc. R. Soc. London **A 341**, 49 (1974).
54. J. M. Bardeen, P. J. Steinhardt and M. S. Turner, Phys. Rev. D **28**, 679 (1983).
55. R. H. Brandenberger and R. Kahn, Phys. Rev. D **29**, 2172 (1984).
56. D. H. Lyth, Phys. Rev. D **31**, 1792 (1985).
57. F. Finelli and R. H. Brandenberger, Phys. Rev. Lett. **82**, 1362 (1999) [arXiv:hep-ph/9809490].
58. B. A. Bassett and F. Viniegra, Phys. Rev. D **62**, 043507 (2000) [arXiv:hep-ph/9909353].
59. F. Finelli and R. H. Brandenberger, Phys. Rev. D **62**, 083502 (2000) [arXiv:hep-ph/0003172].
60. J. H. Traschen, Phys. Rev. D **29**, 1563 (1984).
61. J. H. Traschen, N. Turok and R. H. Brandenberger, Phys. Rev. D **34**, 919 (1986).
62. G. Dvali, A. Gruzinov and M. Zaldarriaga, Phys. Rev. D **69**, 023505 (2004) [arXiv:astro-ph/0303591].
63. L. Kofman, arXiv:astro-ph/0303614.
64. F. Vernizzi, Phys. Rev. D **69**, 083526 (2004) [arXiv:astro-ph/0311167].
65. M. Axenides, R. H. Brandenberger and M. S. Turner, Phys. Lett. B **126**, 178 (1983).
66. S. Weinberg, Phys. Rev. D **67**, 123504 (2003) [arXiv:astro-ph/0302326].
67. W. B. Lin, X. H. Meng and X. M. Zhang, Phys. Rev. D **61**, 121301 (2000) [arXiv:hep-ph/9912510].
68. V. F. Mukhanov and G. V. Chibisov, JETP Lett. **33**, 532 (1981) [Pisma Zh. Eksp. Teor. Fiz. **33**, 549 (1981)].
69. A. H. Guth and S. Y. Pi, Phys. Rev. Lett. **49**, 1110 (1982).
70. A. A. Starobinsky, Phys. Lett. B **117**, 175 (1982).
71. S. W. Hawking, Phys. Lett. B **115**, 295 (1982).
72. V. N. Lukash, Pisma Zh. Eksp. Teor. Fiz. **31**, 631 (1980);
V. N. Lukash, Sov. Phys. JETP **52**, 807 (1980) [Zh. Eksp. Teor. Fiz. **79**, (1980)].
73. W. Press, Phys. Scr. **21**, 702 (1980).
74. K. Sato, Mon. Not. Roy. Astron. Soc. **195**, 467 (1981).
75. V. F. Mukhanov, JETP Lett. **41**, 493 (1985) [Pisma Zh. Eksp. Teor. Fiz. **41**, 402 (1985)].
76. V. F. Mukhanov, Sov. Phys. JETP **67**, 1297 (1988) [Zh. Eksp. Teor. Fiz. **94N7**, 1 (1988 ZETFA,94,1-11.1988)].
77. M. Sasaki, Prog. Theor. Phys. **76**, 1036 (1986).
78. N. Birrell and P.C.W. Davies: *Quantum Fields in Curved Space*, (Cambridge Univ. Press, Cambridge, 1982).
79. R. H. Brandenberger, Nucl. Phys. B **245**, 328 (1984).
80. R. H. Brandenberger and C. T. Hill, Phys. Lett. B **179**, 30 (1986).
81. D. Polarski and A. A. Starobinsky, Class. Quant. Grav. **13**, 377 (1996) [arXiv:gr-qc/9504030].
82. L. P. Grishchuk, *Sov. Phys. JETP* **40**, 409 (1975) [*Zh. Eksp. Teor. Fiz.* **67**, 825 (1974)].

83. F. C. Adams, K. Freese and A. H. Guth, Phys. Rev. D **43**, 965 (1991).
84. B. A. Bassett, D. I. Kaiser and R. Maartens, Phys. Lett. B **455**, 84 (1999) [arXiv:hep-ph/9808404].
85. J. P. Zibin, R. H. Brandenberger and D. Scott, Phys. Rev. D **63**, 043511 (2001) [arXiv:hep-ph/0007219].
86. A. Borde and A. Vilenkin, Phys. Rev. Lett. **72**, 3305 (1994) [arXiv:gr-qc/9312022].
87. R. H. Brandenberger and C. Vafa, Nucl. Phys. B **316**, 391 (1989).
88. W. G. Unruh, *Phys. Rev.* **D51**, 2827 (1995).
89. S. Corley and T. Jacobson, *Phys. Rev.* **D54**, 1568 (1996) [arXiv:hep-th/9601073].
90. J. Martin and R. H. Brandenberger, *Phys. Rev.* **D63**, 123501 (2001) [arXiv:hep-th/0005209].
91. R. H. Brandenberger and J. Martin, *Mod. Phys. Lett.* **A16**, 999 (2001) [arXiv:astro-ph/0005432].
92. J. C. Niemeyer, *Phys. Rev.* **D63**, 123502 (2001) [arXiv:astro-ph/0005533].
93. C. S. Chu, B. R. Greene and G. Shiu, Mod. Phys. Lett. A **16**, 2231 (2001) [arXiv:hep-th/0011241].
94. R. Easther, B. R. Greene, W. H. Kinney and G. Shiu, *Phys. Rev.* **D64**, 103502 (2001) [arXiv:hep-th/0104102].
95. R. Easther, B. R. Greene, W. H. Kinney and G. Shiu, Phys. Rev. D **67**, 063508 (2003) [arXiv:hep-th/0110226].
96. F. Lizzi, G. Mangano, G. Miele and M. Peloso, *JHEP* **0206**, 049 (2002) [arXiv:hep-th/0203099].
97. S. F. Hassan and M. S. Sloth, Nucl. Phys. B **674**, 434 (2003) [arXiv:hep-th/0204110].
98. R. Brandenberger and P. M. Ho, *Phys. Rev.* **D66**, 023517 (2002) [arXiv:hep-th/0203119].
99. A. Kempf and J. C. Niemeyer, *Phys. Rev.* **D64**, 103501 (2001) [arXiv:astro-ph/0103225].
100. C. P. Burgess, J. M. Cline, F. Lemieux and R. Holman, JHEP **0302**, 048 (2003) [arXiv:hep-th/0210233];
C. P. Burgess, J. M. Cline and R. Holman, JCAP **0310**, 004 (2003) [arXiv:hep-th/0306079].
101. U. H. Danielsson, Phys. Rev. D **66**, 023511 (2002) [arXiv:hep-th/0203198];
U. H. Danielsson, JHEP **0207**, 040 (2002) [arXiv:hep-th/0205227].
102. V. Bozza, M. Giovannini and G. Veneziano, JCAP **0305**, 001 (2003) [arXiv:hep-th/0302184].
103. K. Schalm, G. Shiu and J. P. van der Schaar, JHEP **0404**, 076 (2004) [arXiv:hep-th/0401164].
104. M. Lemoine, M. Lubo, J. Martin and J. P. Uzan, *Phys. Rev.* **D65**, 023510 (2002) [arXiv:hep-th/0109128].
105. T. Jacobson and D. Mattingly, *Phys. Rev.* **D63**, 041502 (2001) [arXiv:hep-th/0009052].
106. N. Kaloper, M. Kleban, A. E. Lawrence and S. Shenker, Phys. Rev. D **66**, 123510 (2002) [arXiv:hep-th/0201158].
107. J. Martin and R. H. Brandenberger, *Phys. Rev.* **D65**, 103514 (2002) [arXiv:hep-th/0201189].
108. W. G. Unruh and R. Schutzhold, arXiv:gr-qc/0408009.
109. R. H. Brandenberger, S. E. Joras and J. Martin, Phys. Rev. D **66**, 083514 (2002) [arXiv:hep-th/0112122].
110. D. Amati, M. Ciafaloni and G. Veneziano, *Phys. Lett.* **B197**, 81 (1987).
111. D. J. Gross and P. F. Mende, *Nucl. Phys.* **B303**, 407 (1988).
112. T. Yoneya, *Mod. Phys. Lett.* **A4**, 1587 (1989).
113. M. Li and T. Yoneya, arXiv:hep-th/9806240.
114. S. Cremonini, Phys. Rev. D **68**, 063514 (2003) [arXiv:hep-th/0305244].
115. S. Hollands and R. M. Wald, Gen. Rel. Grav. **34**, 2043 (2002) [arXiv:gr-qc/0205058].
116. J. Martin and R. Brandenberger, Phys. Rev. D **68**, 063513 (2003) [arXiv:hep-th/0305161].
117. T. Tanaka, arXiv:astro-ph/0012431.
118. A. A. Starobinsky, *Pisma Zh. Eksp. Teor. Fiz.* **73**, 415 (2001) [*JETP Lett.* **73**, 371 (2001)] [arXiv:astro-ph/0104043].
119. M. Porrati, Phys. Lett. B **596**, 306 (2004) [arXiv:hep-th/0402038].
120. R. H. Brandenberger and J. Martin, arXiv:hep-th/0410223.
121. A. A. Starobinsky and I. I. Tkachev, JETP Lett. **76**, 235 (2002) [Pisma Zh. Eksp. Teor. Fiz. **76**, 291 (2002)] [arXiv:astro-ph/0207572].
122. A. J. Tolley, N. Turok and P. J. Steinhardt, Phys. Rev. D **69**, 106005 (2004) [arXiv:hep-

th/0306109].
123. T. J. Battefeld, S. P. Patil and R. Brandenberger, Phys. Rev. D **70**, 066006 (2004) [arXiv:hep-th/0401010].
124. C. van de Bruck, M. Dorca, R. H. Brandenberger and A. Lukas, Phys. Rev. D **62**, 123515 (2000) [arXiv:hep-th/0005032].

TOWARDS A TOPOLOGICAL $\mathcal{G}_2$ STRING

JAN DE BOER, ASAD NAQVI
Instituut voor Theoretische Fysica
Valckenierstraat 65, 1018XE Amsterdam, The Netherlands
AND
ASSAF SHOMER
Santa Cruz Institute for Particle Physics
1156 High Street, Santa Cruz, 95064 CA, USA

Abstract. We define new topological theories related to sigma models whose target space is a 7 dimensional manifold of G_2 holonomy. We show how to define the topological twist and identify the BRST operator and the physical states. Correlation functions at genus zero are computed and related to Hitchin's topological action for three-forms. We conjecture that one can extend this definition to all genus and construct a seven-dimensional topological string theory. In contrast to the four-dimensional case, it does not seem to compute terms in the low-energy effective action in three dimensions.

1. Introduction

Topological strings on Calabi-Yau manifolds describe certain solvable sectors of superstrings and as such provide simplified toy models of string theory. There are two inequivalent ways to twist the Calabi-Yau sigma model. This yields topological theories known as the A-model and the B-model, which at first sight depend on different degrees of freedom: the A-model apparently only involves the Kähler moduli and the B-model only the complex moduli. However, this changes once branes are included, and it has been conjectured that there is a version of S-duality which maps the A-model to the B-model [1]. Subsequently, several authors found evidence for the existence of seven and/or eight dimensional theories that unify and extend the A and B-model [2, 3, 4, 5, 6]. This was one of our original motivations to take a closer look at string theory on seven-dimensional manifolds of G_2 holonomy, and to see whether it allows for a topological twist, though we were motivated by other issues as well, such as applications to M-theory compactifications on G_2-manifolds, and as a possible tool to improve our understanding of the relation between supersymmetric gauge theories in three and four dimensions.

The outline of this note is as follows. We will first review sigma-models on target spaces of G_2 holonomy, and the structure of chiral the algebra of these theories. The

L. Baulied et al. (eds.), String Theory: From Gauge Interactions to Cosmology, 117–127.

latter is a non-linear extension of the $c = \frac{21}{2}$ $\mathcal{N} = 1$ superconformal algebra that contains an $\mathcal{N} = 1$ subalgebra with $c = \frac{7}{10}$. This describes a minimal model, the tricritical Ising model, which plays a crucial role in the twisting. We then go on to describe the twisting, the BRST operator, the physical states, and we end with a discussion of topological G_2 strings. Here we briefly summarize our findings. A more detailed discussion will appear elsewhere [7].

There is extensive literature about string theory and M-theory compactified on G_2 manifolds. The first detailed study of the world-sheet formulation of strings on G_2 manifolds appeared in [8]. The world-sheet chiral algebra was studied in some detail in [8, 9, 10, 11]. For more about type II strings on G_2 manifolds and their mirror symmetry, see e.g. [12, 13, 14, 15, 16, 17, 18, 19, 20, 21]. A review of M-theory on G_2 manifolds with many references can be found in [22].

2. G_2 sigma models

We start from an $\mathcal{N} = (1, 1)$ sigma model describing d chiral superfields $X^\mu = \phi^\mu(z) + \theta\psi^\mu(z)$

$$S = \int d^2z\, d^2\theta\, (G_{\mu\nu} + B_{\mu\nu}) D_\theta \mathbf{X}^\mu D_{\bar{\theta}} \mathbf{X}^\nu. \tag{2.1}$$

The super stress-energy tensor is given by $T(z, \theta) = G(z) + \theta T(z) = -\frac{1}{2} G_{\mu\nu} D_\theta X^\mu \partial_z X^\nu$. This $\mathcal{N} = (1, 1)$ sigma model can be formulated on an arbitrary target space. However, the target space theory will have some supersymmetry only when the manifold has special holonomy. This condition ensures the existence of covariantly constant spinors which are used to construct supercharges. The existence of a covariantly constant spinor on the manifold also implies the existence of covariantly constant *p-forms* given by

$$\phi_{(p)} = \epsilon^T \Gamma_{i_1 \dots i_p} \epsilon\, dx^{i_1} \wedge \dots \wedge dx^{i_p}. \tag{2.2}$$

This formal expression may be identically zero. The details of the target space manifold dictate which p-forms are actually present. If the manifold has special holonomy $H \subset SO(d)$, the non-vanishing forms (2.2) are precisely the forms that transform trivially under H.

The existence of such covariantly constant p-forms on the target space manifold implies the existence of extra elements in the chiral algebra. For example, given a covariantly constant p form, $\phi_{(p)} = \phi_{i_1 \cdots i_p} dx^{i_1} \wedge \cdots \wedge dx^{i_p}$ satisfying $\nabla \phi_{i_1 \cdots i_p} = 0$, we can construct a holomorphic superfield current given by

$$J_{(p)}(z, \theta) = \phi_{i_1 \cdots i_p} D_\theta X^{i_1} \cdots D_\theta X^{i_p},$$

which satisfies $D_{\bar{\theta}} J_{(p)} = 0$ on shell. In components, this implies the existence of a dimension $\frac{p}{2}$ and a dimension $\frac{p+1}{2}$ current. For example, on a Kahler manifold, the existence of a covariantly constant Kahler two form $\omega = g_{i\bar{j}}(d\phi^i \wedge d\phi^{\bar{j}} - d\phi^{\bar{j}} \wedge d\phi^i)$ implies the existence of a dimension 1 current $J = g_{i\bar{j}} \psi^i \psi^{\bar{j}}$ and a dimension $\frac{3}{2}$ current $G'(z) = g_{i\bar{j}}(\psi^i \partial_z \phi^{\bar{j}} - \psi^{\bar{j}} \partial_z \phi^i)$, which add to the $(1, 1)$ superconformal currents $G(z)$ and $T(z)$ to give a $(2, 2)$ superconformal algebra.

A generic seven dimensional Riemannian manifold has $SO(7)$ holonomy. A G_2 manifold has holonomy which sits in a G_2 subgroup of $SO(7)$. Under this embedding, the eight dimensional spinor representation **8** of $SO(7)$ decomposes into a **7** and a singlet of G_2, and the latter corresponds to the covariantly constant spinor. The p-form (2.2) is non-trivial only when $p = 3, 4$. In other words, there is a covariantly constant 3-form $\phi^{(3)} = \phi^{(3)}_{ijk} dx^i \wedge dx^j \wedge dx^k$. The hodge dual 4-form is then also automatically covariantly constant.

By the above discussion, the 3-form implies the existence of a superfield current $J_{(3)}(z, \theta) = \phi^{(3)}_{ijk} D_\theta X^i D_\theta X^j D_\theta X^k \equiv \Phi + \theta K$. Explicitly, Φ is a dimension $\frac{3}{2}$ current $\Phi = \phi^{(3)}_{ijk} \psi^i \psi^j \psi^k$ and K is its dimension 2 superpartner $K = \phi^{(3)}_{ijk} \psi^i \psi^k \partial \phi^k$. Similarly, the 4-from implies the existence of a dimension 2 current X and its dimension $\frac{5}{2}$ superpartner M. The chiral algebra of G_2 sigma models thus contains 4 extra currents on top of the two G, T that constitute the $\mathcal{N} = 1$ superconformal algebra. These six generators form a closed quantum algebra which appears explicitly e.g. in [9, 8, 10] (see also [11]).

An important fact, which will be crucial in almost all the remaining analysis, is that the generators Φ and X form a closed sub-algebra: if we define the supercurrent $G_I = \frac{i}{\sqrt{15}} \Phi$ and stress-energy tensor $T_I = -\frac{1}{5} X$ we recognize that this is the unique $\mathcal{N} = 1$ superconformal algebra of the minimal model with central charge $c = \frac{7}{10}$ known as the *Tri-critical Ising Model*. This sub-algebra plays a role similar to the $U(1)$ R-symmetry of the $\mathcal{N} = 2$ algebra in compactifications on Calabi-Yau manifolds.

In fact, with respect to the conformal symmetry, the full Virasoro algebra decomposes in two commuting[1] Virasoro algebras: $T = T_I + T_r$ with $T_I(z) T_r(w) = 0$. This means we can classify conformal primaries by two quantum numbers, namely its tri-critical Ising model highest weight and its highest weight with respect to T_r: $|\text{primary}\rangle = |h_I, h_r\rangle$.

Perhaps it is worth emphasizing the logic here: classically, we find a conformal algebra with six generators in sigma-models on manifolds of G_2 holonomy. In the quantum theory we expect, in the absence of anomalies other than the conformal anomaly, to find a quantum version of this classical algebra. In [9] all quantum extensions were analyzed, and a two-parameter family of quantum algebras was found. Requiring that the quantum algebra has the right central charge (necessary to have a critical string theory) and that it contains the tricritical Ising model (necessary for space-time supersymmetry) fixes the two-parameters. This motivates why this is the appropriate definition for string theory on G_2 manifolds.

3. Tri-Critical Ising Model

Unitary minimal models are labeled by a positive integer $p = 2, 3, \ldots$ and occur only on the "discrete series" at central charges $c = 1 - \frac{6}{p(p+1)}$. The Tri-Critical Ising Model is the second member ($p = 4$) which has central charge $c = \frac{7}{10}$. It is at the same time also a minimal model for the $\mathcal{N} = 1$ superconformal algebra.

[1]This decomposition only works for the Virasoro part of the corresponding $\mathcal{N} = 1$ algebras. The full $\mathcal{N} = 1$ structures do not commute. For example the superpartner of Φ with respect to the full $\mathcal{N} = 1$ algebra is K whereas its superpartner with respect to the $\mathcal{N} = 1$ of the tri-critical Ising model is X.

The conformal primaries of unitary minimal models are labeled by two integers $1 \leq n' \leq p$ and $1 \leq n < p$. The weights in this range are arranged into a "*Kac table*". The conformal weight of the primary $\Phi_{n'n}$ is $h_{n'n} = \frac{[pn'-(p+1)n]^2-1}{4p(p+1)}$. In the Tri-critical Ising model ($p = 4$) there are 6 primaries of weights $0, \frac{1}{10}, \frac{6}{10}, \frac{3}{2}, \frac{7}{16}, \frac{3}{80}$. Below we write the Kac table for the tricritical Ising model. Beside the Identity operator ($h = 0$) and the $\mathcal{N} = 1$ supercurrent ($h = \frac{3}{2}$) the NS sector (first and third columns) contains a primary of weight $h = \frac{1}{10}$ and its $\mathcal{N} = 1$ superpartner ($h = \frac{6}{10}$). The primaries of weight $\frac{7}{16}, \frac{3}{80}$ are in the Ramond sector (middle column).

$n' \backslash n$	1	2	3
1	0	$\mathbf{\frac{7}{16}}$	$\frac{3}{2}$
2	$\mathbf{\frac{1}{10}}$	$\mathbf{\frac{3}{80}}$	$\frac{6}{10}$
3	$\mathbf{\frac{6}{10}}$	$\frac{3}{80}$	$\frac{1}{10}$
4	$\mathbf{\frac{3}{2}}$	$\frac{7}{16}$	0

The Hilbert space of the theory decomposes in a similar way, $\mathcal{H} = \oplus_{n,n'} \mathcal{H}_{n',n}$. A central theme in this work relies on the fact that since the primaries $\Phi_{n'n}$ form a closed algebra under the OPE they can be decomposed into *conformal blocks* which connect two Hilbert spaces. Conformal blocks are denoted by $\Phi^{l',l}_{n',n,m'm}$ which describes the restriction of $\Phi_{n',n}$ to a map that only acts from $\mathcal{H}_{m',m}$ to $\mathcal{H}_{l',l}$.

An illustrative example, which will prove crucial in what follows, is the block structure of the primary $\Phi_{2,1}$ of weight $1/10$. General arguments show that the fusion rule of this field with any other primary $\Phi_{n'n}$ is $\phi_{(2,1)} \times \phi_{(n',n)} = \phi_{(n'-1,n)} + \phi_{(n'+1,n)}$. The only non-vanishing conformal blocks in the decomposition of $\Phi_{2,1}$ are those that connect a primary with the primary right above it and the primary right below in the Kac table, namely[2], $\phi^{n'-1,n}_{2,1,n',n}$ and $\phi^{n'+1,n}_{2,1,n',n}$. This can be summarized formally by defining the following decomposition[3]

$$\Phi_{2,1} = \Phi^{\downarrow}_{2,1} \oplus \Phi^{\uparrow}_{2,1}. \tag{3.1}$$

Similarly, the fusion rule of the Ramond field $\Phi_{1,2}$ with any primary is $\phi_{(1,2)} \times \phi_{(n',n)} = \phi_{(n,n-1)} + \phi_{(n',n+1)}$ showing that it is composed of two blocks, which we denote as follows $\Phi_{1,2} = \Phi^{-}_{1,2} \oplus \Phi^{+}_{1,2}$. Conformal blocks transforms under conformal transformations exactly like the primary field they reside in but are usually not single-valued functions of $z(\bar{z})$.

[2]Note the confusing notation where *down* the Kac table means *larger* n' and vice-versa.

[3]We stress that this decomposition is special to the field $\Phi_{2,1}$ and does not necessarily hold for other primaries which may contain other blocks.

3.1. CHIRAL PRIMARY STATES

The chiral-algebra associated with manifolds of G_2 holonomy[4] allows us to draw several conclusions about the possible spectrum of such theories. It is useful to decompose the generators of the chiral algebra in terms of primaries of the tri-critical Ising model and primaries of the remainder. The commutation relations of the G_2 algebra imply that the some of the generators of the chiral algebra decompose as [8]: $G(z) = \Phi_{2,1} \otimes \psi_{\frac{14}{10}}$, $K(z) = \Phi_{3,1} \otimes \psi_{\frac{14}{10}}$ and $M(z) = a\Phi_{2,1} \otimes \chi_{\frac{24}{10}} + b[X_{-1}, \Phi_{2,1}] \otimes \psi_{\frac{14}{10}}$, with ψ, χ primaries of the indicated weights in the T_r CFT and a, b constants.

Ramond ground states of the full $c = \frac{21}{2}$ SCFT are of the form $|\frac{7}{16}, 0\rangle$ and $|\frac{3}{80}, \frac{2}{5}\rangle$. The existence of the $|\frac{7}{16}, 0\rangle$ state living just inside the tricritical Ising model plays a crucial role in the topological twist. Coupling left and right movers, the only possible RR ground states compatible with the G_2 chiral algebra[5] are a single $|\frac{7}{16}, 0\rangle_L \otimes |\frac{7}{16}, 0\rangle_R$ ground state and a certain number of states of the form $|\frac{3}{80}, \frac{2}{5}\rangle_L \otimes |\frac{3}{80}, \frac{2}{5}\rangle_R$. By studying operator product expansions of the RR ground states we get the following "special" NSNS states $|0,0\rangle_L \otimes |0,0\rangle_R$, $|\frac{1}{10}, \frac{2}{5}\rangle_L \otimes |\frac{1}{10}, \frac{2}{5}\rangle_R$, $|\frac{6}{10}, \frac{2}{5}\rangle_L \otimes |\frac{6}{10}, \frac{2}{5}\rangle_R$ and $|\frac{3}{2}, 0\rangle_L \otimes |\frac{3}{2}, 0\rangle_R$ corresponding to the 4 NS primaries $\Phi_{n',1}$ with $n' = 1, 2, 3, 4$ in the tri-critical Ising model. Note that for these four states there is a linear relation between the Kac label n' of the tri-critical Ising model part and the total conformal weight $h_{total} = \frac{n'-1}{2}$. In fact, it can be shown that, similar to the BPS bound in the $\mathcal{N} = 2$ case, primaries of the G_2 chiral algebra satisfy a (non-linear) bound of the form

$$h_I + h_r \geq \frac{1 + \sqrt{1 + 80h_I}}{8}. \tag{3.2}$$

which is precisely saturated for the four NS states listed above. We will therefore refer to those states as "chiral primary" states. Just like in the case of Calabi-Yau, the $\frac{7}{16}$ field maps Ramond ground states to NS chiral primaries and is thus an analogue of the "spectral flow" operators in Calabi-Yau.

4. Topological Twist

To construct a topologically twisted CFT we usually proceed in two steps. First we define a new stress-energy tensor, which changes the quantum numbers of the fields and operators of the theory under Lorentz transformations. Secondly, we identify a nilpotent scalar operator, usually constructed out of the supersymmetry generators of the original theory, which we declare to be the BRST operator. Often this BRST operator can be obtained in the usual way by gauge fixing a suitable symmetry. If the new stress tensor is exact with respect to the BRST operator, observables (which are elements of the BRST cohomology) are metric independent and the theory is called topological. In particular, the twisted stress tensor should have a vanishing central charge.

[4]We loosely refer to it as "the G_2 algebra" but it should not be confused with the Lie algebra of the group G_2.

[5]Otherwise the spectrum will contain a 1-form which will enhance the chiral algebra. Geometrically this is equivalent to demanding that $b_1 = 0$.

In practice [23, 24], for the $\mathcal{N}=2$ theories, an n-point correlator on the sphere in the twisted theory can conveniently be defined[6] as a correlator in the *untwisted* theory of the same n operators plus two insertions of a spin-field, related to the space-time supersymmetry charge, that serves to trivialize the spin bundle. For a Calabi-Yau 3-fold target space there are two $SU(3)$ invariant spin-fields which are the two spectral flow operators $\mathcal{U}_{\pm\frac{1}{2}}$. This discrete choice in the left and the right moving sectors is the choice between the $+(-)$ twists [25] which results in the difference between the topological A/B models.

In [8] a similar expression was written down for sigma models on G_2 manifolds, this time involving the single G_2 invariant spin field which is the unique primary $\Phi_{1,2}$ of weight $\frac{7}{16}$. It was proposed that this expression could be a suitable definition of the correlation functions of a putative topologically twisted G_2 theory. In other words the twisted amplitudes are defined as[7]

$$\langle V_1(z_1)\dots V_n(z_n)\rangle_{\texttt{twist}} \equiv \langle\Sigma(\infty)V_1(z_1)\dots V_n(z_n)\Sigma(0)\rangle_{\texttt{untwist}}. \tag{4.1}$$

In [8] further arguments were given, using the Coulomb gas representation of the minimal model, that there exists a twisted stress tensor with vanishing central charge. This argument is however problematic, since the twisted stress tensor proposed there does not commute with Felder's BRST operators [26] and therefore it does define a bona fide operator in the minimal model. In addition, a precise definition of a BRST operator was lacking.

We will proceed somewhat differently. We will first propose a BRST operator, study its cohomology, and then use a version of (4.1) to compute correlation functions of BRST invariant observables. We will then comment on the extension to higher genus and on the existence of a topologically twisted G_2 string.

5. The BRST Operator

Our basic idea is that the topological theory for G_2 sigma models should be formulated not in terms of local operators of the untwisted theory but in terms of its (non-local)[8] conformal blocks.

By using the decomposition (3.1) into conformal blocks, we can split any field whose tri-critical Ising model part contains just the conformal family $\Phi_{2,1}$ into its *up* and *down* parts. In particular, the $\mathcal{N}=1$ supercurrent $G(z)$ can be split as

$$G(z) = G^{\downarrow}(z) + G^{\uparrow}(z). \tag{5.1}$$

We claim that $G^{\downarrow}$ is the BRST current and $G^{\uparrow}$ carries many features of an anti-ghost. The basic $\mathcal{N}=1$ relation

$$G(z)G(0) = \left(G^{\downarrow}(z)+G^{\uparrow}(z)\right)\left(G^{\downarrow}(0)+G^{\uparrow}(0)\right) \sim \frac{2c/3}{z^3} + \frac{2T(0)}{z} \tag{5.2}$$

[6] Up to proper normalization.

[7] Up to a coordinate dependent factor that we omit here for brevity and can be found in [7].

[8] It should be stressed that this splitting into conformal blocks is non-local in the simple sense that conformal blocks may be multi-valued functions of $z(\bar{z})$.

proves the nilpotency of this BRST current (and of the candidate anti-ghost) because the RHS contains descendants of the identity operator only and has trivial fusion rules with the primary fields of the tri-critical Ising model and so $(G^{\downarrow})^2 = (G^{\uparrow})^2 = 0$.

More formally, denote by $P_{n'}$ the projection operator on the sub-space $\mathcal{H}_{n'}$ of states whose tri-critical Ising model part lies within the conformal family of one of the four NS primaries $\Phi_{n',1}$. The 4 projectors add to the identity

$$P_1 + P_2 + P_3 + P_4 = 1, \tag{5.3}$$

because this exhaust the list of possible highest weights in the NS sector of the tri-critical Ising model[9]. We can now define our BRST operator in the NS sector more rigorously

$$Q = G^{\downarrow}_{-\frac{1}{2}} \equiv \sum_{n'} P_{n'+1} G_{-\frac{1}{2}} P_{n'}. \tag{5.4}$$

The nilpotency $Q^2 = 0$ is easily proved

$$Q^2 = \sum_{n'} P_{n'+2} G^2_{-\frac{1}{2}} P_{n'} = \sum_{n'} P_{n'+2} L_{-1} P_{n'} = 0, \tag{5.5}$$

where we could replace the intermediate $P_{n'+1}$ by the identity because of property 5.1 and the last equality follows since L_{-1} maps each $\mathcal{H}_{n'}$ to itself.

Q does not commute with the local operator $\mathcal{O}_{\Delta(1),0}$, $\mathcal{O}_{\Delta(2),\frac{2}{5}}$, $\mathcal{O}_{\Delta(3),\frac{2}{5}}$ and $\mathcal{O}_{\Delta(4),0}$ corresponding to the chiral states $|0,0\rangle$, $|\frac{1}{10},\frac{2}{5}\rangle$, $|\frac{6}{10},\frac{2}{5}\rangle$ and $|\frac{3}{2},0\rangle$ (for brevity we will denote those 4 local operators juts by their minimal model Kac label $\mathcal{O}_i$, $i = 1,2,3,4$). However, it can be checked that the following blocks,

$$\mathcal{A}_{n'} = \sum_{m} P_{n'+m-1} \mathcal{O}_{n'} P_m, \tag{5.6}$$

which pick out the maximal "down component" of the corresponding local operator, do commute with Q and are thus in its operator cohomology. Thus the chiral *operators* of the twisted model are represented in terms of the blocks 5.6 of the local operators corresponding to the chiral *states*. Furthermore, it can be shown easily that those chiral operator form a ring under the OPE.

By doing a calculation at large volume, we see that the BRST cohomology has one operator of type $\mathcal{O}_1 \otimes \mathcal{O}_1$, $b_2 + b_3$ of type $\mathcal{O}_2 \otimes \mathcal{O}_2$, $b_4 + b_5$ of type $\mathcal{O}_3 \otimes \mathcal{O}_3$, and one of type $\mathcal{O}_4 \otimes \mathcal{O}_4$. The total BRST cohomology is thus precisely given by $H^*(M)$. Also, one finds that the b_3 operators of type $\mathcal{O}_2 \otimes \mathcal{O}_2$ are precisely the geometric moduli of the G_2 target space[10].

In the topological G_2 theory, genus zero correlation functions of chiral primaries between BRST closed states are position independent. Indeed, the generator of translations on the plane, namely L_{-1}, is BRST exact

[9]For simplicity, we will set $P_{n'} = 0$ for $n' \leq 0$ and $n' \geq 5$, so that we can simply write $\sum_{n'} P_{n'} = 1$ instead of (5.3).

[10]In this work we ignore the b_2 moduli corresponding to the B-field.

$$\{G^{\downarrow}_{-\frac{1}{2}}, G^{\uparrow}_{-\frac{1}{2}}\} = \sum_{n'} P_{n'} G_{-\frac{1}{2}} (P_{n'-1} + P_{n'+1}) G_{-\frac{1}{2}} P_{n'} = \sum_{n'} P_{n'} L_{-1} P_{n'} = L_{-1}. \tag{5.7}$$

This is a crucial ingredient of topological theories.

Moreover, it can be shown [8] that the upper components $\tilde{G}_{-\frac{1}{2}}|\frac{1}{10}, \frac{2}{5}\rangle_L \otimes G_{-\frac{1}{2}}|\frac{1}{10}, \frac{2}{5}\rangle_R$ correspond to exactly marginal deformations of the CFT preserving the G_2 chiral algebra, completely in agreement with the identification of them as the geometric moduli of the theory. Focusing momentarily on the left movers, we can show that $[Q, \{G_{-\frac{1}{2}}, \mathcal{O}_2\}] = \partial \mathcal{A}_2$ [11] so that the very same deformation is physical, namely Q exact, also in the topological theory. Note that the deformation is given by a conventional operator that does not involve any projectors. Combining with the right-movers, we find that the deformations in the action of the topological string are exactly the same as the deformations of the non-topological string as is expected because both should exist on an arbitrary manifold of G_2 holonomy.

Most correlation functions at genus zero vanish. The most interesting one is the three-point function of three operators $Y = \mathcal{O}_2 \otimes \mathcal{O}_2$. These correspond to geometric moduli. If we introduce coordinates t_i on the moduli space of G_2 metrics, then we obtain from a large volume calculation

$$\langle Y_i Y_j Y_k \rangle = \int_M d^7x \sqrt{g} \phi_{abc} \frac{\partial g^{aa'}}{\partial t_i} \frac{\partial g^{bb'}}{\partial t_j} \frac{\partial g^{cc'}}{\partial t_k} \phi_{a'b'c'}. \tag{5.8}$$

One might expect, based on general arguments, that this is the third derivative of some prepotential if suitable 'flat' coordinates are used. We do not know the precise definition of flat coordinates for the moduli space of G_2 metrics, but if we take for example $M = T^7$ and take coordinates such that ϕ is linear in them, then we can verify

$$\langle Y_i Y_j Y_k \rangle = -\frac{1}{21} \frac{\partial^3}{\partial t_i \partial t_j \partial t_k} \int \phi \wedge *\phi. \tag{5.9}$$

The prepotential appearing on the right hand side is exactly the same as the action functional introduced by Hitchin in [27, 28]. A similar action was also used as a starting point for topological M-theory in [3] (see also [2]). This strongly suggests that our topological G_2 field theory is somehow related to topological M-theory.

6. Topological G_2 Strings

In the case of $N = 2$ theories, the computation of correlation functions at genus zero outlined above can be generalized to higher genera [23, 24]. An n-point correlator on a genus-g Riemann surface in the twisted theory can be defined as a correlator in the untwisted theory of the same n operators plus $(2 - 2g)$ insertions of the spin-field that is related to the space-time supersymmetry charge. For a Calabi-Yau 3-fold target space on a Riemann surface with $g > 1$, the meaning of the above prescription is to insert $2g - 2$ of the conjugate spectral flow operator.

[11] Recall that $\mathcal{A}_2$ was defined in (5.6).

To generalize this to the G_2 situation, we would like to have something similar. However, there is only one G_2 invariant spin-field. This is where the decomposition in conformal blocks given in section 3 is useful: the spin-field $\Phi_{2,1}$ could be decomposed[12] in a block $\Phi^+_{2,1}$ and in a block $\Phi^-_{2,1}$. At genus zero we needed two insertions of $\Phi^+_{2,1}$, so the natural guess is that at genus g we need $2g-2$ insertions of $\Phi^-_{2,1}$. However, this is not the full story. We also need to insert $3g-3$ copies of the anti-ghost and integrate over the moduli space of Riemann surfaces to properly define a topological string theory. The anti-ghost is very close to $G^\uparrow$, and the fusion rules of the tri-critical Ising model tell us that there is indeed a non-vanishing contribution to correlation functions of $2g-2$ $\Phi^-_{2,1}$'s and $3g-3$ $G^\uparrow$.

This prescription would therefore work very nicely if we would have found the right anti-ghost. The candidates we tried so far all seem to fail in one way or another. One possible conclusion might be that a twisted stress tensor does not exist and that there is only a sensible notion of topological G_2 sigma models but not of topological G_2 strings. The fact that so far so many properties of the $\mathcal{N}=2$ topological theories appeared to hold also in our G_2 model leads us to believe that a sensible extension to higher genera indeed exists. Identifying the correct twisted stress tensor remains an open problem. Barring this important omission the coupling to topological gravity goes pretty much along the same lines as for the $\mathcal{N}=2$ topological string (details can be found in [7]).

7. Conclusions

An important application of topological strings stems from the realization [23, 24, 25] that its amplitudes agree with certain amplitudes of the physical superstring. Just like $\mathcal{N}=2$ topological strings compute certain F-terms in four dimensional $\mathcal{N}=2$ gauge theories, one might wonder whether the G_2 topological string similarly computes F-terms in three dimensional $\mathcal{N}=2$ gauge theories.

Since G_2 manifolds are Ricci-flat we can consider compactifying the type II superstring on $R^{1,2}\times\mathcal{N}_7$ where $\mathcal{N}_7$ is a 7 dimensional manifold of G_2 holonomy. This reduces the supersymmetry down to two real supercharges in 3 dimension from each worldsheet chirality so we end up with a low energy field theory in 3 dimensions with $\mathcal{N}=2$ supergravity. By studying amplitudes in some detail we observe that, except perhaps at genus zero, the amplitudes do involve a sum over conformal blocks, and the topological G_2 string therefore seems to compute only one of many contributions to an amplitude. This is in contrast to the four dimensional case where the possibility to look only at the (anti)-self dual gravitons allowed to isolate the topological contributions.

Nevertheless, we believe that topological G_2 strings are worthwhile to study. They may possibly provide a good definition of topological M-theory, and a further study may teach us many things about non-topological G_2 compactifications as well. We leave this, as well as the generalization to $spin(7)$ compactifications [29] and the study of branes in these theories to future work.

[12] In terms of the Coulomb gas representation, one of these can be represented as an ordinary vertex operator, the other one involves a screening charge.

Acknowledgements

It is a pleasure to thank Nathan Berkovits, Volker Braun, Lorenzo Cornalba, Robbert Dijkgraaf, Pietro Antonio Grassi, Wolfgang Lerche, Hiroshi Ooguri, Volker Shomerus, Annamaria Sinkovics, Kostas Skenderis and Cumrun Vafa for useful discussions. We would also like to thank the other organizers of the Cargèse 2004 summer school. This work was partly supported by the stichting FOM.

References

1. A. Neitzke and C. Vafa, "N = 2 strings and the twistorial Calabi-Yau," arXiv:hep-th/0402128; N. Nekrasov, H. Ooguri and C. Vafa, "S-duality and topological strings," JHEP **0410**, 009 (2004) [arXiv:hep-th/0403167].
2. A. A. Gerasimov and S. L. Shatashvili, "Towards integrability of topological strings. I: Three-forms on Calabi-Yau manifolds," JHEP **0411**, 074 (2004) [arXiv:hep-th/0409238].
3. R. Dijkgraaf, S. Gukov, A. Neitzke and C. Vafa, "Topological M-theory as unification of form theories of gravity," arXiv:hep-th/0411073.
4. N. Nekrasov, "A la recherche de la m-theorie perdue. Z theory: Chasing m/f theory," arXiv:hep-th/0412021.
5. P. A. Grassi and P. Vanhove, "Topological M theory from pure spinor formalism," arXiv:hep-th/0411167.
6. L. Anguelova, P. de Medeiros and A. Sinkovics, "On topological F-theory," arXiv:hep-th/0412120.
7. J. de Boer, A. Naqvi and A. Shomer, "The Topological G_2 String", to appear.
8. S. L. Shatashvili and C. Vafa, "Superstrings and manifold of exceptional holonomy," arXiv:hep-th/9407025.
9. R. Blumenhagen, "Covariant construction of N=1 superW algebras," Nucl. Phys. B **381**, 641 (1992).
10. J. M. Figueroa-O'Farrill, "A note on the extended superconformal algebras associated with manifolds of exceptional holonomy," Phys. Lett. B **392**, 77 (1997) [arXiv:hep-th/9609113].
11. D. Gepner and B. Noyvert, "Unitary representations of SW(3/2,2) superconformal algebra," Nucl. Phys. B **610**, 545 (2001) [arXiv:hep-th/0101116]; B. Noyvert, "Unitary minimal models of SW(3/2,3/2,2) superconformal algebra and manifolds of G(2) holonomy," JHEP **0203**, 030 (2002) [arXiv:hep-th/0201198];
12. B. S. Acharya, "N=1 M-theory-Heterotic Duality in Three Dimensions and Joyce Manifolds," arXiv:hep-th/9604133.
13. B. S. Acharya, "On mirror symmetry for manifolds of exceptional holonomy," Nucl. Phys. B **524**, 269 (1998) [arXiv:hep-th/9707186].
14. T. Eguchi and Y. Sugawara, "CFT description of string theory compactified on non-compact manifolds with G(2) holonomy," Phys. Lett. B **519**, 149 (2001) [arXiv:hep-th/0108091].
15. R. Roiban and J. Walcher, "Rational conformal field theories with G(2) holonomy," JHEP **0112**, 008 (2001) [arXiv:hep-th/0110302].
16. T. Eguchi and Y. Sugawara, "String theory on G(2) manifolds based on Gepner construction," Nucl. Phys. B **630**, 132 (2002) [arXiv:hep-th/0111012].
17. R. Blumenhagen and V. Braun, "Superconformal field theories for compact G(2) manifolds," JHEP **0112**, 006 (2001) [arXiv:hep-th/0110232]; "Superconformal field theories for compact manifolds with Spin(7) holonomy," JHEP **0112**, 013 (2001) [arXiv:hep-th/0111048].
18. R. Roiban, C. Romelsberger and J. Walcher, "Discrete torsion in singular G(2)-manifolds and real LG," Adv. Theor. Math. Phys. **6**, 207 (2003) [arXiv:hep-th/0203272].
19. K. Sugiyama and S. Yamaguchi, "Coset construction of noncompact Spin(7) and G(2) CFTs," Phys. Lett. B **538**, 173 (2002) [arXiv:hep-th/0204213].
20. T. Eguchi, Y. Sugawara and S. Yamaguchi, "Supercoset CFT's for string theories on non-compact special holonomy manifolds," Nucl. Phys. B **657**, 3 (2003) [arXiv:hep-th/0301164].

21. M. R. Gaberdiel and P. Kaste, "Generalised discrete torsion and mirror symmetry for G(2) manifolds," JHEP **0408**, 001 (2004) [arXiv:hep-th/0401125].
22. B. S. Acharya and S. Gukov, "M theory and Singularities of Exceptional Holonomy Manifolds," Phys. Rept. **392**, 121 (2004) [arXiv:hep-th/0409191].
23. M. Bershadsky, S. Cecotti, H. Ooguri and C. Vafa, "Kodaira-Spencer theory of gravity and exact results for quantum string amplitudes," Commun. Math. Phys. **165**, 311 (1994) [arXiv:hep-th/9309140].
24. I. Antoniadis, E. Gava, K. S. Narain and T. R. Taylor, "Topological amplitudes in string theory," Nucl. Phys. B **413**, 162 (1994) [arXiv:hep-th/9307158].
25. E. Witten, "Mirror manifolds and topological field theory," arXiv:hep-th/9112056.
26. G. Felder, "Brst Approach To Minimal Methods," Nucl. Phys. B **317**, 215 (1989) [Erratum-ibid. B **324**, 548 (1989)].
27. N. Hitchin, "The geometry of three-forms in six and seven dimensions," arXiv:math.dg/0010054.
28. N. Hitchin, "Stable forms and special metrics," arXiv:math.dg/0107101.
29. J. de Boer, A. Naqvi and A. Shomer, "Topological Strings on Exceptional Holonomy Manifolds", to appear.

CHALLENGES OF MATRIX MODELS

ALEXEI MOROZOV
ITEP, Moscow, Russia

1. Introduction

Matrix models appear again and again at the front line of theoretical physics, and every new generation of scientists discovers something new in this seemingly simple subject. It is getting more and more obvious that matrix models capture the very essence of general quantum field theory and provide the crucial representative example for the **string theory** [1]: matrix models may play the same role for string theory as the harmonic oscillator for quantum mechanics. An essential difference is, however, that we have a nearly exhaustive understanding of harmonic oscillator, while nothing like a complete description of matrix models is yet available. Probably, the time is coming to begin a systematic analysis with the goal of building up a theory of matrix model partition functions as the first special functions of string theory [2]. Various applications of matrix model techniques should use these functions as building blocks for the formulation of their results, thus separating the physical content of different applications from a common mathematical formalism. The goal can be to build up an analogue of the powerful free-field formalism, developed in 1980's, which allowed to reduce many problems in **perturbative string theory** [3] (and related issues in representation theory of Kac-Moody algebras, $2d$ conformal field theory and finite-zone solutions in integrable-systems theory) to an almost classical set of special functions: Riemann's theta-functions [4, 5], associated with Riemann surfaces. The problem is that matrix model partition functions are non-trivial generalization of Riemann's theta-functions, which were not fully investigated in XIX-th century, therefore the theory should involve essential new ideas, both in physics and in mathematics. This is the reason why the progress in this field – actually known under the nick-name of **non-perturbative string theory** – has been considerably slower. More and more people begin to realize that generic problems of non-perturbative string theory, i.e. the general theory of phase transitions, have a lot in common, and these most interesting common properties are clearly represented at the simplest level of matrix models. Complications introduced by the sophisticated context of particular applications can and should be separated from the important content, captured at the matrix model level.

In the development of matrix model theory one can distinguish several stages. In fact, these stages are typical for every chapter of the string theory, and they characterize not

L. Baulied et al. (eds.), String Theory: From Gauge Interactions to Cosmology, 129–162.

so much the history, but the kind of questions that are posed and the level of abstraction analyzing these questions. Whenever a new problem arises (or is re-addressed again), it is quite useful to understand its place in this hierarchy.

2. Classical period: introduction of models

During the *classical period* the main task was to *study* concrete phenomena, involving matrix models. Since in these notes we are not going to discuss applications, we just mention a few crucial theoretical methods, developed at this stage.

The most important idea was to reduce the N^2-fold matrix integrals like

$$Z(t|N) = \frac{1}{\mathrm{Vol}(U(N))} \int_{N\times N} d\Phi \exp\left(\sum_k t_k \mathrm{Tr}\, \Phi^k\right) \tag{2.1}$$

to N-fold integrals over eigenvalues ϕ_i of matrices $\Phi = U^\dagger \mathrm{diag}(\phi_i) U$. This idea, closely related in general context to the problem of gauge invariance, is technically based on the possibility to explicitly integrate over "angular variables" in the simplest situations, of which (2.1) is a representative example. The single known exactly solvable generalization, which goes slightly beyond this example, involves the Itzykson-Zuber integral [7]

$$\int_{N\times N} [dU] \exp(\mathrm{Tr} A U^\dagger B U), \tag{2.2}$$

which is used in construction of two important classes of matrix theories: generalized Kontsevich model [8, 9],

$$Z_{GKM}(L|W) \sim \int_{m\times m} dM \exp\left[\mathrm{tr}\left(W(M) + LM\right)\right] \tag{2.3}$$

and even further generalized Kazakov-Migdal-Kontsevich model [10, 11],

$$Z_{G(KM)^2}(\{L_\mu\}|\{W_\alpha\}) \sim \prod_{\mu,\alpha} \int_{m\times m} dM^{(\alpha)} [dU^{(\alpha\beta)}][dU^{(\alpha\mu)}] \cdot$$
$$\cdot \exp\left[\mathrm{tr}\left(W_\alpha(M_\alpha) + M_\alpha U^\dagger_{\alpha\beta} M_\beta U_{\alpha\beta} + L_\mu U^\dagger_{\alpha\mu} M_\alpha U_{\alpha\mu}\right)\right], \tag{2.4}$$

defined for any graph with Hermitian fields M_α and background fields L_μ standing in the vertices and unitary fields $U_{\alpha\beta}$ and $U_{\alpha\mu}$ on the links. In more general situations (sometime quite important for applications, e.g. involving non-trivial multi-brane backgrounds), angular integration is highly non-trivial and remains an open problem. The simplest of such difficulties are echoed in the Gribov copies problem in Yang-Mills theory, while treatment of more sophisticated unitary-matrix integrals face problems, typical for adequate treatment of quantum gravity. The class of theories where gauge (angular) variables can be integrated out in effective way is referred to as **eigenvalue models** [6]. Most of further development concerns eigenvalue models, which until recently remained implicitly synonymous to matrix models.

The next idea of the *classical period* used the fact, that angular integration usually provides non-trivial, but very special measures on the space of eigenvalues, which is often made from Van-der-Monde determinants [12]. For example, (2.1) can be transformed into

$$Z(t|N) = \prod_{i=1}^{N} \int d\phi_i \exp\left(\sum_k t_k \phi_i^k\right) \prod_{i<j}^{N} (\phi_i - \phi_j)^2, \tag{2.5}$$

where $\prod_{i<j}^{N} (\phi_i - \phi_j)^2 = \left(\det_{i,j} \phi_i^{j-1}\right)^2$ can be also considered as discriminant of auxiliary polynomial $P_N(z) = \prod_{i=1}^{N}(z - \phi_i)$. Occurrence of determinants suggested two alternative technical methods of investigation of eigenvalue models: technique of orthogonal polynomials [13] and free-fermion representation [13, 14]. Occurrence of discriminants was not fully exploited yet, though it opens a set of interesting possibilities, both technically and conceptually.

The last idea of the *classical period* which needs to be mentioned, is the idea of **continuum limit**, when the matrix size N tends to infinity. In fact there are infinitely many different continuum limits – relevant for different particular applications, – and only very few have been analyzed so far. However, from the very beginning the main fact was broadly realized: continuum limits of matrix models posses description in terms of Riemann surfaces [15, 16]. Today we know that occurence of spectral surfaces is a general phenomenon, and they occur already at finite values of N [2].

3. First stringy period: generalizations and hunt for structures

The *first stringy period* is characterized by the change of interests: from tools to theory. Instead of studying various applications and developing adequate technical methods to answer the problems, posed by these applications, attention gets concentrated on the search and understanding of internal structures. Instead of *"study"* the main slogan becomes *"deform and generalize"* – this is the standard string theory method of revealing hidden structures. It is at this stage that the three main inter-related structures were discovered behind eigenvalue matrix models: rich Ward identities, integrability and CFT representations.

3.1. WARD IDENTITIES

Occurence of **Ward identities** is the pertinent property of every integral (and thus of *quantum mechanics* and all its generalizations, like quantum field and string theory): they reflect invariance of the integral under the change of integration variable – an archetypical example of auxiliary field. However, this obvious hidden symmetry manifests itself in a rather sophisticated manner: as relation between various correlators in the theory. String theory normally deals with partition functions: the generating functions of all the correlators, summed up with the coefficients like t_k in (2.1), which have the meaning of extra coupling constants, and can be considered as providing the deformation of the bare action. This formalism allows to treat Ward identities as *equations* for partition function, because the change of integration variables can be compensated by the change of the coupling constants, if there are many enough [6]. In particular case of the integral (2.1) the equations are known as **Virasoro constraints** [17] or **loop equations** [15],

$$\hat{L}_-(z)Z(t|N) = 0 \tag{3.1}$$

where

$$\hat{L}_-(z) = \sum_{m\geq -1} z^{-m-2}\hat{L}_m = \mathcal{P}_-\left(\sum_{m=-\infty}^{\infty} z^{-m-2}\hat{L}_m\right) = \mathcal{P}_-\left(\frac{1}{2}(\partial\hat{\phi}(z))^2\right)$$

$$\hat{\phi}(z) = \sum_{k\geq 0}\left(t_k z^k - \frac{1}{2kz^k}\frac{\partial}{\partial t_k}\right)$$

For the integral (2.3) the Ward identities can be written in two forms: either as a Gross-Newman equation [18],

$$\left(W'\left(\partial/\partial L_{tr}\right) + L\right) Z_{GKM}(L|W) = 0, \tag{3.2}$$

which generalizes a similar equation [6] for the Itzykson-Zuber integral (2.2), or as a set of peculiar $\mathcal{W}_n$-constraints (where $n = \deg W'(z)$) [19]. A particular case of these equations in the case of $n = 2$ is **continuum Virasoro constraint** [20, 21],

$$\hat{\mathcal{L}}_{2m} Z_{GKM}\left(L\left|\frac{1}{3}M^3\right.\right) = 0, \quad m \geq -1, \tag{3.3}$$

$$\hat{\mathcal{L}}_{2m} = \frac{1}{2}\sum_{\text{odd } k=1}^{\infty} kt_k\frac{\partial}{\partial t_{k+2m}} + \frac{1}{4}\sum_{\text{odd } k=1}^{2m-1}\frac{\partial^2}{\partial t_k \partial t_{2m-k}} + \frac{1}{16}\delta_{m,0} + \frac{1}{4}t_1^2\delta_{m,-1}$$

where $t_k = r_k + \frac{1}{k}\mathrm{tr}L^{-k/n}$ and $r_k = \frac{n}{k(n-k)}\mathrm{res}\left(W'(z)\right)^{1-k/n}dz = -\frac{2}{3}\delta_{k,3}$. See [6] for detailed discussion of Ward identities for eigenvalue models and [22] for more technicalities, related to generalized Kontsevich model.

3.2. INTEGRABILITY AND RG EVOLUTION

Integrability of matrix models means that partition functions satisfy not only linear equations (Ward identities), but also bilinear Hirota-type equations,

$$\Delta(Z \otimes Z) = 0.$$

Technically, the proofs rely upon determinant representations of eigenvalue models, which are, in turn, immediate corollaries of determinant structures of integration measures, see [6] and references therein. However, the true meaning of integrability remains obscure. Several inter-related ideas should be somehow unified to clarify the issue.

First, integrability should express the fact that the system of Ward identities is rich enough: enough to specify the partition function almost unambiguously – up to some easily controllable degrees of freedom, a sort of *zero modes* with some clear cohomological interpretation. See [23] for the relevant notions of *strong* and *week completeness*.

Second, bilinear equations normally involve the matrix size N in non-trivial way: for example, the simplest bilinear equation for partition function (2.1) – the lowest term of the Toda chain hierarchy – states [24] that

$$Z_N \frac{\partial^2 Z_N}{\partial t_1^2} - \left(\frac{\partial Z_N}{\partial t_1}\right)^2 = Z_{N+1} Z_{N-1} \tag{3.4}$$

This means that bilinear equations involve not only variations of the coupling constant t_k, but also those of conjugate variables, like N. However, in (2.1) only N – a conjugate variable for t_0 (in the sense that $\partial \log Z_N / \partial t_0 = N$) – is present, while there is nothing like conjugate parameters for all other t's. This can suggest that Hirota equations for such restricted partition function are too special to reveal their general structure. In particular, (3.4) holds only for the simplest phase of the model (2.1), other phases belong to multi-component generalizations of Toda chain hierarchy.

Third, while coupling constants parameterize the **bare action**, i.e. the weight of summation over paths in functional integral, the conjugate variables should rather characterize the integration domain (range of integration, boundary conditions, the target space – whatever formalism one prefers to use in introducing them). In this sense conjugate variables are intimately related to generalized renormalization group flows, which are supposed to do exactly the same: describe the dependence of functional integral on the integration domain.

Relation between **integrability** and **renormalization group** is one of the main open problems in modern theoretical physics [1, 6, 25]. To get fully related, both these concepts should be considerably modified. To become a pertinent feature of all partition functions, integrability should not be restricted to ordinary τ-functions [26], associated with free fermions and single-loop Kac-Moody algebras at level $k = 1$. Indeed, τ-functions and bilinear Hirota-like equations can be defined for arbitrary Kac-Moody and even more general Lie algebras and quantum groups [27]. However, the theory of such τ-functions is not reducible to that of Plucker (free-fermion) determinants and requires the full-scale application of the free-field formalism [3] and more sophisticated determinant formulas. Expressions of this type are expected to arise in the study of unitary matrix integrals, but only first steps are being done in this direction [28, 29].

Renormalization group theory also requires considerable generalizations, of which we put especially emphasize on two. First, as already mentioned, it should allow arbitrary variations of integration domain, not just a one-parameter cut-off procedure: renormalization group should study the changes of the *shape* of integration domain, not just of its *volume*. This seems to be already a widely accepted generalization. Second, renormalization group theory should be made applicable to the study of fractal structures, which often reproduce themselves (exactly or approximately) at different scales. This means that the theory should allow non-trivial, periodic and perhaps even chaotic, renormalization group flows [30, 31], and this should not contradict the obvious uni-directional nature of these flows, generated by integrating out degrees of freedom. This entropy-like feature of renormalization group is usually expressed through the **c-theorem** [32], but in the case of systems with infinitely many degrees of freedom the c-function can easily have angular nature, and monotonic decrease of c does not contradict the existence of periodic motions. Besides the obvious example of self-similar fractals, today we already have some field-theory examples of such behaviour [33] – but only at the level of *flows*. It is not so easy to provide examples of *partition functions* (not just flows) with such properties, and the reason for this is simple: we usually consider as *solvable* the systems where answers can be expressed through conventional special functions, and these – almost by definition – have oversimplified branching structure. We do not posses any language to describe more general – and more realistic situations. At the same time such language should exist, because what seems to be sophisticated (fractal and/or chaotic) phase structure is in fact a highly ordered and pure algebraic pattern [34], not so much different from the algebraic-geometry background under the theory of conventional special functions. The jump between *order* (integrability) and *chaos* (infinite-tree structure of bifurcations or phase transitions) can be not so big, and most probably the careful study of partition functions already at the matrix-model level will reveal the deep inter-relation (duality) between the two concepts.

This duality should play a big role in **landscape theory**, which studies the distributions of various algebro-geometric quantities on moduli spaces (either of coupling constants or on the dual space of different branches) and their interplay with renormalization group flows. The crucial hidden double-loop structure, which is supposed to become the basis of such theory [35], should be seen already at the level of matrix models. For old and new attempts to apply – the yet undeveloped – landscape theory to phenomenology see [1] and [36] respectively.

3.3. RELATION TO CONFORMAL THEORY

The third structure behind eigenvalue matrix models is that of $2d$ conformal field theories: partition functions of matrix models coincide with certain correlators in conformal models. This fact is of course related with the free-fermion and KP τ-function representations of simplest partition functions, but it was quickly realized to have a more general meaning [37]: CFT representations are also closely related to Ward identities. In the simplest example of (3.1) the natural observation is that $\frac{1}{2}(\partial\hat{\phi})^2(z)$ looks like the stress tensor $T(z)$ of the $2d$ free scalar field, and therefore one can look for an intertwiner, relating operator $\hat{L}(z)$ in (3.1) with $T(z) = \frac{1}{2}(\partial\varphi)^2(z)$ in the free field theory. The intertwiner in this case is just

$$\langle 0|e^{\oint v(x)\partial\varphi(x)},$$

where $\varphi(z)$ is the $2d$ scalar, $\langle 0|$ – the left vacuum, annihilated by $\varphi_+(z)$, $\langle 0|\varphi_+(z) = 0$, and $v(z) = \sum_k t_k z^k$. Obviously,

$$\hat{L}(z)\langle 0|e^{\oint v(x)\partial\varphi(x)} = \langle 0|e^{\oint v(x)\partial\varphi(x)}\frac{1}{2}(\partial\varphi)^2(z),$$

and acting on the right vacuum $(\partial\varphi)^2_-(z)|0\rangle = 0$, so that

$$\hat{L}_-(z)\langle 0|e^{\oint v(x)\partial\varphi(x)}Q|0\rangle = 0. \tag{3.5}$$

Here Q is inserted in order to make matrix element non-trivial, and the only requirement is that Q commutes with $T(z) = \frac{1}{2}(\partial\varphi)^2(z)$. Such Q can be made out of the *screening charges* $Q_\pm = \oint e^{\pm\sqrt{2}\varphi(x)}$, and $Z(t|N)$ in (2.1) is reproduced if $Q = Q_+^N$. Many generalizations are straightforward, among them the important class of quiver matrix models [37, 38]. An immediate desire is to find a representation of this kind for Kontsevich τ-function – solution of continuum Virasoro constraint (3.3), but this problem appears surprisingly complicated and still remains open, see [39, 40] for important but still insufficient steps towards its solution.

4. Transcendental period: absolutization of structures

The *transcendental* or *second stringy* period is characterized by *absolutization of structures*, revealed at the previous stage. This means that the logic is inverted: now the *structures* are given, and we look for generic objects, possessing such structures – expecting in advance that some of these objects can appear different from original matrix integrals. Today we are mostly at this stage, at least in the theory of matrix models, and only the first careful steps are being done in attempts to understand the hidden meaning of the structures that we observed in applications – and continue to observe again and again in new physical systems when they come to our attention.

Discussion of generic CFT correlators as well as of generic τ-functions is – old and important – but still not a very constructive problem. It lies at the very core of landscape theory, but existing theoretical methods are too week to address the problem directly. The structure which can be rather effectively analyzed by available tools or by their straightforward modifications is that of Ward identities. Working on this structure one can also hope to develop stronger methods, applicable to analysis of the other two structures. To mention just one open question, illustrating the degree of ignorance in this field, it remains unclear to what extent the Virasoro constraints (3.1) and (3.3) *per se*, without explicit integral representations like (2.1), imply integrable structure of partition function.

4.1. WARD IDENTITIES

Defining partition function as solution to Ward identities is a typical D-module-style definition, i.e. basically a problem from *linear algebra*, since Ward identities are *linear* equations (this is what makes examination of this structure much simpler than integrable one, related to quadratic equations). Since we do not expect any mysteries in linear algebra (or, better to say, we think that we already know all of them), the theory building is straightforward, but by no means trivial: we know what to search for, but exhaustive and explicit solution can be quite tedious. We are not yet at the stage when solution of a linear-algebra problem, especially infinite-dimensional, and all its properties can be predicted *a priori*.

Normally, the set of questions one asks about the solution consists of several groups.

4.1.1. *Domain of definition*

First of all, one should decide what class of functions one wants the solution to belong to. Basically, there are two principal alternatives: formal series and globally defined functions. However, even if one chooses formal series, it is still necessary to specify, where *singularities* are allowed to occur. The simplest alternative is between series in positive powers of a variable (singularities are allowed at infinity of the Riemann sphere) and in all integer powers (singularities allowed at infinity and at zero). However, there are many more possibilities: singularities can be allowed at other points (then negative powers of $[z - a]$ can be allowed), the underlying *bare* Riemann surface need not be a sphere (then certain fractional powers of $[z - a]$ are allowed) etc. More than this, different variables can have different kinds of singularities and we have infinitely many such variables. In different applications different requirements are imposed, and one and the same quantity – unique partition function = generic solution to Ward identities – can show up in absolutely different way: different **branches** are relevant for different applications. What can look like a singularity from the point of view of particular application, can be nothing but a branching point or **phase transition** to another branch, which acquires a natural interpretation in terms of an absolutely different application. Once again this emphasizes that the problem of partition functions – even one particular partition function, one particular matrix model D-module,– is almost undistinguishable from the central problem of the string theory [1], which can be formulated as a search of unique universal partition function (a *universal object* of quantum field theory), of which all other thinkable partition functions are particular branches and sub-families. Returning closer to the Earth, even speaking about formal series one – explicitly or implicitly – imposes certain requirements on allowed singularities of solutions, which often do not follow from the Ward identities themselves. Actually, the "rich" Ward identities, that we spoke about in the previous section, fix everything but the behaviour at singularities, they are rich enough to fully constrain the dependence on the choice of the action (on coupling constants), but not rich enough to fix conjugate dependencies, e.g. boundary conditions at possible singularities: in certain sense they fix exactly one half of possible freedom. This is a typical thing for the equation on a wave function in quantum mechanics to do (constrain q-dependence but ignore p, or in other words, there are many solutions of the same Shroedinger equation, differing, say, by the values of energy): this observation is a starting point on the road, leading to identification of Ward identities with Shroedinger-type equation and the partition function with a (string-field theory) wave function.

Convenient way to speak about emerging freedom is in terms of "globally defined" functions: to fix it one can ask to solve Ward identities for functions on Riemann sphere or on torus or anywhere else. The problem is that in most interesting cases – and matrix model partition functions are not an exception – singularities are unavoidable. Moreover, small relaxations of compactness, like "functions with simple poles", which were enough for the formalism of free fields (and thus in dealing with arbitrary $2d$ conformal models), now are not sufficient: τ-functions are different from conformal blocks because they have essential singularities, even if we take those defined on punctured Riemann surfaces of finite genus. Actually, things are even worse: interesting partition functions require this genus to be infinite, if at all defined. Therefore this language – though widely used – by itself does not provide unambiguous classification of solutions.

4.1.2. *Generating functions*

The next question – after the set of allowed functions is somehow specified – is how to represent the answer. There are few chances that the full answer will be any known special function: as already stated, the most important thing is that matrix models produce *new* special functions. What is important, however, these new functions are not *too* new, they are just one step forward more involved than the known special functions, namely the Riemann's theta-functions, which occupy the previous level of complexity. We expect that all kinds of simplifications of matrix model partition functions reduce them to quantities, already expressible through Riemann's theta functions. What can be these simplifications? Instead of considering a generating function of *all* correlators, one can select some one-parametric family, and such reduced generating function can be a candidate. One can take one or another limit (say, one or another continuum limit: naive, double-scaling, fixed genus, ...). One can take a ratio of different branches (a kind of monodromy matrix) or some more sophisticated combination of those.

4.1.3. *Different realizations*

As usual for a D-module, one can look for solutions of linear equations in different forms.

Integral formulas like (2.1) are particular examples of one possible – integral – representation (or *realization*, to avoid confusion with group-theory representations) of solutions. For integral realizations the Ward identities play a role of Picard-Fucks equations: solutions are periods of the form which is converted into a full derivative by application of the corresponding operator (which generates Ward identity). Of course, in such situation one can integrate along any closed contour and different non-homological contours give rise to different solutions. Therefore, when one calls eq.(2.1) Hermitian one matrix model, "Hermitian" actually refers to the measure, $d\Phi = \prod_{i,j=1}^{N} d\Phi_{ij}$ (dictated, in its turn, by the norm $||\delta\Phi||^2 = \mathrm{Tr}\,\delta\Phi^2$), but not to the integration contour: when one defines partition function as solution of the Virasoro constraints, there is no reason to integrate in (2.1) over the contour $\Phi_{ij}^* = \Phi_{ji}$, associated with Hermitian matrices. Moreover, since D-module is defined by linear equations, the superposition principle is applicable, and a sum of any two solutions is again a solution. If matrix integrals like (2.1) with different N satisfy the same Ward identities like Virasoro constraints (3.1), then linear combinations of integrals with different N still are solutions. This means that the matrix size N can be also interpreted as characteristic of integration contour, moreover, this characteristic is not obligatory a positive integer, but can be also negative, rational and even complex-valued.

Differences between possible integral realizations of one and the same partition function are not exhausted by freedom in the choice of integration contours, even with extended interpretation of the term "contour". Partition function can be represented by absolutely different classes of matrix models. The simplest important example is provided by the model (2.1): it can be represented not only by (2.1), but also by a Gaussian integral [41] from Kontsevich family (2.3) with $t_k = \frac{1}{k}\mathrm{tr}L^{-k}$ [42, 6]:

$$Z(t|N) \sim Z_{GKM}\left(L\,\middle|\,\frac{1}{2}M^2 + N\log M\right) \sim$$
$$\sim \int_{m\times m} dM (\det M)^N \exp\left[\mathrm{tr}\left(\frac{1}{2}M^2 + LM\right)\right] \tag{4.1}$$

In this representation analytic continuation to non-integer values of N is even more straightforward. Again, linear combinations of solutions with different values of N, are still solutions to the Virasoro constraints (3.1), and all of them should be considered as particular branches of the 1-matrix model partition function. This provides additional information about the problem of conjugate variables: N itself can serve as one of them, or the coefficients of above-mentioned linear combinations can – changing the former for the latter is a kind of Fourier transform in the space of conjugate variables.

Possibility to represent one and the same partition function by members of two different matrix model families, by (2.1) and by (4.1), is a typical example of *duality* between different families of quantum field theories. One more duality follows from existence of CFT representations for the same partition function, it relates these matrix models to $2d$ free scalar theory and – through one more duality – to free fermions. These dualities illustrate the fact that the nature and the number of integration variables can differ in the most radical way, still partition functions coincide: in absolutely different theories one can find families of identical correlators. Once again, this is an illustration of the general stringy idea: *everything is the same*, one can fully describe one theory (phase) in terms of another. Matrix models provide a nice framework for testing this principle and can help to transform it into constructive and reliable procedure.

4.1.4. *Limiting procedures and asymptotics*

Different realizations of partition functions are adequate or at least convenient for descriptions of different phases of the system (branches of partition function), or at least different *classes* of phases. As usual, phases become pronouncely different in particular asymptotics. Systematic approach implies that all kinds of asymptotics should be investigated. In particular, in the case of matrix models, all kinds of continuum limits should be examined, not only naive or t'Hooft's or double-scaling. Obviously, all kinds of multi-cut solutions and all kinds of multi-scaling limits require attention, but this is only the beginning: by no means all possible asymptotics are exhausted in this way, and also there is a lot of interesting beyond continuum limits. Unfortunately, not too much is known about this variety of limits and nothing like classification of asymptotics is available (this was a trivial thing in the case of one variable and becomes a pretty sophisticated issue in the case of infinitely many variables).

There are two different possibilities to perform limiting procedures: at the level of defining equations (linear Ward identities or bilinear Hirota-style equations) and at the level of particular realizations. Of course, one and the same limit can look very different in different, though equivalent, representations. Because the subject is under-investigated, the set of known realizations is poor and many potentially interesting asymptotics either do not attract attention or are difficult to handle. We mention just two old problems, which still remain unresolved.

In the class of generalized Kontsevich models (2.3) and (2.4) one naturally distinguishes [28, 11] between **Kontsevich phase** – the asymptotics of large L – and the **characters phase** – the asymptotics of small L. In these phases the "time-variables" $t_k = r_k + \frac{1}{k}\mathrm{tr}L^{-k/n}$ with $n = \deg W'(z)$ are respectively close to $r_k = \frac{n}{k(n-k)}\mathrm{res}\left(W'(z)\right)^{1-k/n} dz$ and to infinity. Kontsevich phase corresponds, at least through the relation (4.1), to perturbative phase of the models like (2.1), while the characters phase is the strong coupling limit of these models. In this sense (4.1) is an example of S-duality, interchanging weak and strong coupling phases. Actually, these phases are closely related to the strong and week coupling limits of Yang-Mills theory and thus to confinement problem. However, this problem was never exhaustively analyzed even at the level of matrix models – despite the existence of Kazakov-Migdal-Kontsevich family (2.4), which – in variance with realizations like (2.1) – provides an effective tool for studying both asymptotics and interpolate between them (for these models Kontsevich phase is the WKB asymptotics, while the characters phase is perturbative limit). A systematic analysis, like suggested in [2, 43, 44] for (2.1) continues to wait for its time for Kontsevich and for more general unitary matrix models.

Another abounded problem would be just a small paragraph in this systematic analysis, still it is quite important by itself. This is the problem of "double-scaling" limit [45] of (2.1), which is one of the simplest non-naive large-N asymptotics of $Z(t|N)$, when $N \to \infty$ together with certain special adjustment of t-variables. For investigation of this limit one can make use of various techniques, of which the most important are two: taking limit of loop equations (Virasoro constraints) (3.1) [46] and exploiting the identity (4.1) and taking limit within the family of Kontsevich models [42, 6]. In these ways one can argue that

$$\lim_{\text{d.s.}} Z(t|N) \sim \lim_{\text{d.s.}} Z_{GKM}\left(L\left|\frac{1}{2}M^2 + N\log M\right.\right) = \\ = Z^2_{GKM}\left(\tilde{L}\left|\frac{1}{3}M^3\right.\right) \tag{4.2}$$

where at the l.h.s. $t_{2k+1} = \frac{1}{2k+1}\mathrm{tr}L^{-2k-1} = 0$ and $t_{2k} = \frac{1}{2k}\mathrm{tr}L^{-2k}$, and the quantity at the r.h.s. (expressed through $\tilde{t}_{2k+1} = \frac{1}{k+1/2}\mathrm{tr}\tilde{L}^{-k-1/2}$) has a special name of **Kontsevich τ-function** $\tau_K(\tilde{t})$ and is an increasingly important matrix-model and string-theory special function. However, even reliable derivation of the principally important result (4.2) and exact relation between t, L and $\tilde{t}$, $\tilde{L}$ is not available, nothing to say about corrections to this formula (describing what happens when one *approaches* the limit) or its generalizations. Of special importance among such generalizations are the limits, when $Z(t|N) \to \tau_K^{2n}(\tilde{t})$ with $n > 1$, because of their obvious relation to Givental-style decomposition formulas [47, 2, 40] and their potential role in applications. It goes without saying that various non-perturbative asymptotics of τ_K should be investigated: this is the next natural task after such program is put on track for the basic special function $Z(t|N)$.

4.2. INTEGRABILITY

Integrability is believed [1, 6] to be the pertinent property of quantum partition functions, reflecting the fact that they are results of integration, which erases most of initial information (roughly speaking, *almost all* forms are exact), leaving only cohomological variables. Therefore integrability is intimately related [48] to topological-cohomological-holographic-stringy theories and all these kinds of ideas should be considered together. However, despite many efforts, spent on investigation of particular models, we are still far from formulating clear concepts. Partly this is because examples are dictated more by applications than by the internal logic of the theory, and they provide somewhat chaotic flow of information. There are already several things that we seem to know, but their complete list and relative significance remain obscure.

4.2.1. *Hirota equations*

The first in the list are bilinear Hirota-style equations, which establish contact with Lie algebra theory [27]. Bilinearity is usually related to the properties of comultiplication and to the basic relation $\Delta(g) = g \otimes g$ for group elements [49, 50]. Connes-Kreimer theory [50]-[52] implies, that quantum partition functions respect these properties, even if they are defined through Feynman diagrams, without explicit reference to functional integrals, and identify the underlying algebraic structure with (generalized) renormalization group. As already mentioned, deeper understanding of relation between integrability and renormalization groups remains among the main open problems of the theory.

4.2.2. *Moduli space of solutions*
The second big problem is understanding, classification and control over the freedom in solving Hirota equations. We mentioned already, that these bilinear equations restrict the dependence on coupling constants t_k, and the freedom remains in the conjugate/dual variables, like zero-modes, or boundary conditions, or choices of vacua, or holographic data, or whatever else name and analogy one prefers to use in connection with one's favorite application. The problem is to find not just a name, but adequate language to speak about this data. Suggestive example is provided by original Hirota equations [53], associated with KP hierarchy: there the freedom (or at least a part of it) can be interpreted in terms of Riemann surfaces. Different KP τ-functions, at least from the class of the *finite-zone solutions*, differ by the choice of Krichever data Σ [54]: a complex curve, a point on it and holomorphic coordinates in the vicinity of a point. Given this data, the KP τ-function is fixed to be

$$\tau(t|\Sigma) = e^{tQt}\theta(\sum_k \vec{B}_k t_k | T_{ij}), \tag{4.3}$$

where T_{ij} is the period matrix of the Riemann surface, $\vec{B}_k = \oint_{\vec{B}} \Omega_k$ and $\Omega_k(\zeta) = \left(\zeta^{-(k+1)} + O(1)\right) d\zeta$ are meromorphic 1-forms with vanishing A-periods. If ansatz (4.3) is substituted into Hirota equations, they become equivalent to Shottky condition for the period matrix T_{ij} [55], i.e. require $\theta(\vec{z}|T_{ij})$ to be Riemannian theta function, not just arbitrary Abelian one. Rather straightforwardly, the simplest – rational and solitonic – KP τ-functions can be treated as particular cases of (4.3) with Riemann surface of genus zero. All remaining, non-finite-zone KP τ-functions are more sophisticated and can be considered as associated with infinite-genus Riemann surfaces, though exact meaning is not yet ascribed to these words. Clearly, there are many different kinds of "infinite-genus" τ-functions. Even the principal example of Kontsevich τ-function $\tau_K(t)$, which belongs to this infinite-genus family, is not well understood and investigated.

Anyhow, example (4.3) remains the main reference point in the theory of integrable systems. It underlies the belief that the right language for description of dual/conjugate variables to times (coupling constants) should be that of Hodge theory, different τ-functions should be parameterized by data, encoded in **spectral surfaces**, not obligatory 2-dimensional. Thus integrability is believed to be a unification of group theory and algebraic geometry, with certain combinatorial flavor, already coming from the studies of adjacent and clearly related subjects [34, 50, 51]. At the same time, even despite these ambitious perspectives, too much remains obscure in the general structure of integrability theory, see [56]-[70] for various important constituents which it should finally unify. All this makes natural the fastly increasing attention to integrability concepts, but decisive conceptual breakthrough is still to come.

4.2.3. *Seiberg-Witten theory, Whitham integrability and WDVV equations*
The third problem is the search for an adequate language, which can help to merge Hodge theory with Lie algebra structures. There are various approaches. The most obvious is to study cohomological (topological) field and string theories [71] and find traces of integrability there. This program is very successful in dealing with various examples, but general concepts appear very hard to extract from it. Among such concepts definitely are the WDVV equations [72, 73] and Batalin-Vilkovisky formalism [74, 75]. Alternative approach, from integrability side usually starts from (4.3), directly or implicitly, and notes that theta-function in that formula is oscillating, therefore one can distinguish between fast (oscillating) and slow variables. Then, as usual, one can take an average over fast variables and consider an effective theory induced on the space of slow variables. Since, according to (4.3), the fast variables are exactly the times t_k, the slow variables of the emerging effective theory should be exactly the dual variables, that we are interested in. This idea, known as Whitham theory, has obvious contacts with those of renormalization and renormalization group, and not-surprisingly appears to have immediate applications. While absolutely un-developed in general, it has a solid very well formulated chapter: the Krichever-Hitchin-Seiberg-Witten theory, describing explicit construction of peculiar special functions – **Seiberg-Witten prepotentials** – from the analogue of Krichever data, associated with peculiar Seiberg-Witten families of spectral surfaces (the best known examples are Calabi-Yau spaces and special families of Riemann surfaces). The prepotential $\mathcal{F}(S_k)$ is defined [76]-[78] by two equations:

$$\begin{aligned} S_k &= \oint_{\mathcal{A}_k} \Omega_{SW}, \\ \frac{\partial \mathcal{F}}{\partial S_k} &= \oint_{\mathcal{B}_k} \Omega_{SW}, \end{aligned} \tag{4.4}$$

where $\Omega_{SW}(z)$ is a linear combination of $\Omega_k(z)$ in (4.3) and holomorphic 1-differentials $\vec{\omega}(z)$ (with non-vanishing A-periods), possessing a *special-geometry* property that $\delta\Omega_{SW}$ is a symplectic form on the "sheaf" of spectral surfaces over the moduli space. In practice this condition states that $\delta\Omega_{SW}$ is holomorphic (has no poles) if the variation is taken along the moduli space of Seiberg-Witten curves. Canonical system of contours $\{\mathcal{A}_k, \mathcal{B}_k\}$ in (4.4) consists of all non-contractable cycles (including those around resolved singularities of Ω_k) and their duals, see [78] for details. The variables S_i define a sort of *flat coordinates* on the moduli space.

It appears that the so defined prepotential always satisfies the (appropriately generalized) WDVV equations [73], namely for matrices

$$\left(\hat{\mathcal{F}}_i\right)_{jk} = \frac{\partial^3 \mathcal{F}}{\partial S_i \partial S_j \partial S_k}$$

made out of prepotential's third derivatives,

$$\hat{\mathcal{F}}_i \hat{\mathcal{F}}_j^{-1} \hat{\mathcal{F}}_k = \hat{\mathcal{F}}_k \hat{\mathcal{F}}_j^{-1} \hat{\mathcal{F}}_i \tag{4.5}$$

for any triple (i, j, k). Moreover, this system of equations for infinitely-large matrices often survives reduction to finite-dimensional matrices when only S-variables, associated with ordinary non-contractable contours, are taken into account. This reasons why such reductions exist are far from obvious, and so is the entire theory of WDVV equations. Technically, it relies upon *residue formulas* for moduli derivatives of period matrices, but conceptually it involves reduction theory of non-trivial *algebra of forms* and is still very unsatisfactory. Additional puzzle is that *reduced* WDVV equations are violated, at least when applied naively, in the case of *elliptic* Calogero system (see the third paper in ref.[73]). Other elliptic examples [79, 70] were not yet analyzed in the context of WDVV equations.

At the same time the theory of WDVV equations can appear even more important than it seems. Today this set of equations is the only candidate for a definition of a prepotential in internal terms, which does not refer to explicit construction procedure. Thus WDVV equations can probably be promoted to the status of *definition* of the prepotential and acquire the status, similar to the one that Hirota equations have for τ-functions. (Referring to its possible origin as a Whitham average of (4.3) – or, more, accurately, of the underlying Hirota equations,– prepotential is often called *Whitham or quasiclassical τ-function*, and the above-mentioned problem is to find the *internal* definition of quasiclassical integrability – a very important theoretical challenge.) The main problem with WDVV equations from this point of view is that they describe only "spherical prepotentials": from the study of topological theories and related algebraic constructions [80] we know that there is a whole hierarchy of prepotentials, associated with the **genus expansion** of Yang-Mills and matrix models partition functions, and Seiberg-Witten prepotential is just the zeroth (genus-zero) term of this sequence. Appropriate generalization of WDVV equations for entire sequence is not yet found, for first steps in the cases of genus one and two see [81], and for relation to Batalin-Vilkovisky theory see [82] (to avoid possible confusion, these papers consider only WDVV equations with additional requirement of "unit metric", which is natural in applications to quantum cohomologies [83] and naive topological models – to "geometric prepotentials",– but is not quite straightforward to relax).

4.3. CFT REPRESENTATIONS, WICK THEOREM AND DECOMPOSITION FORMULAS

The theory of $2d$ conformal systems [84], once at the front-line of attention in string theory, was suddenly abandoned before an exhaustive theory was formulated, and many problems, including effective description of entire variety of conformal models and equivalencies (dualities) between them, are left unresolved. Partly, this happened because attempts to glue the entire subject together – in the framework of landscape theory [35] – naturally caused a desire to interpolate between different conformal models, and this unavoidably embeds [32] the world of $2d$ conformal theories into that of $2d$ integrable systems – still a very badly understood subject, intimately related to the theory of quantum Kac-Moody algebras [85, 86, 87], which attracts much less attention than it deserves.

As already mentioned, there is no doubt that every class of matrix models have CFT representations, i.e. partition function of a given family of matrix models can be represented as a correlator of some operator in $2d$ conformal theory. The questions are: what is the way to build this operator – so far it is matter of art rather than a systematic procedure;– what kind of information from the much bigger phase space of $2d$ free fields is lost (projected out) in transition to the matrix models (in particular, even models which are not dual themselves can still produce representations of one and the same matrix model partition function); how the data of matrix model is mapped into the data of conformal model etc. Today we do not know anything similar to answers to any of these questions.

The crucial property of free field (and thus of $2d$ conformal) theories is straightforward realization of the Wick theorem: all correlators are decomposed into multi-linear combinations of pair correlators (see [88] for a rather fresh discussion of the issue). In principle, for topological and integrable theories naive Wick theorem is not applicable: even after all possible simplifications there are contact terms (and prepotential $\mathcal{F}(S)$ is almost never quadratic). In interacting theory transformation to angle-action variables which would explain to what kind of correlators the naive Wick theorem is applicable, is highly non-linear and probably not very useful. A very important question is what substitutes the naive Wick theorem for generic partition functions. As we know, generic partition functions are not at all in general position in the space of all functions, they satisfy many constraints, like linear Ward identities, bilinear Hirota equations, highly non-linear WDVV equations. Thus we can expect that above question about Wick theorem can make sense. Of course, one can try to interpret as avatars of the Wick theorem all of above-mentioned relations on partition functions. Remarkably, there is also a much more profound candidate: decomposition formulas. It appears that partition functions can often be represented as poly-linear combinations of simpler building blocks, like Gaussian matrix models and Kontsevich τ-functions τ_K. So far this fact, which we call Givental-style decomposition, was explicitly formulated in restricted number of cases [47, 2, 40], but we believe that this is just the tip of the iceberg, and further work will confirm the universality and important role of decomposition phenomenon: a probable substitute of Wick theorem for generic partition functions.

5. Towards exhaustive theory of 1-matrix model: the Dijkgraaf-Vafa theory

Dijkgraaf-Vafa theory [89]-[103] deserves special attention, because for the first time an application was found, which requires understanding of a whole variety of phases of a single matrix model, namely the model (2.1). Despite it is still not the *entire* variety – only peculiar phases are considered interesting, where partition function is consistent with the **genus expansion**, – it is still a very big step forward, and it stimulated fast progress in matrix model theory. In this section we briefly characterize some directions of this progress As everywhere in these notes, we concentrate on pure theoretical issues and ignore the – sometime very interesting – results in applications to Yang-Mills, quantum gravity and model building.

5.1. GENUS EXPANSION

Genus expansion, i.e. the t'Hooft's $1/N$ expansion [104], attracts constant attention since the early studies of matrix models [15]. While it has clear meanings both in diagram technique (genus expansion for fat graphs) and in WKB approach to the integrals like (2.1), specification of partition functions possessing genus expansion among generic solutions of Ward identities [17] is not straightforward and looks somewhat artificial. As explained in [2, 44], it involves several steps.

First of all, by rescaling of all time-variables $t_k \to \frac{1}{g}t_k$ one introduces the new parameter g. It appears in Virasoro constraints (3.1) as a coefficient g^2 in front of the double-derivative terms. Second, solution of these Ward identities is looked for among rather special functions, such that

$$g^2 \log Z(t/g|N) = \sum_{p\geq 0} g^{2p} F^{(p)}(t|gN) \tag{5.1}$$

is a series in non-negative powers of g^2, provided the coefficients of this expansion, $F^{(p)}(t|gN)$, depend on the t'Hooft's coupling constant $S = gN$. Though consistent with (3.1), this is actually a tricky requirement. It implies that Z itself is a series in all integer, positive and negative, powers of g^2, just coefficients in front of negative powers are strongly correlated. Moreover, there is no N in Ward identities (3.1). N can be introduced as $S = gN = \partial(g^2 \log Z)/\partial t_0$, but this requirement breaks superposition principle: once it is imposed, the sum of two solutions is not a solution (this is nearly obvious: a sum of two exponentials is not an exponential). This means that genus expansion can be at best a property of the elements of a linear *basis* in the space of solutions, while generic element of this space does not respect it. In other words, genus expansion can be a property of particular integrals, like (2.1), but there is no way to impose this requirement on linear combinations of integrals with different N, which will still be solutions of the Virasoro constraints.

There are at least two ways to deal with this problem (the obvious possibility to ignore it, as is usually done, is not counted). First, one can look for the ways to define special basises – the problem is to define them in invariant way, by specifying their properties, not by explicit construction, like matrix integrals. Second, one can try to formulate genus expansion without explicit reference to N – this road leads to introduction of **check operators** [44].

5.2. GAUSSIAN AND NON-GAUSSIAN PARTITION FUNCTIONS

Requirement of genus expansion is not enough to specify the phase (branch) of partition function completely, it regulates only dependence on the common scaling factor of all couplings t_k and says nothing about their ratios. The next step [2] is to select combinations of t-variables which can appear in denominators. A way to do this is to make a shift, $t_k \to t_k - T_k$, and consider $W(\phi) = \sum_k T_k\phi^k$ as a **bare action** and $t_k\phi^k$ as perturbations. In other words, the branch

$$Z_W(t|N) = \frac{1}{\mathrm{Vol}(U(N))} \int_{N\times N} d\Phi \exp\left(-\mathrm{Tr}W(\Phi) + \sum_k t_k \mathrm{Tr}\,\Phi^k\right) \tag{5.2}$$

of partition function is defined as a formal series in non-negative powers of t_k, while $W(z)$, i.e. T_k's, are allowed to appear in denominator. Once $W(z)$ is introduced, one immediately distinguishes between the Gaussian (with $W(\phi) = \frac{M}{2}\phi^2$) and non-Gaussian phases.

Next, it turns out that while the variable N is naturally introduced in the Gaussian phase, in non-Gaussian case the situation is more sophisticated: there are naturally $n =$

deg $W'(z)$ parameters like N: the phase is still split into a n-parametric family of phases. Additional variables can be interpreted as numbers of eigenvalues concentrated near the n extrema (maxima and minima) of the bare potential $W(\phi)$, a more reliable interpretation is as integration constants in solution of shifted (by $W(z)$) Virasoro constraints. See [2] for further details.

Gaussian partition function can be represented as

$$Z_G^{(M)}(t|S|g) = \exp\left(\frac{1}{g^2}\left(-ST_0 - \frac{M}{2}S^2\right)\right) \exp\left(\sum_{p\geq 0} g^{2p-2}\mathcal{F}_G^{(p)}(t|S)\right) \tag{5.3}$$

and information about the t-dependence of prepotentials $\mathcal{F}_G^{(p)}(t|S)$ can be represented in terms of **multi-densities**, the generating functions of m-point correlators, which – for given p and m – are ordinary poly-differentials on the Riemann sphere.

Multi-densities can be defined in different ways (for different families of correlators) [2], one of the ways consistent with genus expansion and with the free-field representation (3.1) of Virasoro constraints makes use of the operator $\hat{\nabla}(z) = 2d\hat{\phi}_-(z) = \sum_{k=0}^{\infty} \frac{dz}{z^{k+1}}\frac{\partial}{\partial t_k}$:

$$\begin{aligned}
&\left[\hat{\nabla}(z_1)\dots\hat{\nabla}(z_m)\left(g^2 \log Z_W(t|\vec{S}|g)\right)\right]\Big|_{t=0} = \\
&= \rho^{(\cdot|m)}(z_1,\dots,z_m|\vec{S}|g) = \sum_{p\geq 0} g^{2p}\rho^{(p|m)}(z_1,\dots,z_m|\vec{S})
\end{aligned} \tag{5.4}$$

The first *Gaussian* multi-densities are:

$$\begin{aligned}
\rho_G^{(0|1)}(z) &= \frac{z - y_G(z)}{2}, \\
\rho_G^{(1|1)}(z) &= \frac{\nu}{y_G^5} = -\frac{y_G''}{4y_G^2}, \\
\rho_G^{(2|1)}(z) &= \frac{5}{16}\frac{(y_G'')^2}{y_G^5} - \frac{1}{8y_G^2}\partial^2\left(\frac{y_G''}{y_G^2}\right) - \frac{1}{8}\frac{y_G''''}{y_G^4}, \\
&\dots
\end{aligned} \tag{5.5}$$

$$\begin{aligned}
&\rho_G^{(0|2)}(z_1,z_2) = \frac{1}{2(z_1-z_2)^2}\left(\frac{z_1z_2 - 4\nu}{y_G(z_1)y_G(z_2)} - 1\right) = \\
&= -\frac{1}{2y_G(z_1)}\frac{\partial}{\partial z_2}\frac{y_G(z_1)-y_G(z_2)}{z_1-z_2} = -\frac{1}{2y_G(z_2)}\frac{\partial}{\partial z_1}\frac{y_G(z_1)-y_G(z_2)}{z_1-z_2},
\end{aligned}$$

$$\rho_G^{(1|2)}(z_1,z_2) = \frac{\nu}{y_G^7(z_1)y_G^7(z_2)}\Big(z_1z_2(5z_1^4+4z_1^3z_2+3z_1^2z_2^2+4z_1z_2^3+5z_2^4)+$$
$$+4\nu\left[z_1^4-13z_1z_2(z_1^2+z_1z_2+z_2^2)+z_2^4\right]+$$
$$+16\nu^2(-z_1^2+13z_1z_2-z_2^2)+320\nu^3\Big)=$$
$$=\frac{1}{y_{G1}}\left[\left(4\frac{1}{4y_{G1}^2}y_{G1}''-\frac{1}{2y_{G1}}\partial_1^2\right)\frac{1}{2y_{G1}}\frac{\partial}{\partial z_2}\frac{y_{G1}-y_{G2}}{(z_1-z_2)}+\right.$$
$$+\frac{\partial}{\partial z_2}\frac{1}{z_1-z_2}\left(\frac{1}{4y_{G2}^2}y_{G2}''-\frac{1}{4y_{G1}^2}y_{G1}''+\right.$$
$$\left.\left.+\frac{1}{y_{G1}}\left(-\frac{1}{4y_{G1}}y_{G1}''+\frac{1}{2y_{G1}}\frac{\partial}{\partial z_2}\frac{y_{G1}-y_{G2}}{(z_1-z_2)}\right)\right)\right],$$

$$\ldots$$

$$\rho_G^{(0|3)}(z_1,z_2,z_3) = \frac{2\nu(z_1z_2+z_2z_3+z_3z_1+4\nu)}{y_G^3(z_1)y_G^3(z_2)y_G^3(z_3)}=$$
$$=\frac{1}{y_{G1}}\left[2\frac{1}{4y_{G1}^2}\left(\frac{\partial}{\partial z_2}\frac{y_{G1}-y_{G2}}{z_1-z_2}\right)\left(\frac{\partial}{\partial z_2}\frac{y_{G1}-y_{G3}}{z_1-z_3}\right)+\right.$$
$$+\frac{\partial}{\partial z_2}\frac{1}{z_2-z_1}\left(\frac{1}{2y_{G1}}\left(\frac{\partial}{\partial z_3}\frac{y_{G1}-y_{G3}}{z_1-z_3}\right)-\frac{1}{2y_{G2}}\left(\frac{\partial}{\partial z_3}\frac{y_{G2}-y_{G3}}{z_2-z_3}\right)\right)+$$
$$\left.+\frac{\partial}{\partial z_3}\frac{1}{z_3-z_1}\left(\frac{1}{2y_{G1}}\left(\frac{\partial}{\partial z_2}\frac{y_{G1}-y_{G2}}{z_1-z_2}\right)-\frac{1}{2y_{G3}}\left(\frac{\partial}{\partial z_2}\frac{y_{G2}-y_{G3}}{z_2-z_3}\right)\right)\right],$$

$$\ldots$$

We assumed here that $M=1$, then $y_G^2=z^2-4\nu$ and $\nu=S=gN$.

Non-Gaussian are somewhat more sophisticated, they are actually made from Riemann theta-functions for a peculiar Seiberg-Witten family of **hyperelliptic complex curves**. For example, the $(p,m)=(0,2)$-density is the 2-point correlator on such surfaces:

$$\rho_W^{(0|2)}(z_1,z_2)=d_{z_1}d_{z_2}\log E(z_1,z_2), \tag{5.6}$$

where $E(z_1,z_2)\sim\frac{\nu_*(z_1)\nu_*(z_2)}{\theta(\vec{z}_1-\vec{z}_2)}$ is the prime form [5, 3].

5.3. DECOMPOSITION FORMULAS

Rewriting (5.2) in terms of the eigenvalues and then separating the eigenvalues into sets, associated with different extrema of $W(\phi)$, one can deduce [92] a decomposition formula [2] for non-Gaussian Z_W:

$$Z_W\sim\frac{\prod_{i=1}^n\mathrm{Vol}(U(N_i))\,e^{-N_iW(\alpha_i)}}{\mathrm{Vol}(U(N))}\times$$
$$\times\prod_{i<j}^n\alpha_{ij}^{2N_iN_j}\hat{O}_{ij}\prod_{i=1}^n\hat{O}_i\prod_{i=1}^n Z_G^{(M_i)}(t^{(i)}|N_i) \tag{5.7}$$

with α_i denoting the n roots of

$$W'(z) = \sum_{k=0}^{n+1} kT_k z^{k-1} = (n+1)T_{n+1}\prod_{i=1}^{n}(z-\alpha_i),$$

$\alpha_{ij} = \alpha_i - \alpha_j$, and operators

$$\hat{\mathcal{O}}_{ij} = \exp\left(2\sum_{k,l=0}^{\infty}(-)^k\frac{(k+l-1)!}{\alpha_{ij}^{k+l}k!l!}\frac{\partial}{\partial t_k^{(i)}}\frac{\partial}{\partial t_l^{(j)}}\right),$$

$$\hat{\mathcal{O}}_i = \exp\left(-\sum_{k\geq 3}\frac{W^{(k)}(\alpha_i)}{k!}\frac{\partial}{\partial t_k^{(i)}}\right)$$

Here $W^{(k)}(x) = \partial_x^k W(x)$.

As explained in the previous section, this Givental-style formula is an example of general phenomenon, generalizing the Wick theorem from free fields and conformal theories to matrix models and non-trivial integrable systems.

Eq.(5.7) requires (and deserves) deep investigation. Today almost nothing is known about it. It is unclear how (5.7) follows directly from Virasoro constraints, without explicit use of matrix integrals. It is unclear what is exact implication of the relation (4.2) for Gaussian functions, standing at the r.h.s. of (5.7). It is clear that the combination of (4.2) and (5.7) should represent Z_W as some operator acting on τ_K^{2n}, but exact formula is unavailable (Greg Moore in [39] and especially Ivan Kostov in [40] came very close to the answer, but it still escapes).

Eq.(5.7) can be also considered as a definition of a linear basis in the space of solutions of Virasoro constraints, which is formed by the functions, respecting the genus expansion. *Arbitrary* solution can be represented as a linear combination,

$$Z_W(t) = \int_{\vec{S}} \mu(\vec{S}) Z_W(t|\vec{S}), \tag{5.8}$$

with *arbitrary* measure $\mu(\vec{S})$.

5.4. CHECK-OPERATORS AND T-EVOLUTION OPERATOR

According to their definition (5.4), multi-densities do not depend on the time variables t_k, only on $T_0, \ldots, T_{n+1}$ and $\vec{S}$. Actually, dependence on T_n and T_{n+1} is fixed by (3.1), and multi-densities depend on $2n$ variables: n of them are $T's$ and n are S's. Therefore one can ask for a procedure which defines multi-densities directly in terms of these variables, without an intermediate introduction of – infinitely many – auxiliary (from the point of view of multi-densities theory) variables t_k. This problem is solved, at least conceptually, in [2, 44] in terms of **check-operators**.

Multi-densities are not arbitrary functions of their $2n$ variables, the Ward identities (3.1) express all of them in terms of a single function of $n+1$ variables, $T_0, \ldots, T_{n-1}$ and g [2], which we call **bare prepotential**. The choice of bare prepotential is actually the choice of particular branch of partition function. Note, that this means that there are not just infinitely many branches, they form a continuous variety – and this is a property of every reasonable partition function and of anything obtained by application of **evolution operators** (for which the functional integral is a particular realization), see [34] for explanation, how continuous phase structure emerges from discrete bifurcations (phase transitions), and for discussion of the adequate formalism. This set of questions is very important, in particular it can help to understand the *relevant* topology and may be even the metric structure on the space of phases, what is the principal problem in landscape theory. Development of such formalism for matrix models is a task for the future, today we know how to proceed within a given phase, when bare prepotential is somehow specified and the question of relative importance of different choices is not addressed.

5.4.1. *Independent variables and bare prepotential*
To understand what it means to specify the bare prepotential, we consider immediate corollary of (3.1) with $t_k \to \frac{1}{g}(t_k - T_k)$ for partition function at vanishing times t_k. It is given by [44]:

$$Z(T|g)|_{t=0} = \int dkz(k|\eta_2, \dots, \eta_n|g^2)e^{\frac{1}{g^2}(kx-k^2w)} \tag{5.9}$$

with an arbitrary function z of n arguments $(k, \eta_2, \dots, \eta_n)$ and g^2. Here the $\hat{L}_{-1}$-invariant variables are used,

$$w = \frac{1}{n+1}\log T_{n+1}, \quad \eta_k = T_{n+1}^{-\frac{nk}{n+1}}\left(T_n^k + \dots\right), \quad x = T_0 + \dots \sim \eta_{n+1}$$

In particular,

$$\eta_2 = \left(T_n^2 - \frac{2(n+1)}{n}T_{n-1}T_{n+1}\right)T_{n+1}^{-\frac{2n}{n+1}},$$

$$\eta_3 = \left(T_n^3 - \frac{3(n+1)}{n}T_{n-1}T_nT_{n+1} + \frac{3(n+1)^2}{n(n-1)}T_{n-2}T_{n+1}^2\right)T_{n+1}^{-\frac{3n}{n+1}},$$

$$\eta_4 = \left(T_n^4 - \frac{4(n+1)}{n}T_{n-1}T_n^2T_{n+1} + \frac{8(n+1)^2}{n(n-1)}T_{n-2}T_nT_{n+1}^2 - \right.$$
$$\left. - \frac{8(n+1)^3}{n(n-1)(n-2)}T_{n-3}T_{n+1}^3\right)T_{n+1}^{-\frac{4n}{n+1}},$$
$$\dots$$

$$\eta_k = \left(T_n^k + \frac{k(k-2)!}{n!}\sum_{l=1}^{k-1}(-)^l\frac{(n+1)^l(n-l)!}{(k-l-1)!}T_{n-l}T_n^{k-l-1}T_{n+1}^l\right)T_{n+1}^{-\frac{kn}{n+1}}$$

The variable x is obtained from η_{n+1} by normalization and it is the only variable which contains T_0. At the same time only T_0 appears in double-derivative item of $\hat{L}_0$, thus in (5.9) the $\hat{L}_0$-constraint links the x- and w-dependencies. Of course, generic (5.9) does not possess genus expansion, i.e. does not guarantee that the bare prepotential $F(T|g) = g^2\log Z(T|g)$, is expanded in non-negative powers of g^2: as explained above, this requirement is somewhat artificial from the perspective of D-module theory. Still, a simple ansatz makes things consistent (it is not quite clear if this ansatz is absolutely necessary for genus expansion to exist): if

$$S = \frac{\partial F}{\partial T_0} = const \tag{5.10}$$

i.e. is independent of $T_0, \ldots, T_{n+1}$ and g, then,

$$z(k|\eta_2, \ldots, \eta_n|g^2) = \delta(k - S)H(\eta_2, \ldots, \eta_n|S|g^2) \tag{5.11}$$

and

$$F = g^2 \log Z = Sx + \frac{S^2}{n+1} \log T_{n+1} + g^2 \log H(\eta_2, \ldots, \eta_n|S|g^2) \tag{5.12}$$

where H is an arbitrary function of $n+1$ variables. Genus expansion occurs, if we request $g^2 \log H$ is expanded in non-negative powers of g^2.

Introduction of the other $\vec{S}$-variables can be considered as ingenious version of Fourier-Radon transform with the help of Dijkgraff-Vafa partition functions (5.7), which converts $H(\eta_2, \ldots, \eta_n)$ into an arbitrary measure $\mu(\vec{S})$ of the $\vec{S}$-variables in (5.8) [2].

5.4.2. *Check-operator multi-densities*

Remaining constraints $\hat{L}_k$ with $k > 0$ express t-dependencies of $Z(t|T)$ through T-dependencies. In particular, they allow to represent multi-densities (both Gaussian and non-Gaussian) by action of T-dependent operators on partition function (5.9). Such operators, involving only derivatives with respect to T-variables, are named *check-operators* in [44], to distinguish them from *hat-operators*, acting on t-variables. The task is to express multi-densities, defined by application of hat-operators to the full prepotential $F(t|T|g)$, through check-operators applied to the bare prepotential $F(T) = F(t = 0|T|g)$. These check operators will predictably be more sophisticated than hat-operators $\hat{\nabla}(z)$, but instead they can be applied to the *independent* (free) function $F(T)$.

Remarkably, the problem of building check-operator multi-densities appears essentially equivalent to the problem of *Gaussian multi-densities* [44]. Though this fact is not fully proved – and even adequately formulated – yet, it is not too surprising: the needed check-operators are universal, in certain sense they do not depend on the phase, thus they should be restorable from information, available in any phase, including the Gaussian one (once again we encounter the main idea of string theory). Anyhow, modulo some details concerning ordering prescriptions, the first check-operators can be just read from formulas (5.5) for Gaussian multi-densities [44]:

$$\begin{aligned}
\check{\rho}_W^{(0|1)}(z|g) &= \frac{W'(z) - \check{y}(z|g)}{2}, \\
\check{\rho}_W^{(1|1)}(z|g) &= -\frac{1}{4\check{y}^2}\check{y}'', \\
\check{\rho}_W^{(2|1)}(z|g) &= \frac{5}{16}\frac{(\check{y}'')^2}{\check{y}^5} - \frac{1}{8\check{y}^2}\partial^2\left(\frac{\check{y}''}{\check{y}^2}\right) - \frac{1}{8}\frac{\check{y}''''}{\check{y}^4}, \\
&\ldots
\end{aligned} \tag{5.13}$$

$$\check\rho_W^{(0|2)}(z_1,z_2|g) = -\frac{1}{2\check y(z_1|g)}\frac{\partial}{\partial z_2}\frac{\check y(z_1|g)-\check y(z_2|g)}{z_1-z_2},$$

$$\check\rho_W^{(1|2)}(z_1,z_2|g) = \frac{1}{\check y_1}\left[\left(4\frac{1}{4\check y_1^2}\check y_1'' - \frac{1}{2\check y_1}\partial_1^2\right)\frac{1}{2\check y_1}\frac{\partial}{\partial z_2}\frac{\check y_1-\check y_2}{(z_1-z_2)}+\right.$$

$$\left.+\frac{\partial}{\partial z_2}\frac{1}{z_1-z_2}\left(\frac{1}{4\check y_2^2}\check y_2'' - \frac{1}{4\check y_1^2}\check y_1'' + \frac{1}{y_1}\left(-\frac{1}{4\check y_1}\check y_1'' + \frac{1}{2\check y_1}\frac{\partial}{\partial z_2}\frac{\check y_1-\check y_2}{(z_1-z_2)}\right)\right)\right],$$

$$\ldots$$

$$\check\rho_W^{(0|3)}(z_1,z_2,z_3|g) = \frac{1}{\check y_1}\left(2\frac{1}{2\check y_1}\left(\frac{\partial}{\partial z_2}\frac{\check y_1-\check y_2}{z_1-z_2}\right)\frac{1}{2\check y_1}\left(\frac{\partial}{\partial z_2}\frac{\check y_1-\check y_3}{z_1-z_3}\right)+\right.$$

$$+\frac{\partial}{\partial z_2}\frac{1}{z_2-z_1}\left(\frac{1}{2\check y_1}\left(\frac{\partial}{\partial z_3}\frac{\check y_1-\check y_3}{z_1-z_3}\right)-\frac{1}{2\check y_2}\left(\frac{\partial}{\partial z_3}\frac{\check y_2-\check y_3}{z_2-z_3}\right)\right)+$$

$$\left.+\frac{\partial}{\partial z_3}\frac{1}{z_3-z_1}\left(\frac{1}{2\check y_1}\left(\frac{\partial}{\partial z_2}\frac{\check y_1-\check y_2}{z_1-z_2}\right)-\frac{1}{2\check y_3}\left(\frac{\partial}{\partial z_2}\frac{\check y_2-\check y_3}{z_2-z_3}\right)\right)\right),$$

$$\ldots$$

Here

$$\check y^2(z|g) = W'(z)^2 - 4g^2\check R_W(z)$$

and

$$\check R_W(z) = -\sum_{a,b=0}(a+b+2)T_{a+b+2}z^a\frac{\partial}{\partial T_b}$$

The check-multidensities are defined to satisfy

$$\check K_W^{(\cdot|m)}(z_1,\ldots,z_m|g)Z_W(T|g) = K_W^{(\cdot|m)}(z_1,\ldots,z_m|g)Z_W(T|g) =$$

$$= \left[\hat\nabla(z_1)\ldots\hat\nabla(z_m)Z_W(t|T|g)\right]\Big|_{t=0}$$

for *full* correlators $K_W^{(\cdot|m)}(z_1,\ldots,z_m|g)$, which are derivatives of $Z(t)$, not of its logarithm, while for *connected* $\rho_W^{(\cdot|m)}(z_1,\ldots,z_m|g)$

$$(Z_W(T|g))^{-1}\,\check\rho_W^{(\cdot|m)}(z_1,\ldots,z_m|g)\,Z_W(T|g) \neq \rho_W^{(\cdot|m)}(z_1,\ldots,z_m|g) =$$

$$= \hat\nabla(z_1)\ldots\hat\nabla(z_m)\left(g^2\log Z_W(T|g)\right)$$

Relation between the full and connected correlators is provided by straightforward – though heavily looking – combinatorial formula [44]

$$K_W^{(\cdot|m)}(z_1,\ldots,z_m|g) = \sum_{\sigma}^{m!}\sum_{k=1}^{m}\sum_{\nu_1,\ldots,\nu_k=1}^{\infty}\sum_{p_1,\ldots,p_\nu=0}^{\infty} g^{2(p_1+\ldots+p_\nu-\nu)}$$

$$\left(\sum_{\substack{m_1,\ldots,m_k\\ m=\nu_1 m_1+\ldots+\nu_k m_k}} \frac{1}{\nu_1!(m_1!)^{\nu_1}\ldots\nu_k!(m_k!)^{\nu_k}}\rho_W^{(p_1|\tilde{m}_1)}(z_{\sigma(1)},\ldots,z_{\sigma(\tilde{m}_1)})\cdot\right.$$

$$\left.\cdot\,\rho_W^{(p_2|\tilde{m}_2)}(z_{\sigma(\tilde{m}_1+1)},\ldots,z_{\sigma(\tilde{m}_2)})\ldots\rho_W^{(p_\nu|\tilde{m}_\nu)}(z_{\sigma(m-\tilde{m}_\nu+1)},\ldots,z_{\sigma(m)})\right)$$

which serves as generalization of the Wick formula from the case when the only connected correlators are 2-point (note that this simplification does not occur even in Gaussian phases of matrix models beyond naive continuum limit). A similar expression relates the check-operators $\check{K}$ and $\check{\rho}$, in particular,

$$\check{K}_W^{(\cdot|1)}(z;g) = \sum_{p=0}^{\infty} g^{2p-2}\check{\rho}_W^{(p|1)}(z;g). \tag{5.14}$$

Note also, that in variance with (5.5), expressions in (5.13) explicitly contain g^2.

Actually, once the check-operators multi-densities are introduced, one can forget about the genus-expansion constraint: operators can be applied to any bare partition function. Instead of being a constraint on partition functions, genus expansion requirement dictates becomes a selection criterion for a basis in check-operators space. The next step is to build up a theory of the t-evolution operator $\check{U}(t)$, which generates the t-dependence of partition function:

$$Z(t|T) = \check{U}(t)Z(0|T). \tag{5.15}$$

Its existence is almost obvious, see [2, 44], but properties and explicit realizations remain to be found.

5.5. SEIBERG-WITTEN THEORY

The Dijkgraaf-Vafa theory is intimately related to Seiberg-Witten theory.

5.5.1. *Genus-zero level*

For detailed discussion of this relation in the particular case of Dijkgraaf-Vafa partition functions *per se*, i.e. with requirements that genus expansion exists and only genus-zero contributions (spherical prepotentials) are considered, see [95]-[99]. Here are the main points of this analysis. The smallest moduli space of Seiberg-Witten structure unifies all phases with all the bare potentials $W(z)$ of a fixed degree n. It has (complex) dimension $2n$, T and S-variables form the set of $2n$ *flat moduli* of Seiberg-Witten structure, defined with the help of $\Omega_{SW} = \sqrt{W'(z)^2 - 4f_W(z)}dz$, where $f_W(z) = \check{R}_W(z)F^{(0)}(T)$

is a polynomial of degree $n-1$, made from arbitrary function $F^{(0)}(T)$ of n variables $T_0, \ldots, T_{n-1}$ (and dependence on T_n and T_{n+1} prescribed by $\check{L}_{-1}$ and $\check{L}_0$-constraints). The space of all functions of n variables can – in the framework of Fourier-Radon transforms – be parameterized by arbitrary functions of n other parameters, and $\vec{S}$ can play the role of such parameters no better, no worse than any other set of variables. What distinguishes $\vec{S}$ is that they are flat Seiberg-Witten moduli, i.e. the symplectic form on the sheaf of Seiberg-Witten curves is

$$\delta\Omega_{SW} = \omega_i(z) \wedge \delta S_i + \Omega_k(z) \wedge \delta T_k \tag{5.16}$$

and

$$\vec{S} = \oint_{\vec{A}} \Omega_{SW}, \quad T_k = \text{res}_\infty z^k \Omega_{SW} \tag{5.17}$$

The CIV-DV prepotential [89, 90] is defined by

$$\frac{\partial \mathcal{F}_{CIV}}{\partial \vec{S}} = \oint_{\vec{B}} \Omega_{SW}, \quad \frac{\partial \mathcal{F}_{CIV}}{\partial T_k} = \frac{1}{k} \text{res}_\infty z^{-k} \Omega_{SW} \tag{5.18}$$

it can be explicitly calculated term by term and possesses pronounced group-theoretical structure [97, 2], which still awaits its adequate interpretation and explanation. No reasonable generic formulas for this prepotential are yet found, despite it is a genus-zero quantity, and despite it can be deduced in at least two dual ways: from (5.18) and from (5.7). Moreover, even direct relation between these equivalent representations is not yet established. Also important is that there are various differently looking forms for the same items in $\mathcal{F}_{CIV}$, their equivalence follows from peculiar identities [98], which are simple to prove, but not so simple to understand.

5.5.2. *WDVV equations*

In [95, 98, 99] the proof is given of the WDVV equations (4.5) for the CIV-DV prepotential. This is an additional check that $2n$ moduli T and S form complete set, at least at the level of genus-zero prepotentials. In [96] a less obvious observation is reported: sometime this set can be reduced further, for example, by elimination of one of the 4 moduli in the case of $n = 2$, however, the meaning of this observation remains obscure.

The analogues of the CIV-DV prepotential for higher genera can also be defined because of its relation to Ward identities (3.1) – this makes the situation different from generic Seiberg-Witten theory, where the problem of lifting from genus zero to all genera remains very important, but unsolved. This makes it possible to ask, if higher genus prepotentials satisfy the (properly generalized) WDVV equations, for example the ones implied by studies of topological models [81]. This problem was not seriously attacked yet.

5.5.3. *Towards the quantum Seiberg-Witten theory of check multi-densities*

The next natural question in the theory of check operators concerns definition of S-variables. Since in particular phases they are periods of $\rho_W^{(0|1)}$, it is natural to treat S-variables as *operators*, defined as periods of $\check{\rho}^{(0|1)}$. Moreover, under certain conditions the A-periods of higher-genus one-point densities are vanishing, and one can even consider the periods of $\check{\rho}^{(\cdot|1)}$ or

$$\check{S}_i = \frac{g^2}{4\pi i}\oint_{\mathcal{A}_i} \check{K}^{(\cdot|1)}(z). \tag{5.19}$$

This is a very perspective direction of research. Among other, obvious and not-so-obvious, questions is developement of the full-scale version of Seiberg-Witten theory for check-operator quantities, i.e. a quantization of Seiberg-Witten structure. Introduction of check operators with the goal to deal with arbitrary bare prepotentials in a uniform way is a nice illustration of string "third-quantization" procedure, which implies that in order to study *families* of models one actually needs to quantize a representative of the family. An example of the simplest relation in this *quantum Seiberg-Witten theory* is provided by the quantum version of (4.4) [44]:

$$\left[\frac{g^2}{4\pi i}\oint_{\mathcal{A}_i} \check{K},\ \frac{g^2}{2}\oint_{\mathcal{B}_j} \check{K}\right] = g^2\delta_{ij}, \tag{5.20}$$

probably, with no corrections at the r.h.s.

The check-operator A-and B-periods in (5.20) are equal respectively to $\check{S}_i = \check{S}_i^{(0)} + g^2\check{S}_i^{(1)} + O(g^4)$ and $\check{\Pi}_i^{(0)} + O(g^2)$, where [97]

$$\check{S}_i^{(0)} = \frac{g^2}{M_i}\check{R}_i, \qquad \check{\Pi}_i^{(0)} = W(\alpha_i), \tag{5.21}$$

$$g^2\check{S}_i^{(1)} \overset{(5.13)}{=} \operatorname{res}_{\alpha_i}\frac{W'''(z)dz}{8W'(z)^2} = \frac{\partial^4 W\partial^2 W - (\partial^3 W)^2}{8(\partial^2 W)^3}(\alpha_i)$$

and α_i, $i = 1, \ldots, n$ are the roots of $W'(z)$, operators $\check{R}_i = \check{R}_W(\alpha_i)$,

$$M_i = W''(\alpha_i) = (n+1)T_{n+1}\prod_{j\neq i}\alpha_{ij},$$

$\alpha_{ij} = \alpha_i - \alpha_j$. Note that in the leading order the operator B-periods and genus-1 contributions to A-periods do not containt T-derivatives, thus non-trivial are only commutators, involving $\check{S}_i^{(0)}$. These commutators can be easily deduced from two general formulas from [44],

$$\left[\check{R}_W(x), W(z)\right] = \frac{W'(x) - W'(z)}{x - z},$$

$$\left[\check{R}_W(x), \check{R}_W(z)\right] = (\partial_z - \partial_x)\frac{\check{R}'_W(x) - \check{R}'_W(z)}{x - z}, \tag{5.22}$$

together with

$$\frac{\partial \alpha_j}{\partial T_k} = -\frac{k\alpha_j^{k-1}}{M_j}, \tag{5.23}$$

from [98], which implies straightforwardly:

$$\left[\check{R}_W(x), \alpha_j\right] = \sum_{a,b=0} (a+b+2) T_{a+b+2} x^a \frac{\partial \alpha_j}{\partial T_b} =$$

$$= -\frac{1}{M_j} \partial_z \frac{W'(x) - W'(z)}{x - z}\bigg|_{z=\alpha_j}$$

and, since $W'(\alpha_i) = W'(\alpha_j) = 0$,

$$\left[\check{R}_i, \alpha_j\right] = \left[\check{R}_W(\alpha_i), \alpha_j\right] = \frac{1}{\alpha_{ij}} \tag{5.24}$$

Eqs.(5.22) need to be supplemented by (5.24), because in commutators of periods (5.21) the arguments α_i also depend on T_k and their derivatives should be taken into account. Thus instead of (5.22) we need:

$$\left[\check{R}_i, M_j\right] = \left[\check{R}_W(\alpha_i), W''(\alpha_j)\right] =$$

$$= \partial_z^2 \frac{W'(\alpha_i) - W'(z)}{\alpha_i - z}\bigg|_{z=\alpha_j} + W'''(\alpha_j)\left[\check{R}_i, \alpha_j\right] = -\frac{2M_j}{\alpha_{ij}^2}, \tag{5.25}$$

and

$$\left[\check{R}_i, \check{R}_j\right] = \left[\check{R}_W(\alpha_i), \check{R}_W(\alpha_j)\right] =$$

$$= (\partial_z - \partial_x) \frac{\check{R}_W(x) - \check{R}_W(z)}{x - z}\bigg|_{x=\alpha_i,\ z=\alpha_j} +$$

$$+ \left[\check{R}_W(\alpha_i), \alpha_j\right] \check{R}'_W(\alpha_j) - \left[\check{R}_W(\alpha_j), \alpha_i\right] \check{R}'_W(\alpha_i) =$$

$$= \frac{2}{\alpha_{ij}^2}\left(\check{R}_i - \check{R}_j\right) \tag{5.26}$$

Now we have everything to check, that

$$\left[\check{S}_i^{(0)}, \check{S}_j^{(0)}\right] = \left[\frac{g^2}{M_i}\check{R}_i, \frac{g^2}{M_j}\check{R}_j\right] =$$

$$= \frac{g^4}{M_i M_j}\left(\left[\check{R}_i, \check{R}_j\right] - \frac{1}{M_j}\left[\check{R}_i, M_j\right]\check{R}_j + \frac{1}{M_i}\left[\check{R}_j, M_i\right]\check{R}_i\right) = 0$$

Direct generalization of (5.25) states, that

$$\left[\check{R}_i, \partial^m W(\alpha_j)\right] = -m! \sum_{k=0}^{m-2} \frac{1}{(k+1)!} \frac{\partial^{k+2} W(\alpha_j)}{\alpha_{ij}^{m-k}},$$

in particular,

$$\left[\check{R}_i, \frac{\partial^4 W \partial^2 W - (\partial^3 W)^2}{(\partial^2 W)^3}(\alpha_j)\right] = -\frac{24}{M_j \alpha_{ij}^2},$$

so that

$$\left[\check{S}_i^{(0)}, g^2 \check{S}_j^{(1)}\right] = -\frac{3}{M_i M_j \alpha_{ij}^2} = \left[\check{S}_j^{(0)}, g^2 \check{S}_i^{(1)}\right]$$

is symmetric under the permutation $i \leftrightarrow j$. Since also $\left[g^2 \check{S}_i^{(1)}, g^2 \check{S}_j^{(1)}\right] = 0$ and $\left[g^2 \check{S}_i^{(1)}, \check{\Pi}_j^{(0)}\right] = 0$, we finally obtain:

$$\left[\check{S}_i^{(0)} + g^2 \check{S}_i^{(1)}, \check{S}_j^{(0)} + g^2 \check{S}_j^{(1)}\right] = 0,$$

$$\left[\check{\Pi}_i^{(0)}, \check{\Pi}_j^{(0)}\right] = [W(\alpha_i), W(\alpha_j)] = 0,$$

$$\left[\check{S}_i^{(0)} + g^2 \check{S}_i^{(1)}, \check{\Pi}_j^{(0)}\right] = \left[\frac{g^2}{M_i}\check{R}_i, W(\alpha_j)\right] =$$

$$= \frac{g^2}{W''(\alpha_i)} \frac{W'(\alpha_i) - W'(\alpha_j)}{\alpha_{ij}} = g^2 \delta_{ij} \tag{5.27}$$

Note that in the last formula there is no need to vary the argument α_j, because this variation gets multiplied by $W'(\alpha_j) = 0$. When $j = i$ one needs to apply l'Hopital rule to resolve the $0/0$ ambiguity, and it provides the non-vanishing answer. This ends the proof of (5.20). It is an open question, if there are higher order corrections $\sim O(g^4)$ to the r.h.s. of that formula.

Given (5.20), of special interest is this kind of quantization of the WDVV equations. At least in principle, such approach opens a possibility to unify WDVV equations for different genera, the spherical equations from [72, 73] with higher genus equations from [81, 82], and thus provide a new deep connection with geometry of the moduli space of punctured Riemann surfaces.

6. Neoclassical period: back to concrete results and back to physics?

The story about the challenges of matrix models is nearly infinite and can be continued further and further. We, however, draw a line here. The only brief comment that deserves being made at the end of these notes, is that the *transcendental period*, when most attention is concentrated on non-obvious structures underlying the subject and thus on high abstractions, is never the final stage of developement of a theory. Sooner or later, deep insights from transcendental studies open a possibility to solve practical problems and, in our case, derive concrete formulas with the left hand side and the right hand side. Remarkably, some of such results are already starting to emerge. Despite they are rather week, it is principally important that they appear, it means that the theory is on the right track and new knowledge continues to have many non-trivial intersections with the old – what is a necessary feature of a healthy research project.

We mention here just three results of this kind, all concern the theory of CIV-DV prepotentials and all establish non-trivial relations with free fields on Riemann surfaces: the previous-level theory, from which one starts from when departing into the world of integrability and matrix models. The first result [103] identifies 1-loop correction to the CIV-DV prepotential with certain combination of determinants in some – yet unidentified – conformal theory. The second result [40] provides additional evidence in a form of a CFT representation (not fully polished yet) of the all-genus matrix-model partition function Z_W. As expected, this representation actually constructs the partition function by a Givental-style procedure from $2n$ (n = deg $W'(z)$) Kontsevich tau-functions τ_K, obtained by the application of star operators [39] at $2n$ points of the bare spectral curve (Riemann sphere), which become ramification points of the hyperelliptic spectral curve, associated with Z_W. The third result [43] formulates a kind of a perturbation theory for correlation functions of $2n$ operators inserted at these points and represents the partition function as a combinatorial "quantum field theory", essentially of Chern-Simons or, better, Batalin-Vilkovisky type. We consider this small set of preliminary results as a clear manifestation that all the ideas, discussed during the transcendental period can indeed be brought together and the head can catch the tail: the seemingly transcendental ideas will finally form a dense network of knowledge, not a shaky road leading into nowhere...

Acknowledgements

I am indebted for the invitation, hospitality and support to L. Baulieu, B. Pioline, E. Rabinovici and J. de Boer. I thank my numerous co-authors and friends and especially A. Alexandrov, A. Gerasimov, H. Itoyama, S. Kharchev, A. Marshakov, A. Mironov and A. Zabrodin for valuable discussions of matrix models. This work is partly supported by the Federal Program of the Russian Ministry of Industry, Science and Technology No 40.052.1.1.1112 and by the RFBR grant 04-02-16880.

References

1. A.Morozov, Sov.Phys.Usp. **35** (1992) 671-714 (Usp.Fiz.Nauk **162** 83 - 176), http://ellib.itep.ru/mathphys/people/morozov/92ufn-e1. ps & 92ufn-e2.ps
2. A.Alexandrov, A.Mironov and A.Morozov, Int. J. Mod. Phys. **A19** (2004) 4127, hep-th/0310113
3. V.Knizhnik, Sov.Phys.Usp. **32** (1989) 945 (Usp.Fiz.Nauk **159** 454), http://ellib.itep.ru/mathphys/people/ knizhnik/kniz1.ps.gz, kniz2.ps.gz & kniz3.ps.gz; Phys. Lett. **B180** (1986) 247; Comm. Math. Phys. **112** (1987) 567;
D.Lebedev and A.Morozov, Nucl. Phys. **B302** (1988) 342, http://ellib.itep.ru/mathphys/people/morozov/88_np302.ps;
A.Morozov, Nucl. Phys. **B303** (1988) 343-372, http://ellib.itep.ru/ mathphys/people/morozov/87-207.ps;
R.Kallosh and A.Morozov, Int. J. Mod. Phys. **A3** (1988) 1943-1958, http://ellib.itep.ru/mathphys/people/morozov/88-29.ps;
A.Gerasimov, A.Marshakov, A.Morozov, M.Olshanetsky and S.Shatashvili, Int. J. Mod. Phys. **A5** (1990) 2495, http://ellib.itep.ru/ mathphys/people/morozov/89-69.ps, 89-70.eps, 89-72.eps & 89-74.ps;
A.Morozov and A.Perelomov, *Complex Geometry and String Theory*, Encyclopedia of Mathematical Sciences (ed. G.M. Khenkin), Springer-Verlag, **54** (1993) 197-280, http://ellib.itep.ru/mathphys/ people/morozov/89-92.ps, 89-99.ps & 89-107.ps

4. D.Mumford, *Tata Lectures on Theta*, Progress in Mathematics, vol.28, Birkhauser, 1983, 1984
5. J.Fay, *Theta Functions on Riemann Surfaces*, LNM, #352, 1973
6. A.Morozov, Sov.Phys.Usp. **37** (1994) 1 (Usp.Fiz.Nauk **164** 3-62), hep-th/9303139; hep-th/9502091;
A.Mironov, Int. J. Mod. Phys. **A9** (1994) 4335, hep-th/9312212; Phys. of Particles and Nuclei **33** (2002) 537
7. Harish-Chandra, Am. J. Math. **79** (1957) 87;
C.Itzykson and J.-B.Zuber, J.Math.Phys. **21** (1980) 411;
J.Duistermaat and G.Heckman, Invent.Math. **69** (1982) 259
8. M.Kontsevich, Funk. Anal. Prilozh. **25** (1991) v.2, p.50; Comm. Math. Phys. **147** (1992) 1;
M.Adler and P. van Moerbeke, Comm.Math.Phys. **147** (1992) 25;
P.Di Francesco, C.Itzykson and J.-B.Zuber, Comm. Math. Phys. **151** (1993) 193
9. S.Kharchev, A.Marshakov, A.Mironov, A.Morozov and A.Zabrodin, Nucl. Phys. **B380** (1992) 181; Phys. Lett. **B275** (1992) 311;
S.Kharchev, hep-th/9810091
10. V.Kazakov and A.Migdal, Nucl. Phys. **B397** (1993) 214
11. S.Kharchev, A.Marshakov, A.Mironov and A.Morozov, Int. J. Mod. Phys. **A10** (1995) 2015, hep-th/9312210
12. E.P.Wigner, Ann.Math. **53** (1951) 36;
F.J.Dyson, J.Math.Phys. **3** (1962) 140
13. M.Mehta, *Random Matrices*, 2nd ed., Acad. Press., N.Y., 1991;
E.Brezin, C.Itzykson, G.Parisi and J.-B.Zuber, Comm. Math. Phys. **59** (1978) 35;
D.Bessis, C.Itzykson and J.-B.Zuber, Adv. Appl. Math. **1** (1980) 109
14. D.Gross and E.Witten, Phys.Rev. **D21** (1980) 446;
T.Eguchi and H.Kawai, Phys.Rev.Lett. **48** (1982) 1063;
P.Di Francesco, P.Ginsparg and J.Zinn-Justin, J. Phys. Rep. **254** (1995) 1
15. A.Migdal, Phys. Rep. **102** (1990) 199;
J.Ambjorn, J.Jurkiewicz and Yu.Makeenko, Phys. Lett. **B251** (1990) 517;
J.Ambjorn, L.Chekhov, C.F.Kristjansen and Yu.Makeenko, Nucl. Phys. **B404** (1993) 127, erratum in Nucl. Phys. **B449** (1995) 681, hep-th/9302014;
G.Akemann, Nucl. Phys. **B482** (1996) 403, hep-th/9606004
16. K.Demeterfi, N.Deo, S.Jain and C.-I.Tan, Phys.Rev. **D42** (1990) 4105-4122;
C.Crnkovic and G.Moore, Phys. Lett. **B257** (1991) 322;
F.David, Phys. Lett. **B302** (1993) 403, hep-th/9212106;
G.Bonnet, F.David and B.Eynard, J.Phys. **A33** (2000) 6739, cond-mat/0003324
17. A.Mironov and A.Morozov, Phys. Lett. **B252** (1990) 47-52, http://ellib.itep.ru/mathphys/people/morozov/90-252.ps;
F.David, Mod. Phys. Lett. **5** (1990) 1019;
J.Ambjorn and Yu.Makeenko, Mod. Phys. Lett. **5** (1990) 1753;
H.Itoyama and Y.Matsuo, Phys. Lett. **B255** (1991) 202
18. D.Gross ans M.Newman, Nucl. Phys. **B380** (1992) 168, hep-th/9112069
19. A.Marshakov, A.Mironov and A.Morozov, Mod. Phys. Lett. **7** (1992) 1345-1360, hep-th/9201010
20. M.Fukuma, H.Kawai and R.Nakayama, Int. J. Mod. Phys. **A6** (1991) 1385;
R.Dijkgraaf, E.Verlinde and H.Verlinde, Nucl. Phys. **B348** (1991) 435
21. E.Witten, in Proc.of the XX-th Int. Conf. on Diff. Geom. Methods in Theor. Phys., Vol.1,2 (1991) 176-216, World Sci. Publ., River Edge, NJ, 1992, Eds. S.Catto and A.Rocha;
A.Marshakov, A.Mironov and A.Morozov, Phys. Lett. **B274** (1992) 280
22. S.Kharchev, A.Marshakov, A.Mironov and A.Morozov, Mod. Phys. Lett. **8** (1993) 1047-1062, Theor.Math.Phys. **95** (1993) 571-582, hep-th/0208046
23. A.Mironov and A.Morozov, Phys. Lett. **B490** (2000) 173-179, hep-th/0005280
24. A.Gerasimov, A.Marshakov, A.Mironov, A.Morozov and A.Orlov, Nucl. Phys. **B357** (1991) 565
25. A.Gorsky, I.Krichever, A.Marshakov, A.Mironov and A.Morozov, Phys. Lett. **B355** (1995) 466, hep-th/9505035;

A.Morozov, hep-th/9903087.

26. M.Jimbo, T.Miwa and M.Sato, Publ. RIMS **14** (1978) 223, **15** (1979) 201, 577, 871, 1531
27. A.Gerasimov, S.Khoroshkin, D.Lebedev, A.Mironov and A.Morozov, Int. J. Mod. Phys. **A10** (1995) 2589-2614, hep-th/9405011;
A. Mironov, A. Morozov and L. Vinet, Theor.Math.Phys. **100** (1995) 890-899 (Teor.Mat.Fiz. **100** (1994) 119-131), hep-th/9312213
28. A.Mironov, A.Morozov and G.Semenoff, Int. J. Mod. Phys. **A11** (1996) 5031, hep-th/9404005
29. A.Morozov, Mod. Phys. Lett. **7** (1992) 3503-3508, hep-th/9209074;
S.Shatashvili, Commun. Math. Phys. **154** (1993) 421-432, hep-th/9209083;
L.D. Paniak Nucl. Phys. **B553** (1999) 583-600, hep-th/9902089;
P. Zinn-Justin and J.-B. Zuber, J.Phys. **A36** (2003) 3173-3194, math-ph/0209019;
M. Bertola and B. Eynard, J.Phys. **A36** (2003) 7733-7750, hep-th/0303161;
S. Aubert, C.S. Lam, J.Math.Phys. **44** (2003) 6112-6131, math-ph/0307012; J.Math.Phys. **45** (2004) 3019-3039, math-ph/0405036
30. G.t'Hooft, hep-th/0208054;
A.Morozov and A.Niemi, *Nucl.Phys.* **B666** (2003) 311-336, hep-th/0304178
31. M.Tierz, hep-th/0308121;
E.Goldfain, *Chaos, Solitons and Fractals* (2004);
S.Franco (MIT, LNS), Y.H.He, C.Herzog and J.Walcher, Phys.Rev. **D70** (2004) 046006, hep-th/0402120;
J.I.Latorre, C.A. Lutken, E. Rico and G. Vidal, quant-ph/0404120;
J.Gaite, J.Phys. **A37** (2004) 10409-10420, hep-th/0404212;
T. Oliynyk, V. Suneeta and E. Woolgar, hep-th/0410001.
32. A.Zamolodchikov, JETP Lett. **43** (1986) 730; Sov. J. Nucl. Phys. **46** (1987) 1090
33. P.F.Bedaque, H.-W.Hammer and U.van Kolck, Phys.Rev.Lett. **82** (1999) 463, nucl-th/9809025;
D.Bernard and A.LeClair, Phys.Lett. **B512** (2001) 78, hep-th/0103096;
S.D.Glazek and K.G.Wilson, Phys.Rev. **D47** (1993) 4657, hep-th/0203088;
A.LeClair, J.M.Roman and G.Sierra, Phys.Rev. **B69** (2004) 20505, cond-mat/0211338; Nucl.Phys. **B675** (2003) 584-606, hep-th/0301042; **B700** (2004), 407-435, hep-th/0312141;
E.Braaten and H.-W.Hammer, cond-mat/0303249
34. V.Dolotin and A.Morozov, hep-th/0501235
35. A.Gerasimov, D.Lebedev and A.Morozov, Int. J. Mod. Phys. **A6** (1991) 977-988, http://ellib.itep.ru/mathphys/people/morozov/90-04.ps;
A.Morozov, Mod. Phys. Lett. **6** (1991) 1525-1532,
http://ellib.itep.ru/mathphys/people/morozov/90-50.ps
36. N.Arkani-Hamed, S.Dimopoulos and S.Kachru, hep-th/0501082
37. S. Kharchev, A. Marshakov, A. Mironov, A. Morozov and S. Pakuliak, Nucl. Phys. **B404** (1993) 717, hep-th/9208044;
A.Mironov and S.Pakuliak, Theor. Math. Phys. **95** (1993) 604-625 (Teor. Mat. Fiz. **95** (1993) 317-340)
38. I.Kostov, Phys. Lett. **B297** (1992) 74-81;
J.Alfaro and I.Kostov, hep-th/9604011
39. G.Moore, Comm.Math.Phys. **133** (1990) 261-304; Prog. Theor. Phys. Suppl. **102** (1990) 255-286
40. I.Kostov, *COnformal Field Theory Techniques in Random Matrix Models*, preprint SPHT/t98/112, 1998
41. L.Chekhov and Yu.Makeenko, Phys. Lett. **B278** (1992) 271
42. S.Kharchev, A.Marshakov, A.Mironov and A.Morozov, Nucl. Phys. **B397** (1993) 339, hep-th/9203043.
43. B.Eynard, JHEP **0411** (2004) 031, hep-th/0407261
44. A.Alexandrov, A.Mironov and A.Morozov, hep-th/0412099, hep-th/0412205
45. E.Brezin and V.Kazakov, Phys. Lett. **B236** (1990) 144;
M.Douglas and S.Shenker, Nucl. Phys. **B335** (1990) 635;
D.Gross amd A.Migdal, Phys.Rev.Lett. **64** (1990) 127

46. Yu.Makeenko, A.Marshakov, A.Mironov and A.Morozov, Nucl. Phys. **B356** (1991) 574-628, http://ellib.itep.ru/mathphys/people/ morozov/mmmm.ps
47. A.Givental. math.AG/0009067;
J.S.Song and Y.S.Song, J.Math.Phys. **45** (2004) 4539-4550, hep-th/0103254;
A.Alexandrov, J.Math.Phys. **44** (2003) 5268-5278, hep-th/0205261
48. A.Morozov, Talk at INTAS-RFBR School, September 98, hep-th/9810031
49. A.Morozov and L.Vinet, Int. J. Mod. Phys. **A13** (1998) 1651, hep-th/9409093
50. A.Gerasimov, A.Morozov and K.Selivanov, Int. J. Mod. Phys. **A16** (2001) 1531-1558, hep-th/0005053
51. D.Kreimer, Adv. Theor. Math. Phys. **2** (1998) 303-334, hep-th/9810022;
A.Connes and D.Kreimer, Comm. Math. Phys. **199** (1998) 203-242; Lett. Math. Phys. **48** (1999) 85-96, hep-th/9904044; JHEP 9909024, hep-th/9909126; Comm. Math. Phys. **210** (2000) 249-273, hep-th/9912092; hep-th/0003188; Annales Henri Poincare **3** (2002) 411-433, hep-th/0201157;
A.Connes and M.Marcolli, math.NT/0409306; hep-th/0411114
52. D.Malyshev, JHEP **0205** (2002) 013, hep-th/0112146; hep-th/0408230
53. R.Hirota, Topics in Current Physics **17** (1980) 157
54. I.Krichever, Russ. Math. Surveys **32** (1977) v.6 p.185 (Usp.Mat.Nauk **32** (1977) v.6 p.183)
55. S.Novikov, *unpublished*;
B.Dubrovin, Sov.Math.Usp. **36** (1981) v.2 p.11-80;
T.Shiota, *Characterization of Jacobian Varieties in Terms of Soliton Equations*, Harvard University, 1984;
M.Mulase, J.Diff.Geom. **19** (1984) 403-430; Invent.Math. **92** (1988) 1-46
56. A.Marshakov, M.Martellini and A.Morozov, Phys. Lett. **B418** (1998) 294-302, hep-th/9706050
57. A.Mironov and A.Morozov, Phys. Lett. **B524** (2002) 217-226, hep-th/0107114
58. V.Kazakov, Phys. Lett. **B237** (1990) 212;
I.Kostov, Phys. Lett. **B238** (1990) 181;
J. de Boer, A.Sinkovics, E.Verlinde and J.-T. Yee, JHEP **0403** (2004) 023, hep-th/0312135;
C.Johnson, hep-th/0408049
59. M. Adler, A. Morozov, T. Shiota and P. van Moerbeke, Nucl. Phys. Proc. Suppl. **49** (1996) 201-212, hep-th/9603066; hep-th/9610137
60. D.B. Fairlie, J. Govaerts, A. Morozov, Nucl. Phys. **B373** (1992) 214-232, hep-th/9110022
61. N.Dorey, T.Hollowood, S.Prem Kumar and A.Sinkovics, JHEP **11** (2002) 039, hep-th/0209089
62. A.Gerasimov, S.Kharchev, A.Marshakov, A.Mironov, A.Morozov and M.Olshanetsky, Int. J. Mod. Phys. **A12** (1997) 2523-2584, hep-th/9601161
63. I.K.Kostov, I.Krichever, M.Mineev-Weinstein, P.Wiegmann and A.Zabrodin, hep-th/0005259;
I.Krichever, A.Marshakov and A.Zabrodin, hep-th/0309010;
I.Krichever, M.Mineev-Weinstein, P.Wiegmann and A.Zabrodin, nlin.SI/0311005
64. A.M. Levin, M.A. Olshanetsky and A. Zotov, nlin.SI/0110045;
A. Levin and M. Olshanetsky, hep-th/0301078
65. J.A.Minahan and K.Zarembo, JHEP **0303** (2003) 013, hep-th/0212208;
V.Kazakov, A.Marshakov, J.A.Minahan and K.Zarembo, JHEP **0405** (2004) 024, hep-th/0402207;
L.Dolan, C.R.Nappi and E.Witten, JHEP **0310** (2003) 017, hep-th/0308089, hep-th/0401243;
N.Beisert, G.Ferretti, R.Heise, and K.Zarembo, hep-th/0412029
66. R.Boels, J. de Boer, R.Duivenvoorden and J.Wijnhout, JHEP **0403** (2004) 010, hep-th/0305189;
R.Boels and Jan de Boer, hep-th/0411110
67. V.Kazakov and A.Marshakov, J.Phys. **A36** (2003) 3107-3136, hep-th/0211236
68. A.V. Belitsky, V.M. Braun, A.S. Gorsky and G.P. Korchemsky, Int. J. Mod. Phys. **A19** (2004) 4715-4788, hep-th/0407232
69. V. Fock, A. Gorsky, N. Nekrasov and V. Rubtsov, JHEP **0007** (2000) 028, hep-th/9906235;
H.Braden, A.Gorsky, A.Odesskii and V.Rubtsov, Nucl. Phys. **B633** (2002) 414-442, hep-

th/0111066
70. H.Braden, A.Marshakov, A.Mironov and A.Morozov, hep-th/9906240;
A.Mironov and A.Morozov, Phys. Lett. **B475** (2000) 71-16, hep-th/9912088; hep-th/0001168
71. A.Schwarz, Lett. Math. Phys. **2** (1978) 247;
E.Witten, Comm.Math.Phys. **117** (1988) 353; **118** (1988) 411; Nucl. Phys. **B340** (1990) 281;
R.Dijkgraaf and E.Witten, Nucl. Phys. **B342** (1990) 486;
D.Birmingham, M.Blau, M.Rakowski and G.Thompson, Phys. Repts. **209** (1991) 129;
R.Dijkgraaf, Nucl. Phys. **B342** (1990) 486; Nucl. Phys. **B352** (1991) 59; hep-th/9201003;
T.Eguchi, K.Hori, C.-S.Xiong, Int. J. Mod. Phys. **A12** (1997) 1743-1782, hep-th/9605225;
A.Losev, hep-th/9801179;
A.Gerasimov and S.Shatashvili, JHEP **0411** (2004) 074, hep-th/0409238
72. E.Witten, Surv.Diff.Geom. **1** (1991) 243;
R.Dijkgraaf, E.Verlinde and H.Verline, Nucl. Phys. **B352** (() 1991) 59;
B.Dubrovin, LNM **1620** (1996) 120-348, hep-th/9407018
73. A.Marshakov, A.Mironov and A.Morozov, Phys. Lett. **B389** (1996) 43-52, hep-th/9607109; Mod. Phys. Lett. **12** (1997) 773-788, hep-th/9701014; Int. J. Mod. Phys. **A15** (2000) 1157-1206, hep-th/9701123;
A.Mironov and A.Morozov, Phys. Lett. **B424** (1998) 48-52, hep-th/9712177;
A.Morozov, Phys. Lett. **B427** (1998) 93-96, hep-th/9711194;
G.Bonelli and M.Matone, Phys.Rev. **D58** (1998) 045006, hep-th/9712025;
G.Bertoldi and M.Matone, Phys. Lett. **B425** (1998) 104-106, hep-th/9712039; Phys.Rev. **D57** (1998) 6483-6485, hep-th/9712109;
B. de Wit and A. Marshakov, Theor.Math.Phys. **129** (2001) 1504-1510 (Teor.Mat.Fiz. **129** (2001) 230-238), hep-th/0105289;
H.W. Braden and A. Marshakov, Phys. Lett. **B541** (2002) 376-383, hep-th/0205308
74. I.Batalin and G.Vilkovisky, Phys. Lett. **B102** (1981) 27-31; Phys. Lett. **B120** (1983) 166-170; Phys. Rev. **D28** (1983) 2567-2582; Nucl. Phys. **B234** (1984) 106-124; J. Math. Phys. **26** (1985) 172-184;
I.Batalin, Phys.Rev. **D28** (1983) 2567; J.Math.Phys. **26** (1985) 172
75. E.Witten, Phys.Rev. **D46** (1992) 5467-5473, hep-th/9208027; Phys.Rev. **D47** (1993) 3405-3410, hep-th/9210065; hep-th/9306122;
M. Bershadsky, S. Cecotti, H. Ooguri and C. Vafa, Comm. Math. Phys. **165** (1994) 311-428, hep-th/9309140; Nucl. Phys. **B405** (1993) 279-304, hep-th/9302103;
W.Siegel, Phys. Lett. **B151** (1984) 391; hep-th/0107094
B.Zwiebach Nucl. Phys. **B390** (1993) 33, hep-th/9206084;
N. Berkovits and W. Siegel, Nucl. Phys. **B505** (1997) 139-152, hep-th/9703154;
N.Berkovits, JHEP **0004** (2000) 018, hep-th/0001035; JHEP **0409** (2004) 047, hep-th/0406055;
N.Berkovits and P.Howe, Nucl. Phys. **B635** (2002) 75-105, hep-th/0112160;
A.Gorodentsev and A.Losev, lectures at ITEP-DIAS School, 2004.
76. N.Seiberg and E.Witten, Nucl. Phys. **B426** (1994) 19, hep-th/9407087; Nucl. Phys. **B431** (1994) 484, hep-th/9408099;
A.Klemm, W.Lerche, S.Theisen and S.Yankielowicz, Phys. Lett. **B344** (1995) 169, hep-th/9411048;
E.Martinec and N.Warner, Nucl. Phys. **B459** (1996) 97-112, hep-th/9509161;
T.Eguchi and S.Yang, Mod. Phys. Lett. **11** (1996) 131-138, hep-th/9510183;
R.Donagi and E.Witten, Nucl. Phys. **B460** (1996) 299, hep-th/9510101;
Eric D'Hoker, I.M. Krichever, D.H. Phong, Nucl.Phys. **B489** (1997) 179-210, 211-222, hep-th/9609041, hep-th/9609145; hep-th/0212313;
R.Carroll, hep-th/9712110, hep-th/9802130, hep-th/9804086, hep-th/9905010;
A.Marshakov, *Seiberg-Witten Theory and Integrable Systems*, World Scientific, Singapore, 1999
77. B.Dubrovin, I.Krichever and S.Novikov, *Integrable Systems I*, VINITI, Dynamical Systems **4** (1985) 179;
I.Krichever, Comm. Math. Phys. **143** (1992) 415, hep-th/9205110;

B.Dubrovin, Nucl. Phys. **B379** (1992) 627;
T.Nakatsu and K.Takasaki, Mod. Phys. Lett. **11** (1996) 417, hep-th/9509162

78. H.Itoyama and A.Morozov, Nucl. Phys. **B477** (1996) 855, hep-th/9511125; Nucl. Phys. **B491** (1997) 529, hep-th/9512161; hep-th/9601168;
A.Gorsky, A.Marshakov, A.Mironov and A.Morozov, Nucl. Phys. **B527** (1998) 690-716, hep-th/9802007
79. A.Gorsky, A.Marshakov, A.Mironov and A.Morozov, hep-th/9604078
80. A. Losev, G. Moore, N. Nekrasov and S. Shatashvili, Nucl.Phys.Proc.Suppl. **46** (1996) 130-145, hep-th/9509151; hep-th/9511185; Nucl. Phys. **B484** (1997) 196-222, hep-th/9606082;
A. Losev, N. Nekrasov and S. Shatashvili, hep-th/9908204; Class. Quant. Grav. **17** (2000) 1181-1187, hep-th/9911099;
N.Nekrasov, Adv. Theor. Math. Phys. **7** (2004) 831-864, hep-th/0206161, hep-th/0306211;
R.Flume and R. Poghossian, Int. J. Mod. Phys. **A18** (2003) 2541, hep-th/0208176;
A.Losev, A.Marshakov and N.Nekrasov, hep-th/0302191;
N.Nekrasov and A.Okounkov, hep-th/0306238;
N.Nekrasov, S.Shadchin, Comm. Math. Phys. **252** (2004) 359-391, hep-th/0404225
81. E.Getzler, Comm. Math. Phys. **163** (1994) 473-489, hep-th/9305013; alg-geom/9612004; math.AG/9801003
82. A.Losev and S.Shadrin, preprint ITEP-TH-67/04
83. M.Kontsevich and Yu.Manin, Comm.Math.Phys. **164** (1994) 525
84. A.Belavin, A.Polyakov and A.Zamolodchikov, Nucl. Phys. **B241** (1984) 333;
V.Dotsenko, Proceedings of ITEP Winter School, **12** (1985) v.3 p.90-140
85. S.Khoroshkin and V.Tolstoy, hep-th/9406194;
G.Felder, hep-th/9412207;
S. Khoroshkin, D. Lebedev and S. Pakuliak, q-alg/9702002;
A.Varchenko and G.Felder, q-alg/9704005
86. V.Bazhanov, S.Lukyanov and A.Zamolodchikov, Comm. Math. Phys. **177** (1996) 381-398, hep-th/9412229; Comm. Math. Phys. **190** (1997) 247-278, hep-th/9604044; Comm. Math. Phys. **200** (1999) 297-324, hep-th/9805008; Adv.Theor.Math.Phys. **7** (2004) 711-725, hep-th/0307108
87. S. Kharchev, D. Lebedev and M. Semenov-Tian-Shansky, Comm. Math. Phys. **225** (2002) 573-609, hep-th/0102180;
A.Gerasimov, S.Kharchev and D.Lebedev, math.QA/0204206; math.QA/0402112;
A. Gerasimov, S. Kharchev, D. Lebedev and S. Oblezin, math.AG/0409031
88. H. W. Braden, A.Mironov and A.Morozov, Phys. Lett. **B514** (2001) 293-298, hep-th/0105169
89. F.Cachazo, K.Intriligator and C.Vafa, Nucl. Phys. **B603** (2001) 3, hep-th/0103067;
F.Cachazo and C.Vafa, hep-th/0206017
90. R.Dijkgraaf and C.Vafa, Nucl. Phys. **B644** (2002) 3, hep-th/0206255; Nucl. Phys. **B644** (2002) 21, hep-th/0207106; hep-th/0208048
91. L.Chekhov and A.Mironov, Phys. Lett. **B552** (2003) 293, hep-th/0209085
92. A.Klemm, M.Marino and S.Theisen, JHEP **0303** (2003) 051, hep-th/0211216
93. A.Gorsky, Phys. Lett. **B554** (2003) 185-189, hep-th/0210281
94. F.Cachazo, M.Douglas, N.Seiberg and E.Witten, JHEP **0212** (2002) 071, hep-th/0211170;
F.Cachazo, N.Seiberg and E.Witten, JHEP **0302** (2003) 042, hep-th/0301006; JHEP **0304** (2003) 01, hep-th/0303207
95. H.Itoyama and A.Morozov, Nucl. Phys. **B657** (2003) 53, hep-th/0211245
96. H.Itoyama and A.Morozov, Phys. Lett. **B555** (2003) 287, hep-th/0211259
97. H.Itoyama and A.Morozov, Progr. Theor. Phys. **109** (2003) 433, hep-th/0212032
98. H.Itoyama and A.Morozov, Int. J. Mod. Phys. **A18** (2003) 5889, hep-th/0301136
99. L.Chekhov, A.Marshakov, A.Mironov and D.Vasiliev, Phys. Lett. **B562** (2003) 323, hep-th/0301071
100. T.Eguchi and Y.Sugawara (Tokyo U.), JHEP **0305** (2003) 063, hep-th/0305050
101. S.Naculich, H.Schnitzer and N.Wyllard, JHEP **0301** (2003) 015, hep-th/0211254;
I.Bena, R.Roiban and R.Tatar, Nucl. Phys. **B679** (2004) 168-188, hep-th/0211271;
B.Feng, Nucl. Phys. **B661** (2003) 113-138, hep-th/0212010; Phys.Rev. **D68** (2003) 025010,

hep-th/0212274;
K.Ohta, JHEP **0302** (2003) 057, hep-th/0212025;
S.Seki, Nucl. Phys. **B661** (2003) 257-272, hep-th/0212079;
I.Bena, S. de Haro and R.Roiban, Nucl. Phys. **B664** (2003) 45-58, hep-th/0212083;
C.Hofman, JHEP **0310** (2003) 022, hep-th/0212095;
H.Suzuki, JHEP **0303** (2003) 036, hep-th/0212121;
Y.Demasure and R.Janik, Nucl. Phys. **B661** (2003) 153-173, hep-th/0212212;
C.Ahn and S.Nam (Kyung Hee U.), Phys. Lett. **B562** (2003) 141-146, hep-th/0212231;
A. Mironov, Fortsch.Phys. **51** (2003) 781-786, hep-th/0301196;
R.Roiban, R.Tatar and J.Walcher, Nucl. Phys. **B665** (2003) 211-235, hep-th/0301217;
Y.Ookouchi and Y.Watabiki, Mod. Phys. Lett. **18** (2003) 1113-1126, hep-th/0301226;
C.Ahn and Y.Ookouchi, JHEP **0303** (2003) 010, hep-th/0302150;
D.Berenstein, JHEP **0306** (2003) 019, hep-th/0303033;
H.Itoyama and H.Kanno, Phys. Lett. **B573** (() 2003) 227-234, hep-th/0304184; Nucl. Phys. **B686** (2004) 155-164, hep-th/0312306;
M.Matone and L.Mazzucato, JHEP **0307** (2003) 015, hep-th/0305225;
M.Alishahiha, J. de Boer, A.Mosaffa and J.Wijnhout, JHEP **0309** (2003) 066, hep-th/0308120;
S.Aoyama and T.Masuda, JHEP **0403** (2004) 072, hep-th/0309232;
R.Argurio, G.Ferretti and R.Heise, Int. J. Mod. Phys. **A19** (2004) 2015-2078, hep-th/0311066;
M.Gomez-Reino, JHEP **0406** (2004) 051, hep-th/0405242;
P.B.Ronne, hep-th/0408103;
K.Fujiwara, Hiroshi Itoyama and M.Sakaguchi, hep-th/0409060

102. M.Matone, Nucl. Phys. **B656** (2003) 78-92, hep-th/0212253;
A.Dymarsky and V.Pestun, Phys.Rev. **D67** (2003) 125001, hep-th/0301135
103. L.Chekhov, hep-th/0401089
104. G.t'Hooft, Nucl. Phys. **B72** (1974) 461

PHENOMENOLOGY OF NEUTRINO OSCILLATIONS

LEV B. OKUN
ITEP, Moscow, Russia

Abstract.
A short review of phenomenological description of neutrino oscillations and recent experimental data on neutrino masses and mixings.

1. A short history of neutrinos [1-16]

It took about 30 years from the discovery of radioactivity of uranium in 1896 by Henry Becquerel to the hypothesis of neutrino by Wolfgang Pauli in 1930. It was conjectured to solve a few puzzles of β-decay, such as continuous β-spectra. The new particle was baptized by Enrico Fermi who also gave in 1933 the first theory of β-decay. In this theory neutron transformed into proton by emitting electron and antineutrino, $\bar{\nu}$, while proton transformed into neutron emitting positron and neutrino, ν.

The difference between neutrino and antineutrino has been questioned in 1937 by Ettore Majorana, who suggested that $\bar{\nu}$ might be identical to ν, or in other words, that neutrino, like photon, could be a genuinely neutral particle. This tantalizing possibility is still the goal of many experiments looking for neutrinoless double beta decay $2\beta 0\nu$.

After the World War 2 neutrino remained a kind of Cinderella among more famous elementary particles. But there were physicists unshakebly faithful to it. For me the first neutrino knight was Bruno Pontecorvo with whom I was in touch for more than forty years.

In his famous 1946 Chalk River Report he suggested radiochemical Cl-Ar method of detecting neutrinos from the sun, nuclear reactors, accelerators, radioactive sources. In 1958 he raised the question of neutrino-antineutrino oscillations and in 1959 suggested an experimental search for muonic neutrinos. In 1969 he and Volodya Gribov derived formulas describing $\nu_\mu \leftrightarrow \nu_e$ oscillations. (The idea of $\nu_\mu \leftrightarrow \nu_e$ mixing was put forward in 1962 by Ziro Maki, Masami Nakagawa and Shoichi Sakata.) The formula for $\nu_e \leftrightarrow \nu_\mu$ oscillations without "extra factor of 2" misprint, which one can find in some of Pontecorvo papers, was published by Samoil Bilenky and Pontecorvo and by Harald Fritsch and Peter Minkowski in 1976. In the subsequent years a number of authors argued in favor of this "extra factor of 2". But the correct formula has no "extra 2".

L. Baulied et al. (eds.), String Theory: From Gauge Interactions to Cosmology, 163–175.

Two important steps in the theory of neutrino oscillations were made by Lincoln Wolfenstein in 1978 and by Stanislav Mikheev and Alexey Smirnov in 1985. In the first step the charged current contribution to the ν_e-scattering in matter was taken into account. In the second step the resonant $\nu_\mu \leftrightarrow \nu_e$ transition due to this scattering in matter was predicted. This is the famous MSW effect.

The most important land-mark experimental discoveries in the second half of the last century were:

The first detection of $\bar{\nu}$ from a nuclear reactor by Frederic Reines and Clyde Cowan in reaction $\bar{\nu}p \to ne^+$ (1953).

The experimental discovery of parity violation in weak interactions (E. Ambler et al.; R.L. Garwin, L.M. Lederman, M. Weinrich; J.J. Friedman, V.L. Telegdi (1957)) and the idea of longitudinal neutrinos (T.D. Lee, C.N. Yang, L. Landau, A. Salam, 1957).

The discovery by Leon Lederman, Melvin Schwartz, Jack Steinberger et al. that ν_μ and ν_e are two different particles (1962).

The discovery of τ-lepton by Martin Perl (1975) and the proof by four collaborations at LEP I Collider at CERN at the end of last century that only three light neutrinos, ν_e, ν_μ, ν_τ are coupled to Z-boson.

2. Simple formulas for oscillations

Let us start by considering an oversimplified example of two flavors: ν_e, ν_μ. Introduce two unit vectors: ν_l, ν_α, where $l = e, \mu$, while $\alpha = 1, 2$ correspond to two masses: m_1, m_2.

Let us define the mixing angle θ: $c \equiv \cos\theta$, $s \equiv \sin\theta$.

$$\nu_e = c\nu_1 + s\nu_2 \quad , \qquad \nu_1 = c\nu_e - s\nu_\mu \quad ;$$
$$\nu_\mu = -s\nu_1 + c\nu_2 \quad , \qquad \nu_2 = s\nu_e + c\nu_\mu \quad .$$

Usually neutrino oscillations are described by wave functions in "reduced" plane wave approximation: $\nu_l(x,t)$ are expressed in terms of $\nu_\alpha(x,t)$ with fixed boundary conditions at $x = 0, t = 0$.

$$\nu_\alpha(x,t) = \nu_\alpha e^{ip_\alpha x - iE_\alpha t} \quad , \quad \alpha = 1, 2$$
$$\nu_e(x,t) = c\nu_1(x,t) + s\nu_2(x,t) \; ; \;\; \nu_e(0,0) = \nu_e \quad .$$
$$\nu_\mu(x,t) = -s\nu_1(x,t) + c\nu_2(x,t) \; ; \;\; \nu_\mu(0,0) = \nu_\mu \quad .$$

Two scenarios are considered in the literature: "same energy" vs "same momentum". "Same energy": $E_1 = E_2 = E$, no oscillations in time at any x, in particular at $x = 0$. Free neutrinos (on mass shell): $p_\alpha = \sqrt{E^2 - m_\alpha^2} = E - \frac{m_\alpha^2}{2E}$.

$$\Delta p = p_2 - p_1 = \frac{\Delta m^2}{2E} \quad , \quad \Delta m^2 = m_1^2 - m_2^2$$

Consider the case when the source emits electron neutrinos: $\nu(0,0) = \nu_e$. Then the evolution is described by

$$\nu_e(x,t) = [c\nu_1 + s\nu_2 e^{i\varphi(x)}]e^{-iEt + ik_1 x} \quad ,$$

where

$$\varphi(x) = \Delta p x = \frac{\Delta m^2}{2E} x \ .$$

In order to find projections of $\nu_e(x,t)$ on ν_μ and ν_e let us rewrite $\nu_e(x,t)$ in terms of these flavor states:

$$\nu_e(x,t) = [c(c\nu_e - s\nu_\mu) + s(s\nu_e + c\nu_\mu)e^{i\varphi(x)}]e^{-iEt+ip_1 x}.$$

Now it is easy to find probabilities of transition $\nu_e \to \nu_\mu$ and of survival $\nu_e \to \nu_e$.

a) Transition $\nu_e \to \nu_\mu$ $\quad (\nu(x) = \nu_\mu)$:

$$P(\nu_e \to \nu_\mu) = (sc)^2 \cdot |-1 + e^{i\varphi}|^2 = (sc)^2 \cdot 2(1 - \cos\varphi) =$$

$$= \sin^2 2\theta \sin^2 \tfrac{\varphi}{2} = \sin^2 2\theta \sin^2 \tfrac{\Delta m^2 x}{4E}$$

b) Survival $\nu_e \to \nu_e$ $\quad (\nu(x) = \nu_e)$:

$$P(\nu_e \to \nu_e) = 1 - P(\nu_e \to \nu_\mu) = 1 - \sin^2 2\theta \sin^2 \frac{\Delta m^2 x}{4E}$$

For applications of these formulas it is convenient to return from $\hbar, c$ units, which we tacitly used before, to units eV for m, GeV for E, and km for x and to replace x by L which is widely used in literature to denote the distance between the source and the detector of neutrino. Then

$$\varphi(L) = \Delta m^2 L/4E\hbar c = 1.27\Delta m^2(\text{eV}^2)L(\text{km})/E(\text{GeV}) \ .$$

Here a comment on dimensions and coefficient 1.27 is in order.

$$[Et] = [pL] = [\hbar]\ , \quad [(\Delta m^2/E) \cdot L] = [EL] = [Ect] = [\hbar c] \ ,$$

where $[Et]$, ... , $[hc]$ means dimension of Et, ... , hc.

m_1, m_2 are measured in eV, Δm^2 is measured in eV^2,
L is measured in kilometers.
E is measured in GeV.

Recall that $\hbar = 6.582 \cdot 10^{-25}$ GeV·s ,
$c = 299792$ km/s ,
$\hbar c = 1.973 \cdot 10^{-19}$ Gev km .
Hence

$$\frac{\text{eV}^2 \cdot \text{km}}{4\text{GeV}\ \hbar c} = \frac{10^{-18}\ \text{GeV} \cdot \text{km}}{4 \cdot 1.973 \cdot 10^{-19}\ \text{GeV} \cdot \text{km}} = 1.27 \ .$$

My PhD adviser Isaak Yakovlevich Pomeranchuk (1913 – 1966) used to quote a Ukranian philosopher Grigory Skovoroda (1722 – 1794):

"Thanks God: All what is relevant is simple, all what is not simple is not relevant."

Most of the experiments done up to now have been analyzed in terms of the "reduced" two neutrino-flavors plane wave approximation, described in this section. I am tempted to dub it "Skovoroda approximation" for its simplicity.

While oscillations of two neutrinos are described by two masses, m_1, m_2, and one mixing angle θ_{12}, the oscillations of three neutrinos, ν_e, ν_μ, ν_τ, are described by three masses, m_1, m_2, m_3, three angles $\theta_{12}, \theta_{13}, \theta_{23}$, which define the 3×3 PMNS (Pontecorvo–Maki–Nakagawa–Sakata) matrix $U_{l\alpha}$, where l is flavor index ($l = e, \mu, \tau$), while α is mass index ($\alpha = 1, 2, 3$). CP-violation is parametrized by one phase δ in case of Dirac neutrinos, like CKM (Cabibbo–Kobayashi–Maskawa) for quarks, and by three phases $(\delta, \alpha_1, \alpha_2)$ in case of Majorana neutrinos.

$$\begin{pmatrix} \nu_e \\ \nu_\mu \\ \nu_\tau \end{pmatrix} = \begin{pmatrix} U_{e_1} & U_{e_2} & U_{e_3} \\ U_{\mu_1} & U_{\mu_2} & U_{\mu_3} \\ U_{\tau_1} & U_{\tau_2} & U_{\tau_3} \end{pmatrix} \begin{pmatrix} \nu_1 \\ \nu_2 \\ \nu_3 \end{pmatrix} ,$$

$$U = \begin{pmatrix} c_{12}c_{13} & s_{12}c_{13} & s_{13}e^{-i\delta} \\ -s_{12}c_{23} - c_{12}s_{23}s_{13}e^{i\delta} & c_{12}c_{23} - s_{12}s_{23}s_{13}e^{e\delta} & s_{23}c_{13} \\ s_{12}s_{23} - c_{12}c_{23}s_{13}e^{i\delta} & -c_{12}s_{23} - s_{12}c_{23}s_{13}e^{i\delta} & c_{23}c_{13} \end{pmatrix}$$

$$\times \mathrm{diag}(e^{i\alpha_1/2}, e^{i\alpha_2/2}, 1) \ .$$

$$c_{ij} \equiv \cos\theta_{ij} \ , \quad s_{ij} \equiv \sin\theta_{ij} \ .$$

The Majorana phases are sometimes defined by a matrix $\mathrm{diag}(1, e^{i\phi_2}, e^{i\phi_3})$. They do not influence the neutrino oscillations.

3. "Same energy" vs "same momentum" [12-14], [17-25]

In the previous section we assumed that the neutrino of given flavor is a superposition of neutrinos of different masses with equal energies. In case of three neutrinos ($l = e, \mu, \tau$; $\alpha = 1, 2, 3$):

$$E_1 = E_2 = E_3 = E \ , \ \ p_\alpha = E - \frac{m_\alpha^2}{2E} \ .$$

The evolution of matrix U with distance L is expressed by multiplication of each matrix element $U_{l\alpha}$ by $e^{ip_\alpha L}$.

The oscillations in space are caused by

$$\Delta p_{\alpha\beta} L = (p_\alpha - p_\beta) L = -\frac{\Delta m_{\alpha\beta}^2}{2E} L \ ,$$

where $\Delta m_{\alpha\beta}^2 = m_\alpha^2 - m_\beta^2$.

Historically, the prototype of neutrino oscillations were oscillations of neutral kaons which traditionally were described as a result of evolution in time. Hence Gribov and Pontecorvo considered evolution of neutrino wave function in time. Accordingly they assumed that three neutrinos of different masses have the same momentum: $p_1 = p_2 = p_3 = p$. Then their energies were $E_\alpha = p + m_\alpha^2/2p$ and evolution of matrix U with time was due to factors $e^{-iE_\alpha t}$. The oscillations in time were caused by

$$\Delta E_{\alpha\beta} t = (E_\alpha - E_\beta) t = \frac{\Delta m_{\alpha\beta}^2}{2p} t \ .$$

By taking into account that for ultrarelativistic neutrinos equations $pc = E$ and $tc = L$ are valid with good accuracy, one derives for the "same momentum" scenario the same result (up to a non-essential sign) as for the "same energy" one. The fact that for the "same momentum" the initial wave function of neutrino has oscillating flavor at $x = 0$ has been traditionally neglected. This was recently stressed by Michail Vysotsky.

The use of "same energy" was first demonstrated by Igor Kobzarev et al. in 1982 by considering the production of neutrinos on heavy nuclei by a monoenergetic beam of electrons. Further arguments in favor of "same energy" were presented by Leo Stodolsky (1998). Especially insightful were remarks by Harry Lipkin (2002) who stressed the role of massive stationary neutrino detector being at rest in the laboratory frame.

The proper way to rigorously treat neutrino oscillations is to go beyond the "reduced" neutrino plane wave approximation and to consider the quantum mechanical amplitude, describing not only propagation of neutrino, but also its production and detection. In all realistic situations the corrections to the "reduced" formulas are negligible.

4. Solar neutrinos [26]

The study of solar neutrinos is of great interest at least for two reasons: 1) to understand neutrinos; 2) to understand and to monitor the sun. For the latter task neutrinos are especially well suited; they cover the distance from the center of the sun to its surface in about two seconds, while for photons it takes about 100 thousand years.

The main source of solar neutrinos are nuclear reactions of the hydrogen cycle in which four protons ultimately transform into a ^{4}He nucleus, two positrons, and two neutrinos. The hydrogen cycle consists of the following stages:

(1) Burning of protons:

$$99.75\%: \quad p + p \to d + e^+ + \nu_e \qquad E_\nu^{\max} = 0.42 \text{ MeV}$$
$$0.25\%: \quad p + e^- + p \to d + \nu_e \ (pep) \qquad E_\nu = 1.44 \text{ MeV}$$

(2) Burning of *d*:

$$d + p \to {}^3\text{He} + \gamma \qquad Q = 5.5 \text{ MeV}$$

(3) Burning of ^{3}He:

$$86\%: \quad {}^3\text{He} + {}^3\text{He} \to {}^4\text{He} + 2p \qquad Q = 12.9 \text{ MeV}$$
$$14\%: \quad {}^3\text{He} + {}^4\text{He} \to {}^7\text{Be} + \gamma \qquad Q = 1.59 \text{ MeV}$$
$$\sim 10^{-7} \quad {}^3\text{He} + p \to {}^4\text{He} + \nu_e + e^+ \ (hep) \qquad E_\nu^{\max} = 18.77 \text{ MeV}$$

(4) Burning of ^{7}Be:

$$90\%: \quad {}^7\text{Be} + e^- \to {}^7\text{Li} + \nu_e \qquad E_\nu = 0.861 \text{ MeV}$$
$$10\%: \quad {}^7\text{Be} + e^- \to {}^7\text{Li}^* + \nu_e \qquad E_\nu = 0.383 \text{ MeV}$$
$$0.015\%: \quad {}^7\text{Be} + p \to {}^8\text{B} + \gamma \qquad Q = 0.133 \text{ MeV}$$

(5) Burning of ^{7}Li:

$$^7\text{Li} + p \to \ ^4\text{He} + \ ^4\text{He} \quad Q = 17.3\,\text{MeV}$$

(6) Decay of ^{8}B:

$$^8\text{B} \to \ ^8\text{Be} + e^+ + \nu_e \quad E_\nu^{\text{max}} = 14.06\,\text{MeV}$$

(7) Decay of ^{8}Be:

$$^8\text{Be} \to \ ^4\text{He} + \ ^4\text{He} \quad Q = 0.92\,\text{MeV}$$

Here E_ν is the energy of neutrinos in reactions with two-particle final state, E_ν^{max} is the maximum energy of neutrinos in three-particle final state, while Q is the total energy release in a reaction or decay without neutrinos. The percentage gives the yield of the relevant reaction.

The first solar neutrino experiment was started in 1960s by Ray Davis et al. in Homestake mine at a depth of 1.6 km. The neutrino detector was a tank filled with 600 tones of perchlorethylene (C_2Cl_4). The neutrinos were captured in a reaction suggested in 1946 by Bruno Pontecorvo:

$$\nu_e + \ ^{37}\text{Cl} \to \ ^{37}\text{Ar} + e^- \ ,$$

$$^{37}\text{Ar} \to \ ^{37}\text{Cl} + e^+ + \nu_e \ .$$

The half-life of ^{37}Ar is 35 days. The decay of ^{37}Ar signaled the capture of neutrino. The threshold of ^{37}Ar production is 0.86 MeV. Hence the experiment was sensitive mainly to boron ν_e. After many years of running it gave two main results: 1) the flux of boron ν_e exists, 2) the flux is approximately four times smaller than predicted on the basis of the Standard Solar Model by J.Bahcall.

The deficit of boron neutrinos made it natural to look for neutrinos produced in other solar reactions, in particular for pp and ^{7}Be neutrinos. For these low-energy neutrinos another radiochemical reaction suggested in 1966 by Vadim Kuzmin was used:

$$\nu_e + ^{71}\text{Ga} \to e + ^{71}\text{Ge}\ , \quad E_{\text{thr}} = 0.233\,\text{MeV} \ ,$$

$$^{71}\text{Ge} \to \ ^{71}\text{Ga} + e^+ + \nu_e\ , \quad t_{1/2} = 11.4\ \text{days} \ .$$

During 1990s two international Gallium-Germanium experiments – SAGE (Baksan, Russia) and GALLEX/GNO (Gran Sasso, Italy) – confirmed the existence of solar neutrino deficit in low energy part of neutrino spectrum. This time the number of detected neutrinos was by a quarter less than expected in the absence of oscillations.

Further study of the solar neutrinos was performed at Kamioka mine in Japan and at Sudbury Neutrino Observatory (SNO) in Canada. The experiments Kamiokande (1 Kt H_2O) and Superkamiokande (SK, 50 Kt H_2O) by Masatoshi Koshiba et al. used Čerenkov radiation to detect the products of neutrino interactions. They confirmed the solar neutrino deficit; the SK discovered also the deficit of ν_μ in atmospheric neutrinos (see below).

All experiments described above were studying charged current neutrino interactions (CC), in which neutrinos transform into charged leptons. The only exception was neutrino-electron scattering (ES) which was observed at Kamiokande and which involved not only charged currents (CC) but neutral currents (NC) as well. But the number of ES-events was smaller than that of CC.

The SNO experiment was aimed at a full-fledged study of neutral current interactions (NC). For that purpose it used 1 Kt of heavy water, D_2O, in which three types of processes took place:

$$\nu_e d \to ppe^- \quad (\mathrm{CC}) \ ,$$

$$\nu_l d \to pn\nu_l \quad (\mathrm{NC}) \ ,$$

$$\nu_l e^- \to \nu_l e^- \quad (\mathrm{ES}) \ .$$

Here index $l = e, \mu, \tau$; D denotes the atom of deuterium, while d – its nucleus – deuteron.

In 2002 it was proved that the total number of events seen in SNO exactly corresponded to the flux predicted by the Standard Solar Model. Thus, the deficit of electronic neutrinos turned out to be due to their transitions into ν_μ and ν_τ. This was a spectacular demonstration of neutrino oscillations! The solar values of $\Delta m_{12}^2 \sim 0.8 \cdot 10^{-4}$ eV^2 and $\sin^2\theta_{12} \sim 0.3$ were in accord with the so called Large Mixing Angle (LMA) variant of the MSW-effect.

The 2002 Nobel Prize was awarded to R. Davis and M. Koshiba for their pioneering experiments on cosmic neutrinos.

5. Reactor antineutrinos [27-30]

There is only one reactor experiment, KamLAND, which detected oscillations of antineutrinos produced by nuclear reactors. All previous numerous reactor experiments failed to detect these oscillations. The most sensitive of these not enough sensitive experiments is CHOOZ.

The experiment CHOOZ is named after a town in France near the border with Belgium. The source of antineutrinos are two twin pressurized-water reactors with thermal power yield 4.25 GW each, at short distance $\Delta L = 117$ m. The antineutrino detector at distance $L = 1.05$ km contained 5 tons of liquid hydrogen-rich scintillator loaded with Gadolinium. The antineutrinos were captured in reaction $\bar{\nu}_e p \to n e^+$; Gadolinium was chosen due to its large neutron cross section and high γ-ray energy released after neutron capture. At $L \simeq 1$ km and average $E_{\bar{\nu}} \simeq 3$ MeV the CHOOZ experiment is not sensitive to $\Delta m_{12}^2 \sim 0.8 \cdot 10^{-4}$ eV^2, but sensitive to $\Delta m_{31}^2 \sim 25 \cdot 10^{-4}$ eV^2. By using the upper line of matrix U it is easy to derive the amplitude of $\bar{\nu}_e$ survival:

$$A(\bar{\nu}_e \to \bar{\nu}_e) = c_{12}^2 c_{13}^2 e^{i\varphi_1} + s_{12}^2 c_{13}^2 e^{i\varphi_2} + s_{13}^2 e^{i\varphi_3} =$$

$$= e^{i\varphi_1}(c_{12} c_{13}^2 + s_{12}^2 c_{13}^2 + s_{13}^2 e^{i\varphi_{31}}) = e^{i\varphi_1}(c_{13}^2 + s_{13}^2 e^{i\varphi_{31}}) \ .$$

Here $\varphi_i = p_i L$, the CP-violating phase δ is neglected.

$$\varphi_{31} = \varphi_3 - \varphi_1 = \frac{m_1^2 - m_3^2}{2E} \ , \quad \varphi_{21} = \varphi_2 - \varphi_1 \simeq 0 \ .$$

The probability of $\bar{\nu}_e$ survival:

$$P(\bar{\nu}_e \to \bar{\nu}_e) = |A(\bar{\nu}_e \to \bar{\nu}_e)|^2 = c_{13}^4 + s_{13}^4 + 2c_{13}^2 s_{13}^2 \cos\varphi_{31} =$$

$$= 1 - \sin^2 2\theta_{13} \sin^2 \frac{\Delta m_{13}^2 L}{4E} \quad .$$

As no $\bar{\nu}_e$-disappearance was detected, CHOOZ agrees with $\theta_{13} = 0$.

The merit of the CHOOZ experiment was that during the periods when reactors were off, it was possible to measure the background in the detector. The demerits were rather low power of the antineutrino source and small mass of the detector. The merits and demerits are reversed in the experiment KamLAND (Kamioka Liquid-scintillator Antineutrino Detector). It contains 1 Kt liquid scintillator in Kamioka mine and receives antineutrinos from all Japan nuclear power plants (180 GW). With many power plants it is impossible to measure background when they are off. But the large flux of antineutrinos and the large mass of the detector allow to measure antineutrinos coming from large distances (80% of the $\bar{\nu}$-flux comes from 140-210 km). Large value of L/E, in its turn, makes KamLAND sensitive to small value of Δm_{12}^2. The result of 515 days of running the detector is 258 events of $\bar{\nu}_e p \to n e^+$ reaction. Without oscillations it was expected about 365 events. By measuring the number of events as a function of positron and hence neutrino energy E it was established that $\Delta m_{12}^2 = 0.8 \cdot 10^{-4}$ eV^2 with accuracy about 10% at 95% CL in accord with solar data. The fitted value $\sin^2 2\theta_{12} = 0.83$ is less precise, because the survival probability

$$P(\bar{\nu}_e \to \bar{\nu}_e) = 1 - \sin^2 2\theta_{12} \sin^2 \Delta m_{21}^2 L/4E$$

is maximal at $L = 160$ km. To become more sensitive to the value of θ_{12} one needs $L \approx 60$ km.

6. Atmospheric neutrinos [26]

There are many detectors of atmospheric neutrinos: in Antarctica, Mediterranean, Baksan, Gran Sasso, Baikal, but only one of them – Superkamiokande (SK) – was up to now sensitive enough to detect their oscillations.

Atmospheric neutrinos are produced in the decays of pions and kaons, created by the collisions of cosmic ray protons with nuclei of air. At the moment of production most of them are muonic neutrinos.

The effect discovered at SK is the up-down asymmetry in the fluxes of atmospheric ν_μ: the number of the ν_μ coming from above (directly from the sky) was larger than number of ν_μ coming from below (after traversing the Earth). Thus part of these ν_μ disappeared on their way through the Earth. Taking into account that the diameter of the Earth is about 12 000 km, while the energy E_{ν_μ} is in the interval from 1 GeV to 10 GeV, the SK collaboration estimated $\Delta m^2 = 25 \cdot 10^{-4}$ eV^2 with uncertainty $\pm 20\%$. The only effect observed in atmospheric neutrinos was disappearance of ν_μ. There was no appearance of ν_e, which means that the effect of $\nu_\mu \leftrightarrow \nu_e$ mixing is negligible. Thus, the effect was caused by strong $\nu_\mu \leftrightarrow \nu_\tau$ mixing. Hence $\Delta m_{23}^2 = 25 \cdot 10^{-4}$, $\theta_{23} \simeq \pi/4$. (τ-leptons were not effectively produced by ν_τ because of moderate energy of atmospheric neutrinos.)

7. Accelerator neutrinos [27]

The most recent accelerator neutrino experiment is K2K (its name being a slang acronym for "KEK to Kamikande: K to K). Here the source of muonic neutrinos is 12 GeV proton accelerator at KEK (Japan). The mean energy of neutrinos $\bar{E}_{\nu_\mu} = 1.3$ GeV. The target was 50 Kt water Čerenkov detector Superkamiokande (SK) 250 km away from the source. The results of K2K are in accord with those of atmospheric neutrinos: pronounced disappearance of ν_μ without appearance of ν_e. This was the first confirmation of atmospheric $\nu_\mu \leftrightarrow \nu_\tau$ oscillations with $\Delta m^2_{23} = (28 + 4 - 7) \cdot 10^{-4}$ eV^2 (November 2004) from a man-made neutrino source. The total number of ν_μ-events observed up to now is 108, while expected without oscillations were 161±11.

Since 1990s there exists the so-called LSND anomaly coming from the liquid scintillator neutrino detector at Los Alamos accelerator according to which $\sqrt{\Delta m^2_{\mu e}} = 0.4$ eV. It was not confirmed by other oscillation data, in particular, by KARMEN at the Rutherford Laboratory (UK, 1990 - 1995). Instead of discussing it, let us wait for the result of MiniBooNE (the first stage of Booster Neutrino Experiment at FNAL).

8. Patterns of mixing angles and masses [25,31]

Let us summarize the results of neutrino oscillation experiments. Most of the values quoted below have uncertainties up to 20%. The CP-violating parameter δ is not known.

$$\sin^2\theta_{\rm atm} = 0.50 \qquad \Delta m^2_{\rm atm} = 25 \cdot 10^{-4}\ {\rm eV}^2$$
$$\sin^2\theta_{\rm sol} \equiv \sin^2\theta_{21} = 0.30 \quad \Delta m^2_{\rm sol} = 0.8 \cdot 10^{-4}\ {\rm eV}^2$$
$$\sin^2\theta_{13} \le 0.005$$

$$\theta_{\rm sol} = \theta_{21} \simeq 37^0\ \ ({\rm Sun + KAMLAND})$$
$$\Delta m^2_{\rm sol} \equiv |\Delta m^2_{21}| \equiv |m_2^2 - m_1^2|$$
$$\Delta m^2_{\rm atm} \equiv |\Delta m^2_{32}| \equiv |m_3^2 - m_2^2|$$

From oscillations we know that neutrinos do have masses, but we know only the absolute values of two differences of their squares. We do not know the signs of these differences.

The case when $m_3 > m_2$ is called normal hierarchy.

Normal hierarchy is analogous to that of quarks and charged leptons (their masses are given in GeV):

$Q = 2/3$	$Q = -1/3$	$Q = -1$	$Q = 0$
t 175	b 5	τ 1.9	ν_3 $\Delta m^2_{\rm atm} = 25 \cdot 10^{-4}$ eV^2
c 1.3	s 0.15	μ 0.1	ν_2
u $5 \cdot 10^{-3}$	d $10 \cdot 10^{-3}$	e $0.5 \cdot 10^{-3}$	ν_1 $\Delta m^2_{\odot} = 0.8 \cdot 10^{-4}$ eV^2

Inverted hierarchy is more promising for solving the Majorana–Dirac alternative in neutrinoless double β-decay: the heavier m_2, the larger $< m >$, see section 9.

There is also a possibility of quasidegenerate masses: $m_1 \simeq m_2 \simeq m_3 \simeq 0.2 - 0.3$ eV.

For the absolute value of m_{ν_e} only the upper limit is known from two experiments on the electron spectrum in tritium β-decay: $m_{\nu_e} < 2.2$ eV (Troitsk, 2001), Mainz (2003).

An experiment KATRIN (Karlsruhe Tritium Neutrino Experiment) is designed to measure the mass of electron antineutrino directly with sensitivity of 0.2 eV.

As for mixing angles, let us note, that for neutrinos two of them are much larger than for quarks.

9. Double beta decay [32]

While in case of Dirac neutrino the emission of electron proceeds either with absorption of neutrino, or with emission of antineutrino, the Majorana neutrino can be both emitted and absorbed. This refers not only to real neutrinos, but to virtual as well. Therefore in the second order of weak interaction two neutrons in a nucleus can decay into two protons and two electrons by exchanging a virtual Majorana neutrino. The amplitude of such process is proportional to a product of two factors:

$$A(\beta\beta)_{0\nu} \sim < m > \cdot M \ .$$

The factor $< m >$ represents the neutrino propagator:

$$< m > = m_1 U_{e1}^2 + m_2 U_{e2}^2 e^{i\alpha_{21}} + m_3 U_{e3}^2 e^{i\alpha_{31}} \ .$$

The coefficients $U_{e\alpha}$ can be read off from the first line of the matrix U at the end of Section 2.

From neutrino oscillations we know that $|U_{e2}| \gg |U_{e3}|$. That is why the probability of $2\beta0\nu$ decay is larger for larger value of m_2 (inverted hierarchy).

The factor M represents the nuclear matrix element estimated on the basis of measurements of the half-life times of the decays $2\beta 2\nu$.

The most advanced seems to be at present the search of $2\beta0\nu$ decay of ^{76}Ge by two collaborations: IGEX and Heidelberg – Moscow. Both give upper limits

$$| < m > | < (0.35 \div 1.05) \text{ eV at } 90\% \text{ CL} \ ,$$

where factor 3 uncertainty reflects the uncertainty of M. (A part of the latter collaboration claims to observe the $2\beta0\nu$ decay and gives a lower limit $| < m > | > 0.11$ eV.)

Collaboration NEMO3 (^{100}Mo; ^{82}Se) and CUORICINO (^{130}Te) expect to lower the upper limit for $| < m > |$ to 0.2 eV.

In a more distant future a number of collaborations promise an order of magnitude better sensitivity: CUORE (^{130}Te), GENIUS (^{76}Ge), EXO (^{136}Xe), MAJORANA (^{76}Ge), MOON (^{100}Mo).

10. Cosmological and stellar inputs [33]

Though neutrinos play crucial role in astrophysics, up to now the observations of the stars have not contributed significantly to our knowledge of neutrino masses and mixings. Even the most spectacular supernova SN 1987 A provided only a rather modest upper limit on neutrino masses.

From the theory of white dwarfs an upper limit on neutrino magnetic moment can be deduced $\mu_{\nu_e} \leq 10^{-10}\mu_B$, where μ_B is Bohr magneton.

The early universe turns out to be more informative than stars. The precision measurements of cosmic microwave background (CMB) by Wilkinson Microwave Anisotropy Probe (WMAP) in conjuction with data on large scale structure, 2*d*FGRS, Lyman Alpha forest and a host of other astronomical measurements allow to find the best fit to a set of cosmological parameters and in particular to the upper limit on Ω_ν – the energy density in neutrinos.

In case of 3 degenerate neutrino species

$$\Sigma = 94\Omega_\nu h^2 \ ,$$

where $\Sigma = m_1 + m_2 + m_3$, Ω_ν is in units of critical density, and Hubble constant $h = H/H_0$, where $H_0 = 100\text{km} \cdot \text{s}^{-1} \cdot \text{Mpc}^{-1}$.

According to the 2003 fit of WMAP, $\Sigma < 0.7$ eV.

11. Challenges and Prospects [31,34]

Challenge #1. Absolute values of neutrino masses (from oscillations we know only the values of $|\Delta m^2_{12}|$ and $|\Delta m^2_{23}|$). The problem could be solved by KATRIN. At present even the sign of Δm^2_{23} is not known.

Challenge #2. Majorana *vs* Dirac. The problem could be solved by observing the neutrinoless double beta decay.

Challenge #3. CP-violation in leptonic sector. This is especially difficult because we know that $\theta_{13} \ll 1$ and $\Delta m^2_{21}/\Delta m^2_{32} \ll 1$. Note that for $\theta_{13} = 0$ and/or $\Delta m^2_{21}/\Delta m^2_{32} = 0$ the CP-violating effects disappear (recall the mechanism by Kobayashi and Maskawa). Of special interest would be the measurement of α_{21} and α_{31} if the neutrinoless double beta decay is discovered.

In most of the cases only disappearance effects have been observed (disappearance of ν_μ in atmospheric neutrinos, disappearance of ν_e and $\bar{\nu}_e$ in solar and reactor neutrinos). What is needed is the appearance effects, such as the appearance of ν_τ in the CERN - Gran Sasso project (detectors Opera, Icarus).

It is necessary to reach much higher statistics, high intensity, high energy, clean beams, such as beta beams from beta decay of relativistic nuclei, or from neutrino factories where neutrinos appear in the decays of stored relativistic muons. In the nearest perspective of great interest is the 50 GeV JHF PS.

It is necessary to reach much higher accuracy in measurements of the key parameters:

$$\delta(\Delta m^2_{32}) = 10^{-4}\ \text{eV}^2 \ ,$$

$$\delta(\sin^2 2\theta_{13}) = 5 \cdot 10^{-3} \ ,$$

$$\delta(\sin^2 2\theta_{32}) = 10^{-2} \ .$$

Solving the puzzles of neutrinos is an absolute must, because these particles are at the focal point of particle physics, astrophysics and cosmology.

Acknowledgements

Many thanks to Laurent Baulieu and Boris Pioline for their warm hospitality.

I am grateful to A.D. Dolgov, O.V. Lychkovsky, A.A. Mamonov, M.V. Rotaev, M.G. Schepkin, V.L. Telegdi, and M.I. Vysotsky for many valuable comments. This work was partly supported by RFBR grant No.04-02-16538 and by A.von Humboldt Award.

References

1. H. Becquerel, Comp. Rend. **12** (1896) 501, 509. Discovery of radioactivity of uranium salts (in fact, of thorium β-rays).
2. E. Rutherford, Phil. Mag., ser.6, **21** (1911) 669.
3. W. Pauli, Letter of 4 December 1930 to a meeting of "radioactive ladies and gentlemen" at Tübingen (reproduced in L. Brown Phys. Today **31**, No.9 (1978) 23.
4. E. Fermi, Zs. f. Phys. **88** (1934) 161. Theory of β-decay.
5. E. Majorana, Nuovo Cimento **14** (1937) 171. Majorana neutrino.
6. B. Pontecorvo, Report PD-205, Chalk River, 1946, National Research Council of Canada.
7. F. Reines, C.L. Cowan, Phys. Rev. **92** (1953) 830.
8. G. Danby, J.-M. Gaillard, K. Goulianos, L.M. Lederman, N. Mistry, M. Schwartz, J. Steinberger, Phys. Rev. Lett. **9** (1962) 36. Experimental discovery of two kinds of neutrinos.
9. B. Pontecorvo, ZhETF **33** (1957) 549. The idea of $\nu_e \leftrightarrow \bar{\nu}_e$ oscillations.
10. B. Pontecorvo, ZhETF **37** (1959) 1751. Proposal to search for ν_μ.
11. Z. Maki, M. Nakagawa and S. Sakata, Progr. Theor. Phys. **28** (1962) 870. The idea of $\nu_e \leftrightarrow \nu_\mu$ mixing.
12. V. Gribov, B. Pontecorvo, Phys. Lett. **28B** (1969) 493. Description of $\nu_e \leftrightarrow \nu_\mu$ oscillations.
13. S. Bilenky, B. Pontecorvo, Lett Nuovo Cim. **17** (1976) 569.
14. H. Fritzsch, P. Minkowski, Phys. Lett. **B62** (1976) 72. Formula for $\nu_e \leftrightarrow \nu_\mu$ oscillations without "factor of 2" misprint.
15. L. Wolfenstein, Phys. Rev. **D17** (1978) 2369. Neutrino oscillations in matter.
16. S. Mikheev, A. Smirnov, Yad. Fiz. **42** (1985) 1441. Resonance neutrino transitions in matter. Sov. J. Nucl. Phys. **42** (1986) 913.
17. I. Kobzarev, B. Martemyanov, L. Okun, M. Schepkin, Sov. J. Nucl. Phys. **35** (1982) 708. Example of equal energy scenario.
18. L. Stodolsky, Phys. Rev. **D58** (1998) 036006. Arguments in favor of equal energy scenario.
19. H. Lipkin, hep-ph/0212093. On the role of static detector in neutrino oscillations.
20. M.I. Vysotsky, *CKM matrix and CP violation in B-mesons*, Surveys in High Energy Physics, 2003, vol. 18 (1-4), pp. 19-54; hep-ph/0307218, p.18.
21. A. Dolgov, Phys. Rep. **370** (2002) 333.
22. M. Beuthe, Phys. Rep. **375** (2003) 105, hep-ph/0109119. On the role of wave packets.
23. M. Nauenberg, Phys. Lett. **B447** (1999) 23, Err. Phys. Lett. **B452** (1999) 439, hep-ph/9812441. Oscillations of neutrinos from pion decay.
24. B. Kayser, Phys. Lett. **B592**, Issues 1-4 (July 2004) 145. Neutrino mass, mixing and flavor change.
25. R.N. Mohapatra, APS Neutrino Study: Theory of neutrinos, hep-ph/0412099.
26. H. Back et al., APS Neutrino Study: Report of the Solar and Atmospheric Neutrino Experiments Working Group, hep-ex/0412016.
27. M. Goodman, Long-Baseline Neutrino News. Monthly selection of news from <mcg@hep.anl.gov>.
28. Neutrino oscillation industry. A website on neutrino experiments.
29. E. Abouzaid et al., APS Neutrino Study: Report of the Reactor working group, www.aps.org/neutrino; http://apsreactor.uchicago.edu/report.html.
30. M. Apolonio et al., CHOOZ, hep-ex/0301017.
31. S.J. Freedman, B. Kayser, APS Neutrino Study: The neutrino matrix, physics/0411216.

32. C. Aalseth et al., APS Neutrino Study: Neutrinoless double beta decay and direct searches for neutrino mass, hep-ph/0412300.
33. S.W. Barwick et al., APS Neutrino Study: Report of the Neutrino Astrophysics and Cosmology Working Group, astro-ph/0412544.
34. C. Albright et al., APS Neutrino Study: Neutrino Factory and Beta Beam Experiments and Development, www.aps.org/neutrino.

CLOSED STRINGS IN MISNER SPACE: A TOY MODEL FOR A BIG BOUNCE ?

BRUNO DURIN
LPTHE, Université Paris 6,
4 place Jussieu, 75252 Paris cedex 05, France

AND

BORIS PIOLINE
LPTHE, Université Paris 6,
4 place Jussieu, 75252 Paris cedex 05, France
and LPTENS, Département de Physique de l'Ecole Normale Supérieure
24 rue Lhomond, 75231 Paris cedex 05, France

Abstract. Misner space, also known as the Lorentzian orbifold $R^{1,1}/boost$, is one of the simplest examples of a cosmological singularity in string theory. In this lecture, we review the semi-classical propagation of closed strings in this background, with a particular emphasis on the twisted sectors of the orbifold. Tree-level scattering amplitudes and the one-loop vacuum amplitude are also discussed.

Thus I was moving along the sloping curve of the time loop towards that place in which the Friday me before the beating would change into the Friday me already beaten.

I. Tichy, [1]

Despite their remarkable success in explaining a growing body of high precision cosmological data, inflationary models, just as the Hot Big Bang Model, predict an Initial Singularity where effective field theory ceases to be valid [2]. As a quantum theory of gravity, String Theory ought to make sense even in this strongly curved regime, possibly by providing an initial quantum state if the Initial Singularity is truely an Origin of Time, or by escaping it altogether if stringy matter turns out to be less prone to gravitational collapse than conventional field-theoretic matter. Unfortunately, describing cosmological singularities and, less ambitiously, time dependence in string theory has been a naggingly difficult task, partly because of the absence of a tractable closed string field theory framework. Unless stringy (α') corrections in the two-dimensional sigma model are sufficient to eliminate the singularity, quantum (g_s) corrections are expected to be important due to

L. Baulied et al. (eds.), String Theory: From Gauge Interactions to Cosmology, 177–200.

the large blue-shift experienced by particles or strings as they approach the singularity, invalidating a perturbative approach. Nevertheless, one may expect cosmological production of particles, strings and other extended states near the singularity to qualitatively alter the dynamics, and it is not unplausible, though still speculative, that their contribution to the vacuum energy be sufficient to lead to a Big Bounce rather than an Big Bang.

In order to make progress on this issue, it is useful to study toy models where at least α' corrections are under control, and study string production to leading order in g_s. Orbifolds, being locally flat, are immune to α' corrections, and thus a good testing playground. One of the simplest examples of time-dependent orbifolds[1] is the Lorentzian orbifold $\mathbb{R}^{1,1}/boost$ [7, 8, 9], formerly known as Misner space [3] in the gravity literature. Introduced as a local model for the cosmological singularity and chronological horizon of Lorentzian Taub-NUT space, Misner space was shown long ago to exhibit divergences from quantum vacuum fluctuations in field theory, at least for generic choices of vacua [22]. Not surprisingly, this is also true in string theory, although less apparent since the local value of the energy-momentum tensor is not an on-shell observable [5]. Similarly, just as in field theory, tree-level scattering amplitudes of field-theoretical (untwisted) states have been found to diverge, as a result of large graviton exchange near the singularity [32].

While these facts ominously indicate that quantum back-reaction may drastically change the character of the singularity, experience from Euclidean orbifolds suggests that twisted states may alleviate the singularities of the effective field theory description, and that it may be worthwhile to investigate their classical behaviour, overpassing the probable inconsistency of perturbation theory. Indeed, in the context of Misner space, twisted states are just strings that wind around the collapsing spatial direction, and become the lightest degrees of freedom near the singularity. In these notes, we review classical aspects of the propagation of closed strings in Misner space, with particular emphasis on twisted states, based on the recent works [4, 5, 6].

The outline is as follows. In Section 2, we describe the semi-classical dynamics of charged particles and winding strings, and compute their cosmological production rate, at tree level in the singular Misner geometry – although our approach is applicable to more general cases. In Section 3, we analyze the imaginary part of the one-loop amplitude, which carries the same information in principle. In Section 4, we review recent results on scattering amplitudes of untwisted and twisted states, and their relation to the problem of classical back-reaction from a "condensate" of twisted states. Section 5 contains our closing remarks.

1. Semi-classics of closed strings in Misner space

1.1. MISNER SPACE AS A LORENTZIAN ORBIFOLD

Misner space was first introduced in the gravity literature as a local model [3] for the singularities of the Taub-NUT space-time [23] . It can be formally defined as the quo-

[1]The orbifold of $R^{1,1}$ under time reversal may be even simpler, but raises further puzzles related to time unorientability [10]. Discussions of other exact cosmological backgrounds in string theory include [11, 12, 13, 14, 15, 16, 17, 18, 19, 20, 21].

tient of two-dimensional[2] Minkowski space $\mathbb{R}^{1,1}$ by the finite boost transformation B : $(x^+, x^-) \to (e^{2\pi\beta}x^+, e^{-2\pi\beta}x^-)$, where $x^\pm$ are the light-cone coordinates. As such, it is a locally flat space, with curvature localized at the fixed locus under the identification, i.e. on the light-cone $x^+x^- = 0$. The geometry of the quotient can be pictured as four Lorentzian cones touching at their apex (See Figure 1), corresponding to the four quadrants of the covering space $\mathbb{R}^{1,1}$. Choosing coordinates adapted to the boost B,

$$x^\pm = Te^{\pm\beta\theta}/\sqrt{2}\,, \qquad x^+x^- > 0 \quad \text{(Milne regions)} \tag{1.1}$$

$$x^\pm = \pm re^{\pm\beta\eta}/\sqrt{2}\,, \qquad x^+x^- < 0 \quad \text{(Rindler regions)} \tag{1.2}$$

where, due to the boost identification, the coordinates θ and η are compact with period 2π, the metric of the quotient can be written as

$$ds^2 = -2\,dx^+dx^- = \begin{Bmatrix} -dT^2 + \beta^2T^2d\theta^2 \\ dr^2 - \beta^2r^2d\eta^2 \end{Bmatrix} \tag{1.3}$$

The two regions $x^+ < 0, x^- < 0$ (P) and $x^+ > 0, x^- > 0$ (F), describe contracting and expanding cosmologies where the radius of the spatial circle parameterized by θ changes linearly in time, and are often called (compactified) Milne regions. The space-like cones $x^+ > 0, x^- < 0$ (R) and $x^+ < 0, x^- > 0$ (L), often termed "whiskers", are instead time-independent Rindler geometries with compact time η [3]. The Milne and Rindler regions, tensored with a sphere of finite size, describe the Taub and NUT regions, respectively, of the Taub-NUT space-time in the vicinity of one of its infinite sequence of cosmological singularities. It also captures the local geometry in a variety of other cosmological string backgrounds [12, 15, 17, 18]. It is also interesting to note that, combining the boost B with a translation on a spectator direction, one obtains the Gott space-time, i.e. the geometry around cosmic strings in four dimensions [25].

Due to the compactness of the time coordinate η, both Misner and Taub-NUT space-times contain closed timelike curves (CTC) which are usually considered as a severe pathology. In addition to logical paradoxes and exciting prospects [1] raised by time-loops, the energy-momentum tensor generated by a scalar field at one-loop is typically divergent, indicating a large quantum back-reaction. According to the Chronology Protection Conjecture, this back-reaction may prevent the formation of CTC altogether [26]. String theorists need not be intimidated by such considerations, and boldly go and investigate whether the magics of string theory alleviate some of these problems.

String theory on a quotient of flat space[4] is in principle amenable to standard orbifold conformal field theory techniques, although the latter are usually formulated for Euclidean orbifolds. While backgrounds with Lorentzian signature can often be dealt with by (often subtle) Wick rotation from Euclidean backgrounds, the real complication stems from the fact that the orbifold group is infinite, and its action non proper [5]. This however need not

[2]Higher dimensional analogues have also been considered [24].

[3]This should not be confused with thermal Rindler space, which is periodic in *imaginary* time.

[4]String theory on Taub-NUT space, which is not flat, has been studied recently using heterotic coset models [20].

[5]Defining $X^+ = Z\,,\quad X^- = -\bar{Z}$ in the Rindler region, one obtains an orbifold of $\mathbb{R}^2$ by a rotation with an irrational angle. A related model has been studied recently in [27].

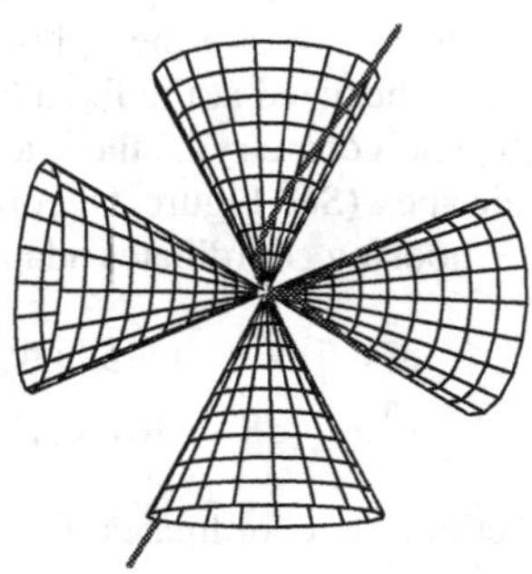

Figure 1. Free particles or untwisted strings propagate from the past Milne region to the future Milne region, with a temporary excursion in the whiskers.

be a problem at a classical level: as we shall see, free strings propagate in a perfectly well-defined fashion on this singular geometry.

1.2. PARTICLES IN MISNER SPACE

As in standard orbifolds, part of the closed string spectrum consists of configurations on the covering space, which are invariant under the orbifold action. Such "untwisted" states behave much like point particles of arbitrary mass and spin. Their trajectory, aside from small-range string oscillations, consist of straight lines on the covering space:

$$X_0^{\pm} = x_0^{\pm} + p^{\pm}\tau \ , \tag{1.4}$$

where $m^2 = 2p^+p^-$ includes the contribution from momentum in the transverse directions to Misner space as well as string oscillators. The momentum along the compact direction is the "boost momentum" $j = x_0^+ p^- + x_0^- p^+$, and is quantized in units of $1/\beta$ in the quantum theory. A massive particle with positive energy ($p^+, p^- > 0$) thus comes in from the infinite past in the Milne region at $\tau = -\infty$ and exits in the future Milne region at $\tau = +\infty$, after wandering in the Rindler regions for a finite proper time. As the particle approaches the light-cone from the past region, its angular velocity $d\theta/dT \sim 1/T$ along the Milne circle increases to infinity by the familiar "spinning skater" effect. It is therefore expected to emit abundant gravitation radiation, and possibly lead to large back-reaction. From the point of view of an observer in one of the Rindler regions, an infinite number of particles of Rindler energy j are periodically emitted from the horizon at $r = 0$ and travel up to a finite radius $r = |j|/M$ before being reabsorbed into the singularity – and so on around the time loop.

Quantum mechanically, the center of mass of a (spinless) untwisted string is described by a wave function, solution of the Klein-Gordon equation in the Misner geometry. Diagonalizing the boost momentum j, the radial motion is governed by a Schrödinger equation

$$\begin{cases} -\partial_x^2 - m^2 e^{2y} - j^2 = 0 \\ -\partial_y^2 + m^2 e^{2y} - j^2 = 0 \end{cases} \quad \text{where} \quad \begin{cases} T &= \pm\sqrt{2x^+x^-} = e^x \\ r &= \pm\sqrt{-2x^+x^-} = e^y \end{cases} \tag{1.5}$$

The particle therefore bounces against an exponentially rising, Liouville-type wall in the Rindler regions, while it is accelerated in a Liouville-type well in the Misner regions. Notice that, in both cases, the origin lies at infinite distance in the canonically normalized coordinate x (y, resp.). Nevertheless, in and out type of wave functions can be defined in each region, and extended to globally defined wave functions by analytic continuation across the horizons at $x^+x^- = 0$.

Equivalently, the wave function for an untwisted string in Misner space may be obtained by superposing a Minkowski plane wave with its images under the iterated boosts B^n, $n \in \mathbb{Z}$. Performing a Poisson resummation over the integer n, one obtains wave functions with a well defined value of the boost momentum j, as a continuous superposition of plane waves

$$f_{j,m^2,s}(x^+, x^-) = \int_{-\infty}^{\infty} dv \, \exp\left(ip^+X^-e^{-2\pi\beta v} + ip^-X^+e^{2\pi\beta v} + ivj + vs\right) \quad (1.6)$$

where s denotes the $SO(1,1)$ spin in $\mathbb{R}^{1,1}$ [9, 5]. This expression defines global wave functions in all regions, provided the v-integration contour is deformed to $(-\infty - i\epsilon, +\infty + i\epsilon)$. In particular, there is no overall particle production between the (adiabatic) in vacuum at $T = -\infty$ and the out vacuum at $T = +\infty$, however there is particle production between the (adiabatic) in vacuum at $T = -\infty$ and the (conformal) out vacuum at $T = 0^-$. This is expected, due to the "spinning skater" infinite acceleration near the singularity, as mentioned above.

1.3. WINDING STRINGS IN MISNER SPACE

In addition to the particle-like untwisted states, the orbifold spectrum contains string configurations which close on the covering space, up to the action of an iterated boost B^w:

$$X^\pm(\sigma + 2\pi, \tau) = e^{\pm 2\pi w\beta} X^\pm(\sigma, \tau) \quad (1.7)$$

In the Milne regions, they correspond to strings winding w times around the compact space-like dimension S^1_θ, which become massless at the cosmological singularity. They are therefore expected to play a prominent rôle in its resolution, if at all. In the Rindler regions, they instead correspond to strings winding around the compact *time-like* dimension S^1_η. Given that a time-loop exist, there is nothing *a priori* wrong about a string winding around time: it is just a superposition of w static (or, more generally, periodic in time) strings, stretched (in the case of a cylinder topology) over an infinite radial distance.

In order to understand the semi-classical aspects of twisted strings [5], let us again truncate to the modes with lowest worldsheet energy, satisfying (1.7):

$$X^\pm_0(\sigma, \tau) = \frac{1}{\nu} e^{\mp\nu\sigma} \left[\pm\alpha^\pm_0 e^{\pm\nu\tau} \mp \tilde{\alpha}^\pm_0 e^{\mp\nu\tau}\right] . \quad (1.8)$$

where $\nu = -w\beta$. As usual, the Virasoro (physical state) conditions determine the mass and momentum of the state in terms of the oscillators,

$$M^2 = 2\alpha^+_0\alpha^-_0 \ , \quad \tilde{M}^2 = 2\tilde{\alpha}^+_0\tilde{\alpha}^-_0 \quad (1.9)$$

where ($M^2 = m^2 + j\nu, \tilde{M}^2 = m^2 - j\nu$) are the contributions of the left-moving (resp. right-moving) oscillators. Restricting to $j = 0$ for simplicity, one may thus choose $\alpha^\pm$ and $\tilde{\alpha}^\pm$ to be all equal in modulus to $m/\sqrt{2}$, up to choices of sign leading to two qualitatively different kinds of twisted strings:

- For $\alpha^+\tilde{\alpha}^- > 0$, one obtains *short string* configurations

$$X_0^\pm(\sigma,\tau) = \frac{m}{\nu\sqrt{2}} \sinh(\nu\tau)\, e^{\pm\nu\sigma} \tag{1.10}$$

winding around the Milne space-like circle, and propagate from infinite past to infinite future (for $\alpha^+ > 0$). When $j \neq 0$, they also extend in the Rindler regions to a finite distance $r_-^2 = (M - \tilde{M})^2/(4\nu^2)$, after experiencing a signature flip on the worldsheet.

- For $\alpha^+\tilde{\alpha}^- < 0$, one obtains *long string* configurations,

$$X_0^\pm(\sigma,\tau) = \frac{m}{\nu\sqrt{2}} \cosh(\nu\tau)\, e^{\pm\nu\sigma} \tag{1.11}$$

propagating in the Rindler regions only, and winding around the time-like circle. They correspond to static configurations which extend from spatial infinity in L or R to a finite distance $r_+^2 = (M + \tilde{M})^2/(4\nu^2)$, and folding back to infinity again.

Notice how, in contrast to Euclidean orbifolds, twisted strings are in no sense localized near the singularity !

Quantum mechanically, the (quasi) zero-modes $\alpha_0^\pm, \tilde{\alpha}_0^\pm$ become hermitian operators with commutation rules [9, 4]

$$[\alpha_0^+, \alpha_0^-] = -i\nu\,, \quad [\tilde{\alpha}_0^+, \tilde{\alpha}_0^-] = i\nu \tag{1.12}$$

Representing α_0^+ as a creation operator in a Fock space whose vacuum is annihilated by α_0^-, introduces an imaginary ordering constant $i\nu/2$ in (1.9) after normal ordering, which cannot be cancelled by any of the higher modes in the spectrum[6]. Thus, in this scheme, there are no physical states in the twisted sector [9]. However, this quantization does not maintain the hermiticity of the zero-mode operators. The analogy of (1.12) to the problem of a charged particle in an electric field will take us to the appropriate quantization scheme in the next section.

1.4. WINDING STRINGS VS. CHARGED PARTICLES

Returning to (1.11), one notices that the complete worldsheet of a twisted closed string can be obtained by smearing the trajectory of the left-movers (i.e. a point with $\tau + \sigma = cste$) under the action of continuous boosts (See Figure 2). In particular, setting $a_0^\pm = \tilde{\alpha}_0^\pm$ and $x_0^\pm = \mp\tilde{\alpha}_0^\pm/\nu$, the trajectory of the left-movers becomes

$$X^\pm(\tau) = x_0^\pm \pm \frac{a_0^\pm}{\nu} e^{\pm\nu\tau}\,. \tag{1.13}$$

[6]Higher excited modes have energy $n \pm i\nu$, and can be quantized in the usual Fock space scheme, despite the Lorentzian signature of the light-cone directions [4].

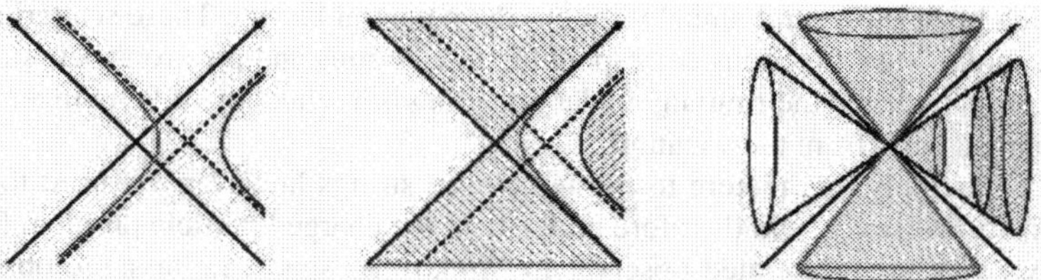

Figure 2. Closed string worldsheets in Misner space are obtained by smearing the trajectory of a charged particle in Minkowski space with a constant electric field. Short (resp. long) strings correspond to charged particles which do (resp. do not) cross the horizon.

which is nothing but the worldline of a particle of charge w in a constant electric field $E = \beta$! Indeed, it is easily verified that the short (long, resp.) string worldsheet can be obtained by smearing the worldline of a charged particle which crosses (does not, resp.) the horizon at $x^+x^- = 0$.

Quantum mechanically, it is easy to see that this analogy continues to hold [4, 5] : the usual commutation relations for a particle in an electric field

$$[a_0^+, a_0^-] = -i\nu\,, \quad [x_0^+, x_0^-] = -\frac{i}{\nu}\,, \tag{1.14}$$

reproduce the closed string relations (1.12) under the identification above. The mass of the charged particle $M^2 = \alpha^+\alpha^- + \alpha^-\alpha^+$ reproduces the left-moving Virasoro generator $m^2 + \nu j$ as well. It is therefore clear that the closed string zero-modes, just as their charge particle counterpart, can be represented as covariant derivatives acting on complex wave functions $\phi(x^+, x^-)$:

$$\alpha_0^\pm = i\nabla_\mp = i\partial_\mp \pm \frac{\nu}{2}x^\pm\,, \quad \tilde{\alpha}_0^\pm = i\tilde{\nabla}_\mp = i\partial_\mp \mp \frac{\nu}{2}x^\pm \tag{1.15}$$

in such a way that the physical state conditions are simply the Klein-Gordon operators for a particle with charge $\pm\nu$ in a constant electric field,

$$M^2 = \nabla_+\nabla_- + \nabla_-\nabla_+\,, \quad \tilde{M}^2 = \tilde{\nabla}_+\tilde{\nabla}_- + \tilde{\nabla}_-\tilde{\nabla}_+ \tag{1.16}$$

Coordinates $x^\pm$ are the (Heisenberg picture) operators corresponding to the location of the closed string at $\sigma = 0$. The radial coordinate $\sqrt{\pm 2x^+x^-}$ associated to the coordinate representation (1.15), should be thought of as the radial position of the closed string in the Milne or Rindler regions.

From this point of view, it is also clear while the quantization scheme based on a Fock space has failed: the Klein-Gordon equation of a charged particle in an electric field is equivalent, for fixed energy p_t, to a Schrödinger equation with an *inverted* harmonic potential,

$$-\partial_x^2 + m^2 - (p_t - Ex)^2 \equiv 0 \tag{1.17}$$

In contrast to the magnetic case which leads to a positive harmonic potential with discrete Landau levels, the spectrum consists of a continuum of delta-normalization scattering states which bounce off (and tunnel through) the potential barrier. These scattering states are the quantum wave functions corresponding to electrons and positrons being reflected by the electric field, and their mixing under tunneling is a reflection of Schwinger production of charged pairs from the vacuum.

In order to apply this picture to twisted closed strings however, we need to project on boost invariant states, and therefore understand the charged particle problem from the point of view of an accelerated observer in Minkowski space, i.e. a static observer in Rindler space.

1.5. CHARGED PARTICLES IN MISNER SPACE

Charged particles in Rindler space have been discussed in [28]. Classical trajectories are, of course, the ordinary hyperbolae from Minkowski space, translated into the Rindler coordinates $(y = e^r, \eta)$. For a fixed value j of the energy conjugate to the Rindler time η, the radial motion is governed by the potential

$$V(y) = M^2 r^2 - \left(j + \frac{1}{2}\nu r^2\right)^2 = \frac{M^2 \tilde{M}^2}{\nu^2} - \left(\frac{M^2 + \tilde{M}^2}{2\nu} - \frac{\nu}{2} r^2\right)^2 \tag{1.18}$$

where, in the last equality, we have translated the charged particle data into closed string data. In contrast with the neutral case ($\nu = 0$), the potential is now unbounded from below at $r = \infty$. For $j < M^2/(2\nu)$ (which is automatically obeyed in the closed string case, where $\tilde{M}^2 > 0$ for non-tachyonic states), the $r = 0$ and $r = \infty$ asymptotic regions are separated by a potential barrier (See Figure 3). Particles on the right ($r \to \infty$) of the barrier correspond to electrons coming from and returning to Rindler infinity, while, for $j > 0$ (resp. $j < 0$), particles on the left ($r \to 0$) correspond to positrons (resp. electrons) emitted from and reabsorbed by the Rindler horizon. Quantum tunneling therefore describes both Schwinger pair production in the electric field (when $j > 0$), and Hawking emission of charged particles from the horizon (when $j < 0$).

Similarly, the trajectories of charged particles in Milne space correspond to other branches of the same hyperbolae, and their motion along the cosmological time T, for a fixed value of the momentum j conjugate to the compact spatial direction θ, is governed by the potential

$$V(T) = -M^2 T^2 - \left(j + \frac{1}{2}\nu T^2\right)^2 = \frac{M^2 \tilde{M}^2}{\nu^2} - \left(\frac{M^2 + \tilde{M}^2}{2\nu} + \frac{\nu}{2} T^2\right)^2 \tag{1.19}$$

The potential is maximal and negative at $T = 0$, although this is at infinite distance in the canonically normalized coordinate x. The classical motion therefore covers the complete time axis $T \in \mathbb{R}$.

Quantum mechanically, the Klein-Gordon equation in the Rindler region is equivalent to a Schrödinger equation in the potential (1.18) or (1.19) at zero-energy, and can be solved in terms of Whittaker functions [28]. Bases of *in* and *out* modes can be defined in each quadrant and analytically continued accross the horizons, e.g. in the right Rindler region

$$\mathcal{V}^j_{in,R} = e^{-ij\eta} r^{-1} M_{-i(\frac{j}{2} - \frac{M^2}{2\nu}), -\frac{ij}{2}}(i\nu r^2/2) \tag{1.20}$$

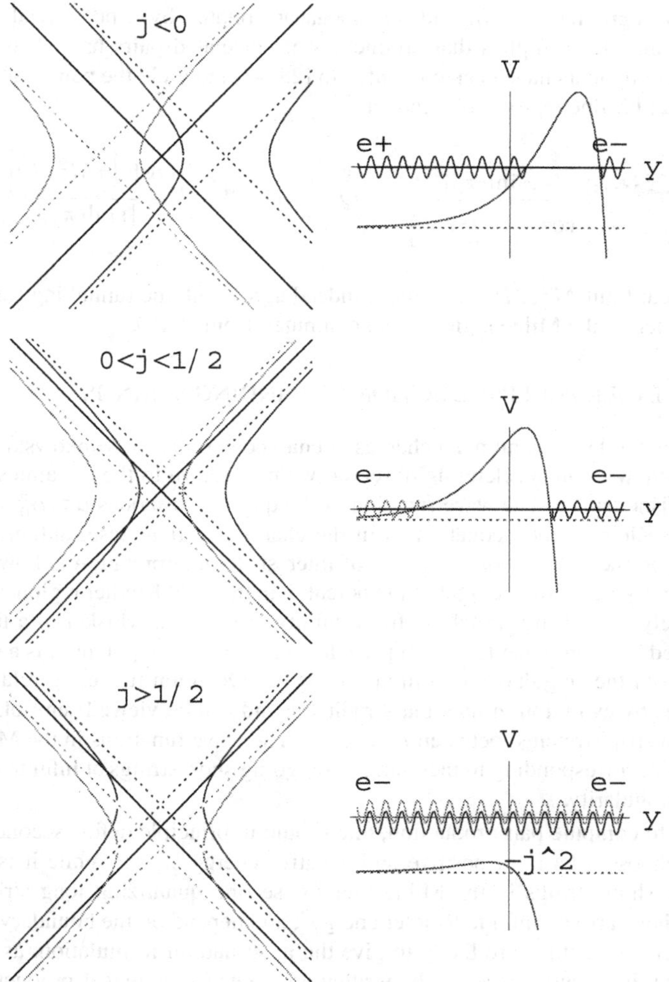

Figure 3.
Left: Classical trajectories of a charged particle in Rindler/Milne space. j labels the Rindler energy or Milne momentum, and is measured in units of M^2/ν. Right: Potential governing the radial motion in the right Rindler region, as a function of the canonical coordinate $y = e^r$.

corresponds to incoming modes from Rindler infinity $r = \infty$, while

$$\mathcal{U}^j_{in,R} = e^{-ij\eta} r^{-1} W_{i(\frac{j}{2}-\frac{M^2}{2\nu}),\frac{ij}{2}}(-i\nu r^2/2) \tag{1.21}$$

corresponds to incoming modes from the Rindler horizon $r = 0$. As usual in time-dependent backgroungs, the *in* and *out* vacua are related by a non-trivial Bogolubov transformation, which implies that production of correlated pairs has taken place. The Bogolubov coefficients have been computed in [28, 4], and yield the pair creation rates in the Rindler and Milne regions, respectively:

$$Q_R = e^{-\pi M^2/2\nu} \frac{|\sinh \pi j|}{\cosh \left[\pi \tilde{M}^2/2\nu\right]}\,, \quad Q_M = e^{-\pi M^2/2\nu} \frac{\cosh \left[\pi \tilde{M}^2/2\nu\right]}{|\sinh \pi j|}\,, \tag{1.22}$$

In the classical limit $M^2, \tilde{M}^2 \gg \nu$, these indeed agree with the tunnelling (or scattering over the barrier, in the Milne regions) rate computed from (1.19).

1.6. SCHWINGER PAIR PRODUCTION OF WINDING STRINGS

Having understood the quantum mechanics of charged particles in Minkowski space from the point of view of an accelerating observer, we now return to the dynamics of twisted strings in Misner space. The wave function of the quasi-zero-modes $\alpha_0^{\pm}, \alpha_0^{\pm}$ is governed by the same Klein-Gordon equation as in the charged particle case, although only the dependence on the radial coordinate r is of interest. Its interpretation is however rather different: e.g, a particle on the right of the potential in the right Rindler region corresponds to an infinitely long string stretching from infinity in the right whisker to a finite radius r_+ and folded back onto itself, while a particle on the left of the potential is a short string stretching from the singularity to a finite radius r_-. Quantum tunneling relates the two type of states by evolution in imaginary radius r, and can be viewed semi-classically as an Euclidean strip stretched between r_+ and r_-. The wave functions in the Milne region are less exotic, corresponding to incoming or outgoing short strings at infinite past, future or near the singularity.

In order to compute pair production, one should in principle define second quantized vacua, i.e. choose a basis of positive and negative energy states. While it is clear how to do so for short strings in the Milne regions, second quantizing long strings is less evident, as they carry an infinite Rindler energy[7], and depend on the boundary conditions at $r = \infty$. However, they are likely to give the most natural formulation, as any global wave function in Misner space can be written as a state in the tensor product of the left and right Rindler regions: the entire cosmological dynamics may thus be described as a state in a time-independent geometry, albeit with time loops !

Fortunately, even without a proper understanding of these issues, one may still use the formulae (1.22) to relate incoming and outcoming components of the closed string wave functions, and compute pair production for given boundary conditions at Rindler infinity. In particular, it should be noted that the production rate in the Milne regions Q_M is infinite for vanishing boost momentum $j = 0$, as a consequence of the singular geometry.

Moreover, although our analysis has borrowed a lot of intuition from the analogy to the charged particle problem, we are now in a position to describe pair production of

[7]The latter can be computed by quantizing the long string worldsheet using σ as the time variable [5].

winding strings in any geometry of the form

$$ds^2 = -dT^2 + a^2(T)d\theta^2 \quad \text{or} \quad ds^2 = dr^2 - b^2(r)d\eta^2 \tag{1.23}$$

(despite the fact that these geometries are not exact solutions of the *tree level* string equations of motion, they may be a useful mean field description of the back-reacted geometry). Neglecting the contributions of excited modes (which no longer decouple since the metric is not flat), the wave equation for the center of motion of strings winding around the compact direction θ or η, is obtained by adding to the two-dimensional Laplace operator describing the free motion of a neutral particle, the contribution of the tensive energy carried by the winding string:

$$\begin{cases} \frac{1}{a(T)}\partial_T\, a(T)\, \partial_T + \frac{j^2}{a^2(T)} + \frac{1}{4}w^2a^2(T) - m^2 = 0 \\ \frac{1}{b(r)}\partial_r\, b(r)\, \partial_r + \frac{j^2}{b^2(r)} + \frac{1}{4}w^2b^2(r) - m^2 = 0 \end{cases} \tag{1.24}$$

Choosing $a(T) = \beta T$ or $a(r) = \beta r$ and multiplying out by $(a^2(T), b^2(r))$, these equations indeed reduce to (1.19) and (1.18)[8]. In particular, for a smooth geometry, the production rate of pairs of winding strings is finite.

2. One-loop vacuum amplitude

In the previous section, we have obtain the production rate of winding strings in Misner space, from the Bogolubov coefficients of the tree-level wave functions. In principle, the same information could be extracted from the imaginary part of the one-loop amplitude. In this section, we start by reviewing the vacuum amplitude and stress-energy in field theory, and go on to study the one-loop amplitude in string theory, both in the twisted and untwisted sectors.

2.1. FIELD THEORY

The one-loop energy-momentum tensor generated by the quantum fluctuations of a free field ϕ with (two-dimensional) mass M^2 and spin s can be derived from the Wightman functions at coinciding points (and derivatives thereof). These depend on the choice of vacuum: in the simplest "Minkowski" vacum inherited from the covering space, any Green function is given by a sum over images of the corresponding one on the covering space. Using a (Lorentzian) Schwinger time representation and integrating over momenta, we obtain

$$G(x^\mu; x'^\mu) = \sum_{l=-\infty}^{\infty} \int_0^\infty d\rho\, (i\rho)^{-D/2} \exp\left[i\rho M^2 - 2\pi s l\right] \tag{2.1}$$
$$\exp\left[-\frac{i}{4\rho}(x^+ - e^{2\pi\beta l}x^{+'})(x^- - e^{-2\pi\beta l}x^{-'})\right]$$

[8] Notice that, in disagreement to a claim in the literature [29], the wave equation for $j = 0$ is *not* regular at the origin.

where s is the total spin carried by the field bilinear. Taking two derivatives and setting $x = x'$, one finds a divergent stress–energy tensor [22]

$$T_{\mu\nu}dx^\mu dx^\nu = \frac{1}{12\pi^2}\frac{K}{T^4}\left(-dT^2 - 3T^2 d\eta^2\right) \tag{2.2}$$

where the constant K is given by

$$K = \sum_{n=1}^{\infty} \cosh[2\pi n s\beta] \frac{2 + \cosh 2\pi n\beta}{(\cosh 2\pi n\beta - 1)^2} \tag{2.3}$$

This divergence is expected due to the large blue-shift of quantum fluctuations near the singularity. Notice that for spin $|s| > 1$, the constant K itself becomes infinite, a reflection of the non-normalizability of the wave functions for fields with spin.

In string theory, the local expectation value $\langle 0|T_{ab}(x)|0\rangle_{ren}$ is not an on-shell quantity, hence not directly observable. In contrast, the integrated free energy, given by a torus amplitude, is a valid observable[9]. In field theory, the free energy may be obtained by integrating the propagator at coinciding points (2.1) once with respect to M^2, as well as over all positions, leading to

$$\mathcal{F} = \sum_{l=-\infty}^{\infty} \int dx^+ dx^- \int_0^\infty \frac{d\rho}{(i\rho)^{1+\frac{D}{2}}} \exp\left(-8i\sinh^2(\pi\beta l)x^+x^- + iM^2\rho\right) \tag{2.4}$$

In contrast to the flat space case, the integral over the zero-modes $x^\pm, x$ does not reduce to a volume factor, but gives a Gaussian integral, centered on the light cone $x^+x^- = 0$. Dropping as usual the divergent $l = 0$ flat-space contribution and rotating to imaginary Schwinger time, one obtains a finite result

$$\mathcal{F} = \sum_{l=-\infty, l\neq 0}^{+\infty} \int_0^\infty \frac{d\rho}{\rho^{1+\frac{D}{2}}} \frac{e^{-M^2\rho - 2\pi\beta s l}}{\sinh^2(\pi\beta l)} \tag{2.5}$$

Consistently with the existence of globally defined positive energy modes for (untwisted) particles in Misner space, $\mathcal{F}$ does not have any imaginary part, implying the absence of net particle production between past and future infinity.

2.2. STRING AMPLITUDE IN THE UNTWISTED SECTOR

We may now compare the field theory result (2.5) to the one-loop vacuum amplitude in string theory with Euclidean world-sheet and Minkowskian target space, as computed in [9, 16]:

$$A_{bos} = \int_{\mathcal{F}} \sum_{l,w=-\infty}^{\infty} \frac{d\rho d\bar\rho}{(2\pi^2\rho_2)^{13}} \frac{e^{-2\pi\beta^2 w^2\rho_2 - \frac{R^2}{4\pi\rho_2}|l+w\tau|^2}}{\left|\eta^{21}(\rho)\,\theta_1(i\beta(l+w\rho);\rho)\right|^2} \tag{2.6}$$

[9] Of course, the spatial dependence of the one-loop energy may be probed by scattering e.g. gravitons at one-loop.

where θ_1 is the Jacobi theta function,

$$\theta_1(v;\rho) = 2q^{1/8}\sin\pi v \prod_{n=1}^{\infty}(1-e^{2\pi i v}q^n)(1-q^n)(1-e^{-2\pi i v}q^n)\,, \quad q = e^{2\pi i\rho} \tag{2.7}$$

In this section, we restrict to the untwisted sector $w = 0$. Expanding in powers of q, it is apparent that the string theory vacuum amplitude can be viewed as the field theory result (2.5) summed over the spectrum of (single particle) excited states, satisfying the matching condition enforced by the integration over ρ_1. As usual, field-theoretical UV divergences at $\rho \to 0$ are cut-off by restricting the integral to the fundamental domain F of the upper half plane.

In contrast to the field theory result, where the integrated free energy is finite for each particle separately, the free energy here has poles in the domain of integration, at

$$\rho = \frac{m}{n} + i\frac{\beta l}{2\pi n} \tag{2.8}$$

Those poles arise only after summing over infinitely many string theory states. Indeed, each pole originates from the $(1-e^{\pm 2\pi i v}q^n)$ factor in (2.7), hence re-sums the contributions of a complete Regge trajectory of fields with mass $M^2 = kn$ and spin $s = k$ ($k \in \mathbb{Z}$). In other words, the usual exponential suppression of the partition function by the increasing masses along the Regge trajectory is overcome by the spin dependence of that partition function. Regge trajectories are a universal feature of perturbative string theory, and these divergences are expected generically in the presence of space-like singularities. Since the pole (2.8) occurs both on the left- and right-moving part, the integral is not expected to give any imaginary contribution, in contrast to the charged open string case considered in [30].

2.3. STRING AMPLITUDE IN THE TWISTED SECTORS

We now turn to the interpretation of the string one-loop amplitude in the twisted sectors ($w \neq 0$), following the analysis in [5]. As in the rest of this lecture, it is useful to truncate the twisted string to its quasi-zero-modes, lumping together the excited mode contributions into a left and right-moving mass squared M^2 and $\tilde{M}^2$. Equivalently, we truncate the path integral to the "mini-superspace" of lowest energy configurations on the torus of modulus $\rho = \rho_1 + i\rho_2$, satisfying the twisted boundary conditions,

$$X^{\pm} = \pm\frac{1}{2\nu}\alpha^{\pm}e^{\mp(\nu\sigma - iA\tau)} \mp \frac{1}{2\nu}\tilde{\alpha}^{\pm}e^{\mp(\nu\sigma + i\tilde{A}\tau)} \tag{2.9}$$

where

$$A = \frac{k}{\rho_2} - i\beta\frac{l+\rho_1 w}{\rho_2}\,, \quad \tilde{A} = \frac{\tilde{k}}{\rho_2} + i\beta\frac{l+\rho_1 w}{\rho_2} \tag{2.10}$$

where $k, \tilde{k}$ are a pair of integers labelling the periodic trajectory, for fixed twist numbers (l, w). Notice that (2.9) is not a solution of the equations of motion, unless ρ coincides with one of the poles. In order to satisfy the reality condition on $X^{\pm}$, one should restrict to configurations with $k = \tilde{k}$, $\alpha^{\pm} = -(\tilde{\alpha}^{\pm})^*$. Nevertheless, for the sake of generality we shall not impose these conditions at this stage, but only exclude the case of a degenerate worldsheet $k = -\tilde{k}$.

We can now evaluate the Polyakov action for such a classical configuration, after rotating $\tau \to i\tau$:

$$\begin{aligned} S = &- \frac{\pi}{2\nu^2 \rho_2} \left(\nu^2 \rho_2^2 - [k - i(\beta l + \nu \rho_1)]^2 \right) R^2 \\ &- \frac{\pi}{2\nu^2 \rho_2} \left(\nu^2 \rho_2^2 - \left[\tilde{k} + i(\beta l + \nu \rho_1) \right]^2 \right) \tilde{R}^2 \\ &- 2\pi i j \nu \rho_1 + 2\pi \mu^2 \rho_2 \end{aligned} \tag{2.11}$$

where the last line, equal to $-i\pi\rho M^2 + i\pi\bar{\rho}\tilde{M}^2$, summarizes the contributions of excited modes, and $\alpha^{\pm} = \pm R e^{\pm\eta}/\sqrt{2}$, $\tilde{\alpha}^{\pm} = \pm \tilde{R} e^{\pm\tilde{\eta}}/\sqrt{2}$.

The path integral is thus truncated to an integral over the quasi-zero-modes $\alpha^{\pm}, \tilde{\alpha}^{\pm}$. Since the action (2.11) depends only on the boost-invariant products R^2 and $\tilde{R}^2$, a first divergence arises from the integration over $\eta - \tilde{\eta}$, giving an infinite factor, independent of the moduli, while the integral over $\eta + \tilde{\eta}$ is regulated to the finite value β after dividing out by the (infinite) order of the orbifold group.

In addition there are divergences coming from the integration over R and $\tilde{R}$ whenever

$$\rho_1 = -\frac{\beta l}{\nu} - i\frac{\nu}{2}(k - \tilde{k}) \,, \qquad \rho_2 = \frac{|k + \tilde{k}|}{2\nu} \tag{2.12}$$

which, for $k = \tilde{k}$, are precisely the double poles (2.8). These poles are interpreted as coming from infrared divergences due to existence of modes with arbitrary size $(R, \tilde{R})$. For $k \neq \tilde{k}$, the double poles are now in the complex ρ_1 plane, and may contribute for specific choices of integration contours, or second-quantized vacua. In either case, these divergences may be regulated by enforcing a cut off $|\rho - \rho_0| > \epsilon$ on the moduli space, or an infrared cut-off on R. It would be interesting to understand the deformation of Misner space corresponding to this cut off, analogous to the Liouville wall in AdS_3 [31].

Rather than integrating over $R, \tilde{R}$ first, which is ill-defined at ρ satisfying (2.12), we may choose to integrate over the modulus ρ first. The integral with respect to ρ_1 is Gaussian, dominated by a saddle point at

$$\rho_1 = -\frac{\beta l}{\nu} + i\frac{\tilde{k}\tilde{R}^2 - kR^2}{\nu(R^2 + \tilde{R}^2)} - 2i\frac{j\nu\rho_2}{R^2 + \tilde{R}^2} \tag{2.13}$$

It is important to note that this saddle point is a local extremum of the Euclidean action, unstable under perturbations of ρ_1. The resulting Bessel-type action has again a stable saddle point in ρ_2, at

$$\rho_2 = \frac{R\tilde{R}|k + \tilde{k}|}{\nu\sqrt{(R^2 + \tilde{R}^2)(4\mu^2 - R^2 - \tilde{R}^2) - 4j^2\nu^2}} \tag{2.14}$$

Integrating over ρ_2 in the saddle point approximation, we finally obtain the action as a function of the radii $R, \tilde{R}$:

$$\begin{aligned} S &= \frac{|k+\tilde{k}|R\tilde{R}\sqrt{(R^2+\tilde{R}^2)(R^2+\tilde{R}^2-4\mu^2)+4j^2\nu^2}}{\nu(R^2+\tilde{R}^2)} \\ &\pm 2\pi j \frac{\tilde{k}\tilde{R}^2-kR^2}{\nu(R^2+\tilde{R}^2)} \pm 2\pi i\beta j l \end{aligned} \tag{2.15}$$

where the sign of the second term is that of $k+\tilde{k}$. S admits an extremum at the on-shell values

$$R^2 = \mu^2 - j\nu, \qquad \tilde{R}^2 = \mu^2 + j\nu \qquad \text{with action} \quad S_{k,\tilde{k}} = \frac{\pi M\tilde{M}}{\nu}|k+\tilde{k}| \tag{2.16}$$

Notice that these values are consistent with the reality condition, since the boost momentum j is imaginary in Euclidean proper time. Evaluating (ρ_1, ρ_2) for the values (2.16), we reproduce (2.12), which implies that the integral is indeed dominated by the region around the double pole. Fluctuations in $(\rho_1, \rho_2, R, \tilde{R})$ directions around the saddle point have signature $(+,+,-,-)$, hence a positive fluctuation determinant, equal to $M^2\tilde{M}^2$ up to a positive numerical constant. This implies that the one-loop amplitude in the twisted sectors does not have any imaginary part, in accordance with the naive expectation based on the double pole singularities. It is also in agreement with the answer in the untwisted sectors, where the globally defined in and out vacua where shown to be identical, despite the occurence of pair production at intermediate times.

Nevertheless, the instability of the Euclidean action under fluctuations of ρ_1 and ρ_2 indicates that spontaneous pair production takes place, by condensation of the two unstable modes. Thus, we find that winding string production takes place in Misner space, at least for vacua such that the integration contour picks up contributions from these states. This is consistent with our discussion of the tree-level twisted wave functions, where tunneling in the Rindler regions implies induced pair production of short and long strings. The periodic trajectories (2.9) describe the propagation across the potential barrier in imaginary proper time, and correspond to an Euclidean world-sheet interpolating between the Lorentzian world-sheets of the long and short strings.

3. Tree-level scattering amplitudes

After this brief incursion into one-loop physics, we now return to the classical realm, and discuss some features of tree-level scattering amplitudes. We start by reviewing the scattering of untwisted modes, then turn to amplitudes involving two twisted modes, which can still be analyzed by Hamiltonian methods. We conclude with a computation of scattering amplitudes for more than 2 twisted modes, which can be obtained by a rather different approach. Our presentation follows [32, 6].

3.1. UNTWISTED AMPLITUDES

Tree-level scattering amplitudes for untwisted states in the Lorentzian orbifold are easily deduced from tree-level scattering amplitudes on the covering space, by the "inheritance principle": expressing the wave functions of the incoming or outgoing states in Misner space as superpositions of Minkowski plane waves with well-defined boost momentum j via Eq. (1.6) (with spin $s = 0$), the tree-level scattering amplitude is obtained by averaging the standard Virasoro-Shapiro amplitude

$$\mathcal{A}_{Mink} = \delta\left(\sum_i p_i\right) \frac{\Gamma\left(-\frac{\alpha'}{4}s\right)\Gamma\left(-\frac{\alpha'}{4}t\right)\Gamma\left(-\frac{\alpha'}{4}u\right)}{\Gamma\left(1+\frac{\alpha'}{4}s\right)\Gamma\left(1+\frac{\alpha'}{4}t\right)\Gamma\left(1+\frac{\alpha'}{4}u\right)} \tag{3.1}$$

under the actions of continuous boosts $p_i^{\pm} \to p_i^{\pm}(v) = e^{\pm\beta v_i} p_i^{\pm}$, with weight $e^{ij_i v_i}$, on all (but one) external momenta. Possible divergences come from the boundary of the parameter space spanned by the v_i, where some of the momenta $p_i^{\pm}(v)$ become large. In a general (Gross Mende, ($s, t, u \to \infty$ with $s/t, s/u$ fixed) high energy regime, the Virasoro-Shapiro amplitude is exponentially suppressed [33] and the integral over the v_i converges. However, there are also boundary configurations with $s, u \to \infty$ and fixed t where the Virasoro-Shapiro amplitude has Regge behavior s^t, in agreement with the fact that the size of the string at high energy grows like $\sqrt{\log s}$. In this regime, using the Stirling approximation to the Gamma functions in (3.1), it is easy to see that the averaged amplitude behaves as

$$A_{Misner} \sim \int^{\infty} dv \, \exp\left[v\left(i(j_2 - j_4) - \frac{1}{2}\alpha'(p_1^i - p_3^i)^2 + 1\right)\right] \tag{3.2}$$

hence diverges for small momentum transfer $(p_1^i - p_3^i)^2 \leq 2/\alpha'$ in the directions transverse to Misner space. There are similar collinear divergences in the other channels as well, both in the bosonic or superstring case.

The situation is slightly improved in the case of Grant space (analogous to the "null brane" considered in [34]), i.e. when the boost identification is combined with a translation of length R on a direction x_2 transverse to the light-cone: in this case, the boost momentum is no longer quantized (although the sum $Rp_2 + \beta j$ still is), and one can construct wave packets which are regular on the horizon, by superposition of states of different boost momentum [34, 16]. Collinear divergences remain, albeit in a reduced range of momentum transfer [6],

$$(\vec{p}_1 + \vec{p}_3)^2 \leq \frac{(\sqrt{1 + 2\alpha' E^2} - 1)^2}{(\alpha' E)^2}, \qquad E = \frac{\beta}{R} \tag{3.3}$$

As $R \to 0$, this reduces to Misner space case as expected.

As a matter of fact, these divergences may be traced to large tree-level graviton exchange near the singularity, or, in the Grant space case, near the chronological horizon [32]. Collinear divergences can in principle be treated in the eikonal approximation, i.e. by resumming an infinite series of ladder diagrams. While a naive application of the flat space result [35] suggests that this resummation may lead to finite scattering amplitudes of untwisted states in Misner space [19], a consistent treatment ensuring that only boost-invariant gravitons are exchanged has not been proposed yet, and prevents us from drawing a definitive conclusion. More generally, it would be extremely interesting to develop eikonal techniques in the presence of space-like singularities, and re-evaluate the claim in [36] that a single particle in Misner space will ineluctably cause the space to collapse.

3.2. TWO-TWIST AMPLITUDES

As we reviewed in Section 1.4, the zero-mode wave functions in the twisted sectors form a continuum of delta-normalizable states with arbitrarily negative worldsheet energy. In contrast to the standard case of twist fields of finite order in Euclidean rotation orbifolds, twisted states in Misner space should thus be described by a continuum of vertex operators with arbitrarily negative conformal dimension. While the conformal field theory of such operators remains ill-understood, amplitudes with two twisted fields only can be computed by ordinary operator methods on the cylinder, in the twisted vacua at $\tau = \pm\infty$ [6].

Stringy fuzziness.
Vertex operators for untwisted states are just a boost-invariant superposition of the ordinary flat space vertex operators. In order to compute their scattering amplitude against a twisted string, it is convenient to write them as a normal ordered expression in the twisted Hilbert space. Since the twisted oscillators have an energy $n \pm i\nu$ with $n \in \mathbb{Z}$, normal ordering gives a different contribution than in the untwisted state,

$$\Delta(\nu) \equiv [X^-_{\succ 0}, X^+_{\prec 0}] - [X^-_{>0}, X^+_{<0}] = \psi(1+i\nu) + \psi(1-i\nu) - 2\psi(1) \tag{3.4}$$

where $\psi(x) = \sum_{n=1}^{\infty}(x+n)^{-1} = d\log\Gamma(x)/dx$. In the above equation, $X^{\pm}_{>,<}$ (resp. $X^{\pm}_{\succ,\prec}$) denote the positive and negative frequency parts (excluding the (quasi) zero-mode contributions) of the embedding coordinates $X^{\pm}(\tau,\sigma)$, as defined by the untwisted (resp. twisted) mode expansion. As a result, the vertex operator for an untwisted tachyon becomes

$$: e^{i(k^+X^- + k^-X^+)} :^{(\text{un.})} = \exp\left[-k^+k^-\Delta(\nu)\right] \; : e^{i(k^+X^- + k^-X^+)} :^{(\nu-\text{tw.})} \tag{3.5}$$

Such a factor is in fact present for all untwisted states, although the normal ordering prescription is slightly more cumbersome for excited states. Since this normal ordering constant depends on the winding number $w = -\nu/\beta$, it cannot be reabsorbed by a field redefinition of the untwisted state, nor of the twisted string. Instead, it can be interpreted as the form factor acquired by untwisted states in the background of a twisted string, due to the zero-point quantum fluctuations of the winding string. The latter polarizes untwisted string states into a cloud of r.m.s. size $\sqrt{\Delta(\nu)}$. which, while proportional to ν at small ν, grows logarithmically with the winding number,

$$\Delta(\nu) = 2\zeta(3)\nu^2 + O(\nu^4) = 2\log\nu - \frac{23}{20} + O(\nu^{-2}) \tag{3.6}$$

Notice that this logarithmic growth winding can be viewed as the T-dual of the Regge growth with energy. It is also interesting to observe the analogy of the form factor in (3.5) with similar factors appearing in non-commutative gauge theories with matter in the fundamental representation – in line with the general relation between twisted strings and charged particles outlined in Section 1.4.

Zero-mode overlaps.
In general, the S-matrix element factorizes into a product of an excited mode contribution, which can be evaluated, just as in flat space, by normal ordering and commutation, and a

(quasi)-zero-mode contribution. In the real space representation (1.15) for the quasi-zero-modes, the latter reduces to an overlap of twisted and untwisted wave functions, e.g. in the three tachyon case,

$$\int dx^+ dx^- \; f_1^*(x^+, x^-) \; e^{i(p_2^- x^+ + p_2^+ x^-)} \; f_3(x^+, x^-) \tag{3.7}$$

where f_1 and f_3 denote eigenmodes of the charged Klein-Gordon equation, and f_2 is an eigenmode of the neutral Klein-Gordon equation, each of which with fixed angular momentum j_i. Considering higher excited modes such as the graviton would introduce extra factors of covariant derivatives $\alpha^\pm$ in (3.7).

In order to evaluate these overlaps, it is convenient to use a different representation and diagonalize half of the covariant derivative operators, e.g.

$$\alpha^- = i\nu\partial_{\alpha^+}, \quad \tilde{\alpha}^+ = i\nu\partial_{\tilde{\alpha}^-} \tag{3.8}$$

acting on functions of the variables $\alpha^+, \tilde{\alpha}^-$ taking values in the quadrant $\mathbb{R}^\epsilon \times \mathbb{R}^{\tilde{\epsilon}}$. On-shell wave functions are now powers of their arguments,

$$f(\alpha^+, \tilde{\alpha}^-) = N_{in} \left(\epsilon\, \alpha^+\right)^{\frac{M^2}{2i\nu} - \frac{1}{2}} \left(\tilde{\epsilon}\, \tilde{\alpha}^-\right)^{\frac{\tilde{M}^2}{2i\nu} - \frac{1}{2}} \tag{3.9}$$

The notation N_{in} for the normalization factor anticipates the fact that this representation is appropriate to describe an *in* state. The choice of the signs ϵ and $\tilde{\epsilon}$ of α^- and $\tilde{\alpha}^+$ distinguishes between short strings ($\epsilon\tilde{\epsilon} = 1$) and long strings ($\epsilon\tilde{\epsilon} = -1$). Of course, the oscillator representation (3.8) can be related to the real-space representation via the intertwiner

$$f(x^+, x^-) = \int d\tilde{\alpha}^+ d\alpha^- \Phi^{in}_{\nu,\tilde{\alpha}^+,\alpha^-}(x^+, x^-) f(\alpha^+, \tilde{\alpha}^-) \tag{3.10}$$

where the kernel is given by

$$\Phi^{in}_{\nu,\alpha^+,\tilde{\alpha}^-}(x^+, x^-) = \exp\left(\frac{i\nu x^+ x^-}{2} - i\alpha^+ x^- - i\tilde{\alpha}^- x^+ + \frac{i}{\nu}\alpha^+\tilde{\alpha}^-\right) \tag{3.11}$$

This kernel may be viewed as the wave function for an off-shell winding state with "momenta" α^+ and $\tilde{\alpha}^-$. Equivalently, one may diagonalize the complementary set of operators,

$$\alpha^+ = -i\nu\partial_{\alpha^-}, \quad \tilde{\alpha}^- = -i\nu\partial_{\tilde{\alpha}^+} \tag{3.12}$$

leading to on-shell wave functions

$$f(\alpha^-, \tilde{\alpha}^+) = N_{out} \left(\epsilon\, \alpha^-\right)^{-\frac{M^2}{2i\nu} - \frac{1}{2}} \left(\tilde{\epsilon}\, \tilde{\alpha}^+\right)^{-\frac{\tilde{M}^2}{2i\nu} - \frac{1}{2}} \tag{3.13}$$

Those are related to the real-space representation by the kernel

$$\Phi^{out}_{\nu,\tilde{\alpha}^+,\alpha^-}(x^+, x^-) = \exp\left(-\frac{i\nu x^+ x^-}{2} - i\tilde{\alpha}^+ x^- - i\alpha^- x^+ - \frac{i}{\nu}\tilde{\alpha}^+\alpha^-\right) \tag{3.14}$$

Replacing $f_1^*(x^+, x^-)$ and $f_3(x^+, x^-)$ by their expression in terms of the *out* and *in* wave functions (3.9), (3.13) respectively, renders the $x^\pm$ Gaussian (albeit with a non-positive definite quadratic form). The remaining $\alpha^\pm, \tilde{\alpha}^\pm$ integrals can now be computed in terms of hypergeometric functions. Including the form factor from the excited modes, we obtain, for the 3-point amplitude,

$$\langle 1| : e^{i(p_2^+ X^- + p_2^- X^+)} : |3\rangle = \frac{g_s}{2\nu}\, \delta_{\sum j_i}\, \delta\left(\sum p_i^{\perp}\right) \exp\left[-p_2^+ p_2^- \tilde{\Delta}(\nu)\right]$$
$$\left(-p_2^+\right)^{\mu-1} \left(-p_2^-\right)^{\tilde{\mu}-1} U\left(\lambda, \mu, i\frac{p_2^+ p_2^-}{\nu}\right) U\left(\tilde{\lambda}, \tilde{\mu}, i\frac{p_2^+ p_2^-}{\nu}\right) \quad (3.15)$$

where the non-locality parameter $\tilde{\Delta}(\nu)$ includes the contribution of the quasi-zero-mode,

$$\tilde{\Delta}(\nu) = \psi(i\nu) + \psi(1 - i\nu) - 2\psi(1) \quad (3.16)$$

The parameters of the Tricomi confluent hypergeometric functions U appearing in (3.15) are given by

$$\lambda = \frac{1}{2} + \frac{M_3^2}{2i\nu} \qquad \tilde{\lambda} = \frac{1}{2} + \frac{\tilde{M}_3^2}{2i\nu}$$
$$\mu = 1 + \frac{M_3^2 - M_1^2}{2i\nu} \qquad \tilde{\mu} = 1 + i\frac{\tilde{M}_3^2 - \tilde{M}_1^2}{2i\nu} \,. \quad (3.17)$$

The amplitude is finite, and it is proportional to the overlap of the zero-mode wave-functions, up to the smearing due to the form factor of the untwisted string in the background of the twisted string. Similar expressions can be obtained for 3-point functions in superstring theory involving an untwisted massless state.

Four-point amplitudes.

The same techniques allow to compute 4-point amplitudes, which now include an integral over the location of the 4-th vertex, as well as on the relative boost parameter v between the two untwisted vertices. The complete expression can be found in [6] and is somewhat abstruse, however it is useful to consider the factorization limit $z \to 0$ where $T(3), T(4)$ (resp. $T(1), T(2)$) come together:

$$\langle 1|T(2)T(3)|4\rangle \to g_s^2\, \delta_{-j_1+j_2+j_3+j_4}\, \delta\left(-\vec{p}_1 + \sum_{i=1}^{3} \vec{p}_i\right)$$
$$\int_{-\infty}^{\infty} dv\, e^{i(j_3-j_1)v} \int dz d\bar{z}\, |z|^{2\vec{p}_3\cdot\vec{p}_4 + \vec{p}_3\cdot\vec{p}_3 - 2} \exp\left[-\left(p_2^+ p_2^- + p_3^+ p_3^-\right)\tilde{\Delta}(\nu)\right]$$
$$(-1)^{\mu+\tilde{\mu}}\, (p_2^+)^{-\tilde{\lambda}}\, (p_2^-)^{-\lambda}\, (p_3^+)^{\mu-\lambda-1}\, (p_3^-)^{\tilde{\mu}-\tilde{\lambda}-1}\, z^{-\frac{1}{2}M_1^2 - \frac{i\nu}{2}}\, \bar{z}^{-\frac{1}{2}\tilde{M}_1^2 - \frac{i\nu}{2}} \quad (3.18)$$

The amplitude diverges whenever $j_3 = j_1$ due to the propagation of winding strings with vanishing boost momentum in the intermediate channel. This result closely parallels the discussion in Ref. [32], where tree-level scattering amplitudes of four untwisted states where found to diverge, due to large graviton exchange near the singularity.

3.3. MORE THAN TWO TWISTED STRINGS

Scattering amplitudes involving three or more twisted states can be obtained by mapping to an analogous problem which is now very well understood: the Wess-Zumino-Witten model of a four-dimensional Neveu-Schwarz plane wave [37, 38, 39], with metric

$$ds^2 = -2dudv + d\zeta d\bar{\zeta} - \frac{1}{4}\zeta\bar{\zeta}du^2 \,, \qquad H = dudxd\bar{x} \tag{3.19}$$

where $\zeta = x_1 + ix_2$ is the complex coordinate in the plane. In the light-cone gauge $u = p\tau$, it is well known that the transverse coordinate X has the mode expansion of a complex scalar field twisted by a real, non rational angle proportional to the light-cone momentum p [39]. In fact, there exists a free-field representation where the vertex operator of a physical state with non-zero p is just the product of a plane wave along the (u, v) light cone coordinates, times a twist field[10] creating a cut z^p on the world-sheet. Correlation functions of physical states have been computed using standard WZW techniques [41, 42], and, by removing the plane wave contribution, it is then possible to extract the correlator of twist fields with arbitrary angle.

Referring the reader to [6] for more details, we simply quote the result for the three twist amplitude: in real-space representation (1.15), the amplitude (hence, the OPE coefficient of 3 twist fields) is given by the overlap

$$\int dx_1^\pm dx_2^\pm \qquad \exp\left[(x_1^+ - x_2^+)(x_1^- - x_2^-)\Xi(\nu_1,\nu_2)\right] \\ [f_1(x_1^\pm)f_2(x_2^\pm)]^* \, f_3\left(x_3^\pm - \frac{\nu_1 x_1^\pm + \nu_2 x_2^\pm}{\nu_1+\nu_2}\right) \tag{3.20}$$

where the characteristic size of the kernel is given by the ratio

$$\Xi(\nu_1,\nu_2) = -i\frac{1 - \frac{i\nu_3}{\nu_1\nu_2}\frac{\gamma(i\nu_3)}{\gamma(i\nu_1)\gamma(i\nu_2)}}{1 + \frac{i\nu_3}{\nu_1\nu_2}\frac{\gamma(i\nu_3)}{\gamma(i\nu_1)\gamma(i\nu_2)}} \tag{3.21}$$

with $\gamma(p) \equiv \Gamma(p)/\Gamma(1-p)$. As $\nu_i \to 0$, $\Xi(\nu_1,\nu_2) \sim 1/(2\zeta(3)\,\nu_3^2)$ so that the interaction becomes local, as expected for flat space vertex operators. For larger ν however, the non-locality scale $1/\sqrt{\Xi}$ diverges when $\nu_1\nu_2\gamma(i\nu_1)\gamma(i\nu_2) = i\nu_3\gamma(i\nu_3)$. The origin of this divergence is not well understood at present.

3.4. TOWARD CLASSICAL BACK-REACTION

While computing the back-reaction from the quantum production of particles and strings remains untractable with the present techniques, the results above give us a handle on a related problem, namely the linear response of closed string fields to a classical (coherent) condensate of winding strings. Indeed, consider deforming Misner space away from the orbifold point, by adding to the free worldsheet action a condensate of marginal twist operators:

$$S_\lambda = \int d^2\sigma\, \partial X^+ \bar{\partial} X^- + \lambda_{-w} V_{+w} + \lambda_{+w} V_{-w} \tag{3.22}$$

While this deformation is marginal at leading order, it implies a one-point function for untwisted fields

$$\langle e^{ikX} \rangle_\lambda \sim \lambda_w \lambda_{-w} \langle w | e^{ikX} | -w \rangle \,, \tag{3.23}$$

[10] For integer p, new "spectrally flowed" states appear describing long strings [40].

which needs to be cancelled by deformating S at order λ^2 by an untwisted field: this is the untwisted field classically sourced by the winding string with vertex operator $V_{\pm w}$. In addition, the same winding string also sources twisted states whose winding number is a multiple of w:

$$\langle V_{-2w} \rangle_\lambda \sim \lambda_w \lambda_w \langle w | V_{-2w} | w \rangle \,, \tag{3.24}$$

The 3-point functions in (3.23), (3.24) are precisely the amplitudes which have been computed the two previous sections. It is thus possible to extract the corrections to the metric and other string fields to leading order in the deformation parameter λ_w. In the Euclidean orbifold case, such a procedure allows to resolve a conical ALE singularity into a smooth Eguchi Hanson gravitational instanton. Whether the same procedure allows to resolve the divergences of the Lorentzian orbifold remains an intriguing open question.

4. Discussion

In this lecture, we have taken a tour of the classical aspects of the propagation of closed strings in a toy model of a cosmological singularity: Misner space, a.k.a. the Lorentzian orbifold $\mathbb{R}^{1,1}/\mathbb{Z}$. Our emphasis has been particularly on twisted sectors, which play such an important rôle in resolving the conical singularities of Euclidean orbifolds. In particular, we have obtained a semi-classical understanding of the pair production of winding strings, as a tunneling effect in the Rindler regions, in close analogy to Schwinger pair creation in an electric field. Despite the fact that the one-loop amplitude remains real, indicating no overall particle production between infinite past and infinite future, it is clear that abundant production of particles and strings takes place near the singularity.

While tree-level scattering amplitudes exhibit severe divergences due to the infinite blue-shift near the singularity, it is quite conceivable that the back-reaction from the cosmological production of particles and winding strings may lead to a smooth cosmology, interpolating between the collapsing and expanding phases. Indeed, winding strings behave much like a two-dimensional positive cosmological constant, and may thus lead to a transient inflation preventing the singularity to occur.

Unfortunately, incorporating back-reaction from quantum production lies outside the scope of current perturbative string technology at present. A second quantized definition of string theory would seem to be a prerequisite to even formulate this question, however, unlike the open string case, a field theory of off-shell closed strings has remained elusive, and may even be excluded on general grounds. A generalization of the usual first quantized approach allowing for non-local deformations of the worldsheet [43] may in principle incorporate emission of correlated pairs of particles, however do not seem very tractable at present.

Instead, the most practical approach seems to consider classical deformations by twisted fields away from the orbifold point. In contrast to the problem of quantum back-reaction, this may be treated in conformal perturbation theory, and we have taken some steps in this direction. It remains to see whether Misner space is a good approximation to the resulting space.

More importantly, Misner space appears to be a very finely tuned example of the space-like singularities which are generically expected to occur in classical Einstein gravity: as shown long ago by Belinsky, Khalatnikov and Lifshitz, and independently by Misner himself (see e.g. [44] for a recent review), the generic approach to a cosmological singularity consists of a chaotic sequence of "Kasner" epochs (of which Milne/Misner space is a special example with zero curvature) and curvature-induced bounces, occuring heterogeneously through space. An outstanding question is therefore to understand string theory in Misner (Mixmaster) space.

Acknowledgements

Both authors are grateful to M. Berkooz, D. Reichman and M. Rozali for a very enjoyable collaboration, and to B. Craps, G. d'Appolonio, E. Kiritsis, and G. Moore for useful discussions. B. D. would like to thank the organizers of the Cargèse 2004 ASI for giving him the opportunity to present part of this work in the Gong Show.

References

1. S. Lem, "The Seventh Voyage", in *The Star Diaries*, Varsaw 1971, english translation New York, 1976.
2. A. Borde and A. Vilenkin, "Eternal inflation and the initial singularity," Phys. Rev. Lett. **72** (1994) 3305 [arXiv:gr-qc/9312022].
3. C. W. Misner, in *Relativity Theory and Astrophysics I: Relativity and Cosmology*, edited by J. Ehlers, Lectures in Applied Mathematics, Vol. 8 (American Mathematical Society, Providence, 1967), p. 160.
4. M. Berkooz, and B. Pioline, "Strings in an electric field, and the Milne universe," JCAP **0311** (2003) 007 [arXiv:hep-th/0307280].
5. M. Berkooz, B. Pioline and M. Rozali, "Closed strings in Misner space: Cosmological production of winding strings," JCAP **07** (2004) 003 [arXiv:hep-th/0405126].
6. M. Berkooz, B. Durin, B. Pioline and D. Reichmann, "Closed strings in Misner space: Stringy fuzziness with a twist," arXiv:hep-th/0407216.
7. G. T. Horowitz and A. R. Steif, "Singular String Solutions With Nonsingular Initial Data," Phys. Lett. B **258**, 91 (1991).
8. J. Khoury, B. A. Ovrut, N. Seiberg, P. J. Steinhardt and N. Turok, "From big crunch to big bang," Phys. Rev. D **65** (2002) 086007 [arXiv:hep-th/0108187].
9. N. A. Nekrasov, "Milne universe, tachyons, and quantum group," arXiv:hep-th/0203112.
10. V. Balasubramanian, S. F. Hassan, E. Keski-Vakkuri and A. Naqvi, "A space-time orbifold: A toy model for a cosmological singularity," Phys. Rev. D **67** (2003) 026003 [arXiv:hep-th/0202187]; R. Biswas, E. Keski-Vakkuri, R. G. Leigh, S. Nowling and E. Sharpe, "The taming of closed time-like curves," JHEP **0401** (2004) 064 [arXiv:hep-th/0304241].
11. I. Antoniadis, C. Bachas, J. R. Ellis and D. V. Nanopoulos, "Cosmological String Theories And Discrete Inflation," Phys. Lett. B **211** (1988) 393; I. Antoniadis, C. Bachas, J. R. Ellis and D. V. Nanopoulos, "An Expanding Universe In String Theory," Nucl. Phys. B **328** (1989) 117; I. Antoniadis, C. Bachas, J. R. Ellis and D. V. Nanopoulos, "Comments On Cosmological String Solutions," Phys. Lett. B **257** (1991) 278.
12. C. R. Nappi and E. Witten, "A Closed, expanding universe in string theory," Phys. Lett. B **293**, 309 (1992) [arXiv:hep-th/9206078].
13. C. Kounnas and D. Lust, "Cosmological string backgrounds from gauged WZW models," Phys. Lett. B **289** (1992) 56 [arXiv:hep-th/9205046].
14. E. Kiritsis and C. Kounnas, "Dynamical topology change in string theory," Phys. Lett. B **331** (1994) 51 [arXiv:hep-th/9404092].
15. S. Elitzur, A. Giveon, D. Kutasov and E. Rabinovici, "From big bang to big crunch and beyond," JHEP **0206**, 017 (2002) [arXiv:hep-th/0204189]; S. Elitzur, A. Giveon and E. Rabinovici, "Removing singularities," JHEP **0301**, 017 (2003) [arXiv:hep-th/0212242].
16. L. Cornalba and M. S. Costa, "A New Cosmological Scenario in String Theory," Phys. Rev.

D **66**, 066001 (2002) [arXiv:hep-th/0203031]; L. Cornalba, M. S. Costa and C. Kounnas, "A resolution of the cosmological singularity with orientifolds," Nucl. Phys. B **637**, 378 (2002) [arXiv:hep-th/0204261]; L. Cornalba and M. S. Costa, "On the classical stability of orientifold cosmologies," Class. Quant. Grav. **20** (2003) 3969 [arXiv:hep-th/0302137];

17. B. Craps, D. Kutasov and G. Rajesh, "String propagation in the presence of cosmological singularities," JHEP **0206**, 053 (2002) [arXiv:hep-th/0205101]; B. Craps and B. A. Ovrut, "Global fluctuation spectra in big crunch / big bang string vacua," Phys. Rev. D **69** (2004) 066001 [arXiv:hep-th/0308057].
18. E. Dudas, J. Mourad and C. Timirgaziu, "Time and space dependent backgrounds from non-supersymmetric strings," Nucl. Phys. B **660**, 3 (2003) [arXiv:hep-th/0209176].
19. L. Cornalba and M. S. Costa, "Time-dependent orbifolds and string cosmology," Fortsch. Phys. **52**, 145 (2004) [arXiv:hep-th/0310099].
20. C. V. Johnson and H. G. Svendsen, "An exact string theory model of closed time-like curves and cosmological singularities," arXiv:hep-th/0405141.
21. N. Toumbas and J. Troost, "A time-dependent brane in a cosmological background," JHEP **0411** (2004) 032 [arXiv:hep-th/0410007].
22. W. A. Hiscock and D. A. Konkowski, "Quantum Vacuum Energy In Taub - Nut (Newman-Unti-Tamburino) Type Cosmologies," Phys. Rev. D **26** (1982) 1225.
23. A. H. Taub, "Empty Space-Times Admitting A Three Parameter Group Of Motions," Annals Math. **53**, 472 (1951); E. Newman, L. Tamburino and T. Unti, "Empty Space Generalization Of The Schwarzschild Metric," J. Math. Phys. **4** (1963) 915.
24. J. G. Russo, "Cosmological string models from Milne spaces and SL(2,Z) orbifold," arXiv:hep-th/0305032.
25. J. R. I. Gott, "Closed Timelike Curves Produced By Pairs Of Moving Cosmic Strings: Exact Solutions," Phys. Rev. Lett. **66**, 1126 (1991); J. D. Grant, "Cosmic strings and chronology protection," Phys. Rev. D **47** (1993) 2388 [arXiv:hep-th/9209102].
26. S. W. Hawking, "The Chronology protection conjecture," Phys. Rev. D **46**, 603 (1992).
27. D. Kutasov, J. Marklof and G. W. Moore, "Melvin Models and Diophantine Approximation," arXiv:hep-th/0407150.
28. C. Gabriel and P. Spindel, "Quantum charged fields in Rindler space," Annals Phys. **284** (2000) 263 [arXiv:gr-qc/9912016].
29. N. Turok, M. Perry and P. J. Steinhardt, "M theory model of a big crunch / big bang transition," Phys. Rev. D **70** (2004) 106004 [arXiv:hep-th/0408083].
30. C. Bachas and M. Porrati, "Pair Creation Of Open Strings In An Electric Field," Phys. Lett. B **296**, 77 (1992) [arXiv:hep-th/9209032].
31. J. M. Maldacena, H. Ooguri and J. Son, "Strings in AdS(3) and the SL(2,R) WZW model. II: Euclidean black hole," J. Math. Phys. **42**, 2961 (2001) [arXiv:hep-th/0005183].
32. M. Berkooz, B. Craps, D. Kutasov and G. Rajesh, "Comments on cosmological singularities in string theory," arXiv:hep-th/0212215.
33. D. J. Gross and P. F. Mende, "The High-Energy Behavior Of String Scattering Amplitudes," Phys. Lett. B **197**, 129 (1987).
34. H. Liu, G. Moore and N. Seiberg, "Strings in a time-dependent orbifold," JHEP **0206**, 045 (2002) [arXiv:hep-th/0204168]; H. Liu, G. Moore and N. Seiberg, "Strings in time-dependent orbifolds," JHEP **0210**, 031 (2002) [arXiv:hep-th/0206182].
35. D. Amati, M. Ciafaloni and G. Veneziano, "Classical And Quantum Gravity Effects From Planckian Energy Superstring Collisions," Int. J. Mod. Phys. A **3** (1988) 1615.
36. G. T. Horowitz and J. Polchinski, "Instability of spacelike and null orbifold singularities," Phys. Rev. D **66**, 103512 (2002) [arXiv:hep-th/0206228].
37. C. R. Nappi and E. Witten, "A WZW model based on a nonsemisimple group," Phys. Rev. Lett. **71**, 3751 (1993) [arXiv:hep-th/9310112];
38. D. I. Olive, E. Rabinovici and A. Schwimmer, "A Class of string backgrounds as a semiclassical limit of WZW models," Phys. Lett. B **321** (1994) 361 [arXiv:hep-th/9311081].
39. E. Kiritsis and C. Kounnas, "String Propagation In Gravitational Wave Backgrounds," Phys. Lett. B **320** (1994) 264 [Addendum-ibid. B **325** (1994) 536] [arXiv:hep-th/9310202]; E. Kiritsis, C. Kounnas and D. Lust, "Superstring gravitational wave backgrounds with space-time

supersymmetry," Phys. Lett. B **331**, 321 (1994) [arXiv:hep-th/9404114].
40. E. Kiritsis and B. Pioline, "Strings in homogeneous gravitational waves and null holography," JHEP **0208**, 048 (2002) [arXiv:hep-th/0204004].
41. G. D'Appollonio and E. Kiritsis, "String interactions in gravitational wave backgrounds," arXiv:hep-th/0305081.
42. Y. K. Cheung, L. Freidel and K. Savvidy, "Strings in gravimagnetic fields," JHEP **0402** (2004) 054 [arXiv:hep-th/0309005].
43. O. Aharony, M. Berkooz and E. Silverstein, "Multiple-trace operators and non-local string theories," JHEP **0108** (2001) 006 [arXiv:hep-th/0105309]; M. Berkooz, A. Sever and A. Shomer, "Double-trace deformations, boundary conditions and spacetime singularities," JHEP **0205** (2002) 034 [arXiv:hep-th/0112264]; E. Witten, "Multi-trace operators, boundary conditions, and AdS/CFT correspondence," arXiv:hep-th/0112258.
44. T. Damour, M. Henneaux and H. Nicolai, "Cosmological billiards," Class. Quant. Grav. **20**, R145 (2003) [arXiv:hep-th/0212256].

PHENOMENOLOGICAL GUIDE TO PHYSICS BEYOND THE STANDARD MODEL

STEFAN POKORSKI
Institute for Theoretical Physics, University of Warsaw
and
Theory Division, CERN, Geneve

Abstract. Various aspects of physics beyond the Standard Model are discussed from the perspective of the fantastic phenomenological success of the Standard Model, its simplicity and predictive power.

1. Introduction

The Standard Model is a successful theory of interactions of quarks and leptons at energies up to about hundred GeV. Despite that success it is widely expected that there is physics beyond the Standard Model, with new characteristic mass scale(s), perhaps up to, ultimately, a string scale.

The expectation is motivated by several fundamental questions that remain unanswered by the Standard Model. The most pressing one is better understanding of the mechanism of the electroweak symmetry breaking. The origin of flavour and of the pattern of fermion masses and of CP violation also remain beyond its scope. Moreover, we know now that the physics of the Standard Model cannot explain the baryon asymmetry in the Universe. And on the top of all that come two recent strong experimental hints for physics beyond the Standard Model, that is very small neutrino masses and the presence of dark matter in the Universe. The list can be continued by including dark energy and inflation.

The Standard Model does not explain the scale of the electroweak symmetry breaking. It is a free parameter of the theory, taken from experiment. Moreover, once we accept the point of view that the Standard Model is only an effective "low energy" theory which is somehow cut-off at a mass scale M, and if $M \gg M_{W,Z}$, the electroweak symmetry breaking mechanism based on use of an elementary Higgs field is unstable against quantum corrections (this is the so-called hierarchy problem).

Many different extensions of the Standard Model have been proposed to avoid the hierarchy problem and, more ambitiously, to calculate the scale of the electroweak breaking in terms of, hopefully, more fundamental parameters. Some extensions give the complete

L. Baulied et al. (eds.), String Theory: From Gauge Interactions to Cosmology, 201–228.

Standard Model, with one Higgs doublet, as their low energy approximation in the sense of the Appelquist - Carazzone decoupling and in some others the mechanism of the electroweak symmetry breaking cannot be decoupled from the bigger theory. The general idea is that the bigger theory has some characteristic mass scale M only order of magnitude bigger than $M_{W,Z}$, which plays the role of a cut-off to the electroweak sector. All those extensions of the Standard Model have distinct experimental signatures. The experiments at the LHC will, hopefully, shed more light on the mechanism of the electroweak symmetry breaking and will support one of those (or still another one ?) directions.

One approach is based on low energy supersymmetry. The scale M is identified with the mass scale of supersymmetric partners of the Standard Model particles. Supersymmetry is distinct in several very important points from all other proposed solutions to the hierarchy problem. First of all, it provides a general theoretical framework which allows to address many physical questions. Supersymmetric models, like the Minimal Supersymmetric Standard Model or its simple extensions satisfy a very important criterion of "perturbative calculability". In particular, they are easily consistent with the precision electroweak data. The Standard Model is their low energy approximation in the sense of the Appelquist - Carazzone decoupling, so most of the successful structure of the Standard Model is built into supersymmetric models. Unfortunately, there are also some troublesome exceptions: there are new potential sources of Flavour Changing Neutral Current (FCNC) transitions and of CP violation, and baryon and lepton numbers are not automatically conserved by the renormalizable couplings. But even those problems can at least be discussed in a concrete way. The quadratically divergent quantum corrections to the Higgs mass parameter (the origin of the hierarchy problem in the Standard Model) are absent in any order of perturbation theory. Therefore, the cut-off to a supersymmetric theory can be as high as the Planck scale and "small" scale of the electroweak breaking is still natural. But the hierarchy problem of the electroweak scale is solved at the price of a new hierarchy problem of the soft supersymmetry braking scale versus the Planck (string) scale. Spontaneous supersymmetry breaking and its transmission to the visible sector is a difficult problem and a fully satisfactory mechanism has not yet been found. Again on the positive side, supersymmetry is not only consistent with Grand Unification of elementary forces but, in fact, makes it very successful. And, finally, supersymmetry is needed for string theory.

All other extensions of the electroweak theory proposed as solutions to the hierarchy problem rely on an onset of some kind of strong dynamics at energy scales not much higher than the electroweak scale. In some of them, like Higgless models with dynamical electroweak symmetry breaking or strong gravity in large extra dimensions, the strong dynamics is simply a cut-off directly to the electroweak sector and appears already at $\mathcal{O}(1\ \text{TeV})$. In models with the Higgs boson as a pseudo-Goldstone boson (e.g. Little Higgs models) and models with gauge fields present in extra dimensions the cut-off scale M is identified with the characteristic scale of new perturbative physics, e.g. with the scale of breaking of some global symmetry or with the radii of extra dimensions. However, since those models are non-renormalizable and, moreover, in the bigger theory the quadratic divergences to the scalar mass parameter are absent typically only at one loop level, new physics itself has to be cut-off by some unknown strong dynamics at a scale one or two orders of magnitude higher than the M. Generally speaking, there is no Appelquist - Carazzone decoupling of new physics and the precision tests of such a version of the electroweak theory are not possible at the same level of accuracy as in the renormalizable Standard Model.

It is clear that models with early onset of strong dynamics cannot be easily, if at all, reconciled with Grand Unification. Also, they are very strongly constrained by precision electroweak data. There have been constructed models that work but simple models are usually ruled out. Moreover, various aspects of flavour physics are often very obscure.

Spontaneous symmetry breaking in the condensed matter physics and in QCD is due to some collective effects. In supersymmetric models, such effects are presumably responsible for spontaneous breaking of supersymmetry and, in consequence, for the generation of soft mass terms. However, the electroweak symmetry breaking is driven by perturbative quantum corrections, generated by the large top quark Yukawa coupling, to the scalar potential. In the Little Higgs models, the Higgs boson is a Goldstone boson of a bigger spontaneously broken global symmetry group. The Higgs potential needed for the electroweak symmetry breaking is also given by quantum corrections, with important contribution from the top quark Yukawa coupling. Thus, one thing many models have in common is that the electroweak symmetry is broken by perturbative quantum effects and linked to the large mass of the top quark.

At present, all extensions of the Standard Model remain speculative and none is fully satisfactory. Remembering the simplicity, economy and success of the Standard Model, one may wonder if in our search for its extensions shouldn't the Hipocrates principle *Primum non nocere* play more important role than it does. Indeed, various new ideas offer surprisingly low ratios of benefits to losses. It is, therefore, appropriate to begin by reviewing the basic structure of the Standard Model that underlies its success. It is likely that it gives us important hints for the physics beyond.

2. The Standard Model

2.1. BASIC STRUCTURE

The underlying principles of the electroweak theory are:

1. local $SU(2)_L \times U(1)_Y$ gauge symmetry and electroweak unification
2. spontaneous breaking of $SU(2)_L \times U(1)_Y$ gauge symmetry to $U(1)_{EM}$, by the Higgs mechanism with one Higgs doublet
3. matter content (chiral fermions)
4. renormalizability

Massless chiral fermions are the fundamental objects of matter: left-handed, with helicity $\lambda = -1/2$, and right-handed, with helicity $\lambda = 1/2$. It is so because parity and charge conjugation are not the symmetries of our world. The left-handed fermions carry different weak charges from the right-handed fermions. Chiral fermion fields are two-component (Weyl) spinors (see e.g. [1]):

$\underline{SU(2)_L \text{ doublets}}$

$$q_1 \equiv \begin{pmatrix} u \\ d \end{pmatrix} \quad q_2 \equiv \begin{pmatrix} c \\ s \end{pmatrix} \quad q_3 \equiv \begin{pmatrix} t \\ b \end{pmatrix} \tag{2.1}$$

$$l_1 \equiv \begin{pmatrix} \nu_e \\ e \end{pmatrix} \quad l_2 \equiv \begin{pmatrix} \nu_\mu \\ \mu \end{pmatrix} \quad l_3 \equiv \begin{pmatrix} \nu_\tau \\ \tau \end{pmatrix} \tag{2.2}$$

with the electric charge and the hypercharge ($Y = Q - T_3$) assigned as below

	u	d	ν	e
Q	2/3	-1/3	0	-1
Y	1/6	1/6	-1/2	-1/2

These are left-handed chiral fields in the representation $(0, 1/2)$ of the $SL(2, C)$, each describing two massless degrees of freedom: a particle with the helicity $\lambda = -1/2$ and its antiparticle with $\lambda = +1/2$. (The chiral fields can also be written as four-component spinors (see e.g. [1]) but in the following we shall be using the Weyl notation).

Right-handed fields $[(1/2, 0)$ of $SL(2, C)]$ in the same representations of $SU(2)_L \times U(1)_Y$ as the left-handed fields (2.1) and (2.2) do not exist in Nature. Instead, we have

<u>$SU(2)_L$ singlets</u>

$$\begin{array}{lll} u_R, & c_R, & t_R \\ d_R, & s_R, & b_R \\ e_R, & \mu_R, & \tau_R \end{array}$$

in $(1, +2/3)$, $(1, -1/3)$ and $(1, -1)$ of $SU(2)_L \times U(1)_Y$, respectively. These are right-handed chiral fields in the $(1/2, 0)$ representation of the group $SL(2, C)$. For constructing a Lorentz invariant Lagrangian, it is more convenient to take as fundamental fields only the left-handed chiral fields. Thus, we introduce left-handed chiral fields, e.g

$$u^c, \quad c^c, \quad t^c \tag{2.3}$$

in $(1, -2/3)$ of $SU(2)_L \times U(1)_Y$, such that

$$\bar{u^c} \equiv CPu^c(CP)^{-1} = u_R \tag{2.4}$$

Indeed, CP transformation results in the simultaneous change of chirality and charges (representation $R \to R^*$ for internal symmetries). Moreover, we see that the electric charge $Q = T_3 + Y$ satisfies, e.g.

$$Q_{u^c} = -2/3 = -Q_u \tag{2.5}$$

and the two left-handed fields u and u^c become charge conjugate to each other when $U_{EM}(1)$ remains the only unbroken symmetry:

$$Cu^cC^{-1} = u \tag{2.6}$$

We note that the matter chiral fields of the SM do not include a right-handed neutrino field ν_R in $(1, 1)$ of $SU(2)_L \times U(1)_Y$ (such a charge assignment preserves the relation $Q = T_3 + Y$) or equivalently, a left-handed field ν^c such that

$$\nu_R = CP\nu^c(CP)^{-1} \tag{2.7}$$

but we can supplement the Standard Model with such a particle, if useful.

The breaking of the electroweak symmetry is generated by the potential of the Higgs doublet $H = \begin{pmatrix} H^+ \\ H^0 \end{pmatrix}$ with the hypercharge $Y = +1/2$:

$$V = m^2 H^\dagger H + \frac{\lambda}{2}(H^\dagger H)^2 \tag{2.8}$$

When $m^2 < 0$ is chosen, the Higgs doublet acquires the vacuum expectation value. Indeed, the minimum of the potential is for

$$\langle H^\dagger H \rangle = -\frac{m^2}{\lambda} \equiv \frac{v^2}{2} \tag{2.9}$$

By $SU_L(2)$ rotation we can always redefine the vacuum so that only the VEV of the lower component of the Higgs doublet is non-zero. The $SU_L(2) \times U_Y(1)$ symmetry is then broken down to $U'(1)$ which is identified with $U_{EM}(1)$ with $Q = T^3 + Y$ because $(T^3 + Y)\begin{pmatrix} 0 \\ v \end{pmatrix} = 0$. The parameters m and λ are free parameters of the Standard Model. Equivalently, the scale of the electroweak symmetry breaking is not predicted by the theory and must be taken from experiment.

2.2. FERMION MASSES

Higgs doublets (and only doublets) have $SU(2)_L \times U(1)_Y$ invariant renormalizable couplings to the chiral fermions of the Standard Model. For the charged fermions, we can write down the following Yukawa couplings:

$$\mathcal{L}_{Yukawa} = -Y_l^{BA} H_i^* l_{iA} e_B^c - Y_d^{BA} H_i^* q_{iA} d_B^c - Y_u^{BA} \epsilon_{ij} H_i q_{iA} u_B^c + hc \tag{2.10}$$

where i is the $SU(2)_L$ index and A, B are generation indices. We use the fact that the two-dim representation of $SU(2)$ is real and $i\tau_2 H$ transforms as H^*, i.e. as $2^*(\equiv 2)$ of $SU(2)$. Therefore, $(i\tau_2 Hq) = \epsilon_{ij} H_i q_j$ is also an invariant of $SU(2)$. After spontaneous breaking of $SU(2)_L \times U(1)_Y$ to $U(1)_{EM}$ by the Higgs boson vacuum expectation value v we obtain the Dirac masses

$$\mathcal{L}_{mass} = -v(Y_l^{BA} e_A e_B^c + Y_d^{BA} d_A d_B^c + Y_u^{BA} u_A u_B^c) + hc \tag{2.11}$$

However, at the level of the full, $SU(2)_L \times U(1)_Y$ invariant theory, there is no renormalizable term that would give neutrino mass. It is so because ν^c is absent from the spectrum of the SM. Thus, in the SM, neutrinos are massless.

The interactions (2.10,2.11) are written in some "electroweak" basis defined by eigenvectors of the $SU(2)_L \times U(1)_Y$ symmetry group. In such a basis, both the fermion masses and the Yukawa couplings are in general non-diagonal in the flavour indices (A, B). However, we can introduce another set of fields (say, primed fields) describing physical particles (mass eigenstates). The flavour of the primed fields is defined in the mass eigenstate basis. The two sets of fields are related to each other by unitary transformations:

$$\begin{aligned} u = U_L u' \qquad d = D_L d' \\ u^c = u'^c U_R^\dagger \quad d^c = d'^C D_R^\dagger \\ e = E_L e' \\ e^c = e'^c E_R^\dagger \end{aligned} \tag{2.12}$$

which, of course, do not commute with the $SU(2)_L \times U(1)_Y$ gauge transformations and can be performed only after the spontaneous breakdown of the gauge symmetry. In eq.(2.12), the fields $u,\ d$, e denote three-dimensional vectors in the flavour space.

The transformations (2.12) diagonalize the mass terms and the Yukawa couplings defined by (2.10). After diagonalization we can combine the chiral fields into Dirac fields. The weak currents can be expressed in terms of the physical (mass eigenstates) fields:

$$\begin{aligned} J_\mu^- &= \sum_{A,B} \bar{u}'_A \bar{\sigma}_\mu (V_{CKM})_{AB} d'_B \\ &+ \sum_A \bar{\nu}'_A \bar{\sigma}_\mu e'_A \end{aligned} \tag{2.13}$$

where the Cabibbo-Kobayashi-Maskawa matrix $V_{CKM} = U_L^\dagger D_L$. Note that the lepton current is diagonal in flavour (defined in the charged lepton mass eigenstate basis) because the massless neutrino field can be redefined by the transformation $\nu_A = E_{AB}^L \nu'_B$ where E_{AB}^L is the transformation diagonalizing the charged lepton mass matrix (see (2.12)). Thus, for the lepton current, $V_{CKM} = E^{L\dagger} E^L = 1$.

It is important to remember that in the SM (with only one Higgs doublet) the Yukawa couplings to the physical Higgs boson (and, in fact, also the couplings to the Z^0) and the mass terms are diagonalized by the same unitary rotations. So they are flavour diagonal. The only source of flavour non-conservation resides in V_{CKM}. In particular, not only the global lepton number but also each flavour lepton number is separately conserved.

2.3. APPROXIMATE CUSTODIAL SYMMETRY OF THE STANDARD MODEL AND THE PRECISION ELECTROWEAK DATA

The Higgs sector of the SM is invariant under global $SO(4)$ symmetry acting on four real components of the complex doublet. The group $SO(4) \simeq SU(2)_L \times SU(2)_R$ and the Higgs doublet can be written as a 2×2 matrix Φ

$$\Phi = \begin{pmatrix} H^+ & H^{0*} \\ H^0 & -H^- \end{pmatrix}, \tag{2.14}$$

which transforms as $(\mathbf{2}, \mathbf{2})$ of the latter group (whose first factor is just the gauged weak isospin group):

$$\Phi \longrightarrow \Phi' = U_L \Phi U_R\,. \tag{2.15}$$

The vacuum expectation value of the Higgs field breaks $SU(2)_L \times SU(2)_R$ to its diagonal subgroup called "custodial" $SU(2)$ acting on the three would-be Goldstone bosons G^a:

$$H = (H^+ \quad H^0) \longrightarrow \frac{1}{\sqrt{2}} e^{\frac{iG^a \tau^a}{v}} (0 \quad v + h^0) \tag{2.16}$$

In the rest of the electroweak Lagrangian the $SU(2)_R$ subgroup and therefore also the custodial $SU(2)$ symmetry is broken by the Yukawa interactions and by the $U(1)_Y$ coupling. However, the custodial symmetry is still seen at the tree level since by the Higgs mechanism it ensures that the gauge bosons W^+, W^- and W^0 of $SU(2)_L$ transform as a custodial triplet. In consequence, the ratio of the strength of charged and neutral current interaction at the tree level is equal to one

$$\rho \equiv \frac{M_W^2}{\cos^2 \theta_W M_Z^2} = 1 , \tag{2.17}$$

where $\cos^2 \theta_W = g_2^2/(g_2^2 + g_1^2)$.

This relation is consistent to a very good approximation with experimental data but this is not the end of the success of the SM. Since the rest of the Lagrangian violates the custodial symmetry, there are quantum corrections to the relation $\rho = 1$. Since the global $SU(2)_L \times SU(2)_R$ symmetry of the Higgs sector fixes the structure of counterterms in the scalar potential, quantum corrections to the relation $\rho = 1$ must be finite! In one-loop approximation one gets

$$\Delta\rho \equiv \rho - 1 = \frac{3g_2^2}{64\pi^2} \frac{m_t^2 - m_b^2}{M_W^2} - \frac{g_1^2}{64\pi^2} \frac{11}{3} \ln \frac{M_h^2}{M_W^2} . \tag{2.18}$$

The first term is of the order of 1% and, is in perfect agrrement with the precision electroweak data. Thus, fits to the data give $M_h \sim \mathcal{O}(M_W)$ although logarithmic dependence of $\Delta\rho$ on M_h does not allow for precise determination of this mass. We shall discuss later on the importance of the Higgs particle mass for various extensions of the SM.

We conclude that the approximate custodial symmetry of the SM is in fantastic agreement with experimental data. Any alternative mechanism of the electroweak symmetry breaking or any extension of the SM must not violate the custodial symmetry of the electroweak interactions. Furthermore, we see that the renormalizable SM with one Higgs doublet, has very strong predictive power which allows for its precision tests at the level of one per mille.

2.4. GIM MECHANISM AND THE SUPPRESSION OF FCNC AND CP VIOLATING TRANSITIONS

It is well established experimentally that the amplitudes for processes such as e.g. $K^0(d\bar{s})$-$\bar{K}^0(\bar{d}s)$ mixing caused by electroweak interactions, $\langle K^0|\mathcal{H}_{\rm weak}|\bar{K}^0\rangle$, are strongly suppressed in comparison with the amplitudes for the charged current transitions like $n \to pe^-\bar{\nu}_e$ or $\mu^- \to e^-\bar{\nu}_e\nu_\mu$. A good measure of the mixing is the mass difference between neutral kaon mass eigenstates $|K_L^0\rangle = \frac{1}{\sqrt{2}}(|K^0\rangle - |\bar{K}^0\rangle)$ and $|K_S^0\rangle = \frac{1}{\sqrt{2}}(|K^0\rangle + |\bar{K}^0\rangle)$ (we neglect here even smaller CP violation): $\Delta M_K = 3 \times 10^{-12}$ MeV and is suppressed by factor 10^6 compared to what one could expect for a generic electroweak transition. This fact finds a very elegant explanation in the Standard Model. The charged current transitions shown in Fig. 1 are present at the tree level whereas it follows from the structure of the theory that the diagrams shown in Fig. 2 are absent. In the diagrams the quark fields are of course mass eigenstate fields and the couplings are obtained by rotating from an electroweak basis, in which the theory is formulated, to the mass eigenstate basis, in which the quark flavour is defined. It is obvious that the Z^0 couplings are flavour di-

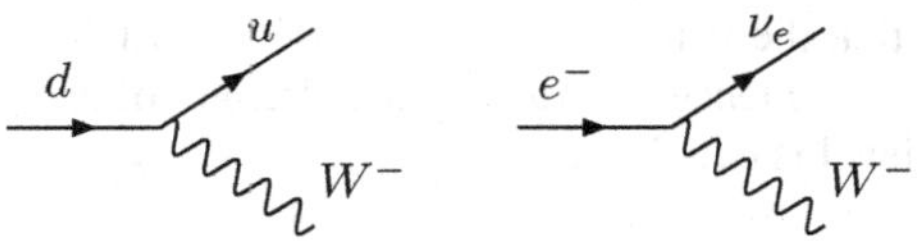

Figure 1. Charged current transitions in the Standard Model.

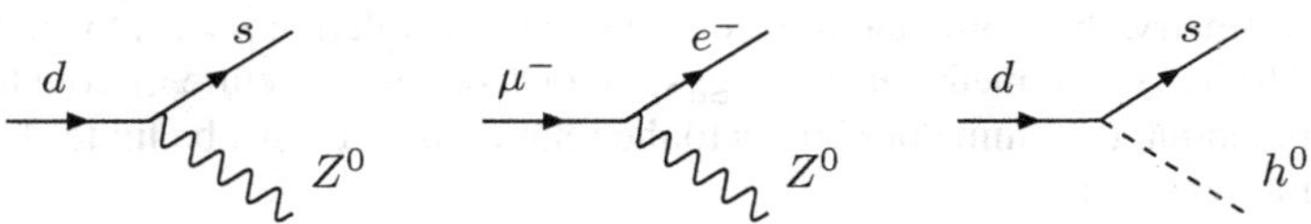

Figure 2. Neutral current transitions absent at the tree level in the SM.

agonal. More interesting is the absence in the SM of the scalar flavour changing neutral currents. This result follows from the fact that in the SM there is only one Higgs doublet. Because of that, diagonalizing the fermion mass matrices one simultaneously obtains also diagonal Yukawa couplings to the physical Higgs boson. In models with more Higgs doublets, additional discrete symmetry has to be imposed to ensure that only one Higgs doublet couples to the quarks of the same charge. In the minimal supersymmetric model two doublets are needed for supersymmetric theory but (by the holomorphicity of the superpotential) only one Higgs doublet can couple to the same charge quarks. The absence of flavour non-diagonal neutral currents at the tree level is not sufficient to account for the observed suppression of processes like kaon mixing or $b \rightarrow s\gamma$. For example the 1-loop diagrams shown in Fig. 3 generate the $K^0(d\bar{s})$-$\bar{K}^0(\bar{d}s)$ transitions and the coefficient C in the effective Lagrangian

$$\mathcal{L}_{\text{eff}} = C\ (\bar{s}_L\gamma_\mu d_L)(\bar{s}_L\gamma_\mu d_L) \tag{2.19}$$

describing their contribution (in the limit of external quark momenta small compared to M_W) is generically of order $C \sim \frac{\alpha^2}{M_W^2} \sim \alpha\ G_F$. However, in the SM the sum of all such contributions is suppressed by a factor $\sim 10^{-4}$ due to the so-called (generalized to 3 generations of quarks) Glashow-Illiopoulos-Maiani mechanism. The coefficient C

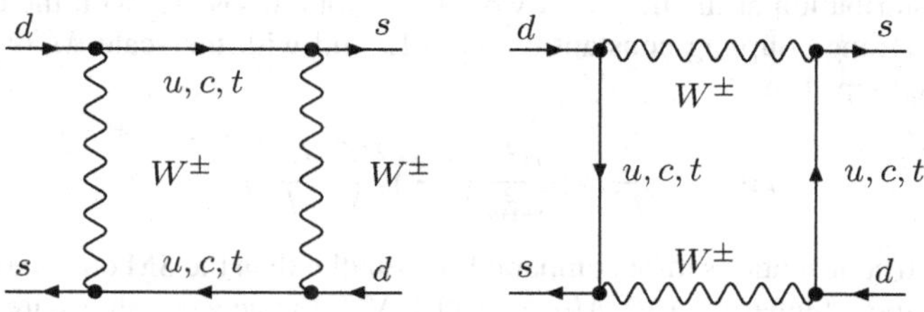

Figure 3. Leading SM contribution to K^0-$\bar{K}^0$ mixing.

generated by diagrams of Fig. 3 (and the diagrams in which one or both $W^\pm$ are replaced by the unphysical would-be Goldstone bosons $G^\pm$) is finite and has dimension mass^{-2}. The whole effective Lagrangian can be written as

$$\begin{aligned}\mathcal{L}_{\text{eff}} &= \frac{1}{2}\left(\frac{g_2}{\sqrt{2}}\right)^4 \sum_{i,j=u,c,t} V^\star_{is}V_{id}V^\star_{js}V_{jd} \\ &\times \int \frac{d^4q}{(2\pi)^4} \frac{[\overline{\Psi}_s\gamma_\mu P_L(\not{q}+m_{q_i})\gamma_\nu P_L\Psi_d][\overline{\Psi}_s\gamma^\nu P_L(\not{q}+m_{q_j})\gamma^\mu P_L\Psi_d]}{(q^2-m^2_{q_i})(q^2-m^2_{q_j})(q^2-M^2_W)^2}\end{aligned} \tag{2.20}$$

The top quark contribution is suppressed by the smallness of the product $V^*_{ts}V_{td}$. The rest contributes to the coefficient C in (2.19)

$$\begin{aligned}A_{uc} &\sim \left(\frac{g_2}{\sqrt{2}}\right)^4 \frac{1}{M^2_W} \sum_{i,j=u,c} V^\star_{is}V_{id}V^\star_{js}V_{jd}\left[1+\mathcal{O}\left(\frac{m^2_{q_i}}{M^2_W},\frac{m^2_{q_j}}{M^2_W}\right)\right] \\ &\sim \alpha\, G_F\left\{(V^\star_{ts}V_{td})^2+\mathcal{O}\left(\sum_{i,j=u,c} V^\star_{is}V_{id}V^\star_{js}V_{jd}\frac{m^2_q}{M^2_W}\right)\right\}\end{aligned} \tag{2.21}$$

where in the last line, unitarity of the CKM matrix has been used: $V^\star_{us}V_{ud}+V^\star_{cs}V_{cd}=-V^\star_{ts}V_{td}$.

From this example it is clear that for the empirical pattern of quark masses and mixing angles there is strong suppression of FCNC in the SM. It is much stronger than "naturally" expected.

The predictions of the SM for the FCNC transitions are in very good agreement with experimental data. This is also true for CP violation. The only source of CP violation in the SM is the phase of the CKM matrix. As a result, the effects of CP violation in the kaon system, in which they were first observed, are proportional to the masses of the light quarks and small CKM mixing angles and hence very small (this is not so for the B-meson systems in which CP violation is probed by present experiments).

The strong suppression of the FCNC and CP violating transitions, so nicely consistent with the SM is a big challenge for any of its extension. This is easy to understand on a qualitative basis. Let us suppose that new physics contributes to such transitions at 1-loop level (any contribution at the tree level would be a total disaster!) with the couplings of order of the strong coupling constant $\alpha_s \approx 0.12$ and with the scale M of the particle masses in the loop. Then

$$\Delta C \sim \frac{\alpha_s^2}{M^2} = \frac{\alpha^2}{M_W^2}\left(\frac{\alpha_s}{\alpha}\right)^2\left(\frac{M_W}{M}\right)^2 \tag{2.22}$$

Thus, for such contributions to be comparable or smaller than the SM one, the new physics scale M has to be higher than $10^3 M_W \sim 100$ TeV. If the new couplings are of the order of α then we get $M \gtrsim 10$ TeV. If the scale of new physics extension of the SM is below these limits, the new physics must somehow control the flavour effects!

2.5. BARYON AND LEPTON NUMBER CONSERVATION

The principles (1) - (4) imply global $U(1)$ symmetries of the theory: baryon and lepton number conservation $\Delta B = \Delta L = 0$. In fact, for leptons the implication is even stronger, namely $U_e(1) \times U_\mu(1) \times U_\tau(1)$ is a global symmetry of the electroweak Lagrangian and the lepton flavour numbers are separately conserved: $\Delta L_e = \Delta L_\mu = \Delta L_\tau$=0. For quarks, quark mixing explicitly breaks quark flavour $U(1)$'s and only the total baryon number is conserved.

The conservation of the baryon and lepton numbers by the renormalizable couplings of the Standard Model is beautifully consistent with experimental limits on the life time of the proton , $\tau_p \gtrsim 10^{33}$ years, and on the branching ratios for the lepton flavour violating decays, e.g. $BR(\mu \to e\gamma) \lesssim 10^{-11}$. Proton decay and lepton flavour violating decays occur, if at all, many orders of magnitude less frequently than generic electroweak processes. Actually, in the Standard Model those conservation laws are violated by chiral anomaly. The diagrams shown in Fig. 4 where $j^a_{\mu L}$'s are $SU(2)_L$ gauge currents and $j^{L,B}_\mu$ is the baryon or lepton current of the $U(1)$ global symmetries gives (insisting on the conservation of the gauge currents [1])

$$\begin{aligned} \partial^\mu j^B_\mu &\propto (\mathrm{Tr}B)\sum_{a=1}^{3} W^a_{\mu\nu}\tilde{W}^{a\mu\nu} \\ \partial^\mu j^L_\mu &\propto (\mathrm{Tr}L)\sum_{a=1}^{3} W^a_{\mu\nu}\tilde{W}^{a\mu\nu} \end{aligned} \tag{2.23}$$

Only $SU_L(2)$ doublets contribute to the traces, so they do not vanish, and $W^a_{\mu\nu}$ is the $SU(2)_L$ field strength. Non-perturbative effects give, in general a non-zero condensate $W^a_{\mu\nu}\tilde{W}^{\mu\nu}$ (topological baryon and lepton number non-conservation) but the effect is totally negligible at zero temperature. At non-zero temperature, the topological baryon and lepton number non-conservation is enhanced and can play important physical role because the $(B-L)$ current is anomaly free: $\partial^\mu j^{B-L}_\mu \propto \mathrm{Tr}(B-L) = 0$ and the quantum number $B-L$ is conserved. Thus in the presence of some hypothetical perturbative lepton number and CP violation, topological effects may convert leptogenesis into baryogenesis.

Incidentally, with the right-handed neutrino included in the spectrum, the diagrams in Fig. 5 do not give any anomaly, neither, and the $(B - L)$ symmetry can, therefore, be gauged.

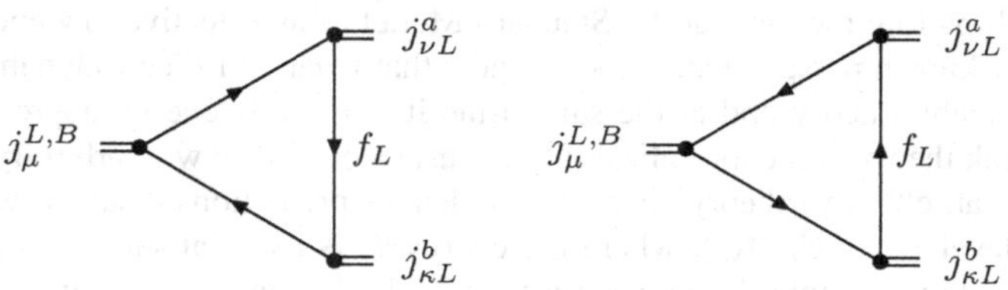

Figure 4. Anomalies in the B and L currents in the SM.

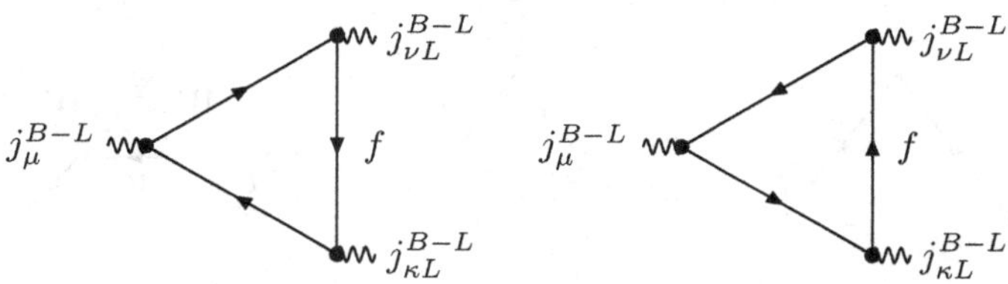

Figure 5. $[U(1)_{B-L}]^3$ anomaly.

3. Hints from the Standard Model for its extensions

3.1. IS THE EFFECTIVE LOW ENERGY ELECTROWEAK THEORY INDEED THE RENORMALIZABLE STANDARD MODEL?

In the construction of the SM Lagrangian we have been, so far, guided by its renormalizability. Accepting the fact that the SM is only an effective theory one may wonder, however, how important is its renormalizability. Unitarity and symmetries are certainly more fundamental requirements and indeed e.g. the physics of pions is described by a non-renormalizable effective low energy theory (non-linear σ-model). It is, therefore, useful to recall the main differences between the two classes of quantum field theories.

In a renormalizable theory its cut-off can be taken to infinity and the whole UV sensitivity is hidden in a finite number of free parameters, the same at any order of perturbation expansion. Calculations with arbitrary precision are, therefore, possible with a fixed number of parameters whose values can be determined from the experimental data. If some theory gives as its low energy approximation a renormalizable theory, then according to the Appelquist-Carazzone decoupling theorem, the effects of heavy degrees of freedom characterized by a mass scale M show up only as corrections in the form of higher dimensional operators allowed by the symmetries of the renormalizable theory:

$$\mathcal{L}_{\text{eff}} = \mathcal{L}_{\text{renormalizable}} + \mathcal{O}\left(\frac{1}{M^n}\right) O_{4+n} \tag{3.1}$$

This is a window to new physics (if such corrections are needed by experiment) even if we do not know the theory at the scale M.

It is worth putting the view at the Standard Model as an effective low energy theory into the better known perspective. We know now that Quantum Electrodynamics (QED) is a renormalizable theory and at the same time it is the low energy approximation to the electroweak theory. Its renormalizability means calculability with arbitrary precision. But it is only an effective theory so we know that its predictions disagree with experiment at the level $\sim \mathcal{O}(E/M_{\mathrm{W}})$, where the energy E is the characteristic energy for a given process. For example, let us have a look at the lepton magnetic moment. It gets contributions from the diagrams depicted in Fig. 6. Thus, for the nonrelativistic effective

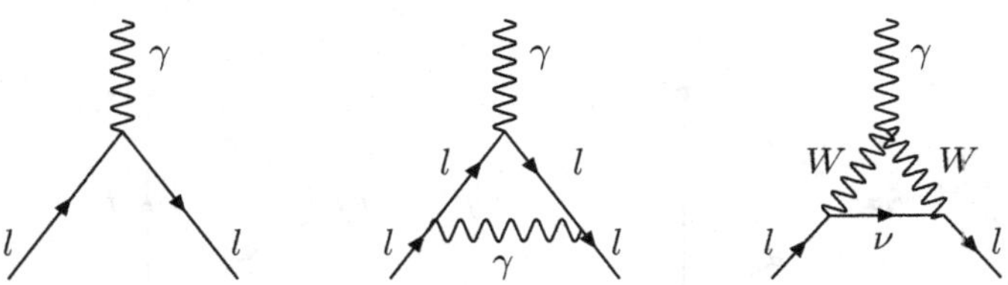

Figure 6. One loop contributions to the anomalous lepton magnetic moment in the SM.

interaction with the magnetic field we get

$$\mathcal{H}_{\text{eff}} = \frac{e}{2m_l}\boldsymbol{\sigma}\cdot\mathbf{B}(1+\frac{\alpha}{2\pi}+\mathcal{O}\left(\alpha\frac{m_l^2}{M_W^2}\right)+\ldots) \tag{3.2}$$

where the role of the energy scale is played by the lepton mass m_l. The "weak" correction is calculable in the full electroweak theory, but at the level of QED as an effective theory it has to be added as a new non-renormalizable (but $U(1)_{EM}$ invariant) interaction

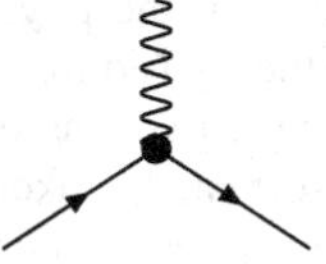

Figure 7. Effective photon-lepton vertex.

$$\mathcal{L}_{\text{eff}} = \frac{m_l}{M^2}\bar{\psi}\sigma_{\mu\nu}\psi F^{\mu\nu} \qquad \text{(dim 5)} \tag{3.3}$$

This would have been a way to discover weak interactions (and to measure the weak scale) in purely electromagnetic processes: we extend QED to a non-renormalizable theory by adding higher dimension operators and look for their experimental manifestation in purely electromagnetic processes once the experimental precision is high enough. Luckily enough for us, effective QED may also contain other than (3.3) non-renormalizable corrections, $U(1)_{EM}$ invariant but violating the conservation of quantum numbers that are accidentally conserved in QED, for instance flavour. Such corrections manifest themselves as different type of interactions - weak interactions - and were easy to discover experimentally. Similarly, among many possible non-renormalizable corrections to the SM which respect the $SU(2)_L \times U(1)_Y$ gauge symmetry there are such that violate e.g. the lepton and/or baryon number conservation or give Majorana masses to neutrinos. We shall discuss them in the following.

In a non-renormalizable theory one either has to keep explicit logarithmic cut-off dependence ($\frac{1}{16\pi^2} \ln \frac{\Lambda}{\mu}$ where μ is some low energy scale) or the number of counterterms (i.e. the number of free parameters of the theory) must increase at each order of perturbation expansion. The value of the cut-off Λ is dictated by the consistency between one-loop calculations and the contribution of the higher dimensional operators. Typically, the theory becomes strongly interacting above the cut-off scale.

A physically important non-renormalizable effective theory is the theory of pions [2, 1]. The pions are pseudo-Goldstone bosons of the (approximate) global chiral symmetry $SU(2)_L \times SU(2)_R$ of strong interactions which is spontaneously broken down to $SU(2)_V$ of isospin. The physics of the light degrees of freedom (pions) is described by a non-linear σ-model. The chiral symmetry $SU(2)_L \times SU(2)_R$, non-linearly realized on the pion fields, requires non-renormalizable interaction. The lowest dimension one is

$$\mathcal{L}_{\text{pions}} = f_\pi^2 \text{Tr}(\partial_\mu U^\dagger \partial^\mu U) \tag{3.4}$$

where the fields

$$U = e^{i\pi^a \tau^a / f_\pi} \tag{3.5}$$

transform under $SU(2)_L \times SU(2)_R$ linearly: $U \to V_L U V_R^\dagger$. The constanf f_π is the pion decay constant. The chiral symmetry cannot be reconciled with an effective renormalizable theory of pions and the Appelquist-Carazzone decoupling does not work.

The important question is: is the true low energy approximation to the more fundamental theory which explains the mechanism of electroweak symmetry breaking the renormalizable SM (like in supersymmetric extensions of the SM) or non-renormalizable electroweak theory (like in higgsless models with dynamical electroweak symmetry breaking and in models with the Higgs boson as a pseudo-Goldstone boson of some spontaneously broken bigger global symmetry)?

The predictive power and the phenomenological success of the SM suggests the first case. On the other hand, one may argue that the second option would more resemble spontaneous symmetry breaking in the condensed matter physics and in strong interactions.

3.2. MATTER CONTENT AND DEEPER UNIFICATION?

There are two striking aspects of the matter spectrum in the Standard Model. One is the chiral anomaly cancellation [2, 1], which is necessary for a unitary (and renormalizable) theory, and occurs thanks to certain conspiracy between quarks and leptons suggesting a deeper link between them. The potential source of chiral anomalies in the Standard Model are the triangle diagrams like the ones shown in Figs. 4 in which now the external lines correspond to all possible triplets of currents coupled to the three types of gauge fields in the electroweak theory: $U(1)_Y$ gauge field B_μ, $SU(2)_L$ gauge fields W_μ^a $(a = 1, 2, 3)$ and/or $SU(3)_C$ gauge fields A_μ^a $(a = 1, \ldots, 8)$ and internal fermion lines correspond to all chiral fermions in the theory. Most of the anomaly coefficients vanish due to the group structure. The most interesting ones are the anomalies with one $U(1)_Y$ current and two $SU(2)_L$ currents and the one with three $U(1)_Y$ currents. They vanish (a necessary condition for consistency of the electroweak theory) provided

$$\sum Q_i = 0\,, \tag{3.6}$$

(where Q_i's are the electric charges of the fields) separately for doublets and singlets of $SU(2)_L$. Incidently, the same condition is sufficient for vanishing of the mixed $U(1)_Y$-gravitational anomaly given by the diagrams like those shown in Figs. 4 but now with two currents corresponding to the energy-momentum tensors and the third one to the $U(1)_Y$ current. The condition (3.6) is satisfied in the SM because quark and lepton contributions cancel each other.

The second striking feature of the matter spectrum in the Standard Model is that it fits into simple representations of the $SU(5)$ and $SO(10)$ groups [3]. Indeed, we have, for $SU(5)$

$$\begin{aligned} \mathbf{5}^* &= \begin{pmatrix} \nu_e \\ e^- \end{pmatrix}_L,\ d_L^c \\ \mathbf{10} &= \begin{pmatrix} u \\ d_L \end{pmatrix},\ u_L^c,\ e_L^c \\ \mathbf{1} &= \nu_L^c \quad \text{(if the right} - \text{handed neutrino is added to the spectrum)} \end{aligned}$$

and, for $SO(10)$,

$$\mathbf{16} = \mathbf{5}^* + \mathbf{10} + \mathbf{1}$$

The assignment of fermions to the $SU(5)$ representations fixes the normalization of the $U(1)_Y$ generator:

$$Q = T_3 + Y = L^{11} + \sqrt{\frac{5}{3}} L^{12}\,, \tag{3.7}$$

where L^{ij} are the $SU(5)$ generators satisfying the normalization condition $[L^{ij},\ L^{kl}] = \frac{1}{2}\delta^{ik}\delta^{jl}$.

Both facts, the anomaly cancellation and the pattern of fermion spectrum, strongly suggest some kind of quark and lepton unification, at least at some very deep level, with some big group and some mechanism of its breaking. In addition, in line with the above conclusion is a well known fact that, with normalization given by eq (3.7), the running gauge couplings of the Standard Model approach each other at high scale of order 10^{13} GeV. Although unification of the gauge couplings in the Standard Model is only very approximate, it is nevertheless a remarkable fact that the strength of strong and electroweak interactions become comparable at certain energy scale.

3.3. NEUTRINO MASSES: EVIDENCE FOR NEW VERY HIGH MASS SCALE?

There is at present strong experimental evidence for neutrino oscillations whose most obvious and most natural explanation is that neutrinos are massive and their mass eigenstates are different from the weak interaction eigenstates. This is the first experimental evidence for physics beyond the Standard Model.

The smallness of the neutrino masses can be easily understood as due to the presence of a new very high mass scale M. If it makes sense to rely on the Appelquist-Carazzone decoupling theorem then the mass scale M would manifest itself via higher dimension operators.

Neutrino mass terms may appear as dimension five operator

$$\frac{1}{M}(Hl_A)\lambda_{AB}(Hl_B) \tag{3.8}$$

where we use the following notation: $(Hl) \equiv \epsilon_{ij}H^i l^j$ denotes $SU(2)_L$ contraction and $ll \equiv \epsilon^{\alpha\beta}l_\alpha l_\beta$ denote Lorentz contraction. After spontaneous SM gauge symmetry breaking by the Higgs boson VEV the operator (3.8) gives indeed a Majorana mass matrix for neutrinos:

$$\mathcal{L}_{\nu\ \mathrm{mass}} = -m_{AB}\ \nu_A\nu_B + \mathrm{H.c.}\ , \qquad m_{AB} = \frac{v^2}{M}\lambda_{AB} \tag{3.9}$$

Small neutrino masses are obtained for big value of M, with the constants $\lambda_{AB} \sim \mathcal{O}(1)$. This is called a see-saw mechanism.

A possible and, in fact quite elegant, origin of the mass scale M would be the existence of another left-handed particle ν^c, a singlet of $SU(2)_L \times U(1)_Y$, i.e. a field such that

$$CP\nu^c(CP)^{-1} \equiv \nu_R \tag{3.10}$$

with a Majorana mass term

$$\mathcal{L}_{\mathrm{Majorana}} = M_{\mathrm{Maj}}^{AB}\ \nu_A^c\nu_B^c B + \mathrm{H.c.} \tag{3.11}$$

It can be interpreted as a right-handed neutrino field. Moreover, we can construct Yukawa interactions

$$\epsilon_{ij}H_i\nu_B^c Y_\nu^{BA} l_j^A + \mathrm{H.c.} \tag{3.12}$$

with a new set of (neutrino) Yukawa couplings Y_ν^{BA}. Both terms are $SU(2)_L \times U(1)_Y$ invariant and even renormalizable. We can consider then the diagram shown in Fig. 8. At

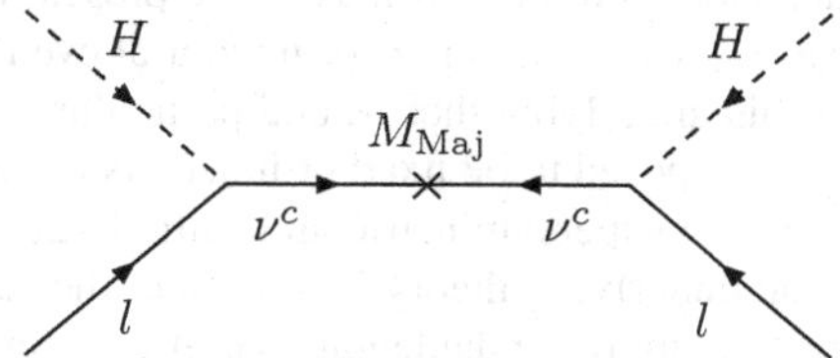

Figure 8. Diagram generating the dimension 5 operator.

the electroweak scale v, if $M_{Maj} \gg v$, we obtain the effective interaction shown in Fig. 9 described by the operator

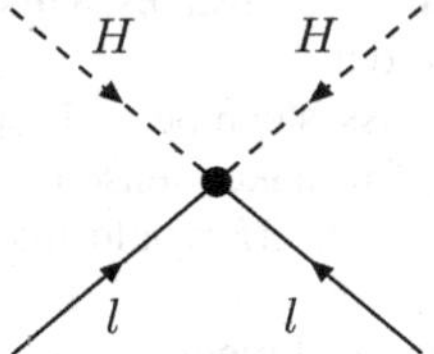

Figure 9. Effective dimension 5 operator.

$$\epsilon_{ij} H_i \, l_j^A \, Y_\nu^{DA} (M_{Maj}^{-1})^{DC} Y^{CB} \, \epsilon_{ij} H_i l_j^B \tag{3.13}$$

We recognize the previously introduced operator (3.8) with

$$\frac{\lambda_{AB}}{M} = (Y_\nu^T M_{MAJ}^{-1} Y_\nu)_{AB} \tag{3.14}$$

The see-saw mechanism with a new mass scale M is the most compelling explanation of the smallness of the neutrino masses [4, 5]. Indeed

i) the smallness of m_ν is then related to its zero electric charge

ii) the smallness of m_ν is also related to lepton number violation at the scale M

iii) with $m_\nu \sim Y^2 \frac{v^2}{M}$, $v = 240$ GeV and for $Y \sim \mathcal{O}(1)$ we get $m_\nu \sim (0.01 \div 0.1)$ eV for $M \sim (10^{15} \div 10^{13})$ GeV So, the scale M is close to the GUT scale.

iv) ν^c completes the spinor representation of S0(10)

v) heavy ν^c can play important role in baryogenesis via leptogenesis.

3.4. HIERARCHY PROBLEM IN THE SM: HINT FOR A NEW *LOW* MASS SCALE?

Quantum corrections to the Higgs potential mass parameter m^2 in eq. (2.8) in the SM are quadratically divergent. If the SM is an effective low energy theory and has a cut-off at some mass scale M of new physics, $\Lambda_{SM} \sim M$, it means then that quantum corrections to m^2 are quadratically dependent on the new mass scale present in the underlying more fundamental theory. When $M \gg M_Z$ this is very unnatural even if m^2, that is M_Z, remains a free parameter of this underlying theory, and particularly difficult to accept if in the underlying theory m^2 is supposed to be fixed or indeed is fixed by some more fundamental considerations (as e.g. in supersymmetric and Little Higgs models, respectively). The latter is necessary if the underlying theory is to predict the scale of the electroweak symmetry breaking in terms of "more fundamental" parameters and, generically, in terms of its own cut-off Λ_{New}. Thus, for naturalness of the Higgs mechanism in the SM there should exist a new mass scale $M \gtrsim M_Z$, say only order of magnitude higher than M_Z and better understanding the mechanism of the electroweak symmetry breaking is, hopefully, a bridge to new physics that will be explored at the LHC.

For a more quantitative discussion of this so-called hierarchy problem we recall that in general in a field theory with a cut-off Λ and some scalar field(s) ϕ that can acquire

VEV(s) the 1-loop effective potential is

$$\Delta V_{1-\text{loop}}(\phi) = \frac{1}{2}\int^{\Lambda} \frac{d^4k}{(2\pi)^4} \,\text{STr}\ln\left[k^2 - \mathcal{M}^2(\phi)\right] = c\Lambda^4 + c'\Lambda^4 \ln \Lambda^2$$
$$+\frac{1}{32\pi^2}\Lambda^2 \text{STr}\mathcal{M}^2(\phi) + \frac{1}{64\pi^2}\text{STr}\mathcal{M}^4(\phi)\ln\frac{\mathcal{M}^2(\phi)}{\Lambda^2} + \ldots \quad (3.15)$$

where Str$\mathcal{M}^2(\phi)$ =Tr$[(-)^F M^2(\phi)]$ with F - the fermion number operator and $\mathcal{M}^2(\phi)$ is the full ϕ-dependent mass matrix for all fields of the theory. The first terms in the expansion are the ϕ-independent contribution to the vacuum energy. We are interested in quantum corrections δm^2 to the mass parameter m^2 of the ϕ field potential. They are obtained by expanding

$$\text{STr}\mathcal{M}^2(\phi) = c_2\phi^2 + \ldots\,, \qquad \text{STr}\mathcal{M}^4(\phi) = c_4\phi^2 + \ldots\,, \quad (3.16)$$

The corrections proportional to c_2 (to c_4) are in general quadratically (logarithmically) dependent on the cut-of scale Λ. In the SM with a cut-of Λ_{SM} we find

$$\delta m^2 = \left(\frac{\partial^2 \Delta V_{1-\text{loop}}(\phi)}{\partial\phi^2}\right)_{\text{min}} = \frac{3}{64\pi^2}\left(3g_2^2 + g_1^2 + \lambda - 8y_t^2\right)\Lambda_{\text{SM}}^2 + \ldots\,, \quad (3.17)$$

If the SM was the correct theory up to the mass scale suggested by the see-saw mechanism, $\Lambda_{\text{SM}} \sim M_{GUT}$

$$|\delta m^2| \sim 10^{30}\ \text{GeV}^2 \sim 10^{26} M_W^2\ !$$

Clearly, this excludes the possibility of understanding the magnitude of Fermi constant $G_F \sim M_W^{-2}$ in any sensible way. We also see that for $|\delta m^2| \sim M_W^2$, $10M_W^2$, $100M_W^2$ one needs $\Lambda_{\text{SM}} \lesssim 0.5$ TeV, $\lesssim 1$ TeV, $\lesssim 6$ TeV, respectively. If the scale of new physics is in the above range it should be discovered at the LHC.

However, for a solution of the hierarchy problem it is not enough to have a low physical cut-off scale of the SM. The deeper theory has its own cut-off scale Λ_{New} and the dependence on it of δm^2 calculated in this deeper theory should be mild enough, in order not to reintroduce the hierarchy problem for the electroweak scale.[1]

Many theoretical ideas have been proposed for solving the hierarchy problem of the electroweak scale. In supersymmetric extensions of the SM the dependence on their own cut-off scales Λ_{New} is only logarithmic because the quadratic divergences cancel out at any order of the perturbation expansion. Since the effective potential $V_{1-\text{loop}}(\phi)$ depends only on $\ln\Lambda_{\text{New}}$ (to any order of the perturbation expansion) the scale Λ_{New} can be as high as the Planck scale. The quadratic dependence on the SM cut-off scale Λ_{SM}, that is on the mass scale M_{SUSY} of the superparticles, shows up in

$$\text{STr}\mathcal{M}^4(\phi) = f(M_{\text{SUSY}}^2)\,\phi^2 + \ldots \quad (3.18)$$

and more explicitly, at the 1-loop as[2]

$$\delta m^2 = \frac{1}{16\pi^2}\left(3g_2^2 + g_1^2 - 12y_t^2\right) M_{\text{SUSY}}^2 \ln\frac{\Lambda_{\text{New}}^2}{M_{\text{SUSY}}^2}\,, \quad (3.19)$$

[1]The hierarchy of some other (new) scales is nevertheless usually present.

[2]The formula (3.19) applies in fact to $m_{H_2}^2$ which for $\tan\beta \gtrsim 5$ is the most important for electroweak symmetry breaking.

where we have replaced all soft supersymmetry breaking mass terms including the Higgs boson mass by $M_{\rm SUSY}$.

Eq. (3.19) shows that in supersymmetric models the electroweak scale is calculable in terms of the known coupling constants and the (unknown) scales $M_{\rm SUSY}$ and $\Lambda_{\rm New}$. For a natural solution to the hierarchy problem of the electroweak scale $M_{\rm SUSY}$ has to be low, say $M_{\rm SUSY} \lesssim \mathcal{O}(10)M_W$. However, a new very difficult question appears about the hierarchy $\Lambda_{\rm New}/M_{\rm SUSY}$. This is the question about the mechanism of supersymmetry breaking. In gravity mediation scenarios $\Lambda_{\rm New} \sim M_{\rm Pl}$. In gauge mediation scenarios $\Lambda_{\rm New}$ is low but it is a new, introduced by hand, scale.

Other ideas for solving the hierarchy problem of the electroweak scale are more "pragmatic". Focusing on the scenarios with some predictive power, their general structure is the following: the low energy electroweak theory (but not necessarily the renormalizable SM) is embedded in a bigger one with a characteristic mass scale $\Lambda_{\rm SM} \sim M \sim \mathcal{O}(1\ {\rm TeV})$. The new physics is under perturbative control up to its cut-off $\Lambda_{\rm New} \gtrsim \mathcal{O}(10\ {\rm TeV})$, high enough to avoid any conflict with precision electroweak data (to be discussed later). For such scenarios with $\Lambda_{\rm New} \gtrsim \mathcal{O}(10\ {\rm TeV})$ to be useful for solving the electroweak hierarchy problem the dependence of δm^2 on $\Lambda_{\rm New}$ calculated in the extended theory has to be weak enough. This is obtained by ensuring that at least 1-loop contribution to the effective potential (3.15) have no quadratic dependence on $\Lambda_{\rm New}$:

$$(\delta m^2)_{1-{\rm loop}} = 0 \cdot \Lambda^2_{\rm New} + \mathcal{O}(\ln \Lambda_{\rm New}) + {\rm const.} \tag{3.20}$$

E.g. in the Little Higgs models the vanishing of the c_2 in ${\rm STr}\mathcal{M}^2(\phi)$ in eq. (3.16) is ensured by cancellation between contributions from particles of the same statistics. Such models predict the existence of new quark-like fermions and gauge bosons with masses $\sim M$. In these models, the quadratic dependence of δm^2 on $\Lambda_{\rm New}$ is present in higher order of the perturbation expansion but it is suppressed by loop factors. The tree level Higgs mass parameter m^2 usually vanishes $m^2(M)_{\rm tree} = 0$ as e.g. the Higgs boson is a Goldstone boson of some bigger (approximate) symmetry spontaneously broken at the scale M, i.e. M is identified with the "decay constant" and $\Lambda_{\rm New} \sim 4\pi M$. The electroweak symmetry is broken by quantum corrections. The electroweak scale is then predicted e.g. $M_W = M_W({\rm couplings}, M, \Lambda_{\rm New})$ with a mild dependence on $\Lambda_{\rm New} \gtrsim \mathcal{O}(10\ {\rm TeV})$ at which new unknown strong interactions set on. The crucial role is played by the new physics parameter M. In judging the plausibility of such ideas it is worth remembering our remarks in 3.1.

3.5. NEW LOW MASS SCALE AND PRECISION ELECTROWEAK DATA

The presence of new physics at low energy scale, $M \sim 1$ TeV raises the question on its contribution to the electroweak observables. We can address this question in a model independent way if again we assume the Appelquist-Carazzone decoupling scenario, i.e. renormalizable SM and corrections to it from new physics as higher dimension operators:

$$\mathcal{L}_{\rm eff} = \mathcal{L}_{\rm SM} + \sum_{\hat{O}_i^{n+4}} \frac{c_i}{M^n}\hat{O}_i^{n+4}\,. \tag{3.21}$$

This time we are interested in operators which contribute to the electroweak observables. Such operators are necessarily of dimension $n \geq 6$. One can classify various contributions from new physics according to the value of the coefficients c_i in the Lagrangian (3.21): $c_i \sim \mathcal{O}(1)$ for new tree-level contributions or contributions from new strong interacting sector; $c_i \sim \mathcal{O}(1/16\pi^2)$ for contributions from perturbative new physics at 1-loop. Fitting the electroweak observables one obtains limits on $\frac{c_i}{M^2}$. Strictly speaking, the limits are applicable to new physics which gives renormalizable SM as its low energy limit but the results are also indicative for new mass scales in models like e.g. Little Higgs, in which the Appelquist-Carazzone decoupling does not work. In any case, the constraints on such models from the electroweak observables can be discussed model by model.

The task of using electroweak data to put limits on the new scale M is greatly facilitated by the expectation that the dominant corrections from new physics will show up as corrections to the gauge boson self-energies (the so-called oblique corrections). This expectation is based on the experience gained in the SM and with its various hypothetical extensions. Several authors have obtained limits on the scale M under plausible assumption that the main corrections to the SM fits appear in gauge boson self-energies

A comment is in order here. Gauge boson self-energies are not gauge invariant objects. The vertex and box contributions contain also pieces which are independent of the external legs. They cancel the gauge dependence of the full gauge boson self-energies and restore Ward-identity. Therefore, strictly speaking, there is an ambiguity in extracting the gauge boson self energies from fits to experimental data. Fortunately, the vertex and box contributions are in the SM in the commonly used gauges much smaller than the gauge boson self energies and the neglection of gauge dependence of the latters in obtaining experimental information about their magnitude seems to be a reasonable approach.

In the SM, with unbroken electromagnetic $U(1)$ gauge symmetry there are four independent gauge boson self-energies $\Pi_{ij}(q^2)$:

$$\int d^4x\, e^{-iq\cdot x}\langle J_i^\mu(x) J_j^\nu(0)\rangle = ig^{\mu\nu}\Pi_{ij}(q^2) + q^\mu q^\nu \text{ term} \tag{3.22}$$

For instance, we can take $\Pi_{\gamma\gamma}$, $\Pi_{3\gamma}$, Π_{33} and Π_{11} as independent quantities ($i = 1, 3$ are $SU(2)_L$ indices; the QED Ward identity implies $\Pi_{11}(q^2) = \Pi_{22}(q^2)$. In the limit of $q^2 \ll M^2$ we can expand

$$\begin{aligned}
\Pi_{\gamma\gamma}(q^2) &\approx q^2\Pi'_{\gamma\gamma}(0)\\
\Pi_{3\gamma}(q^2) &\approx q^2\Pi'_{3\gamma}(0)\\
\Pi_{33}(q^2) &\approx \Pi_{33}(0) + q^2\Pi'_{33}(0)\\
\Pi_{11}(q^2) &\approx \Pi_{11}(0) + q^2\Pi'_{11}(0)
\end{aligned} \tag{3.23}$$

($\Pi_{\gamma\gamma}(0) = 0$ by QED Ward identity; the only non-zero contribution to $\Pi_{3\gamma}(0)$ comes from the $W^\pm$-charged Goldston boson loop). Thus, oblique corrections to the electroweak observables are to a good approximation parametrized by six constants. Three of them (or three linear combinations) are fixed in terms of α, M_Z and G_F by the renormalization procedure. In the remaining three combinations the UV divergences must cancel.[3] One

[3]The finitness of the gauge sector contribution to S, T and U requires the inclusion of the terms with $\Pi_{3\gamma}(0)$.

usually defines [6, 7]

$$\begin{aligned}\alpha T &\equiv \frac{e^2}{s^2c^2M_Z^2}\left[\Pi_{11}(0)-\Pi_{33}(0)\right] = \frac{\Pi_{WW}(0)}{M_W^2} - \frac{\Pi_{ZZ}(0)}{M_Z^2}\\ \alpha S &\equiv 4e^2\left[\Pi'_{33}(0)-\Pi'_{3\gamma}(0)\right] \propto \Pi'_{3Y}(0)\\ \alpha U &\equiv 4e^2\left[\Pi'_{11}(0)-\Pi'_{33}(0)\right]\end{aligned} \tag{3.24}$$

It is clear from their definition that the parameters S, T and U have important symmetry properties: T and U vanish in the limit of unbroken custodial $SU(2)_V$ symmetry. The parameter S vanishes when $SU(2)_L$ is unbroken; unbroken $SU(2)_V$ is not sufficient for vanishing of S because $S \propto \Pi'_{3Y}(0) = \Pi'_{3L,3R}(0) + \Pi'_{3L,B-L}$ (the decomposition is labelled by the $SU(2)_L \times SU(2)_R \times U(1)_{B-L}$ quantum numbers) and $\mathbf{3}_L \times \mathbf{3}_R = \mathbf{1}+\mathbf{5}$ under $SU(2)_V$.

It turns out that in the SM the quantum corrections to the ρ parameter defined in section 2 as a ratio of physical (and measured) observables are to a very good approximation given by αT:

$$\Delta\rho \equiv \rho - 1 = \frac{M_W}{M_Z\cos\theta} - 1 \approx \alpha T \tag{3.25}$$

and, according to the eq. (2.18), depend quadratically on the top quark mass and logarithmically on the SM Higgs boson mass. Eq. (3.25) is a good approximation because other corrections to ρ (vertex corrections) are in the SM negligibly small. Similarly, in the SM

$$\alpha S = \frac{e^2}{48\pi^2}\left(-2\ln\frac{m_t^2}{M_Z^2} + \ln\frac{M_h^2}{M_Z^2}\right) . \tag{3.26}$$

It was discussed in section 2.3 that the SM quantum corrections agree excellently with electroweak data. The only free parameter in the fits of the SM to these data is the Higgs boson mass. The main part of this dependence enters through the ρ parameter and the data favour negligible contribution to ρ from $\ln(M_h/M_W)$ (see eq. (2.18)). The fits give $M_h \approx \mathcal{O}(100$ GeV), with a big error since the dependence of the fits on the Higgs boson mass is only logarithmic. Such fits determine the values of parameters S_{SM}, T_{SM} and U_{SM} for the best fitted value of M_h.

We can discuss the room for new physics contribution to the electroweak fits by writing more generally:

$$\begin{aligned}T &= T_{SM}(M_h) + \Delta T\ ,\\ S &= S_{SM}(M_h) + \Delta S\ ,\\ U &= U_{SM}(M_h) + \Delta U\ ,\end{aligned} \tag{3.27}$$

where $T_{SM}(M_h)$ etc. is the SM contribution for some fixed value of the Higgs boson mass.

A fit to the data gives now some values for ΔT, ΔS and ΔU as a function of the assumed value of M_h and shows how much room we have for new physics for different values of M_h. It is clear that for, say, $M_h = 115$ GeV such fits give ΔT, ΔS and ΔU consistent with zero and the only room for new physics is in the errors of the fitted values of these quantities. For larger values of M_h we have more room for new physics contributions. As we can see from eqs. (2.18) and (3.26) it must be positive to T and negative to S to balance the contribution of the larger Higgs boson mass. The fitted values of ΔT, ΔS and ΔU for different values of M_h give limits on the coefficients c_i/M of the dimension six operators that contribute to ΔT, ΔS and ΔU.

Several interesting conclusions have been reached in such studies [8]. First of all, independently of the assumed value of M_h, for $c_i \sim \mathcal{O}(1)$ the fits to the electroweak data give a lower limit $M \gtrsim \mathcal{O}(4$ TeV) This limit is reached only for very correlated signs of the coefficients c_i. Thus, qualitatively speaking, any new physics with $M \lesssim \mathcal{O}(10$ TeV) must be perturbative and cannot contribute at the tree level to be consistent with electroweak data. This strongly suggests a perturbative solution to the hierarchy problem.

Secondly, if new physics is indeed perturbative and shows up only at loop level, then the fits give $M_h \lesssim 240$ GeV independently of the value of the scale M.

Finally, if $M_h \gtrsim 300$ GeV then new physics with the scale $M \sim 10 \div 30$ TeV and with $c_i \sim \mathcal{O}(1)$ (i.e. strongly interacting) is actually needed! Thus, experimental discovery of the Higgs boson and the determination of its mass will be a strong hint about the kind of new physics one may expect.

4. Supersymmetric extensions of the Standard Model

In the rest of these lectures we discuss supersymmetric models as at present the most complete theoretical framework going beyond the Standard Model [9]. It has quite a few attractive features and also a number of difficulties, so that the full success of the SM is not automaticaaly recovered. We have already mentioned some of them in several places in these lectures but here we collect them together and extend our discussion. We shall not discuss one fundamental unresolved issue for supersymmetry which is the mechanism of spontaneous supersymmetry breaking and its transmission to the the SM sector. On the phenomenological side, the related problem is that of new and potentially dangerous sources in the soft supersymmetry breaking parameters of the FCNC and CP violating transitions. For a review of all these aspects of supersymmetric and supergravity models see e.g. [10, 11, 12].

4.1. PRECISION ELECTROWEAK DATA

As discussed in section 3.4 supersymmetric models like the Minimal Supersymmetric Standard Model (MSSM) or its simple extensions satisfy a very important criterion of calculability. Most of the structure of the Standard Model is built into them, so the renormalizable Standard Model is their low energy approximation. Supersymmetric models are easily consistent [14] with the electroweak data since the supersymmetric quantum corrections to the Standard Model fits are suppressed by powers of the mass scale M_{SUSY} of supersymmetric particles and for $M_{SUSY} > \mathcal{O}(500)$ GeV are well below experimental errors (in particular, the custodial symmetry breaking by the sfermion masses is sufficiently suppressed). Thus, the predictive power of the Standard Model remains intact and its success is not accidental.

4.2. THE ELECTROWEAK SYMMETRY BREAKING

Supersymmetric models solve the hierarchy problem of the electroweak scale. In the limit of unbroken supersymmetry the quadratically divergent quantum corrections to the Higgs mass parameter are absent in any order of perturbation theory. When supersymmetry is softly broken by a mass scale M, the superpartners get their masses from the electroweak breaking and from the supersymmetry breaking mass terms $\sim M$. They decouple at energies smaller than M and the quadratically divergent Standard Model contribution to the Higgs mass parameter is cut-off by M and, therefore, depends quadratically on M. Thus, the hierarchy problem of the electroweak scale disappears if $M \lesssim \mathcal{O}(1)$ TeV. The cut-off to a supersymmetric theory can be as high as the Planck scale and "small" scale of the electroweak symmetry breaking is still natural.

The electroweak symmetry breaking may be triggered by radiative corrections to the Higgs potential:

$$(\delta m_{H_2}^2)_{1-\mathrm{loop}} \sim -\mathcal{O}(0.1)\, M^2 \ln \frac{\Lambda}{M} \tag{4.1}$$

This formula follows from eq. (3.19). If we assume that $m_0^2 \sim M^2$, i.e. that the tree level Higgs mass parameter is approximately equal to the soft supersymmetry breaking scale the radiative electroweak symmetry breaking ($(m_{H_2}^2)_{1-\mathrm{loop}} < 0$) is triggered by the large top quark Yukawa coupling, hidden in the numerical factors of eq. (4.1). With the Higgs boson self-interactions fixed by the gauge couplings of the Standard Model

$$\lambda\phi^4 \to g^2\phi^4 \tag{4.2}$$

one obtains the correct prediction for the electroweak scale for $\Lambda \sim M_{GUT,\,Pl}$. This nicely fits with unification of the gauge couplings.

4.3. THE MASS OF THE LIGHTEST HIGGS BOSON

Supersymmetric models typically restrict the couplings in the Higgs potential and give strong upper bounds on the mass of the lightest Higgs particle [13]. In the minimal model the Higgs boson self-coupling comes from the D-terms and its self-coupling is the gauge coupling, eq.(4.2). Therefore, at the tree level

$$M_{\mathrm{Higgs}} < M_Z \approx 91 \text{ GeV} \tag{4.3}$$

There are large quantum corrections to this result. They depend quadratically on the top quark mass and logarithmically on the stop mass scale $M_{\tilde{t}} \sim M_{\mathrm{SUSY}}$:

$$M_{\mathrm{Higgs}}^2 = \lambda v^2 \tag{4.4}$$

where λ is given by

$$\lambda = \frac{1}{8}(g_2^2 + g_1^2)\cos^2 2\beta + \Delta\lambda\,, \qquad \text{with} \quad \Delta\lambda = \frac{3g_2^2}{8\pi^2}\frac{m_t^4}{v^2 M_W^2}\ln\frac{M_{\tilde{t}}^2}{m_t^2}\,. \tag{4.5}$$

The present experimental limit $M_{\mathrm{Higgs}} > 114$ GeV requires $M_{\tilde{t}} \gtrsim 500$ GeV and for $M_{\tilde{t}} < 1$ TeV, $M_{\mathrm{Higgs}} < 130$ GeV. The closer the Higgs mass would be to the present experimental limit, the better it would be for the "naturalness" of the electroweak scale. Clearly, in the MSSM, the tunning in the Higgs potential depends exponentially on the Higgs mass and one may eventually have some tension here (see eqs. (4.5), (3.19)).

One can depart from the minimal model and relax the bound on the Higgs mass. For instance, with an additional chiral superfield which is a Standard Model singlet, one may couple the singlet to the Higgs doublets and get additional contributions to the Higgs self-coupling. Explicit calculations show that in such and other models, with $M \lesssim 1$ TeV, the bound on the Higgs mass cannot be raised above $\sim$ 150 GeV if one wants to preserve perturbative gauge coupling unification.

4.4. GAUGE COUPLING UNIFICATION

It is well known [15] that in the framework of the MSSM with degenerate sparticle spectrum characterized by $M_{SUSY} \approx 1$ TeV the three experimentally measured gauge couplings unify with high precision at the scale $M_{GUT} \sim 10^{16}$ GeV. This gives support to perturbative new physics at $\mathcal{O}(1$ TeV). Supersymmetry and the idea of grand unification (see section 3.2) mutually strengthen their attractiveness.

A closer look at the unification is interesting. One may ask how precise is the unification when the superpartner masses are not degenerate and different from 1 TeV. It has been understood that even for nondegenerate superpartner spectrum the superpartner mass dependence of the RG evolution of the gauge couplings can be described to a very good approximation by a single effective parameter T. The superparticle threshold effects are correctly included in the supersymmetric 1-loop RGE whose running starts at T, with the SM RG equations used below the scale T. For consistency, 2-loop running should also be included.

T depends strongly on the higgsino (μ) and gaugino (M_i) mass parameters and much weaker on the sfermion masses. Exact unification of the measured gauge couplings requires $T \approx 1$ TeV, i.e. the higgsino and the gaugino physical masses ~ 1 TeV if degenerate. However, a more plausible assumption that the parameters μ and M_i are approximately degenerate and ~ 1 TeV at the GUT scale gives $T \approx 100$ GeV because of strong renormalization effects. Thus, a realistic spectrum does not give exact unification and one may wonder about the accuracy of unification in the MSSM.

In order to define what we understand by 'successful unification' let us first recall the one-loop renormalization group equations in the SM and MSSM. At one-loop the gauge couplings $\tilde{\alpha}_i$ of the three group factors of $G_{\rm SM}$ run according to the equations:

$$\frac{1}{\tilde{\alpha}_i(Q)} = \frac{1}{\tilde{\alpha}_i(M_Z)} - \frac{b_0^{(i)}}{2\pi} \ln\left(\frac{Q}{M_Z}\right) + \delta_i \tag{4.6}$$

Here, $1/\tilde{\alpha}_i(M_Z) = (58.98 \pm 0.04,\ 29.57 \pm 0.03,\ 8.40 \pm 0.14)$ are the experimental values of the gauge couplings at the Z^0-pole and $b_0^{(i)}$ are the one-loop coefficients of the relevant beta-functions. They read $b_0 = (\frac{1}{10} + \frac{4}{3}N_g, -\frac{43}{6} + \frac{4}{3}N_g, -11 + \frac{4}{3}N_g)$ in the SM and $b_0 = (\frac{3}{5} + 2N_g, -5 + 2N_g, -9 + 2N_g)$ in the MSSM, where N_g is the number of generations. Threshold corrections (e.g. from heavy GUT gauge bosons) are represented by the parameters δ_i. As explained earlier using the MSSM RG equations directly from the electroweak scale for $T \approx M_Z$ means that the supersymmetric threshold corrections corresponding to a realistic mass spectrum are properly included.

In the bottom-up approach one can speak about the gauge coupling unification if in some range of scales Q the couplings defined by eq. (4.6) with, in general, Q-dependent $\delta_i(Q)$ can take a common value $\alpha_i(Q) = \alpha_{\rm GUT}$ for reasonably small values of $\delta_i(Q)$ (compared to $\alpha_{\rm GUT}^{-1}$).[4] The condition for the unification can be succintly written as

$$\epsilon_{ijk}\left(\frac{1}{\tilde{\alpha}_i(M_Z)} + \delta_i\right)(b_0^{(j)} - b_0^{(k)}) = 0 \tag{4.7}$$

Putting in the experimental values for $\alpha_i(M_Z)$ and the beta-function coefficients we get:

$$\begin{aligned} -41.1 + 3.8\delta_1 - 11.1\delta_2 + 7.3\delta_3 = 0 &\qquad \text{(SM)} \\ -0.9 + 4\delta_1 - 9.6\delta_2 + 5.6\delta_3 = 0 &\qquad \text{(MSSM)} \end{aligned} \tag{4.8}$$

We see, that to achieve the gauge coupling unification at the one-loop level we need the threshold corrections δ_i to be of order 10% $\alpha_{\rm GUT}^{-1}$ in the SM, while in the MSSM we need only δ_i ~1% $\alpha_{\rm GUT}^{-1}$. In the MSSM once the two loop effects are inluded one needs δ_i's by factor 2 larger. The unification of the gauge cuplings in the MSSM is indeed very precise: it admits (and requires) only 2% threshold corrections from the GUT physics. These 2% corrections give 10% effect on α_s at M_Z scale, but the precision of unification in the MSSM should be judged by the necessary for exact unification threshold corrections at the GUT scale.

Unification of the gauge couplings does not necessarily imply the standard GUT theories with all their problems, like spontaneous breaking of the GUT gauge group by VEVs of some Higgs fields, the doublet-triplet splitting problem, etc. Many different solutions have been proposed.

With the threshold corrections of the right order of magnitude, the unification scale can be estimated from the equation:

$$\frac{1}{\alpha_1(M_Z)} - \frac{1}{\alpha_2(M_Z)} - \frac{1}{2\pi}(b_0^{(1)} - b_0^{(2)}) \ln\left(\frac{M_{\rm GUT}}{M_Z}\right) + (\delta_1 - \delta_2) = 0 \tag{4.9}$$

For the sake of concreteness, we assume here that $\delta_1 = \delta_2 = 0$ and that all threshold corrections are accounted for by δ_3 (thus, the unification point is assumed to be where α_1 and α_2 intersect). Putting in the experimental numbers and the beta-function coefficients we get $M_{\rm GUT} \approx 1 \times 10^{13}$ GeV in the SM and $M_{\rm GUT} \approx 2 \times 10^{16}$ GeV in the MSSM.

The scale of unification in the MSSM is determined very precisely to be in the range $(2 \div 4) \times 10^{16}$ GeV. This is interesting because it is very close to the reduced Planck scale $M_{\rm Pl} = 2 \times 10^{18}$ GeV and could be considered as evidence for unification including gravity.[5] But one to two orders of magnitude difference between the two scales needs some explanation. Of course, new particles in incomplete $SU(5)$ representations would

[4]Whether there exists a unified model able to provide such values of $\delta_i(Q)$'s is a different question.

[5]In string theories without the stage of Grand Unification below the compactification scale M_S the couplings unify at M_S which e.g. in weakly coupled heterotic string is about factor 5 below $M_{\rm Pl}$.

alter the running and could push the unification scale closer to the Planck scale. However, one must not destroy the precision of the unification by new threshold corrections, so this possibility looks very fine-tuned and *ad hoc*. An interesting possibility would be to unify the three gauge interactions with gravity at $M_{GUT} \sim 10^{16}$ GeV by changing the energy dependence of the gravity coupling. This is possible if gravitational interactions (and only they) live in more than 3 spatial dimentions. The effective four-dimensional Planck constant is then

$$M_{\rm Pl}^2 = M_{\rm Pl\,(4+n)}^{2+n} R^n \,, \tag{4.10}$$

where n is the number of extra dimensions, R is their compactification radius and $M_{\rm Pl\,(4+n)}^{2+n}$ is the Planck scale in $4+n$ dimensions which we would like to take equal to $M_{GUT} \sim 10^{16}$ GeV. For $n = 1$ (like in the M-theory of Horava and Witten) we get $1/R \sim 10^{14}$ GeV.

4.5. PROTON DECAY

In the SM the baryon number is (perturbatively) conserved since there are no renormalizable couplings violating this symmetry (see section. 2.5). Experimental search for proton decay, e.g $p \rightarrow e^+\pi^0, p \rightarrow K^+\nu$ is one of the most fundamental tasks for particle physics. The present limit is $\tau_p > 10^{33}$ years. If the SM is only an effective low energy theory the remnants of new physics should show up as non-renormalizable corrections to the SM. The lowest dimension operators for the proton decay (with the particle spectrum of the SM) is the set of dimension 6 operators of the form

$$\hat{O}_i^{(6)} = \frac{c_i^{(6)}}{M_{(6)}^2} qqql \tag{4.11}$$

For such operators for the proton lifetime we get

$$\frac{1}{\tau_p} = \frac{[c_i^{(6)}]^2}{16\pi} \frac{m_p^5}{M_{(6)}^4}$$

The limit $\tau_p > 10^{33}$ years gives then $M_{(6)} > \sqrt{c_i^{(6)}} \times 10^{16}$ GeV (this is only a very rough estimate which neglects strong interaction effects). Any new physics with lower mass scale that could lead to proton decay should be coupled with $c_i^{(6)} \ll 1$. For instance, for $c_i^{(6)} \sim \alpha_{\rm GUT} \approx 1/25$ we get $M_{(6)} \gtrsim \times 10^{15}$ GeV which is still too high for the SM unification ($M_{\rm GUT} \approx 10^{13}$ GeV).

In supersymmetric extensions of the SM, with softly broken low energy supersymmetry there are low mass scalars in the spectrum with masses $M \sim \mathcal{O}(1$ TeV), which may have renormalizable couplings to quarks and leptons.

Indeed, the most general renormalizable superpotential in the minimal supersymmetric model is

$$\begin{aligned} w = &\hat{U}^c\hat{Q}\hat{H}_u + \hat{D}^c\hat{Q}\hat{H}_d + \hat{E}^c\hat{L}\hat{H}_d + \hat{H}_d\hat{H}_u \\ &+\hat{D}^c\hat{Q}\hat{L} + \hat{E}^c\hat{L}\hat{L} + \hat{U}^c\hat{D}^c\hat{D}^c + \hat{L}\hat{H}_u \,, \end{aligned} \tag{4.12}$$

(the coupling constants and the flavour indices are suppressed). The second line is also consistent with the SM gauge symmetry but these interactions do not conserve baryon and lepton numbers and give renormalizble couplings of scalars to fermions. After integrating the scalars out one gets dimension 6 operators as in (4.11) with $M_{(6)} \sim M_{\rm SUSY}$ from diagrams shown in Fig. 10 and to be consistent with the limit on the proton lifetime we

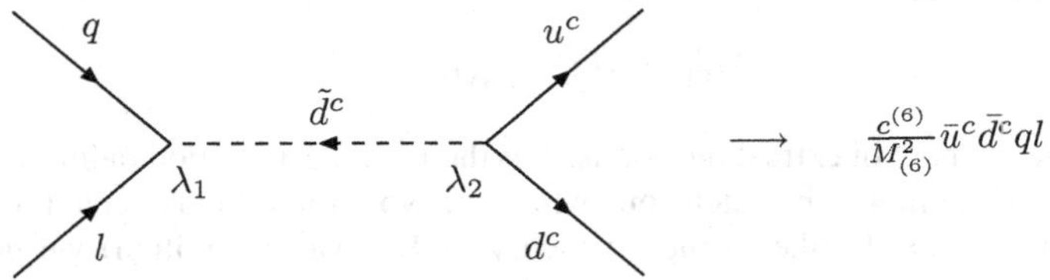

Figure 10. Diagram generating the dimension 6 operator.

need $c^{(6)} = \lambda_1 \lambda_2 < 10^{-26}$.

One can forbid the terms in the second line of eq. (4.12) by imposing a discrete symmetry, the so-called matter parity $R_p = (-1)^{3(B-L)}$. Such a symmetry could for instance, be a discrete remnant of the gauged $U(1)_{B-L}$ in the $SO(10)$ theory [17]. Matter parity is equivalent to R-parity $R = (-1)^{2S+3(B-L)}$ acting on the component fields, where S is their spin, since Lorentz-invariant interactions preserve $(-1)^{2S}$. We get then a stable LSP - candidate for dark matter in the Universe.

In supersymmetric GUT models, even with $R-$parity imposed, there is still another source of dangerous contributions to the proton decay amplitudes. These are the dimension 5 operators

$$\hat{O}_i^{(5)} = \frac{c_i^{(5)}}{M^2_{(5)}} qq\tilde{q}\tilde{l} \tag{4.13}$$

which when inserted into one loop diagrams with gaugino exchanges give rise via diagrams shown in Fig. 11 to dimension 6 operators. In the effective dimension 6 oper-

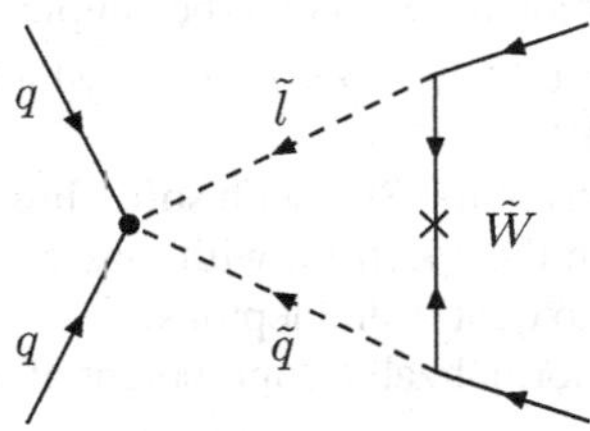

Figure 11. Loop diagram generating the dimension 6 operator form the dimension 5 operator.

ator of the form (4.11) one then gets $c^{(6)} = \alpha_{\rm GUT} c^{(5)}$, $M^2_{(6)} = M_{(5)} M_{\rm SUSY}$. From $\tau_p > 10^{33}$ years and for $M_{\rm SUSY} \sim \mathcal{O}(1~\text{TeV})$, $M_{(5)} \sim M_{\rm GUT} \sim 10^{16}$ GeV, one gets $c^{(5)} \lesssim 10^{-7}$. Thus we need small couplings in the amplitudes generating the dimension 5 operators. In SUSY GUT's dimension 5 operators originate from the exchange of the colour triplet scalars present in the Higgs boson GUT multiplets as shown in Fig. 12, so $c^{(5)} \sim Y^2$ (Yukawa couplings for the quarks of the first two generations)

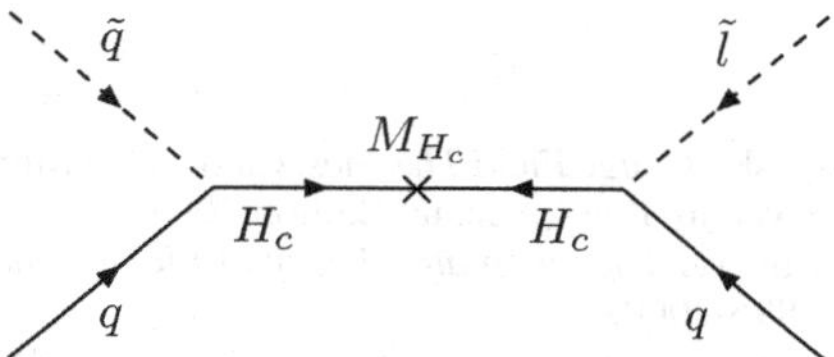

Figure 12. Diagram generating the dimension 5 operator for proton decay.

and $M_{(5)} \sim M_{H_c} \sim M_{\rm GUT}$.

Given various uncertainties, e.g. in the unknown squark, gaugino and heavy Higgs boson mass spectrum, such contributions in supersymmetric GUT models are marginally consistent with the experimental limits on the proton lifetime, particularly in models more complicated than the minimal supersymmetric $SU(5)$ model. Concrete classical GUT models are, however, not very attractive and plagued with various problems like e.g. doublet-triplet splitting problem for the Higgs boson multiplets. There are several interesting other ideas like unification in (small) extra dimensions or in string theory [18], which offer the possibility of avoiding those difficulties and simultaneously preserving the attractive features of GUT's. In some of such models proton is stable or its lifetime makes its decays unobservable experimentally. An interesting question is: what if proton after all decays but slow enough to rule out classical GUT models?

5. Summary

Thinking about new physics from the perspective of the extremely succesful Standard Model is very challenging. At present there is no approach that fully and convincingly incorporates this succes into its structure. Focusing on the electroweak symmetry breaking alone, one may wonder if the high predictive power of the renormalizable SM for the electroweak observables and its perfect agreement with experimental data is significant or partly accidental. If significant - it supports supersymmetry; if partly accidental - we have more room for various speculations about the mechanism of electroweak symmetry breaking. Experiments are needed to put us on the right track, and hopefully, experiments at the LHC will give us the necessary insight.

Acknowledgements

I thank Emilian Dudas, Gordon Kane, Jan Louis, Hans Peter Nilles, Jacek Pawełczyk, Fernando Quevedo, Zurab Tavartkiladze and especially my close colaborator Piotr Chankowski for many useful conversations about physics beyond the Standard Model. My special thanks go to Jihad Mourad who patiently reviewed with me the content of these lectures during my visit to the Astroparticle and Cosmology Insititute (Paris VII) in June 2004, from the perspective of a string theorist. I thank also the Organizers of the Cargese School for a very enjoyable time.

References

1. See for instance, S. Pokorski, *Gauge Field Theories*, Cambridge Monographs on Mathematical Physics, Cambridge University Press, Second Edition 2000
2. See for instance, S. Weinberg, *The Quantum Theory of Fields* vols. I-III, Cambridge University Press, Cambridge 1996-2000.
3. See for instance, G.G. Ross, *Grand Unified Theories*, Addison-Wesley, Redwood City CA, 1985.
4. For a review on quantum corrections to neutrino masses and mixing see P.H. Chankowski and S. Pokorski, *Int. J. Mod. Phys.* **A17** (2002) 575.
5. For a review of the works on the neutrino mass theory see G. Altarelli and F. Feruglio, Theoretical Models of Neutrino Masses and Mixing, hep-ph/0206077.
6. G. Altarelli and R. Barbieri, *Phys. Lett.* **B253** (1991) 161.
7. M. E. Peskin and T. Takeuchi, *Phys. Rev.* **D46** (1992) 381.
8. See for instance, L. J. Hall and C. F. Kolda, **Phys. Lett. B459** (1999) 213; C. F. Kolda and H. Murayama, *JHEP* **0007** (2000) 035.
9. *Perspectives in supersymmetry*, G.L. Kane ed., World Scientific, Singapore 1998.
10. H.-P. Nilles, *Phys. Rept.* **C110** (1984) 1.
11. F. Gabbiani, E. Gabrielli, A. Masiero and L. Silvestrini, *Nucl. Phys.* **B477** (1996) 321.
12. D. J. H. Chung, L. L. Everett, G. L. Kane, S. F. King, J. Lykken and L. T. Wang, hep-ph/0312378.
13. *Perspectives on Higgs Physics II*, G.L. Kane ed., World Scientific, Singapore 1998.
14. See for instance, S. Pokorski, in Proceedings of the *International Workshop on Supersymmetry and Unification of Fundamental Interactions (SUSY 95)*, Palaiseau, France, (hep-ph/9510224).
15. S. Raby, in *Review of Particle Physics, Phys. Lett.* **B592** 1.
16. *Heavy flavour Physics*, Adv. Ser. Direct. High Energy Phys. **15** A.J. Buras and M. Lindner eds., World Scientific, Singapore 1998.
17. S.P. Martin, *Phys. Rev.* **D42** (1992) R2769.
18. See for instance, K. R. Dienes, *Phys. Rept.* **287** (1997) 447.

INTRODUCTION TO COSMIC F- AND D-STRINGS

JOSEPH POLCHINSKI
KITP, University of California, Santa Barbara, CA USA

Abstract. In these lectures I discuss the possibility that superstrings of cosmic length might exist and be observable. I first review the original idea of cosmic strings arising as gauge theory solitons, and discuss in particular their network properties and the observational bounds that rule out cosmic strings as the principal origin of structure in our universe. I then consider cosmic superstrings, including the 'fundamental' F-strings and also D-strings and strings arising from wrapped branes. I discuss the conditions under which these will exist and be observable, and ways in which different kinds of string might be distinguished. We will see that each of these issues is model-dependent, but that some of the simplest models of inflation in string theory do lead to cosmic superstrings. Moreover, these could be the first objects seen in gravitational wave astronomy, and might have distinctive network properties.

1. Introduction

Seeing superstrings of cosmic size would be a spectacular way to verify string theory. Witten considered this possibility in the context of perturbative string theory, and found that it was excluded for several reasons [1]. Perturbative strings have a tension close to the Planck scale, and so would produce inhomogeneities in the cosmic microwave background far larger than observed. The scale of this tension also exceeds the upper bound on the energy scale of the inflationary vacuum, and so it would have been difficult to produce these strings after inflation, while any strings produced earlier would have been diluted beyond observation. Ref. [1] also identified instabilities that would prevent long strings from surviving on cosmic time scales.

In recent years we have understood that there are much more general possibilities for the geometry of the compact dimensions of string theory, including localized branes, and this allows the string tension to be much lower, anything between the Planck scale and the weak scale. Also, we have found new kinds of extended object in string theory. Thus the question of cosmic superstrings (and branes) must be revisited, and this has been done beginning in [2, 3]. A necessary set of conditions is:

1. The strings must be *produced* after inflation.

L. Baulied et al. (eds.), String Theory: From Gauge Interactions to Cosmology, 229–253.

2. They must be *stable* on cosmological timescales.
3. They must be *observable* in some way, but not already excluded.

Ref. [1] thus showed that perturbative strings fail on all three counts. If we do find models that satisfy these three conditions then there is one more condition that we would hope to satisfy:

4. Cosmic superstrings should be *distinguishable* from other kinds of cosmic string, in particular gauge theory solitons.

We will see that each of these four issues is separately model-dependent but that some fortuitous things do happen, and there are simple models in which all these conditions are met, including some of the most fully-developed models of inflation in string theory.

These lectures give a pedagogical introduction to this subject. Section 2 reviews the story of cosmic strings that arise as gauge theory solitons. I discuss the network properties of these strings, and their observational signatures and bounds. Section 3 discusses the cosmic string candidates in various compactifications of superstring theory. Sections 4 through 7 go through each of the conditions on the list above, explaining the issues and their dependence on the details of the compactification. Section 8 presents brief conclusions.

The outline of these lectures (and some of the prose!) follows [4], but the treatment is more detailed and pedagogical.

2. Cosmic string review

2.1. STRING SOLITONS

Cosmic strings might also arise as solitons in a grand unified gauge theory, and for some time these were a candidate for the source of the inhomogeneities that led to the structure in our universe. I start with a review of this subject, since almost all of the ideas and results carry over to superstrings. A complete treatment of this subject would require at least a full volume, and the interested reader is directed to the book by Vilenkin and Shellard [5] and the review by Hindmarsh and Kibble [6]; for a recent look at the subject see [7]. I will have space only to give an overview of the most important ideas.

In any field theory with a broken $U(1)$ symmetry, there will be classical solutions that are extended in one dimension [8, 9].[1] Consider the standard symmetry-breaking potential for a complex scalar field,

$$V(\Phi) = \lambda(|\Phi|^2 - v^2)^2 \,. \tag{2.1}$$

This has a ring of degenerate minima, $\Phi = ve^{i\psi}$ for any phase ψ. Suppose we have a configuration in two space dimensions such that the potential energy is nonzero only in a localized region, falling off rapidly at infinity. At sufficiently large radius R the field must then be approximately at a minimum of the potential, but possibly a different minimum in different directions:

$$\Phi(R, \theta) \approx ve^{i\psi(\theta)} \,. \tag{2.2}$$

[1] I will make a more precise statement, in terms of homotopy groups, in section 4.

As we circle the origin at long distance, the phase $\psi(\theta)$ must make some integer number n of circuits of the ring of minima. This winding number is conserved in time because it is determined by the topology of the field at arbitrarily long distance.

Thus there is a conserved topological quantum number, and the minimum energy configuration in each sector is a stable topological soliton. By continuity, whenever the winding number n is nonzero the field must somewhere pass through zero, so a nonzero winding number implies that there must be a nontrivial 'lump' of potential energy. The size of this core region is set by the Higgs mass $m_H \sim a\sqrt{\lambda}$. Quantum corrections shift the minimum energy in each sector, but do not destabilize the soliton unless they actually restore the $U(1)$ symmetry. Assuming rotational symmetry gives the form

$$\Phi(r,\theta) = f(r)e^{in\theta}\,, \quad f(0) = 0\,, \quad f(\infty) \to v\,. \tag{2.3}$$

We can add in the third dimension simply by extending the polar coordinates to cylindrical coordinates (r, θ, z) with the field independent of z, so that the lump becomes a long straight string.

The broken $U(1)$ can be either a global or a gauge symmetry. In the global case the Hamiltonian density at long distance is

$$\mathcal{H} = |\Pi|^2 + |\vec{\partial}\Phi|^2 + V(\Phi) \to \frac{v^2 n^2}{r^2}\,, \quad r \to \infty\,, \tag{2.4}$$

because the gradient of Φ falls as $1/r$. The string tension, $\int dr\, 2\pi r\mathcal{H}$, is logarithmically divergent at long distance. Nevertheless these solutions are of interest. For example, for two oppositely oriented straight strings separated by a distance L, the divergence is cut off at L. This configuration is topologically trivial and unstable to annihilation of the long strings, but the force between them falls as $1/R$ so in fact the strings can be quite long-lived.

When the $U(1)$ is gauged, the gradient in the covariant derivative $\vec{D}\Phi = \vec{\partial}\Phi - i\vec{A}\Phi$ can be cancelled by an appropriate gauge field,

$$\vec{A}(r,\theta) = \hat{\theta}g(r)\,, \quad g(0) = 0\,, \quad g(\infty) \to \frac{n}{r}\,. \tag{2.5}$$

The energy is then finite. The core of the gauge string contains a magnetic flux

$$\int F = \oint_C A = 2\pi n\,, \tag{2.6}$$

where C is a large circle around the origin. Most of the literature on cosmic strings deals with gauge strings. Most of the superstrings that we will encounter will indeed be analogous to gauge strings, and so this is what we assume in the discussion of network properties and signatures below, but we will see that there is also a possibility of global strings.

2.2. FORMATION OF STRINGS

These classical string solutions exist whenever there is a broken $U(1)$ symmetry, and whenever a $U(1)$ symmetry *becomes* broken during the evolution of the universe a network of strings must actually form. Such a symmetry-breaking transition can occur as a result either of coupling to a time-dependent scalar field or of thermal effects. In the first case the potential for Φ might be of the form

$$V(\Phi) = \lambda(|\Phi|^2 - g\chi)^2 \tag{2.7}$$

with χ a scalar field that is slowly rolling from negative values, for which the potential is minimized at $\Phi = 0$, to positive values, where the symmetry is broken. In the thermal case, the vacuum is found by minimizing the free energy $E - TS$. The entropy is usually larger in the unbroken phase, because it has a greater number of massless degrees of freedom. Thus the unbroken phase might be favored at high temperature and then the symmetry break as the universe cools and we approach the zero-temperature potential (2.1).

The Higgs field Φ thus starts at zero, and then when the symmetry breaks it rolls down to one of the vacua. However, it will not role in the same direction everywhere: by causality, it cannot be correlated on distances greater than the horizon scale. In practice, the correlation length is usually less than this, determined by the intrinsic length scales of the field theory. Thus it chooses random directions in different places, and inevitably there will be some trapped winding, so that strings are left over at the end. This is the Kibble argument [10], and indeed it is what simulations show. A fraction of the string is in the form of finite sized loops, and a fraction is in the form of infinite strings; the latter enter and leave the boundaries of the simulation volume no matter how large this is taken to be. These populations are cleanly separated because the distribution of lengths of finite loops falls rapidly for long loops, giving a convergent integral for the total amount of string in loops. Presumably the existence of the infinite strings is implied by the causality argument. They are important because they begin to stretch with the expansion of the universe, while the small loops quickly decay away.

2.3. EVOLUTION OF STRING NETWORKS

To get some understanding of the evolution of the string networks, let us first make the crude assumption that the string network just expands along with the growth of the universe. If we look at a given comoving region, the length of string within, and therefore the total energy, grows as the scale factor $a(t)$. The comoving volume scales as $a(t)^3$, so the energy density ρ_s scales as $a(t)^{-2}$. This would dominate over matter, $\rho_m \propto a(t)^{-3}$, and radiation, $\rho_r \propto a(t)^{-4}$.

However, there are several processes that reduce the string energy density relative to this estimate:[2]

1. The long strings, which form as random walks, expand uniformly only on scales greater than the horizon length t. On shorter scales they tend to straighten over time.
2. When two strings collide they can either pass through each other or they can reconnect (intercommute), as in Fig. 1. For gauge theory solitons, the value of P is essentially 1. In an adiabatic collision, gauge theory strings always reconnect, $P = 1$,

[2]For reasons of space I will focus on strings whose only long-distance interactions are gravitational. Global strings would also have a long-ranged Goldstone boson field, and could decay via Goldstone boson emission. There are also superconducting strings [11], which have strong couplings to gauge fields. These are perhaps less likely to arise in the superstring case, for reasons that I will explain.

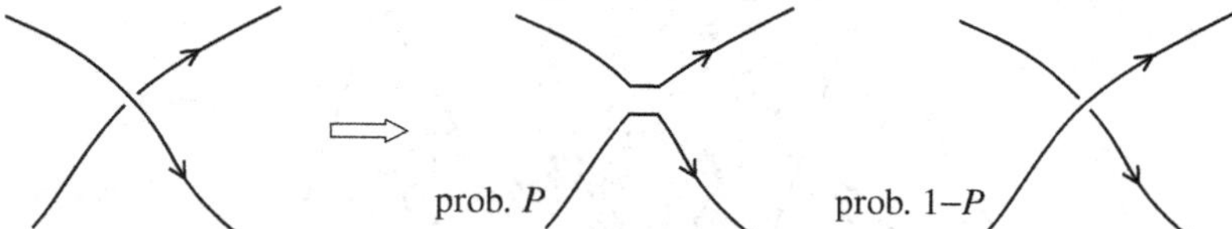

Figure 1. When two strings of the same type collide, they either reconnect, with probability P, or pass through each other, with probability $1 - P$. For classical solitons the process is deterministic, and $P = 1$ for the velocities relevant to the string network. When the strings reconnect, a sharp kink is left on each new string, and each of these separates into a right-moving and a left-moving kink: the result is four kinks, as shown.

because reconnection allows the flux in the string core to take an energetically favorable shortcut. Simulations show that this remains true at the moderately relativistic velocities that are present in string networks [12, 13, 14].[3]

(a) When a long string intersects itself, a loop breaks off. Loops on scales shorter than the horizon do not expand and so behave like massive matter.

(b) When two different long strings reconnect they produce two long kinked strings. The kinks tend to straighten in time, as in point 1, which reduces the total string length. (This effect is not enough to produce a scaling solution without also closed loop formation from self-intersection).

3. The closed loops eventually decay through gravitational radiation.

Simulations show that these processes act with the maximum efficiency allowed by causality, so that the appearance of the string network at any time looks the same when viewed at the horizon scale t, with a few dozen strings spanning the horizon volume and a gas of loops of various sizes [15, 16, 17], as in Fig. 2. This is known as the *scaling solution*. The scaling solution is an attractor: if we start with too much string, the higher collision rate will reduce it, while if we start with too little then there will be few collisions until the amount of string per horizon volume (which is increasing relative to the comoving volume) approaches the scaling value. Thus the details of the initial distribution do not matter, as long as there are some infinite strings.

The total length of string within a horizon is then a numerical constant times t, and the horizon volume is t^3, so the energy density ρ_s is a constant times μt^{-2}, where μ is the string tension. The scale factor is proportional to $t^{1/2}$ in the radiation-dominated era and $t^{2/3}$ in the matter dominated era, so it works out that the ratio of the string energy density to the dominant energy density in each era is a constant. The simulations show that $\rho_s/\rho_m \sim 60G\mu$ during the matter-dominated era and $\rho_s/\rho_r \sim 400G\mu$ during the radiation dominated era. This is our first encounter with the dimensionless combination $G\mu$, about which we will have much more to say, but for now we just assert that we are interested in values of $G\mu$ less than 10^{-6}. The strings would thus be a small but fixed fraction of the total density. In particular, it would take some very exotic network scenario for strings to be the dark matter.

[3]Cosmic strings move at moderately relativistic speeds. In flat spacetime the virial theorem implies that v^2 has a mean value of $\frac{1}{2}$; this is somewhat reduced by the expansion of the universe.

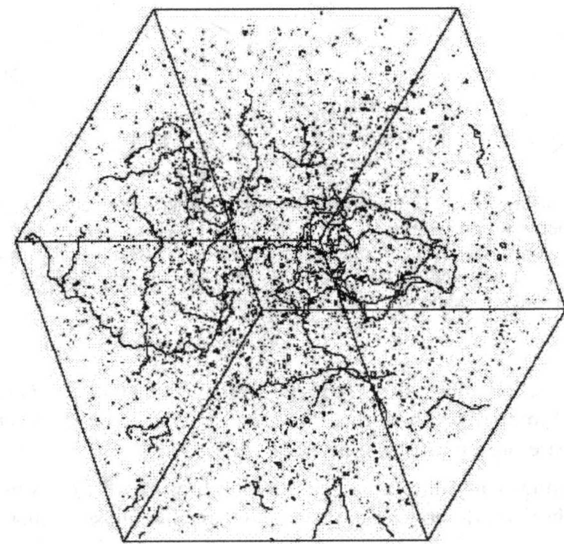

Figure 2. Cosmic string simulation. A side of the cube is around a third of the horizon length. Strings that appear to end are just leaving the simulation volume. From Allen and Shellard [17].

The horizon distance, approximately t, sets the overall scale of the string network, but there are important features on smaller length scales. In particular, there is a lower cutoff on the size of loops, coming from their decay via gravitational radiation. Dimensionally, the gravitational wave power emitted by a loop of size l is $G\mu^2$. Equating the loop mass μl to the power times t determines the typical lifetime for a loop of size l; equivalently it determines the size l, as a function of the time t, below which the decay becomes rapid. Inserting the constants from a more careful treatment, the result is

$$l \sim 50G\mu t\,. \tag{2.8}$$

Taking $G\mu = 10^{-9}$, which will turn out to be a typical value, gives $l \sim 500$ lightyears today. Thus there is large hierarchy of scales in the network.

Another important property of the network is the existence of a kinked structure at short distance [16]. We have noted that reconnection of loops produces kinks. As t increases the number of kinks per horizon length necessarily increases, and this effect is stronger than the tendency of the kinks to straighten as the string stretches. The study of this structure in simulations is limited by the resolution of the simulations, but it has generally been assumed that it is cut off by gravitational radiation at the same scale (2.8) (see however [18]). The short distance structure is important because the simulations indicate that it determines the typical size of the loops that break off the infinite strings.

In summary, we should note that the behavior of string networks is challenging to understand in detail: the large ratio of scales makes numerical simulation difficult, and while there exist various analytic models, the problem is very nonlinear and existing models only capture part of the physics. If cosmic strings are ever detected, it will become important to work toward a much more precise understanding.

Finally, it is interesting to contrast the properties of one-dimensional defects with those of two-dimensional and zero-dimensional defects. If a network of domain walls forms and subsequently annihilates to the maximum extent allowed by causality, then the typical spacing between walls will be of order t. The energy density in domain walls will then scale as $1/t$, as compared to the $1/t^2$ of strings, and this would come to dominate over the matter and radiation densities. Such a network is therefore excluded. For point particles, annihilation is not able to reduce the density to the scaling value: points have a harder time finding each other than strings. The spacing then scales as $a(t)^{-1}$ and the density as $a(t)^{-3}$, just like other massive matter; such defects would therefore come to dominate in the radiation-dominated era and are excluded.

Thus, whereas strings represent a cosmic opportunity, domain walls and point defects are forbidden. In section 4 we will discuss potential instabilities that could eliminate cosmic strings, and the analogous decays would also provide solutions to the domain wall and point defect problems. However, there is one kind of point defect that provides a serious challenge, namely the magnetic monopole [19, 20]. These exist in all theories that unify electromagnetism with the other interactions (for a recent overview see [21]). They cannot decay (magnetic charge conservation) and they are not confined (at least today), so they present a vexing cosmological problem. However, if the phase transition in which they are produced took place before inflation, they would have been diluted the point that a monopole would probably never be seen; this was one of the original motivations for inflation [22]. By the same token, cosmic strings will only be of interest if they are produced at the end of inflation, or later.

2.4. STRING SIGNATURES AND BOUNDS

Since we are considering strings whose only interactions are gravitational, all of their effects are controlled by the dimensionless product of Newton's constant and the string tension, $G\mu$. One can think of this in two ways: it is the string tension in Planck units, and it sets the size of the typical metric perturbation produced by a string. For example, the geometry around a long straight string is conic, with a deficit angle $8\pi G\mu$.

The string network produces inhomogeneities proportional to $G\mu$, and because of the scaling property of the string network these are scale invariant. A value $G\mu \sim 10^{-5.5}$ would give scale-invariant perturbations of the right magnitude to produce the galaxies and CMB fluctuations.[4] This was a viable theory for some time [23, 24], but it is now excluded. For example, the CMB power spectrum is wrong: the actual spectrum shows a pattern of peaks and dips, whereas the spectrum from strings would be smooth. There is a simple reason for this. The fluctuations produced in inflation have a definite phase. This phase is maintained from the end of inflation until the perturbations go nonlinear, and is imprinted as oscillations of the power spectrum. Strings, on the other

[4] I am going to quote values of $G\mu$ to the nearest half order of magnitude. This is less precise than most numbers given in the literature, but it is all I will need and it roughly reflects the uncertainties in the understanding of the string network. To give more precise numbers would require a more detailed discussion, and so the reader should consult the references.

hand, each keep their own time, there is no common phase. Fitting cosmological constant plus cold dark matter *plus* strings to the CMB power spectrum gives an upper limit $G\mu < 10^{-6}$ [25, 26]. Beyond the power spectrum, strings will produce nongaussianities in the CMB. Recent limits are somewhat stronger than those from the power spectrum, around $G\mu < 10^{-6.5}$ [27, 28].[5] Incidentally, strings affect the CMB in more than one way. Strings at early times will induce the matter inhomogeneities that are reflected in the CMB, but even if the CMB were uniform, strings at the present time would produce apparent inhomogeneities. We have noted that a static string bends light. If a string moves with velocity v transverse to the observer there will be a differential redshift of order $8\pi v G\mu$ between the two sides.

Another limit on $G\mu$ comes from pulsar timing. Because the energy in the strings eventually goes into gravitational waves, strings produce a large stochastic gravitational wave background. The classic reference on this subject [29] quotes a limit on the energy density in stochastic gravitational waves, per logarithmic frequency range, as $\Omega_{\rm GW} < 1.2 \times 10^{-7}$, using $h^2 = 0.5$ for the Hubble parameter. In the frequency range of interest one can estimate the stochastic background from a network of strings as $\Omega_{\rm GW} = 0.04 G\mu$ [5], meaning that $G\mu < 10^{-5.5}$. Different analyses of *the same* data, with different statistical methods, give a (controversial) bound a factor of 6 stronger [30] and another a factor of 1.6 weaker [31]. The limits from pulsars should increase rapidly with greater observation time, and recent work [32] quotes a bound 30 times stronger than [29], meaning that $G\mu < 10^{-7}$. However, this paper again obtains much weaker limits using other methods and so this should not be regarded as a bound until there is agreement on the analysis.[6] The bounds here are from gravitational waves with wavelengths comparable to the size of the emitting string loop, and do not include a potentially substantial enhancement due to high-frequency cusps; see section 6.

Thus far we have quoted upper bounds, but there are possible detections of strings via gravitational lensing. A long string will produce a pair of images symmetric about an axis, very different from lensing by a point mass. Such an event has been reported recently [33, 34]. The separation of around two arc-seconds corresponds to $G\mu$ equal to 4×10^{-7} times a geometric factor of order 1. I have heard varying opinions on how seriously to take such an observation, as similar pairs in the past have turned out to coincidental. There is further discussion in the review [7], which also discusses a possible time-dependent lens, as from an oscillating loop, with $G\mu \sim 10^{-7.5}$ [35].

3. Cosmic strings from string theory

Now let us consider the candidates for cosmic superstrings that might arise in various vacua of string theory, starting with the perturbative heterotic string. The gravitational and gauge couplings in this case have a common origin from the closed string interaction, and so by calculating the trilinear amplitudes one finds the relation (e.g. Eq. 18.2.4 of [36])

$$4\kappa^2 = \alpha' g^2 \,. \tag{3.1}$$

[5]This limit is for a network of strings. For a single stray cosmic string, the limit is an order of magnitude weaker.

[6]I would like to thank E. Flanagan and H. Tye for communications on this point.

The gauge and gravitational couplings are equal up to a factor of α' required by the dimensions, and a factor of 4 which arises from various conventions. This holds for the ten-dimensional couplings, but because the gauge and gravitational fields both live in the bulk they reduce in the same way, $\kappa_4^2 = \kappa_{10}^2/V$ and $g_4^2 = g_{10}^2/V$ where V is the compactification volume. Expressing the relation in terms of $G = \kappa_4^2/8\pi$, $\mu = 1/2\pi\alpha'$, and the heterotic gauge coupling $\alpha_{\rm h} = g_4^2/4\pi$ gives

$$G\mu = \frac{\alpha_{\rm h}}{16\pi} . \tag{3.2}$$

The heterotic gauge coupling should be at or above the minimal GUT value $1/20$, so $G\mu$ is at least of order 10^{-3}. The existence of such cosmic strings is therefore excluded [1].

We cannot turn this around and use it as evidence against perturbative heterotic string theory, because in fact a cosmic string network of this type could not form. The CMB limit on tensor perturbations implies an upper limit on the vacuum energy during inflation, $G^2 V_{\rm inf} \lesssim (\delta T/T)^2 \sim 10^{-10}$. In most cases this bounds the tension of any strings that might subsequently form, $G\mu \lesssim (G^2 V_{\rm inf})^{1/2} \lesssim 10^{-5}$. Thus, perturbative heterotic strings could only have been produced in a phase transition before inflation, and the many e-foldings of inflation then make it unlikely that any would be found within our horizon. There is a second reason that heterotic strings would not reach cosmic size, as we see in section 5.

For the perturbative type I string the dimensional analysis is a bit different but the conclusions are the same. The *strongly* coupled $E_8 \times E_8$ heterotic string can have a lower tension [37]. This is because the gauge fields live on the nine-dimensional boundary while gravity lives in the ten-dimensional bulk. More generally, if the gauge fields are confined to a brane while gravity propagates in the bulk, the string tension is suppressed by some power of $R/L_{\rm P}$ where R is the size of the dimensions transverse to the brane [38, 39].

Even without large compact dimensions, the string tension can be reduced by a gravitational redshift (warp factor) [40]

$$ds^2 = e^{2A(y)}\eta_{\mu\nu}dx^\mu\, dx^\nu + \ldots \tag{3.3}$$

where y are the compact coordinates. The string is localized at some point y_0 in the compact space, so the string worldsheet action is proportional to

$$\mu = \frac{e^{2A(y_0)}}{2\pi\alpha'} . \tag{3.4}$$

In interesting compactifications there are often *throats* where the warp factor e^{2A} is much less than 1. The string will fall to the bottom of such a throat, so the four-dimensional tension μ measured in four dimensions can be reduced by a large factor relative to the tension $1/2\pi\alpha'$ seen by a local ten-dimensional observer. This is the RS idea, that different four-dimensional scales arise from a single underlying scale through gravitational redshifting.

Thus the tension is essentially a free parameter, which can lie anywhere from the Planck scale down to the experimental limit near the weak scale. For example, in the warped models it is determined by the depth of the throat, which comes out as the exponential of a function of flux integers that can take a wide range of values [41].

The range becomes much smaller when we focus on *brane-antibrane inflation* models [42, 43, 44, 45]. This is a nice geometric idea for obtaining a slow-roll inflationary potential, and the basis for most current attempts to describe inflation in string theory. That is, the early universe could have contained an extra brane-antibrane pair, separated in the compact directions. The potential energy of these branes would drive inflation. The inflaton field is then the separation between the branes: this has a potential which is rather flat when the branes are separated and steepens as they approach, until at some point a field becomes tachyonic and the brane and antibrane annihilate rapidly.

To understand inflationary cosmology in detail one needs to know the scalar potential. This potential has long been problematic, especially for states of positive vacuum energy, because of potential instabilities in the moduli directions. Recently the tools have been developed to identify a large class of stable solutions [46, 47]. Thus we will focus on the resulting *KLMT model* [48], since it is the most detailed model of inflation in string theory. This is based on a strongly warped compactification of the IIB string theory, with inflation arising from a D3/$\overline{\rm D3}$ pair at the bottom of a throat.

In models of brane inflation, the value of $G\mu$ can be deduced from the observed magnitude of the CMB fluctuations $\delta T/T$. That is, one assumes that $\delta T/T$ arises from the quantum fluctuations of the inflaton; this is natural given the flat form of the inflaton potential. For any given brane geometry the inflaton potential has a definite functional form. For example, in the KLMT model the D3/$\overline{\rm D3}$ potential is $V \sim V_0[1 - O(\phi^{-4})]$. One then finds (Eq. C.12 of [48])

$$G^2 V_0 \sim 0.05(\delta T/T)^3 N_e^{-5/2} \sim 1.5 \times 10^{-20}\ ; \tag{3.5}$$

here $N_e \sim 60$ is the number of e-foldings at the wavelengths responsible for structure formation. The vacuum energy from a D3/$\overline{\rm D3}$ in the warped space is $2T_3 e^{4A_0}$ where A_0 refers to the bottom of the inflationary throat. The F-strings produced after inflation (the mechanism for which is the subject of the next section) will sit in the same throat and so $\mu = e^{2A_0}/2\pi\alpha'$. Since $T_3 = T_{\rm F}^2/2\pi g_s$ in ten dimensions [36] we have for the tension of the IIB string in the inflationary throat

$$\mu_{\rm F} \sim 2 \times 10^{-10} \sqrt{g_{\rm s}}\ , \tag{3.6}$$

where $g_{\rm s}$ is the string coupling.

One might take $g_{\rm s}$ to have a representative value of 0.1. There is one very specific class of models in which it is related to observed quantities. If the matter fields live on D3's or $\overline{\rm D3}$'s, then $g_{\rm s}$ turns out to be exactly $\alpha_{\rm YM}$. However, $g_{\rm s}$ is really e^{Φ} at the position of the brane, and the dilaton Φ in most solutions varies in the compact space and so will not have the same value at the position of the strings as it has at the Standard Model brane. In the special case that it is constant, we can identify $g_{\rm s}$ in the brane tension with the unified coupling $\alpha_{\rm GUT}$. Unification in brane models is generally nonstandard, but a likely value for $\alpha_{\rm GUT}$ is still around 0.1.

Note that these warped manifolds can have many throats and therefore strings of many different tensions. One possibility is that the Standard Model fields live in a throat whose depth corresponds to the weak scale, whereas the depth (3.5) of the inflationary throat gives something closer to the GUT scale. Thus there would be the possibility to see TeV-scale strings and extra dimensions at accelerators, and at the same time GUT-scale strings in the sky.

In addition to the fundamental F-strings, string theory has many other extended objects. For example the IIB theory has a D-string whose tension in ten dimensions is $T_{\rm D} = T_{\rm F}/g_{\rm s}$, and so in the KLMT model

$$\mu_{\rm D} \sim 2 \times 10^{-10}/\sqrt{g_{\rm s}}; \tag{3.7}$$

the geometric mean $(\mu_{\rm F}\mu_{\rm D})^{1/2}$ is independent even of $g_{\rm s}$. There are also bound states of p F-strings and q D-strings with a distinctive tension formula

$$\mu_{(p,q)} = \mu_{\rm F}\sqrt{p^2 + q^2/g_{\rm s}^2}\,. \tag{3.8}$$

There are also a variety of higher dimensional D-, NS- and M-branes in the various regimes of string theory. Any of these will look like a string if all but one of its spatial directions is wrapped on a compact cycle. In the KLMT model, any brane not contained fully in a throat will have a Planck-scale tension because a portion of it passes through the region where the warp factor is close to 1, so we will get additional interesting strings only if there are nontrivial cycles in the throat. The most generic throat, the Klebanov-Strassler throat [2] based on the conifold singularity [50], has no cycles that give additional strings, but there are many less generic singularities with a wide variety of nontrivial cycles; these have been extensively studied in the context of gauge/string duality.

For models based on large compact dimensions, no fully stabilized examples are known. One can fix the bulk moduli by hand and study inflation based on the brane moduli. This is done in [3, 51], which consider a variety of wrapped branes in type II compactifications, and find values in the range

$$10^{-12} \lesssim G\mu \lesssim 10^{-6}\,. \tag{3.9}$$

We should emphasize again that it is the assumption that the inhomogeneities arise from the quantum fluctuation of the inflaton, rather than some other mechanism, that allow us to tie the tension to observed quantities in brane inflation. If another mechanism is responsible, the inflation scale and the tension could be much lower.

It may still turn out to be the case that our vacuum is well-described by weakly coupled heterotic string theory. If so, inflation and cosmic strings might simply arise in the effective low energy field theory (this would be true for other compactifications as well). The possible strings include both the magnetic flux tubes discussed in section 2, and also electric flux tubes that exist in strongly coupled confining theories [1]. The heterotic string, whether weakly or strongly coupled, has the advantage that it more readily makes contact with standard grand unification. A recent paper [52] argues that cosmic strings should be present in a wide class of grand unified inflationary models.

4. Production of cosmic F- and D-strings

If these branes are D-branes, then there is a $U(1)$ gauge symmetry on each of the brane and antibrane, and this $U(1) \times U(1)$ disappears when the branes annihilate. One linear combination of the $U(1)$'s is Higged. The Kibble argument then applies, so that a network of strings must be left over when the branes annihilate [2, 3]. These are D-strings, as one can see by studying the conserved charges [53, 54]. More precisely, if the branes that annihilate are $D(3+k)$-branes, extended in the three large dimensions and wrapped on k small dimensions, then the result is $D(1+k)$-branes that extend in one large dimension and are wrapped on the same small dimensions. The simplest case is $k=0$, where D3/$\overline{\text{D3}}$ annihilation produces D1-branes.

The second linear combination of $U(1)$'s is confined. We can think of confinement as dual Higgsing, by a magnetically charged field, and so we would expect that again the Kibble argument implies production of strings [55, 56]. These are simply the F-strings, the 'fundamental' superstrings whose quantization defines the theory, at least perturbatively.

It is notable that this process produces only strings, and not the cosmologically dangerous monopoles or domain walls [2, 3]. The breaking of a $U(1)$ produces defects of codimension two, and the Kibble argument requires that the codimension be in the large directions: the small directions are in causal contact.[7]

It is striking that what appears to be the most natural implementation of inflation in string theory produces strings and not dangerous defects, but we should now ask how generic this is. Even these models are not, in their current form, completely natural: like all models of inflation they require tuning at the per cent level to give a sufficiently flat potential [48, 58]. There might well be other flat regions in the large potential energy landscape of string theory. An optimistic sign is that there are arguments entirely independent of string theory to indicate that inflation terminates with a symmetry-breaking transition: this is known as hybrid inflation [59], and leads to efficient reheating as well as production of strings [60, 61, 62]. However, not every symmetry-breaking pattern produces strings. For example, in strongly coupled heterotic string theory [37], there are M5-branes and M2-branes. It would seem natural to use M5-branes with two wrapped directions to implement brane inflation. Since the M2-branes have codimension three relative to the M5-branes it will not be so easy to produce strings; this is currently under investigation.[8]

In [69] it was proposed that cosmic fundamental strings could have been produced in a Hagedorn transition in the early universe. The Hagedorn transition corresponds to the formation of strings of infinite length, and so some percolating strings would survive as the universe cooled below the Hagedorn temperature. This idea has a simple realization in the warped models.[9] We have noted that the effective string tension, and so the Hagedorn temperature, is different in different throats. It is possible that after inflation a deeper throat reheats above its Hagedorn temperature, leading to string formation as the universe cools (a black hole horizon would then form at the bottom of the throat, corresponding to the Hawking-Page transition [70, 71]). In fact, this is essentially equivalent to the Kibble mechanism. The throat degrees of freedom have a dual gauge description, in terms of which the Hagedorn transition corresponds to deconfinement. The transition to a confining phase as the universe cools is the electric-magnetic dual of a symmetry-breaking transition. The strings produced in this way would necessarily have a lower tension than

[7]Ref. [55] identifies processes that might suppress production of D-strings. Ref. [57] argues that in some cases there can be disorder in the compact dimensions and so monopoles and domain walls would be produced. We believe that, at least for the KLMT model, neither of these should be relevant.

[8]There has also been recent discussion of a more exotic symmetry-breaking pattern in D-brane inflation [63, 64, 65]. For other discussions of string production in brane inflation see [66, 67, 68]

[9]Similar ideas are being considered by A. Frey and R. Myers.

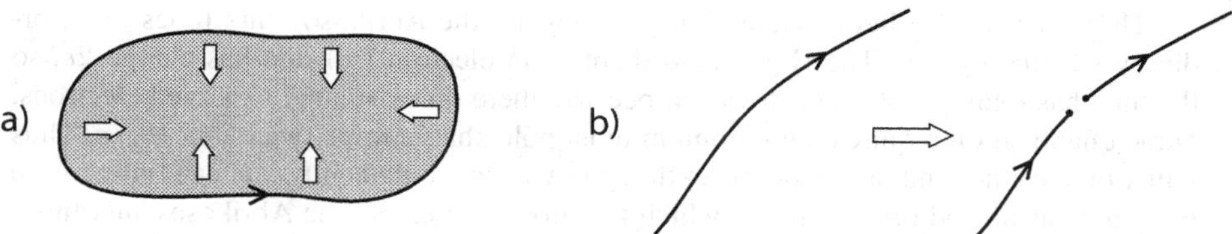

Figure 3. Instabilities of macroscopic strings: a) Confinement by a domain wall. b) Breakage.

those produced directly in brane annihilation, because of the inefficiency of the extra thermal step.

5. Stability of strings

5.1. FIELD THEORY STRINGS

Ref. [1] identified two instabilities that would prevent superstrings from growing to cosmic size. Actually, these same two instabilities exist for field theory soliton strings [72] — one for global strings and the other for gauge strings — so let us first discuss them in this context.

In the case of global strings, we have noted that the long-ranged Goldstone boson has gradient energy. It does not have potential energy at long distance as long as the broken $U(1)$ symmetry is exact: the broken vacua are then exactly degenerate. However, there are general arguments in string theory that there are no exact global symmetries [73, 36]. More generally, the no-hair theorems imply that black holes can destroy global charges, so in any theory of gravity these can not be exactly conserved. Thus the degeneracy of the vacua will not be exact; the trough in the potential is tilted and so there is a potential energy cost even at long distance for a field configuration that circles the trough. This cost is minimized by have the scalar field make its excursion in a kink of finite width, rather than uniformly in angle as in the exact case (2.3). Thus there is a domain wall, with energy proportional to its area, bounded by the string.

The wall exerts a transverse force on the string and forces it to collapse, as in Fig. 3. This is clear for a loop of string bounding a domain wall, but is less intuitive for a network with infinite strings. One can picture the network, with the domain walls, as a complicated shape formed from strips whose width is the typical transverse separation between strings. The timescale for the collapse of the strips, and the disappearance of the strings, is set by the width of the strips, not their (potentially unbounded) length [74].

For gauge strings, the $U(1)$ symmetry is exact because it is a gauge symmetry, and all energies fall exponentially with distance from the string. A magnetic flux runs along the core of the string, and it is the conservation of this flux that prevents the string from breaking. However, in any unified theory one expects that there will be electric and magnetic sources for every flux [21], so that the string can break by creation of a monopole/antimonopole pair. If this is possible, it will happen not only once but everywhere along the length of the string, and so the string breaks up into short segments rather thans growing to cosmic length.

There can be absolutely stable strings. Consider the Abrikosov flux tubes in an ordinary superconductor. The Higgs field there is an electron pair and has charge $2e$, so the tube has total flux $2\pi/2e$. However, because there do exist singly charged electrons, Dirac quantization requires the minimum monopole charge to be twice this, $2\pi/e$. Thus a flux tube cannot end on a monopole, though two can; equivalently one can think of the monopole as a bead on a string, at which the flux reverses. So the Abrikosov flux tubes (in an infinite system) are absolutely stable. One can think of this in terms of an unbroken discrete gauge symmetry $(-1)^{Q/e}$, which acts as -1 on the electron and $+1$ on the BCS condensate. As one circles the string, fields come back to themselves only up to this transformation. Thus the string can be detected by an Aharonov-Bohm experiment at arbitrarily long distance, and so it can never just disappear. We will refer to these absolutely stable strings as 'discrete' strings. Gauge strings without a discrete gauge charge are truly invisible at long distance, and so there is no obstacle to their breaking.

As an aside, we explain this in terms of the homotopy classification of defects (see e.g. Ref. [75] for an introduction). For gauge group G broken to H there is an exact sequence

$$\begin{aligned}\Pi_2(G/H) \quad &\to \quad \Pi_1(H) \to \Pi_1(G) \to \Pi_1(G/H) \\ &\to \quad \Pi_0(H) \to \Pi_0(G) \to \Pi_0(G/H)\,. \end{aligned} \tag{5.1}$$

In the discussion of strings we started with $G = U(1)$, for which $\Pi_1(G) = Z$, and broke to $H = I$ for which $\Pi_1(H) = \Pi_0(H) = I$. Thus the strings are charged under $\Pi_1(G/H) = Z$. Now however let us embed the $U(1)$ into a semisimple group; we are always free to assume this to be simply connected by going to the covering group, so $\Pi_1(G) = I$ and the stable strings are classified by $\Pi_1(G/H) = \Pi_0(H)_G$ — that is, by unbroken discrete symmetries that lie in the connected component of the identity in G.

It can also be that strings that are not absolutely stable have decays that are slow on cosmic timescales. For monopole pair creation, for example, the classic Schwinger calculation gives a pair production rate suppressed by $e^{-\pi M^2/\mu}$, where M is the monopole mass. If the monopole mass is an order of magnitude larger than the scale of the string tension, then the decay will be slow on cosmological time scales. One can think of a succession of symmetry breakings such as $SU(2) \to U(1) \to 1$. There are no stable strings for this pattern, but if the scale of the first symmetry breaking is higher than that of the second then there will be string solutions in the effective field theory describing the second breaking, and these can be long-lived. Similarly the tension of the domain walls might be suppressed by a hierarchy of scales.

5.2. STRING THEORY STRINGS

5.2.1. *Perturbative strings*

Now let us turn to string theory. Consider first perturbative strings [1], compactified for example on a torus or a Calabi-Yau manifold. The type I string couples to no form field, and of course it can break. The heterotic and type II strings are effectively global strings, because they couple to the long-distance form field $B_{\mu\nu}$ which in four dimensions is dual to a scalar. A confining force is produced by instantons, which are magnetic sources for $B_{\mu\nu}$ (thus the completeness of the magnetic sources [21] again enters). It is simplest to analyze this in terms of the dual scalar Θ, defined by $d\Theta = *dB$ modulo Chern-Simons terms; Θ is the phase of the earlier Φ. The instantons become electric sources, meaning that they give rise to explicitly Θ-dependent terms in the effective action: the instanton and anti-instanton amplitudes are weighted by $e^{\pm i\Theta}$.

It is an interesting exercise to work out the decay rates for the various strings. The time scale for the breaking of Type I strings is the string scale unless one tunes the string coupling essentially to zero. For the heterotic string, the one definite contribution comes from QCD instantons. The field strength for $B_{\mu\nu}$ is of the form

$$H = dB + \omega(A)\,, \tag{5.2}$$

where $d\omega(A) = \mathrm{Tr}\,F^2$ and the trace runs over all gauge fields. Thus the magnetic source dH contains the QCD instanton number. The scale of the potential is $m_u m_d m_s \Lambda_{\mathrm{QCD}} \sim 10^{-7}\mathrm{GeV}^4 \equiv V_{\mathrm{inst}}$. The potential energy

$$\int M_{\mathrm{P}}^2(\partial\Theta)^2 + V_{\mathrm{inst}}\cos\Theta \tag{5.3}$$

is of order $M_{\mathrm{P}}^2\tau^{-1} + V_{\mathrm{inst}}\tau$ for a domain wall of thickness τ. This is minimized for $\tau \sim M_{\mathrm{P}}/\sqrt{V_{\mathrm{inst}}}$ (the inverse axion mass), giving tension

$$T_{\mathrm{dw}} \sim M_{\mathrm{P}}\sqrt{V_{\mathrm{inst}}}\,. \tag{5.4}$$

The acceleration of a string due to the wall is then $a = T_{\mathrm{dw}}/\mu$ which is roughly $\sqrt{V_{\mathrm{inst}}}/M_{\mathrm{P}}$. The string network will collapse when this is comparable to the Hubble expansion rate, which is just when the energy density in matter is of order $\sqrt{V_{\mathrm{inst}}}$, near the QCD scale. This is too early to leave observable traces. Further, the estimate (5.4) is just a lower bound, because it is likely that there are other gauge groups becoming strong at higher energy. Indeed, it is a challenge to find an scalar (axion) in string theory whose potential is dominated by QCD instantons and so would solve the strong-CP problem.

NS5-brane instantons, wrapped on the whole compact space, are also a magnetic source for $B_{\mu\nu}$ [76]. The NS5-brane instantons have action

$$T_5 V_6 = \frac{4\pi^2\alpha'}{2\kappa_{10}^2}V_6 = \frac{2\pi}{\alpha_{\mathrm{h}}}\,. \tag{5.5}$$

This is the same as the action for a gauge theory instanton at the string scale. In fact the heterotic NS5-branes are just the string-scale limit of instantons, whereas above we considered QCD-scale instantons. If we take the standard GUT value $\alpha_{\mathrm{h}} \sim 0.05$ then the contribution to V_{inst} is of order $10^{-50}M_{\mathrm{P}}^4$. This is larger than the QCD contribution; it might be suppressed by fermion masses (from zero modes), or enhanced if the GUT coupling is increased.

The same estimates apply to perturbative (non-brane) compactifications of the Type II theory, with the gauge fields arising from world-sheet current algebras, although it is known that the Standard Model cannot be obtained from these [77].

5.2.2. *Brane models*

In the KLMT model [48] we have noted that the candidate cosmic strings are the F- and D1-strings. In ten dimensions these strings couple to form fields $B_{\mu\nu}$ and $C_{\mu\nu}$, but these form fields have no four-dimensional massless modes in this model. They transform non-trivially under the orientifold projection or F theory monodromy that the model requires, and this removes their zero modes. Thus the strings should be unstable to breakage, and indeed this can occur in several ways.

First, the projection that removes the form field produces an oppositely oriented image string on the covering space of the compactification, and breakage occurs through a segment of the string annihilating with its image. If the image is coincident with the string, breakage will be rapid. If the image is not coincident, then the string must fluctuate to find its image. Due to the warp factor the strings feel a potential

$$\mu = e^{2A(y)}\mu_{10} \, . \tag{5.6}$$

The strings are normally confined to the throat where inflation occurs, which as discussed in section 3 has a warp factor e^{A_0} of order 10^{-4}. When the string annihilates with its image, the role of the monopole is played by a bit of string that stretches out of the inflationary throat over to the image throat. Since this passes through bulk region where the warp factor is of order one, the breakage rate is proportional to $e^{-\pi e^{-2A_0}}$ and so totally negligible even on cosmological timescales [56]. Thus the strings will be stable to self-annihilation if there is no orientifold fixed point (or F-theory equivalent) in the inflationary throat. There is no particular reason for the throat to be coincident with any fixed point. Their relative positions are fixed by the complex structure moduli, which depend on flux integers, and these are expected to take rather generic values [78, 79, 80].

The strings can also break on a brane. The model must include branes on which the Standard Model (SM) fields live. If these are D3-branes in the inflationary throat then the strings will break; if they are D7-branes that pass through the inflationary throat then all but the D1-string will break [56]. If they are outside the throat then the strings are stable for the same reason as above. In the simplest implementation of the RS idea the SM branes must be in a different throat: the depth of the inflationary throat is something of order the GUT scale, while the depth of the SM throat should be of order the weak scale. One must ask whether other branes might still be in the inflationary throat, but this is not possible: these would have low energy degrees of freedom which would receive most of the energy during reheating, rather than the SM fields. So this gives a simple scenario in which the strings are stable, but much more work is needed to see whether it is viable, and whether it is generic. An important question is whether reheating through the bulk can transfer energy efficiently to the SM throat; a recent paper gives an affirmative answer [81].

More complicated geometries with additional cycles in the inflationary throat have additional strings from wrapped branes. These strings have not been studied in full detail, but in examples one finds that they can be gauge, global, or discrete, depending on the details of the topology (both in the throat and globally). Similarly their production and stability depends on details.

The stabilization of moduli is not been as fully developed in models based on large compact dimensions, but as a prototype we can consider the T^6/Z_2 orientifold which is T-dual to the Type I string. In the Type I string there are BPS D1- and D5-branes [82], which couple electrically and magnetically to the sole Type I RR form $C_{\mu\nu}$. There are also non-BPS D3- and D7-branes [54]. All of these give rise to strings when all but one direction are wrapped on the T^6. After T-duality these become BPS D7- and D3-branes, and non-BPS D5- and D1-branes, all of which are strings in four dimensions. The BPS strings couple to RR form fields and so are global. The relevant instantons come from a Euclidean D-branes. The calculation is similar to that for the NS5-brane instanton above, with the brane volume replacing V_6, but now the compact dimensions can be much larger and so we get essentially stable strings with a long-range RR axion field.

The non-BPS strings separate into two images as in the earlier discussion. Now, it is important to note that the T^6/Z_2 example is nongeneric in an important sense: because of its high degree of symmetry the strings have moduli and move freely in the compact dimensions. Even if there is not a deep warp factor, supersymmetry breaking will lead to a potential at some scale that will localize the strings. Just like a large warping, a large separation leads to a large action for the decay and so these strings can again be stable if separated from their images.

Adding the SM and other branes to the large dimension models, the stability of the strings depends on the exact geometry. In some cases it might appear that a string has both instabilities, in that it couples to a form field but can break on a brane [56, 83]. The point is that breakage on a brane implies the existence of a gauge field on the brane to ensure continuity of the form source, and this Higgses the form field so that the result is a gauge string, subject to breakage but not confinement [56]. We expect that in string theory as in field theory these strings willl be exactly stable only if stabilized by a discrete gauge symmetry visible in the Aharanov-Bohm effect: anything that can happen will [21].

In summary, it is encouraging to see that strings can be stabilized as a side effect of certain generic properties such as warping and/or large dimensions, which are needed to lower the inflationary scale below the Planck scale in these models.

6. Seeing Cosmic Strings

We will consider below strings that have only gravitational interactions. However, the possibility, noted above, of global strings with a long-ranged scalar field should not be overlooked. Another possibility is a strong coupling to SM or other light fields, and in particular superconducting strings [11] which carry massless degrees of freedom charged under an unbroken gauge symmetry. Generically the strings that one finds are rather boring. In ten dimensions they have massless collective coordinates in the transverse directions, and massless fermions, but these are all gapped by symmetry-breaking effects leaving only the minimal collective coordinates. Light four-dimensional fields generally live on branes, and we have seen that stability requires cosmic strings to be separated from most other branes, implying interactions of gravitational strength. An example of an exception would be a D7-brane coincident with a D1 string; the SM fields might live on or couple strongly to the D7-brane. However, a superconducting string requires that the D1 and D7 be exactly coincident; to ensure this would require a discrete symmetry (which might be orbifolded). More interesting strings are thus possible in special cases.

The CMB and pulsar bounds on $G\mu$ quoted in section 2 are at the upper end of the brane inflation range, ruling out the highest-tension models. Both bounds will improve in the coming decade by at least one or two orders of magnitude. An additional exciting prospect comes from LIGO [84, 85]. Under most circumstances LIGO is at a disadvantage looking for cosmological backgrounds because these fall with increasing frequency: LIGO is looking at frequencies that are 10^{10} times those of the pulsar measurements

(100 Hz versus years^{-1}). For example, the stochastic background coming from the low harmonics of cosmic strings has a constant energy density per logarithmic scale.[10] This translates into $\omega^2 \tilde{h}(\omega)^2 \propto \omega^{-1}$ for the Fourier transform $\tilde{h}$ of the strain (fractional change in the metric). For a stochastic background (different frequencies uncorrelated) the strain from approximate frequency ω is of order $\omega^{1/2}\tilde{h}(\omega)$ which is of order ω^{-1} here. Pulsar timing is sensitive to strains around 10^{-14} or 10^{-15} while LIGO is sensitive to much smaller strains 10^{-22} or 10^{-23}, but the frequency penalty of 10^{10} more than offsets this.

However, something unexpectedly nice happens. When a loop of string in three space dimensions oscillates as governed by the Nambu action, it typically forms a cusp several times per oscillation [86]. To see this go to conformal gauge, where the solution is just a sum or right- and left-movers

$$X^\mu(\sigma, \tau) = f^\mu(\sigma - \tau) + g^\mu(\sigma + \tau) , \tag{6.1}$$

and the gauge conditions imply that f' and g' are both null. Classically there is enough residual gauge freedom to set $X^0 = 2\tau$ so that $f^0 = \tau - \sigma$, $g^0 = \tau + \sigma$. Then $\vec{f}'$ and $-\vec{g}'$ both have unit norm, and as functions of their arguments they trace out closed curves on the unit sphere as σ goes around the strings; also, each averages to zero by periodicity. It follows that these curves will typically intersect (an even number of times).[11] Each intersection represents an event that occurs once per period, and when it occurs the σ-parameterization becomes singular. A representative example for the leading behavior at the singularity (where for simplicity we put the intersection at $\sigma = \tau = 0$) is

$$\begin{aligned} f^x &= (\sigma - \tau) - c^2(\sigma - \tau)^3/2 , \quad f^y = c(\sigma - \tau)^2 , \\ g^x &= -(\sigma + \tau) + c'^2(\sigma + \tau)^3/2 , \quad g^y = c'(\sigma + \tau)^2 , \end{aligned} \tag{6.2}$$

with c and c' constants of order the inverse size of the loop. It follows (for the generic case that $c \neq c'$) that at $\tau = 0$ the string forms a cusp, $y \propto |x|^{2/3}$.

The instantaneous velocity $(\vec{f}' - \vec{g}')/2$ approaches the speed of light at the cusp. Like the crack of a whip, a great deal of energy is concentrated in the tip, but this whip is perhaps hundreds of light-years long and with a tension not so far below the Planck scale. It thus emits an intense beam of gravitational waves in the direction of its motion [84]. The Fourier transform of a cusp singularity is much larger at high frequency than for a smooth function, so that even when the suppression for being off-axis is included this comes within reach of LIGO.

This is shown in Fig. 4, reproduced from [85]. Under optimistic assumptions (but not, I think, too optimistic), even LIGO I is close to discovery sensitivity of one event per year over much of the range of interesting tensions, including the narrow range of

[10] In making this statement we must consider the relative redshifting of modes at different frequencies. At both the pulsar and the LIGO frequencies the current bounds come from waves emitted during the radiation dominated era, and in this case the effective red shifts are the same due to the scaling property. (One can deduce the time of emission for a current freqency ω by solving $\omega = l^{-1}(t)/(1 + z(t))$ with $l(t)$ the loop size (2.8) and $z(t)$ the redshift.) Lower frequency waves emitted during the matter-dominated era will experience less redshifting.

[11] A kink implies a discontinuity in f' or g' and so a gap in one of the curves. Since cosmic strings have a short distance structure with many kinks, there will be many gaps and this may reduce the number of intersections.

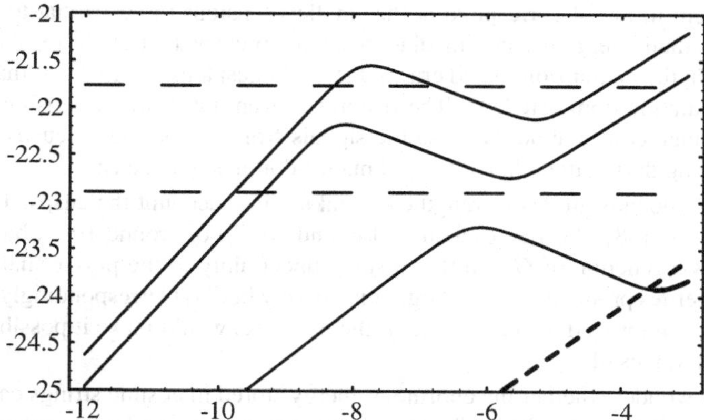

Figure 4. Gravitational wave cusp signals, taken from Damour and Vilenkin [85]. The horizontal axis is $\log_{10}\alpha$ where $\alpha = 50G\mu$. Thus the brane inflation range $10^{-12} \lesssim G\mu \lesssim 10^{-6}$ becomes $-10.3 < \log_{10}\alpha < -4.3$. The vertical axis is $\log_{10} h$ where h is the gravitational strain in the LIGO frequency band. The upper and lower dashed horizontals are the sensitivities of LIGO I and Advanced LIGO at one event per year. The upper two curves are the cusp signal under optimistic and pessimistic network assumptions; the pessimism is that a large number of kinks may suppress the cusps. The lowest solid curve is the signal from kinks, which form whenever strings reconnect. The dashed curve is the stochastic signal.

the KLMT model. This is remarkable: cosmic superstrings might be the brightest objects in gravitational wave astronomy, and the first discovered! LIGO I to date has around 0.1 design-year of data, but it is supposed to begin a new science run in January 2005 at close to design sensitivity and with a good duty cycle. Advanced LIGO is sensitive over almost the whole range, and with a higher event rate; it has a target date of 2009. LISA, a few years later, is even more sensitive. In magnitude of strain it is comparable to LIGO I, but it is looking at a frequency 10,000 times lower and so the typical cusp strains are 1000 times greater [85]. The cusp events might be seen in a search for unmodeled bursts. The shape is not as complex as for stellar and black hole inspirals, but modeling the specific frequency dependence will increase the signal-to-noise ratio. The power-law frequency dependence of the cusp is distinctive. See also [87] for a discussion of the form of the cusp signal.

The dependence of the sensitivity on string tension in Fig. 4 is interesting because it is not monotonic. This comes about as follows [85]. As the string tension decreases, the coupling to gravity becomes weaker and so does the gravitational wave burst from a given cusp. However, since gravitational radiation is the only decay channel for string loops they will live longer.[12] Thus as we decrease $G\mu$ there are more but smaller cusps. With smaller intrinsic events one cannot see as far, but there are more events to see; the competition between these effects depends on the geometry of the universe. Thus, the three regimes

[12]There is another effect as well, the dependence of the short-distance network structure on the gravitational radiation, but we can overlook this for simplicity.

that are evident in Fig. 4 correspond at the smallest tensions to cusps that took place at redshifts less than one, at the intermediate tensions to cusps that are at redshifts greater than one but in the matter-dominated era, and at the largest tensions to cusps that occurred during the radiation-dominated era. The rise in the event rate with decreasing tension in the middle range comes about because the signals from smaller, later, cusps suffer less from redshifting during the relatively rapid matter-dominated expansion.

The pulsar bounds are also strengthed by taking into account the cusps. This is analyzed in detail in [88]; this quotes a current bound on $G\mu$ of around 10^{-7}, but the curve is very flat as a function of $G\mu$ so that a small uncertainty in the pulsar analysis or the network properties produces a much larger uncertainty in $G\mu$. Correspondingly, improved pulsar data and improved understanding of the networks would make it possible to reach much smaller values of $G\mu$.

One may wonder whether the enormous energy stored in cosmic strings can manifest itself in other spectacular effects. The cusps appear to be the best source potential source of high energy particles and radiation, but [89] argues that this is still not observable. Note that short distance physics affects the string only very near the cusp, and even LIGO frequency range is looking on a much longer scale.

7. Distinguishing Superstrings

If gravitational wave cusps or some other signature of cosmic strings are seen, this will be just the beginning of the story. Detailed observations will be able to determine some of the microscopic properties of the strings. For example, it is possible to cleanly distinguish weakly coupled fundamental strings from cosmic strings in perturbative GUTs.

7.1. RECONNECTION

The microscopic structure of the string core does not affect the evolution of strings that are light-years in length, except when two strings cross and their cores interact. We have noted that gauge theory solitons will always reconnect. For F-strings, reconnection is a quantum process, and takes place with a probability P of order g_s^2. The numerical factors are worked out in [90]. To be precise, P is a function of the relative angle and velocity in the collision, but it is simplest to the value averaged over collision parameters.

An important issue is the motion of the string in the compact dimensions. For many supersymmetric compactifications, strings can wander over the whole compact space. Thus they can miss each other, leading to a substantial suppression of P [51, 55]. However, we have noted that in realistic compactifications strings will always be confined by a potential in the compact dimensions. Even if the scale of the potential is low, the fluctuations of the strings are only logarithmic in the ratio of scales — this is characteristic of one-dimensional objects [56]. Thus there is no suppression by powers of the size of the compact dimensions, but the logarithm can be numerically important — it tends to offset powers of π that appear in the numerator. The value of g_s, and the scale of the confining potential, are not known, but in a variety of models [90] finds $10^{-3} \lesssim P \lesssim 1$. For D-D collisions the situation is more complicated, and in the same models one finds $10^{-1} \lesssim P \lesssim 1$. For F-D collisions, P can vary from 0 to 1.

To determine the observational effect of changing P one must feed the given value into the network evolution. A simple argument suggests that the signatures scale as $1/P$: the amount of string in the network must be increased by this factor in order for an increased number of collisions per unit length of string to offset the reduced P in each collision ([88] and references therein). This is a bit oversimplified, because there are issues connected with the sub-horizon scale structure in the string network [16, 91] that can work in either direction, but is supported by simulations as well ([92] and reference therein).[13] If we take this $1/P$ as a model, we see that for the smaller values of P discussed above there can be a substantial increase in the signal even above the encouraging values found in the last section, so that LIGO might soon begin to see *many* cusps. Of course, the existing bounds become stronger.

To first approximation there are two relevant parameters, μ and P. Each individual cusp event has only a single parameter to measure, its overall strength h: because it is a power law there is no characteristic frequency scale. (There is a high frequency cutoff, determined by the alignment of the cusp with the detector [85], but this gives no information about the cusp itself.) After $O(10)$ cusps are seen one can begin to plot a spectrum, $dN \sim Ah^{-B}dh$, and from the two parameters A and B constrain μ and P. There are degeneracies — B depends primarily on the epoch in which the cusp took place — but with a more detailed spectrum, and ultimately with data from kink events and pulsars, this degeneracy will be resolved. Thus μ and P will be overdetermined, and nonstandard network behavior (such as we are about to discuss) will be detectable.

If P is only slightly less than one, say 0.5, then it will require precision simulation of the networks and good statistics on the signatures to distinguish this from 1.0. It should be noted that even with given values of μ and P there are still substantial uncertainties in the current understanding of the behavior of string networks. This has recently been discussed in [88], which concludes that the sensitivities given in Fig. 3 are only weakly dependent on the unknowns.

The discussion of cosmic strings from grand unified theories has focussed on perturbative unification, as suggested by the successful prediction of the Weinberg angle. In this case the only strings are the classical solitons, and $P = 1$ is a fairly robust signature. Thus these can be differentiated from fundamental strings unless we are very unlucky and g_s is very close to 1, or we have a very unusual field theory.[14]

Strongly coupled confining gauge theories can have electric flux tube strings. For these the reconnection probability is of order $1/N^2_{\rm color}$, so finding a small value for P would not rule out a field-theoretic origin entirely, just a perturbative one. We should note that perturbative string theory gives a prediction for the functional dependence on the collision parameters, $P(v, \theta)$ [95, 90]. It would seem very difficult to determine this from observations. If it were possible to map out the string network in detail, through lensing, then perhaps this function might be studied.

[13] Refs. [93, 94] discuss other network evolution issues. We should note that the second of these is concerned with higher-dimensional excitations which, according to our discussion, should be massive and so decoupled except at early times.

[14] One example: a soliton string with a massless internal scalar mode (besides its normal collective coordinates) would be like a string moving in a higher dimensional space and so two such strings could 'miss' each other in field space. However, such modes will always get a mass from symmetry breaking, and I expect that this effect will be significant only in rather contrived models.

7.2. NETWORKS

The second potentially distinguishing feature of the superstring networks is the existence of both F- and D-strings [55, 56, 90], and moreover bound states of p F-strings and q D-strings with the distinctive tension formula (3.8). In this case, when strings of different types collide, rather than reconnecting they form more complicated networks with trilinear vertices. It is then possible that the network does not scale, but gets into a frozen phase where it just stretches with the expansion of the universe [10, 96]. If so, its density would come to dominate at the tensions that we are considering. The F-D networks have not yet been simulated, but simulations of comparable networks suggest that they scale, possibly with an enhanced density of strings [97, 98, 99]. From the discussion above, it follows that one will not directly read off the spectrum (3.8) from the observations, but there should eventually be enough information to distinguish F-D string networks from other types.

Warped models with more complicated throats, as well as models with large compact dimensions, can give rise to a richer spectrum of strings [3, 51]. Still, the (p, q) spectrum (3.8) is worth particular attention: it is an attractive possibility that inflation takes place in a warped throat, and this is the spectrum that arises in the most generic throat. Thus, while the landscape of string theory may be vast, this particular local geometric feature may be common to a large swath of it.

Networks with multiple types of string can also arise in field theories, though I do not know any *perturbative* field theory that gives the particular spectrum (3.8). However, because of duality there will be gauge theory strings that are very hard to differentiate from the F- and D-strings that we are discussing. In particular, in the KLMT model the strings exist in a Klebanov-Strassler throat [2], which has a dual description as a cascading gauge theory, in which there are (p, q) strings.[15]

Indeed, the existence of dualities between string theories and field theories raises the issue, what really is string theory? This is beyond our current scope, but I note that in the present case there is a quantitative question. The Klebanov-Strassler throat has a parameter $g_s M$; when this is large the string description is the valid one, and when it is small the gauge description is the valid one. If we can get enough information about the string network, perhaps combined with information from the CMB, then there might well be enough information to test the hypothesis that we are seeing a Klebanov-Strassler theory, and to measure $g_s M$.

8. Conclusions

As we have seen, each of the four conditions that we discussed at the beginning is independently model-dependent. However, quite a number of things have worked out surprisingly well: the production of strings in brane inflation, the possible stabilization of the strings as a side effect of other properties of the models (in particular, of the stabilization of the vacuum itself), the possibility to see strings over many interesting orders of magnitude of tension, and the existence of properties that distinguish different kinds of string so that after the strings are discovered we can do a lot of science with them. In any case, searching

[15]For discussions of relations between field theory strings and F- and D-strings see [100, 101, 102, 103, 104, 105, 106, 107, 108]; see [109, 110] for some related recent bounds.

for cosmic strings is a tiny marginal cost on top of experiments that will already be done, and it is great that string theorists will have a stake in these experiments over the coming decade and more.

Acknowledgements

I would like to thank E. Copeland, M. Jackson, N. Jones, and R. Myers for collaborations, and N. Arkani-Hamed, L. Bildsten, G. Dvali, A. Filippenko, T. Kibble, A. Linde, A. Lo, A. Lommen, J. Preskill, S. Trivedi, H. Tye, T. Vachaspati, and A. Vilenkin for discussions and communications. This work was supported by National Science Foundation grants PHY99-07949 and PHY00-98395.

References

1. E. Witten, Phys. Lett. **B153**, 243 (1985).
2. N. Jones, H. Stoica, and S. H. H. Tye, JHEP **07**, 051 (2002), hep-th/0203163.
3. S. Sarangi and S. H. H. Tye, Phys. Lett. **B536**, 185 (2002), hep-th/0204074.
4. J. Polchinski, arXiv:hep-th/0410082.
5. A. Vilenkin and E. Shellard, *Cosmic strings and other topological defects* (Cambridge Univ. Press 1994).
6. M. B. Hindmarsh and T. W. B. Kibble, Rept. Prog. Phys. **58**, 477 (1995), hep-ph/9411342.
7. T. W. B. Kibble, arXiv:astro-ph/0410073.
8. A. A. Abrikosov, Sov. Phys. JETP **5**, 1174 (1957).
9. H. B. Nielsen and P. Olesen, Nucl. Phys. **B61**, 45 (1973).
10. T. W. B. Kibble, J. Phys. **A9**, 1387 (1976).
11. E. Witten, Nucl. Phys. **B249**, 557 (1985).
12. E. P. S. Shellard, Nucl. Phys. **B283**, 624 (1987).
13. R. A. Matzner, Computers in Physics **2**, 51 (1988).
14. K. J. M. Moriarty, E. Myers, and C. Rebbi, Phys. Lett. **B207**, 411 (1988).
15. A. Albrecht and N. Turok, Phys. Rev. D **40**, 973 (1989).
16. D. P. Bennett and F. R. Bouchet, Phys. Rev. D **41**, 2408 (1990); PUPT-90-1162, *Based on Invited talks given at Cosmic String Workshop, Cambridge, England, Jul 2-7, 1989.*
17. B. Allen and E. P. S. Shellard, Phys. Rev. Lett. **64**, 119 (1990).
18. X. Siemens, K. D. Olum and A. Vilenkin, Phys. Rev. D **66**, 043501 (2002) [arXiv:gr-qc/0203006].
19. Y. B. Zeldovich and M. Y. Khlopov, Phys. Lett. B **79**, 239 (1978).
20. J. P. Preskill, Phys. Rev. Lett. **43**, 1365 (1979).
21. J. Polchinski, Int. J. Mod. Phys. **A19S1**, 145 (2004), hep-th/0304042.
22. A. H. Guth, Phys. Rev. D **23**, 347 (1981).
23. Y. B. Zeldovich, Mon. Not. Roy. Astron. Soc. **192**, 663 (1980).
24. A. Vilenkin, Phys. Rev. Lett. **46**, 1169 (1981).
25. L. Pogosian, S. H. H. Tye, I. Wasserman, and M. Wyman, Phys. Rev. **D68**, 023506 (2003), hep-th/0304188.
26. L. Pogosian, M. C. Wyman, and I. Wasserman, astro-ph/0403268.
27. E. Jeong and G. F. Smoot, astro-ph/0406432.
28. A. Lo and E. Wright, in preparation (unpublished).
29. V. M. Kaspi, J. H. Taylor and M. F. Ryba, Astrophys. J. **428**, 713 (1994).
30. S. E. Thorsett and R. J. Dewey, Phys. Rev. D **53**, 3468 (1996).
31. M. P. McHugh, G. Zalamansky, F. Vernotte and E. Lantz, Phys. Rev. D **54**, 5993 (1996).
32. A. N. Lommen, astro-ph/0208572.
33. M. Sazhin *et al.*, Mon. Not. Roy. Astron. Soc. **343**, 353 (2003), astro-ph/0302547.
34. M. V. Sazhin *et al.*, (2004), astro-ph/0406516.

35. R. E. Schild, I. S. Masnyak, B. I. Hnatyk, and V. I. Zhdanov, (2004), astro-ph/0406434.
36. J. Polchinski, *String theory. Vol. 2: Superstring theory and beyond* (Cambridge, UK: Univ. Pr., ADDRESS, 1998), p. 531.
37. E. Witten, Nucl. Phys. **B471**, 135 (1996), hep-th/9602070.
38. N. Arkani-Hamed, S. Dimopoulos, and G. R. Dvali, Phys. Lett. **B429**, 263 (1998), hep-ph/9803315.
39. I. Antoniadis, N. Arkani-Hamed, S. Dimopoulos, and G. R. Dvali, Phys. Lett. **B436**, 257 (1998), hep-ph/9804398.
40. L. Randall and R. Sundrum, Phys. Rev. Lett. **83**, 3370 (1999), hep-ph/9905221.
41. S. B. Giddings, S. Kachru, and J. Polchinski, Phys. Rev. **D66**, 106006 (2002), hep-th/0105097.
42. G. R. Dvali and S. H. H. Tye, Phys. Lett. **B450**, 72 (1999), hep-ph/9812483.
43. S. H. S. Alexander, Phys. Rev. **D65**, 023507 (2002), hep-th/0105032.
44. C. P. Burgess *et al.*, JHEP **07**, 047 (2001), hep-th/0105204.
45. G. R. Dvali, Q. Shafi, and S. Solganik, hep-th/0105203.
46. S. Kachru, R. Kallosh, A. Linde and S. P. Trivedi, Phys. Rev. D **68**, 046005 (2003) [arXiv:hep-th/0301240].
47. E. Silverstein, arXiv:hep-th/0405068.
48. S. Kachru *et al.*, JCAP **0310**, 013 (2003), hep-th/0308055.
49. I. R. Klebanov and M. J. Strassler, JHEP **08**, 052 (2000), hep-th/0007191.
50. P. Candelas and X. C. de la Ossa, Nucl. Phys. B **342**, 246 (1990).
51. N. T. Jones, H. Stoica, and S. H. H. Tye, Phys. Lett. **B563**, 6 (2003), hep-th/0303269.
52. R. Jeannerot, J. Rocher and M. Sakellariadou, Phys. Rev. D **68**, 103514 (2003) [arXiv:hep-ph/0308134].
53. A. Sen, JHEP **09**, 023 (1998), hep-th/9808141.
54. E. Witten, JHEP **12**, 019 (1998), hep-th/9810188.
55. G. Dvali and A. Vilenkin, JCAP **0403**, 010 (2004), hep-th/0312007.
56. E. J. Copeland, R. C. Myers, and J. Polchinski, JHEP **06**, 013 (2004), hep-th/0312067.
57. N. Barnaby, A. Berndsen, J. M. Cline and H. Stoica, arXiv:hep-th/0412095.
58. N. Iizuka and S. P. Trivedi, Phys. Rev. D **70**, 043519 (2004) [arXiv:hep-th/0403203].
59. A. D. Linde, Phys. Rev. **D49**, 748 (1994), astro-ph/9307002.
60. J. Yokoyama, Phys. Rev. Lett. **63**, 712 (1989).
61. L. Kofman, A. D. Linde, and A. A. Starobinsky, Phys. Rev. Lett. **76**, 1011 (1996), hep-th/9510119.
62. I. Tkachev, S. Khlebnikov, L. Kofman, and A. D. Linde, Phys. Lett. **B440**, 262 (1998), hep-ph/9805209.
63. J. Urrestilla, A. Achucarro, and A. C. Davis, Phys. Rev. Lett. **92**, 251302 (2004), hep-th/0402032.
64. T. Watari and T. Yanagida, Phys. Lett. **B589**, 71 (2004), hep-ph/0402125.
65. K. Dasgupta *et al.*, JHEP **08**, 030 (2004), hep-th/0405247.
66. E. Halyo, hep-th/0402155.
67. T. Matsuda, Phys. Rev. **D70**, 023502 (2004), hep-ph/0403092.
68. T. Matsuda, hep-ph/0406064.
69. F. Englert, J. Orloff and T. Piran, Phys. Lett. B **212**, 423 (1988).
70. S. W. Hawking and D. N. Page, Commun. Math. Phys. **87**, 577 (1983).
71. E. Witten, Adv. Theor. Math. Phys. **2**, 505 (1998) [arXiv:hep-th/9803131].
72. J. Preskill and A. Vilenkin, Phys. Rev. **D47**, 2324 (1993), hep-ph/9209210.
73. T. Banks and L. J. Dixon, Nucl. Phys. **B307**, 93 (1988).
74. A. Vilenkin and A. E. Everett, Phys. Rev. Lett. **48**, 1867 (1982).
75. S. Coleman, *Aspects of Symmetry* (Cambridge Univ. Press 1988).
76. C. G. . Callan, J. A. Harvey and A. Strominger, Nucl. Phys. B **359**, 611 (1991).
77. L. J. Dixon, V. Kaplunovsky and C. Vafa, Nucl. Phys. B **294**, 43 (1987).
78. R. Bousso and J. Polchinski, JHEP **06**, 006 (2000), hep-th/0004134.
79. M. R. Douglas, JHEP **05**, 046 (2003), hep-th/0303194.
80. S. Ashok and M. R. Douglas, JHEP **01**, 060 (2004), hep-th/0307049.
81. N. Barnaby, C. P. Burgess and J. M. Cline, arXiv:hep-th/0412040.

82. J. Polchinski and E. Witten, Nucl. Phys. B **460**, 525 (1996) [arXiv:hep-th/9510169].
83. L. Leblond and S. H. H. Tye, JHEP **03**, 055 (2004), hep-th/0402072.
84. T. Damour and A. Vilenkin, Phys. Rev. Lett. **85**, 3761 (2000), gr-qc/0004075.
85. T. Damour and A. Vilenkin, Phys. Rev. **D64**, 064008 (2001), gr-qc/0104026.
86. N. Turok, Nucl. Phys. **B242**, 520 (1984).
87. X. Siemens and K. D. Olum, Phys. Rev. **D68**, 085017 (2003), gr-qc/0307113.
88. T. Damour and A. Vilenkin, arXiv:hep-th/0410222.
89. J. J. Blanco-Pillado and K. D. Olum, Phys. Rev. D **59**, 063508 (1999) [arXiv:gr-qc/9810005].
90. M. G. Jackson, N. T. Jones, and J. Polchinski, (2004), hep-th/0405229.
91. D. Austin, E. J. Copeland, and T. W. B. Kibble, Phys. Rev. **D48**, 5594 (1993), hep-ph/9307325.
92. M. Sakellariadou, arXiv:hep-th/0410234.
93. C. J. A. Martins, Phys. Rev. D **70**, 107302 (2004) [arXiv:hep-ph/0410326].
94. A. Avgoustidis and E. P. S. Shellard, arXiv:hep-ph/0410349.
95. J. Polchinski, Phys. Lett. **B209**, 252 (1988).
96. A. Vilenkin, Phys. Rev. Lett. **53**, 1016 (1984).
97. T. Vachaspati and A. Vilenkin, Phys. Rev. **D35**, 1131 (1987).
98. D. Spergel and U.-L. Pen, Astrophys. J. **491**, L67 (1997), astro-ph/9611198.
99. P. McGraw, Phys. Rev. **D57**, 3317 (1998), astro-ph/9706182.
100. K. Becker, M. Becker and A. Strominger, Phys. Rev. D **51**, 6603 (1995) [arXiv:hep-th/9502107].
101. J. D. Edelstein, C. Nunez and F. A. Schaposnik, Nucl. Phys. B **458**, 165 (1996) [arXiv:hep-th/9506147].
102. G. Dvali, R. Kallosh, and A. Van Proeyen, JHEP **01**, 035 (2004), hep-th/0312005.
103. E. Halyo, JHEP **03**, 047 (2004), hep-th/0312268.
104. P. Binetruy, G. Dvali, R. Kallosh, and A. Van Proeyen, Class. Quant. Grav. **21**, 3137 (2004), hep-th/0402046.
105. S. S. Gubser, C. P. Herzog, and I. R. Klebanov, (2004), hep-th/0405282.
106. A. Achucarro and J. Urrestilla, JHEP **08**, 050 (2004), hep-th/0407193.
107. S. S. Gubser, C. P. Herzog, and I. R. Klebanov, (2004), hep-th/0409186.
108. A. Lawrence and J. McGreevy, (2004), hep-th/0409284.
109. J. Rocher and M. Sakellariadou, arXiv:hep-ph/0405133.
110. J. Rocher and M. Sakellariadou, arXiv:hep-ph/0412143.

LONG TIME SCALES AND ETERNAL BLACK HOLES

JOSE L. F. BARBÓN
Instituto de Física Teórica IFT UAM/CSIC
Facultad de Ciencias C-XVI
C.U. Cantoblanco, E-28049 Madrid, Spain

AND

ELIEZER RABINOVICI
Racah Institute of Physics, The Hebrew University
Jerusalem 91904, Israel

Abstract. We discuss the various scales determining the temporal behaviour of correlation functions in the presence of eternal black holes. We point out the origins of the failure of the semiclassical gravity approximation to respect a unitarity-based bound suggested by Maldacena. We find that the presence of a subleading (in the large-N approximation involved) master field does restore the compliance with one bound but additional configurations are needed to explain the more detailed expected time dependence of the Poincaré recurrences and their magnitude.

1. Introduction

Hawking's semiclassical analysis of black hole evaporation suggests that most of the information contained in initial scattering states is shielded behind the event horizon, never to return back to the asymptotic region far from the evaporating black hole [1]. In this picture, the singularity is capable of absorbing all the infalling information, which is then destroyed or transmitted to other geometrical realms, depending on one's hypotheses about the microphysics of the singularity. From the point of view of measurements on the Hawking radiation, the evaporation is not described by a unitary S-matrix. Rather, quantum coherence is violated and the linear evolution in Hilbert space takes pure states into mixed states. Still, probability is conserved, since density matrices ρ remain Hermitian, $\rho^\dagger = \rho$, positive, $\rho > 0$ and normalized, $\text{Tr}\rho = 1$ under time evolution.

The AdS/CFT correspondence [2] is not consistent with this picture. In this construction, quantum gravity in a $(d+1)$-dimensional asymptotically Anti-de Sitter spacetime (AdS) of curvature radius R is *defined* in terms of a conformal field theory (CFT) on a spatial sphere $\mathbf{S}^{d-1}$ of radius R. The effective expansion parameter in the gravity side $1/N^2 \sim G_{\rm N}/R^{d-1}$, maps to an appropriate large N limit of the CFT. For example, for

L. Baulied et al. (eds.), String Theory: From Gauge Interactions to Cosmology, 255–263.

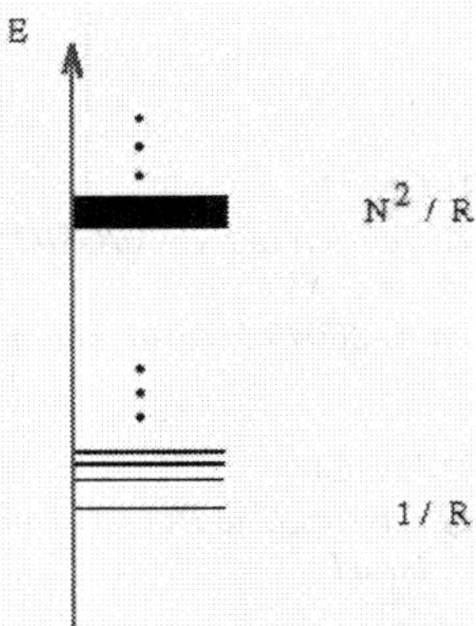

Figure 1. The energy spectrum of a CFT representing AdS_{d+1} quantum gravity. The spectrum is discrete on a sphere of radius R, with gap of order $1/R$. The asymptotic energy band of very dense "black hole" states sets in beyond energies of order N^2/R. The corresponding density of states is that of a conformal fixed point in d spacetime dimensions.

two-dimensional CFT's N^2 is the central charge. When the CFT is a gauge theory, the AdS side is a string theory, N is the rank of the gauge group, and the string perturbative expansion in powers of $g_s \sim 1/N$ is identified with 't Hooft's $1/N$ expansion in the gauge theory side.

According to this definition, the formation and evaporation of small black holes with Schwarschild radius $R_S \ll R$, should be described by a unitary process in terms of the CFT Hamiltonian. Thus, there is no room for violations of coherence as a matter of principle. Unfortunately, the CFT states corresponding to small black holes are hard to describe, and it remains a challenge to put the finger on the precise error in Hawking's semiclassical analysis in that case.

For large AdS black holes with Schwarschild radius $R_S \gg R$ one may attempt to rise to the challenge, since they are thermodynamically stable and can exist in equilibrium at fixed (high) temperatures $1/\beta \gg 1/R$. Indeed, the corresponding Bekenstein–Hawking entropy scales like that of N^2 conformal degrees of freedom at high energy,

$$S \sim \sqrt{N}\,(E\,R)^{\frac{d-1}{d}} \sim N^2\,(R/\beta)^{d-1}\,. \tag{1.1}$$

Therefore, large AdS black holes with inverse Hawking temperature $\beta \ll R$ describe the leading approximation to the thermodynamical functions of the canonical CFT state

$$\rho_\beta = \frac{e^{-\beta H}}{Z(\beta)}\,, \qquad Z(\beta) = \mathrm{Tr}\,\exp(-\beta H)\,. \tag{1.2}$$

This suggests that we can test the semiclassical unitarity argument by careful analysis of slight departures from thermal equilibrium, rather than studying a complete evaporation instability in the vacuum. Ref. [3] proposes to look at the very long time structure of correlators of the form

$$G(t) = \mathrm{Tr}\,[\,\rho\,A(t)\,A(0)\,]\,, \tag{1.3}$$

for appropriate Hermitian operators A. In the semiclassical approximation, one expects these correlators to decay as $\exp(-\Gamma\, t)$ with $\Gamma \sim \beta^{-1}$. However, because the CFT lives in finite volume, the spectrum is actually discrete (c.f. Fig 1), and the correlator must show nontrivial long time structure in the form of Poincaré recurrences (see [4, ?]). This result, which is straightforward from the boundary theory point of view, has far reaching consequences as far as the bulk physics is concerned.

Hence, the failure of $G(t)$ to vanish as $t \to \infty$ can be used as a criterion for unitarity preservation beyond the semiclassical approximation. This argument can be made more explicit by checking the effect of coherence loss on the long-time behaviour of $G(t)$. Using the results of [6] one can simulate the required decoherence by coupling an ordinary quantum mechanical system to a random classical noise. It is then shown in [7] that this random noise forces $G(t)$ to decay exponentially for large t, despite having a discrete energy spectrum. This shows that the long-time behaviour of correlators probes the strict quantum coherence of the bounded system.

At the same time, one would like to identify what kind of systematic corrections to the leading semiclassical approximation are capable of restoring unitarity. A proposal was made in [3] in terms of topology-changing fluctuations of the AdS background. Our purpose here is to investigate these questions and offer an explicit estimate of the instanton effects suggested in [3] (see also [8]). Ultimately, this analysis should provide information about the nature of the black hole singularity.

2. Long-time details of thermal quasi-equilibrium

Poincaré recurrences occur in general bounded systems. Classically they follow from the compactness of available phase space, plus the preservation of the phase-space volume in time (Liouville's theorem). Quantum mechanically, they follow from discreteness of the energy spectrum (characteristic of spatially bounded systems) and unitarity, since

$$G_\beta(t) = \frac{1}{Z(\beta)} \sum_{i,j} e^{-\beta E_i} \, |A_{ij}|^2 \, e^{i(E_i - E_j)t} \tag{2.1}$$

defines a quasiperiodic function of time (we have chosen the canonical density matrix for the initial state). After initial dissipation on a time scale Γ^{-1}, where Γ measures the approximate width of matrix elements of A in the energy basis, the correlator will show $O(1)$ "resurgences" when most of the relevant phases complete a period (c.f. Fig 2). The associated time scale is $t_H \equiv 1/\langle\omega\rangle$, with $\langle\omega\rangle = \langle E_i - E_j\rangle$ an average frequency in (2.1). We can estimate $\langle\omega\rangle$ as $\Gamma/\Delta n_\Gamma$, where Δn_Γ is the number of energy levels in the relevant band of width Γ. Introducing the microcanonical entropy in terms of the level-number function as $n(E) \equiv \exp S(E)$, we have

$$\Delta n_\Gamma \approx \int_{\langle E\rangle - \Gamma/2}^{\langle E\rangle + \Gamma/2} dE \, \frac{dn}{dE} = \int_{\langle E\rangle - \Gamma/2}^{\langle E\rangle + \Gamma/2} dE \, \beta(E) \, e^{S(E)} \approx \Gamma \, \beta \, e^{S(\beta)} \,. \tag{2.2}$$

where we have introduced the microcanonical inverse temperature as $\beta(E) \equiv dS/dE$. From this analysis we obtain an estimate

$$t_H \sim \beta \, e^{S(\beta)} \,. \tag{2.3}$$

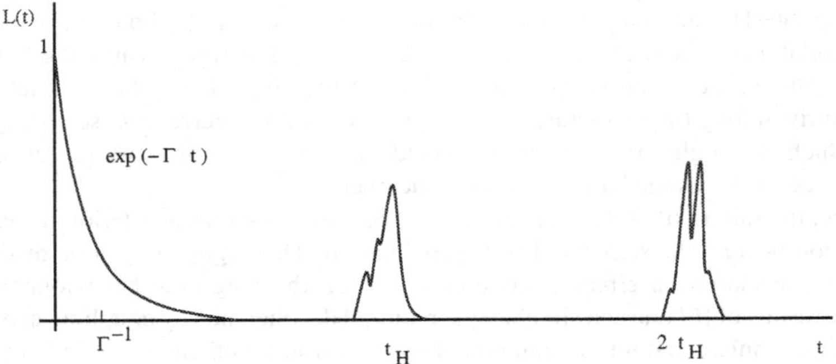

Figure 2. Schematic representation of the very long time behaviour of the normalized time correlator $L(t)$ in bounded systems. The initial decay with lifetime of order Γ^{-1} is followed by O(1) "resurgences" after the Heisenberg time $t_H \sim \beta \, \exp(S)$ has elapsed. Poincaré recurrence times can be defined by demanding the resurgences to approach unity with a given *a* priori accuracy, and scale like a double exponential of the entropy.

Following [9] we call this the Heisenberg time scale. Poincaré times can be defined in terms of quasiperiodic returns of $G_\beta(t)$ with a given *a priori* accuracy. In a sense, the Heisenberg time is the smallest possible Poincaré time.

A more quantitative criterion can be used by defining a normalized positive correlator, $L(t)$, satisfying $L(0) = 1$, and its infinite time average,

$$L(t) \equiv \left| \frac{G(t)}{G(0)} \right|^2 , \qquad \overline{L} \equiv \lim_{T\to\infty} \frac{1}{T} \int_0^T dt \, L(t) \, . \tag{2.4}$$

The profile of $L(t)$ is sketched in Fig 2. The time average can be estimated by noticing that the graph of $L(t)$ features positive "bumps" of height ΔL and width Γ, separated a time t_H, so that

$$\overline{L} \sim \frac{\Delta L}{\Gamma \, t_H} \, . \tag{2.5}$$

For the case at hand $\Delta L \sim 1$, $t_H \sim \beta \, e^S$, and we obtain (c.f. [5, 7])

$$\overline{L} \sim \frac{e^{-S(\beta)}}{\beta \Gamma} \, . \tag{2.6}$$

Since both β and Γ scale as N^0 in the large-N limit of the dual CFT, the "recurrence index" $\overline{L} \sim \exp(-N^2)$ scales as a nonperturbative effect in the semiclassical approximation.

Indeed, one finds $\overline{L} = 0$ in gravity perturbation theory in the AdS black hole background. The reason is that the relevant eigenfrequencies ω (the so-called normal modes of the black hole) form a continuous spectrum to all orders in the $1/N$ expansion. For a static metric of the form

$$ds^2 = -g(r) \, dt^2 + \frac{dr^2}{g(r)} + r^2 \, d\Omega_{d-2}^2 \, , \tag{2.7}$$

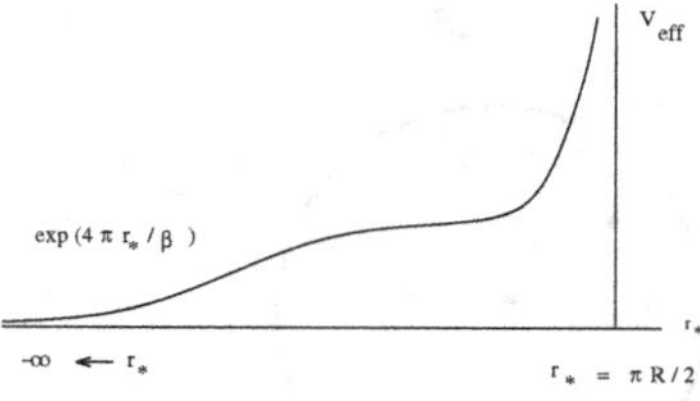

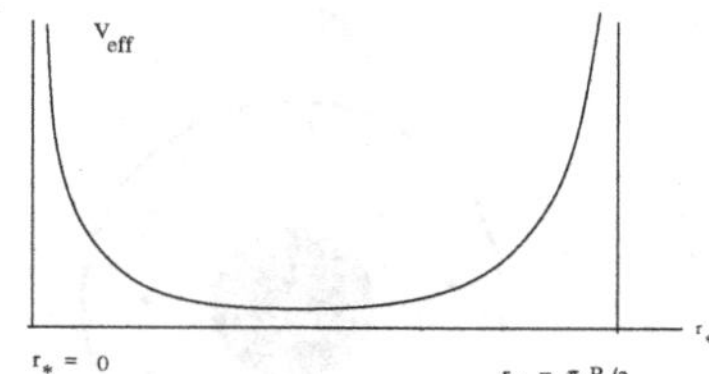

Figure 3. The effective potential determining the semiclassical normal frequency modes in a large AdS black hole background (left). In Regge–Wheeler coordinates the horizon is at $r_* = -\infty$, whereas the boundary of AdS is at $r_* = \pi R/2$ (only the region exterior to the horizon appears). There is a universal exponential behaviour in the near-horizon (Rindler) region. The effective one-dimensional Schrödinger problem represents a semi-infinite barrier with a continuous energy spectrum. This contrasts with the analogous effective potential in vacuum AdS with global coordinates (right). The domain of r_* is compact and the spectrum of normal modes is discrete with gap of order $1/R$.

the normal frequency spectrum follows from the diagonalization of a radial Schrödinger operator

$$\omega^2 = -\frac{d^2}{dr_*^2} + V_{\text{eff}}(r_*) \,, \tag{2.8}$$

with

$$V_{\text{eff}} = \frac{d-2}{2}\, g(r) \left(\frac{g'(r)}{r} + \frac{d-4}{2r^2}\, g(r) \right) + g(r) \left(-\frac{\nabla_\Omega^2}{r^2} + m^2 \right) \tag{2.9}$$

for a scalar field of mass m (analogous potentials can be deduced for higher spin fields). Here we have defined the Regge–Wheeler or "tortoise" coordinate $dr_* = dr/g(r)$. We have shown in Fig. 3 the form of the resulting effective potentials for large AdS black holes, compared with the case of the vacuum AdS manifold. The vacuum AdS manifold, corresponding to the choice $g(r) = 1 + r^2/R^2$ in (2.7), behaves like a finite cavity, as expected. The distinguishing feature of the black-hole horizon is a a non-degenerate zero, $g(r_0) = 0$, which induces the universal scaling

$$V_{\text{eff}}(r_*) \propto \exp(4\pi r_*/\beta) \quad \text{as} \quad r_* \to -\infty \,, \tag{2.10}$$

with $1/\beta = g'(r_0)/4\pi$ the Hawking temperature and the horizon $r = r_0$ appearing at $r_* = -\infty$. The spectrum is discrete in pure AdS, and continuous in the AdS black hole. Physically, this just reflects the fact that the horizon is an infinite redshift surface, so that we can store an arbitrary number of modes with finite total energy, provided they are sufficiently red-shifted by approaching the horizon [10]. Since the thermal entropy of perturbative gravity excitations in the vacuum AdS spacetime scales as $S(\beta)_{\text{AdS}} \sim N^0$, we see that the perturbative Heisenberg time of the AdS spacetime is of $O(1)$ in the large-N limit, leading to $\overline{L}_{\text{AdS}} = O(1)$. On the other hand, we have $\overline{L}_{\text{bh}} = 0$ in this approximation.

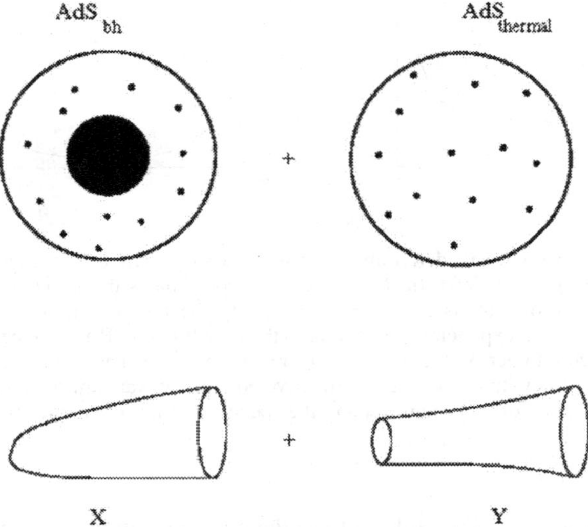

Figure 4. Summing over large-scale fluctuations of the thermal ensemble in which a black hole spontaneously turns into radiation (and viceversa) is represented in the Euclidean formalism as the coherent sum of thermal saddle points of different topology. The "cigar-like" geometry X represents the black-hole master field (in the CFT language) and the cylindrical topology Y represents the thermal gas of particles.

3. Topological diversity and unitarity

It is instructive to understand these perturbative results in the Euclidean formalism, obtained by $t = -i\tau$ in (2.7), followed by an identification $\tau \equiv \tau + \beta$. The resulting metric for the vacuum AdS spacetime has a non-contractible $\mathbf{S}^1$ given by the τ compact direction. We call Y this Euclidean manifold. On the other hand, the black hole spacetime with $g(r_0) = 0$ has different topology, since the thermal $\mathbf{S}^1$ shrinks to zero size at $r = r_0$. The choice $1/\beta = g'(r_0)/4\pi$ ensures smoothness at $r = r_0$. We call this Euclidean black hole manifold X.

The real-time correlation functions in the black hole background, $G(t)_X$, follow by analytic continuation from their Euclidean counterparts. Since X is a completely smooth manifold in the $1/N$ expansion, so is the Euclidean correlator $G(it)_X$ for $t \neq 0$. The continuous spectrum arising in the spectral decomposition of $G(t)_X$ is a consequence of the contractible topology of X, since the Hamiltonian folliation by τ = constant surfaces is singular at $r = r_0$. Therefore, it seems that improving on the semiclassical prediction for $\overline{L}$ requires some sort of topology-change process. The proposal of [3] is precisely that: instead of evaluating the semiclassical correlators on X, one should sum coherently the contribution of X and Y. Normally one neglects the contribution of Y on a quantitative basis (at high temperatures $R \gg \beta$). However, here the contribution of X to $\overline{L}$ vanishes and one is forced to consider the first correction. Since Y has a discrete spectrum in perturbation theory, the net result for $\overline{L}$ should be non-vanishing in

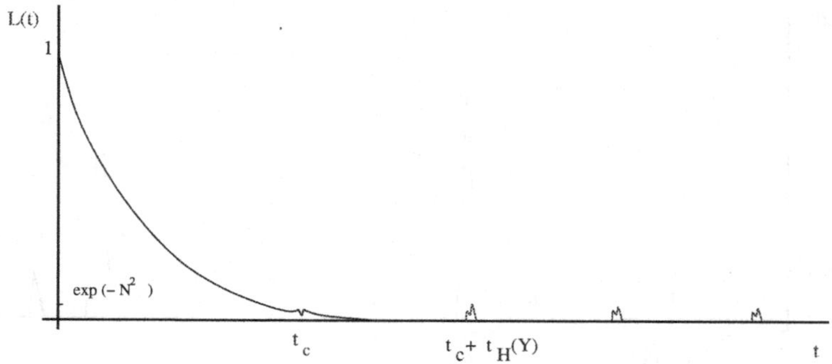

Figure 5. The instanton approximation to the correlator $L(t)_{\rm inst}$ features the expected exponential decay $\exp(-\Gamma\, t)$ induced by the contribution of the X-manifold, whereas the resurgences are entirely due to the interference with the Y-manifold, leading to small bumps of order $\exp(-2\Delta I) \sim \exp(-N^2)$, separated a time $t_H(Y) \sim N^0$. These bumps are noticeable against the background of the X-manifold after a time $t_c \sim \Delta I/\Gamma$.

this approximation. Physically, this superposition of Euclidean saddle points (or master fields, in the language of the CFT) corresponds to large-scale fluctuations in which the AdS black hole is converted into a graviton gas at the same temperature and viceversa.

A more detailed estimate of this "instanton" approximation to $\overline{L}$ yields (c.f. [7])

$$\overline{L}_{\rm inst} \approx C\, e^{-2\Delta I}\ , \tag{3.1}$$

where $C = O(N^0)$, $\Delta I = I_Y - I_X$ and $I = -\log\, Z(\beta)$, calculated in the classical gravity approximation. Since $I_Y \sim -N^0$ and $I_X \sim -N^2$, the exponential suppression factor is of order $\exp(-2|I_X|) \sim \exp(-N^2)$, reproducing the expected scaling (2.6), at least in order of magnitude (however, in general $S_X \neq -2|I_X|$, even at high temperature).

However, the apparent success of (3.1) turns out to be somewhat coincidental. If we consider the full time profile of $L(t)$ rather than the infinite time average, we find

$$L(t)_{\rm inst} \approx L(t)_X + C\, e^{-2\Delta I}\ L(t)_Y\ . \tag{3.2}$$

The resulting structure is shown in Figs. 5 and 6. The instanton approximation to the normalized correlator features the normal dissipation with lifetime $\Gamma^{-1} \sim \beta$ coming from the contribution of X. However, the resurgences are controlled by $L(t)_Y$, damped by a factor $\exp(-2\Delta I) \sim \exp(-N^2)$, and separated a time $t_H(Y) \sim N^0$. Hence, the very long time behaviour as shown in Fig. 6 is very different from the expected one, although the infinite time average comes out right in order of magnitude:

$$\overline{L} \sim \frac{\Delta L}{\Gamma\, t_H} \sim \frac{e^{-N^2}}{\Gamma\cdot\beta} \sim \frac{1}{\Gamma\cdot\beta\, e^{N^2}}\ . \tag{3.3}$$

We can also find the time scale t_c for which the large-scale instantons considered here are quantitatively important on the graph of $L(t)$. This is shown in Fig. 5 and yields $t_c \sim \Delta I/\Gamma \sim N^2$.

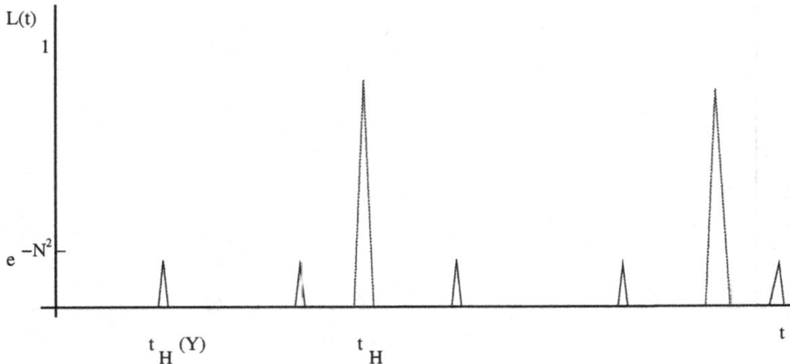

Figure 6. Schematic representation of the very long time behaviour of $L(t)_{\rm inst}$ (dark line) compared to the expected pattern for the exact quantity $L(t)$. The resurgences of $L(t)_{\rm inst}$ occur with periods of order $t_H(Y) = O(N^0)$ and have amplitude of order $\exp(-N^2) \ll 1$. The expectations for the exact CFT, in the dashed line, are $O(1)$ resurgences with a much larger period $t_H \sim \exp(N^2) \gg t_H(Y)$, corresponding to tiny energy spacings of order $\exp(-N^2)$. Despite the gross difference of both profiles, the infinite time average is $O(e^{-N^2})$ for both of them.

A more complicated set of Euclidean saddle points can be analyzed for the three-dimensional case of BTZ black holes. The authors of [11] conclude that resummation of an infinite family of $SL(2, \mathbf{Z})$ saddle points is unlikely to alter the conclusions presented here on the basis of the leading instanton approximation. They also point out that the semiclassical approximation breaks down for times longer than $t_c \sim c$, where c is the central charge of the dual two-dimensional CFT.

4. Conclusions

The study of very long time features of correlators in black hole backgrounds is a potentially important approach towards unraveling the mysteries of black hole evaporation and the associated physics at the spacelike singularity. We have seen that large scale topology-changing fluctuations proposed in [3] begin to restore some of the fine structure required by unitarity, but fall short at the quantitative level. Presumably the appropriate instantons occur on microscopic scales and involve stringy dynamics. While semiclassical black holes do faithfully reproduce "coarse grained" inclusive properties of the system such as the entropy (c.f. [12]), additional dynamical features of the horizon may be necessary to resolve finer details of the information loss problem. Roughly, one needs a systematic set of corrections that could generate a "stretched horizon" of Planckian thickness [13]. The crudest model of such stretched horizon is the brick-wall model of 't Hooft [10]. In this phenomenological description one replaces the horizon by a reflecting boundary condition at Planck distance $\epsilon \sim \ell_P$ from the horizon. This defines a "mutilated" X_ϵ manifold, of cylindrical topology, leading to a discrete spectrum of the right spacing in order of magnitude.

We have also seen that the characteristic time for large topological fluctuations to be important is $t_c \sim O(N^2)$ in the semiclassical approximation. In [14] it was argued that semiclassical two-point functions probe the black hole singularity on much shorter characteristic times, thereby justifying the analysis on the single standard black hole manifold. However, we have seen that detailed unitarity is only restored on time scales of order $t_H \sim \exp(N^2)$. Thus $t_c \ll t_H$ and we conclude that such semiclassical analysis of the singularity is bound to be incomplete, as it misses whatever microphysics is responsible for the detailed unitarity restoration in the quantum mechanical time evolution.

Acknowledgements

E. R. would like to thank the KITP at Santa Barbara for hospitality during the completion of this work, under grant of the National Science Foundation No. PHY99-07949. The work of J.L.F.B. was partially supported by MCyT and FEDER under grant BFM2002-03881 and the European RTN network HPRN-CT-2002-00325. The work of E.R. is supported in part by the Miller Foundation, the BSF-American Israeli Bi-National Science Foundation, The Israel Science Foundation-Centers of Excellence Program, The German-Israel Bi-National Science Foundation and the European RTN network HPRN-CT-2000-00122.

References

1. S.W. Hawking, *Phys. Rev.* **D14** (1976) 2460.
2. J. Maldacena, *Adv. Theor. Math. Phys.* **2** (1998) 231 hep-th/9711200. S.S. Gubser, I.R. Klebanov and A.M. Polyakov, *Phys. Lett.* **B428** (1998) 105 hep-th/9802109. E. Witten, *Adv. Theor. Math. Phys.* **2** (1998) 253 hep-th/9802150.
3. J. Maldacena, *J. High Energy Phys.* **0304** (2003) 021 hep-th/0106112.
4. L Susskind, Contribution to Stephen Hawking's 60th birthday celebration. hep-th/0204027. L. Dyson, J. Lindesay and L. Susskind, *J. High Energy Phys.* **0208** (2002) 045 hep-th/0202163. N. Goheer, M. Kleban and L. Susskind, *J. High Energy Phys.* **0307** (2003) 056 hep-th/0212209.
5. L. Dyson, M. Kleban and L. Susskind, *J. High Energy Phys.* **0210** (2002) 011 hep-th/0208013.
6. T. Banks, L. Susskind and M. E. Peskin, *Nucl. Phys.* **B244** (1984) 125.
7. J.L.F. Barbón and E. Rabinovici, *J. High Energy Phys.* **0311** (2003) 047 hep-th/0308063.
8. D. Birmingham, I. Sachs, S. N. Solodukhin, *Phys. Rev.* **D67** (2003) 104026 hep-th/0212308. I. Sachs, hep-th/0312287.
9. M. Srednicki, cond-mat/9809360.
10. G. 't Hooft, *Nucl. Phys.* **B256** (1985) 727.
11. M. Kleban, M. Porrati and R. Rabadan, *J. High Energy Phys.* **0410** (2004) 030 hep-th/0407192.
12. G.W. Gibbons and S.W. Hawking, *Phys. Rev.* **D15** (1977) 2752.
13. K.S. Thorne, R.H Price and D.A. McDonald, *"Black Holes: The Membrane Paradigm"*. Yale University Press 1986. L. Susskind, L. Thorlacius and J. Uglum, *Phys. Rev.* **D48** (1993) 3743 hep-th/9306069.
14. P. Kraus, H. Ooguri and S. Shenker, *Phys. Rev.* **D67** (2003) 124022 hep-th/0212277. L. Fidkowski, V. Hubeny, M. Kleban and S. Shenker, hep-th/0306170.

SEMICLASSICAL STRINGS AND ADS/CFT

ARKADY A. TSEYTLIN
Physics Department, The Ohio State University,
Columbus OH43210-1106, USA and
at Imperial College, London and Lebedev Physics Institute, Moscow

Abstract. We discuss AdS/CFT duality in the sector of "semiclassical" string states with large quantum numbers. We review the coherent-state effective action approach, in which similar 2d sigma model actions appear from the $AdS_5 \times S^5$ string action and from the integrable spin chain Hamiltonian representing the N=4 super Yang-Mills dilatation operator. We consider mostly the leading-order terms in the energies/anomalous dimensions which match but comment also on higher-order corrections.

1. Introduction

The $\mathcal{N} = 4$ SYM theory with $SU(N)$ gauge group is a family of conformal theories parametrized by the two numbers – N and $g_{\rm YM}$. Four-dimensional conformal theories have apparently much less symmetry than their two-dimensional cousins and thus should be much harder to solve (i.e. to determine their spectrum of dimensions of conformal primary operators and their correlation functions). There are strong indications that this problem may simplify in the planar ($N \to \infty$, $\lambda \equiv g^2_{\rm YM} N$=fixed) limit. In the large N limit there are no formal objections to integrability of a 4-d quantum field theory, and, indeed, the AdS/CFT duality [1, 2, 3] implies the existence of a hidden integrable 2-d structure corresponding to that of $AdS_5 \times S^5$ string sigma model (on a sphere or a cylinder). A major first step towards the solution of the SYM theory would be to determine the spectrum of anomalous dimensions $\Delta(\lambda)$ of the primary operators built out of products of local gauge-covariant fields.

The AdS/CFT duality implies the equality between the AdS_5 energies E of quantum closed string states (as functions of the effective string tension $T = \frac{R^2}{2\pi} = \frac{\sqrt{\lambda}}{2\pi}$ and quantum numbers like S^5 angular momenta J_i) and the dimensions Δ of the corresponding local SYM operators (see, e.g., [4]). To check the duality one would like also to understand how strings "emerge" from the field theory, in particular, which (local, single-trace) gauge theory operators [5] correspond to which "excited" string states and how one may verify the matching of their dimensions/energies beyond the well-understood BPS/supergravity sector. One would then like to use the duality as a guide to deeper level of the structure

L. Baulied et al. (eds.), String Theory: From Gauge Interactions to Cosmology, 265–290.

of quantum SYM theory. For example, the results motivated by comparison to string theory may allow one to "guess" the general structure of the SYM anomalous dimension matrix and may also suggest new methods of computing anomalous dimensions in less supersymmetric gauge theories.

Below we shall review some recent progress in checking AdS/CFT correspondence in a subsector of non-BPS string/SYM states with large quantum numbers.

1.1. GENERALITIES

Let us start with brief remarks on the SYM and the string sides of the duality. The SYM theory contains a gauge field, 6 scalars ϕ_m and 4 Weyl fermions, all in adjoint representation of $SU(N)$. It has global conformal and R-symmetry, i.e. is invariant under $SO(2,4) \times SO(6)$. To determine (in the planar limit) scaling the dimensions of local gauge-invariant operators one, in general, needs to find the anomalous dimension matrix to all orders in λ and then to diagonalize it. The special case is that of chiral primary or BPS operators (and their descendants) $\mathrm{tr}(\phi_{\{m_1}...\phi_{m_k\}})$ whose dimensions are protected, i.e. do not depend on λ. The problem of finding dimensions appears to simplify also in the case of "long" operators containing large number of fields under the trace. One example is provided by "near-BPS" operators [6] like $\mathrm{tr}(\Phi_1^J \Phi_2^n...) + ...$, where $J \gg n$, and $\Phi_k = \phi_k + i\phi_{k+3}$, $k = 1, 2, 3$. Below we shall consider "far-from-BPS" operators like $\mathrm{tr}(\Phi_1^{J_1} \Phi_2^{J_2}...) + ...$, with $J_1 \sim J_2 \gg 1$.

The type IIB string action in $AdS_5 \times S^5$ space has the following structure

$$I = -\frac{1}{2}T \int d\tau \int_0^{2\pi} d\sigma \, (\partial^p Y^\mu \partial_p Y^\nu \eta_{\mu\nu} + \partial^p X^m \partial_p X^n \delta_{mn} + ...) \ , \qquad (1.1)$$

where $Y^\mu Y^\nu \eta_{\mu\nu} = -1, \ \ X^m X^n \delta_{mn} = 1 \ , \ \eta_{\mu\nu} = (-++++-)$, $T = \frac{\sqrt{\lambda}}{2\pi}$ and dots stand for the fermionic terms [7] that ensure that this model defines a 2-d conformal field theory. The closed string states can be classified by the values of the Cartan charges of the obvious symmetry group $SO(2,4) \times SO(6) - E, S_1, S_2; J_1, J_2, J_3$, i.e. by the AdS_5 energy, two spins in AdS_5 and 3 spins in S^5. The mass shell (conformal gauge constraint) condition then gives a relation $E = E(Q,T)$. Here T is the string tension and $Q = (S_1, S_2, J_1, J_2, J_3; n_k)$ where n_k stand for higher conserved charges (analogs of oscillation numbers in flat space). The BPS (chiral primary) string states are point-like (supergravity modes), near-BPS (BMN) states are nearly pointlike, while generic semi-classical far-from-BPS states are represented by extended closed string configurations.

According to the AdS/CFT duality quantum closed string states in $AdS_5 \times S^5$ should be dual to quantum SYM states at the boundary, i.e. in $R \times S^3$ or, via radial quantization, to local single-trace operators at the origin of R^4. Such operators have the following structure
$\mathrm{tr}(D_{1+i2}^{S_1} D_{3+i4}^{S_2} \Phi_1^{J_1} \Phi_2^{J_2} \Phi_3^{J_2}...) + ...$ (where scalars and covariant derivatives may be also replaced by gauge field strength factors and fermions). The energy of a string state should then be equal to the dimension of the corresponding SYM operator, $E(Q,T) = \Delta(Q,\lambda)$, where on the SYM side the charges Q should characterise the eigen-operator of the anomalous dimension matrix. By analogy with the flat space case and ignoring α' corrections (i.e. assuming $R \to \infty$ or $\alpha' \to 0$) the excited string states are expected to have energies

$E \sim m \sim \frac{1}{\sqrt{\alpha'}} \sim \lambda^{1/4}$ [2]. This represents a non-trivial prediction for strong-coupling asymptotics of SYM dimensions. The asymptotics may, however, be different in the limits where the charges Q are also large, e.g., of order $\lambda^{1/2}$ as in the semiclassical limit [8].

In general, the natural (inverse-tension) perturbative expansion on the string side will be given by $E = \sum_{n=-1}^{\infty} \frac{c_n}{(\sqrt{\lambda})^n}$, while on the SYM side the usual planar perturbation theory will give the eigenvalues of the anomalous dimension matrix as $\Delta = \sum_{n=0}^{\infty} a_n \lambda^n$. The AdS/CFT duality implies that the two expansions should be the strong-coupling and the weak-coupling asymptotics of the same function. To check the relation $E = \Delta$ is then a non-trivial problem, except in the case of 1/2 BPS (single-trace chiral primary) operators which are dual to the supergravity states when the energies/dimensions are protected from corrections [4] and thus can be matched on the symmetry grounds.

1.2. SEMICLASSICAL STRING STATES: BMN AND BEYOND

For generic non-BPS states the situation with checking the duality looked hopeless until the remarkable suggestion of [6][1] and then of [8] that a progress can be made by (i) concentrating on a subsector of states with large or "semiclassical" values of quantum numbers, $Q \sim T \sim \sqrt{\lambda}$ and (ii) considering a new limit

$$Q \to \infty\,, \qquad \tilde{\lambda} \equiv \frac{\lambda}{Q^2} = \text{fixed}\,. \tag{1.2}$$

On the string side $\frac{Q}{\sqrt{\lambda}} = \frac{1}{\sqrt{\tilde{\lambda}}}$ plays the role of a semiclassical parameter (like rotation frequency) which can then be taken to be large. The energy of such states is $E = Q + f(Q, \lambda)$, where $f \to 0$ in the $\lambda \to 0$ limit. The duality implies that such semiclassical string states (as well as states represented by small fluctuations near them) should be dual to "long" SYM operators with a large canonical dimension, i.e. containing large number of fields or derivatives under the trace. In this case the duality map becomes more explicit.

[1]The history of the BMN limit is somewhat non-trivial. It started with the important observation that the Penrose limit of the $AdS_5 \times S^5$ space [9] leads to a maximally supersymmetric plane wave geometry supported by the Ramond-Ramond 5-form flux. Remarkably, the $AdS_5 \times S^5$ string theory action [7] in this limit (i.e. Green-Schwarz string action in plane-wave background) becomes essentially quadratic and thus its spectrum can be found explicitly [10]. Motivated by that, ref. [6] gave a dual SYM interpretation of the corresponding string states, suggested that string energies can be compared to perturbative SYM dimensions computed in the same limit, and indeed directly checked this to the first non-trivial order. Ref. [8] then explained that, on the string-theory side, the BMN limit is nothing but a semiclassical expansion near a particular (point-like) string solution. This interpretation was further clarified and extended in [11, 12], suggesting, in particular, how one can in principle compute corrections to the BMN limit (which was later done in full detail in [13] and especially in [14]). This made it clear that the plane-wave connection is not fundamental but rather is a special feature of semiclassical expansion near a certain string configuration (represented by massless $AdS_5 \times S^5$ geodesic wrapping S^5). Expanding near other string solutions leads to other special "Penrose-type" limits of string geometry (geometry "seen" by a fundamental string probe) which are described by the corresponding quadratic fluctuation actions. Expansions near a class of "fast" string configurations (discussed below) for which the world sheet becomes null in the effective zero-tension limit [15, 16, 17] may be interpreted as a stringy generalization of the Penrose limit (keeping in mind, of course, that this analogy applies only in the strict zero-tension limit).

The simplest possibility is to start with a BPS state that carries a large quantum number and consider small fluctuations near it, i.e. a set of *near-BPS* states characterised by a large parameter [6]. The only non-trivial example of such a BPS state is represented by a point-like string moving along a geodesic in S^5 with a large angular momentum $Q = J$. Then $E = J$ and the dual operator is trΦ^J, $\Phi = \phi_1 + i\phi_2$. The small closed strings representing near-by fluctuations are ultrarelativistic, i.e. their kinetic energy is much larger than their mass. They are dual to SYM operators of the form tr$(\Phi^J ...)$ where dots stand for a small number of other fields and/or covariant derivatives (one needs to sum over different orders of the factors to find an eigenstate of the anomalous dimension matrix). The energies of the small fluctuations happen to have the structure [10, 6]

$$E = J + \sqrt{1 + n^2 \tilde{\lambda}}\, K + O(\frac{1}{J}) \ = J + K + k_1 \tilde{\lambda} + k_2 \tilde{\lambda}^2 + \dots . \qquad (1.3)$$

One can argue in general [11, 12] and check explicitly [13, 14] that higher-order quantum string sigma model corrections are indeed suppressed in the limit (1.2), i.e. in the large J, fixed $\tilde{\lambda} \equiv \frac{\lambda}{J^2} = \lambda'$ limit. A remarkable feature of this expression is that E is analytic in $\tilde{\lambda}$, suggesting direct comparison with perturbative SYM expansion in λ.

Indeed, it can be shown that the first three $\tilde{\lambda}$, $\tilde{\lambda}^2$ and $\tilde{\lambda}^3$ terms in the expansion of the square root agree precisely with the one [6], two [19] and three [20, 21, 23] loop terms in the anomalous dimensions of the corresponding operators. There is also (for a 2-impurity $K = 2$ case) an argument [24] suggesting how the full $\sqrt{1 + n^2 \tilde{\lambda}}$ expression may appear on the perturbative SYM side (for reviews of various aspects of the BMN limit see also [25]). However, the general proof of the consistency of the BMN limit in the SYM theory (i.e. that the usual perturbative expansion can be rewritten as an expansion in $\tilde{\lambda}$ and $\frac{1}{J}$) remains to be given. Also, to explain why the string and the SYM expressions match one should show that the string limit (first $J \to \infty$, then $\tilde{\lambda} = \frac{\lambda}{J^2} \to 0$) and the SYM limit (first $\lambda \to 0$, then $J \to \infty$) produce exactly the same expressions for the energies/dimensions, even though, in general, the two limits may not commute, cf. [26, 27, 28].

If one moves away from the near-BPS limit and considers, e.g., a non-supersymmetric state with a large angular momentum $Q = S$ in AdS_5 [8], a similar direct quantitative check of the duality is no longer possible: here the classical energy is not analytic in λ and quantum corrections are no longer suppressed by powers of $\frac{1}{S}$ (but as usual are suppressed by powers of $\frac{1}{\sqrt{\lambda}}$). However, it is still possible to demonstrate a remarkable qualitative agreement between S-dependence of the string energy and of the SYM anomalous dimension. The energy of a folded closed string rotating at the center of AdS_5 which is dual [8] to twist 2 operators on the SYM side (tr$(\Phi_k^* D^S \Phi_k)$, $D = D_1 + iD_2$, and similar operators with spinors and gauge bosons that mix at higher loops [32, 33]) has the following form when expanded at large S :

$$E = S + f(\lambda) \ln S + O(S^0) \qquad (1.4)$$

On the string side

$$f(\lambda)_{\lambda \gg 1} = c_0 \sqrt{\lambda} + c_1 + \frac{c_2}{\sqrt{\lambda}} + \dots , \qquad (1.5)$$

where $c_0 = \frac{1}{\pi}$ is the classical [8] and $c_1 = -\frac{3}{\pi} \ln 2$ is the 1-loop [11] coefficient. On the gauge theory side one finds the *same* S-dependence of the anomalous dimension with the perturbative expansion of the $\ln S$ coefficient being

$$f(\lambda)_{\lambda \ll 1} = a_1 \lambda + a_2 \lambda^2 + a_3 \lambda^3 + \ldots \tag{1.6}$$

where $a_1 = \frac{1}{2\pi^2}$ [31], $a_2 = -\frac{1}{96\pi^2}$ [32], and $a_3 = \frac{11}{360 \times 64\pi^2}$ [33]. Like in the case of the SYM entropy [34], here one expects the existence of a smooth interpolating function $f(\lambda)$ that connects the two perturbative expansions. In fact, observing that the factor $\frac{1}{\pi}$ in (1.5) and factor $\frac{1}{\pi^2}$ in (1.6) seem to factorize, one can suggest a simple square root type interpolating formula for $f(\lambda)$ that seem indeed to give a good fit [32, 33] (cf. also the discussion end of section 4).

1.3. MULTISPIN STRING STATES

One may wonder still if examples of quantitative agreement between string energies and SYM dimensions found for the near-BPS (BMN) states exist also for more general non-BPS string states. Indeed, it was noticed already in [11] that a string state that carries large spin in AdS_5 as well as large spin J in S^5 has, in contrast to the above $J = 0$ case, an analytic expansion of its energy in $\tilde{\lambda} = \frac{\lambda}{J^2}$, just as in the BMN case with a large oscillation number $K \sim S$. It was observed in [35] that semiclassical string states carrying several large spins (with at least one of them being in S^5) have a regular expansion of their energy E in powers of $\tilde{\lambda}$ and it was then suggested, by analogy with the near-BPS case, that the expansion of E in small effective tension or $\tilde{\lambda}$ may be possible to match with the perturbative expansion of the SYM dimensions.

For a classical rotating closed string solution in S^5 one has $E = \sqrt{\lambda}\mathcal{E}(w_i)$, $J_i = \sqrt{\lambda} w_i$ so that $E = E(J_i, \lambda)$. The required key property is that (in contrast to the case of a single spin in AdS_5) there should be no $\sqrt{\lambda}$ factors in the expansion of the classical energy E in small λ

$$E = J + c_1 \frac{\lambda}{J} + c_2 \frac{\lambda^2}{J^3} + \ldots = J \left[1 + c_1 \tilde{\lambda} + c_2 \tilde{\lambda}^2 + \ldots\right] . \tag{1.7}$$

Here $J = \sum_{i=1}^3 J_i$, $\tilde{\lambda} \equiv \frac{\lambda}{J^2}$ and $c_n = c_n(\frac{J_i}{J})$ are functions of ratios of the spins which are finite in the limit $J_i \gg 1$, $\tilde{\lambda}$ =fixed.

The simplest example of such solution is provided by a circular string rotating in two orthogonal planes in S^3 part of S^5 with the two angular momenta being equal, $J_1 = J_2$ [35]:

$$\mathrm{X}_1 \equiv X_1 + iX_2 = \cos(n\sigma)\, e^{iw\tau} , \quad \mathrm{X}_2 \equiv X_3 + iX_4 = \sin(n\sigma)\, e^{iw\tau} ,$$

and with the global AdS_5 time being $t = \kappa\tau$. The conformal gauge constraint implies $\kappa^2 = w^2 + n^2$ and thus

$$E = \sqrt{J^2 + n^2\lambda} = J(1 + \frac{1}{2} n^2 \tilde{\lambda} - \frac{1}{8} n^4 \tilde{\lambda}^2 + \ldots) , \tag{1.8}$$

where $J = J_1 + J_2 = 2J_1$. For fixed J the energy thus has a regular expansion in string tension (in contrast to what happens in flat space where $E = \sqrt{\frac{2}{\alpha'} J}$).

Similar expressions (1.7) are found also for more general rigid multispin closed strings [35, 36, 37, 38, 39, 40]. In particular, for a folded string rotating in one plane of S^5 and with its center of mass orbiting along big circle in another plane the coefficients c_n are transcendental functions (expressed in terms of elliptic integrals) [37]. More generally, the 3-spin solutions are described by an integrable Neumann model [38, 39] and the coefficients c_n in the energy are expressed in terms of genus two hyperelliptic functions. The reason why choosing a particular string ansatz one gets an integrable effective 1-d model lies in the integrability of the original $S^5 = SO(6)/SO(5)$ classical sigma model [30] (see also [44]).

To be able to hope to compare the classical energy to the SYM dimension one should be sure that higher string α' corrections are suppressed in the limit $J \to \infty$, $\tilde{\lambda}$ =fixed. Formally, this is of course the case since $\alpha' \sim \frac{1}{\sqrt{\lambda}} \sim \frac{1}{J\sqrt{\tilde{\lambda}}}$; what is more important, the $\frac{1}{J}$ corrections are again analytic in $\tilde{\lambda}$ [36], i.e., as in the BMN case, the expansion in large J and small $\tilde{\lambda}$ is well-defined on the string side,

$$E = J \left[1 + \tilde{\lambda}(c_1 + \frac{d_1}{J} + ...) + \tilde{\lambda}^2 (c_2 + \frac{d_2}{J} + ...) + ... \right] , \tag{1.9}$$

with the classical energy (1.7) being the $J \to \infty$ limit of the exact expression.

The reason for this particular form of the energy (1.9) can be explained as follows [11, 12, 40]. We are computing the $AdS_5 \times S^5$ superstring sigma model loop corrections to the mass of a stationary solitonic solution on a 2-d cylinder (no IR divergences). This theory is conformal (due to the crucial presence of the fermionic fluctuations) and thus it does not depend on a UV cutoff. The relevant 2d fluctuations are massive and their masses scale as $w \sim \frac{1}{\sqrt{\tilde{\lambda}}}$. As a result, the inverse mass expansion is well-defined and the quantum corrections should be proportional to positive powers of $\tilde{\lambda}$. This was explicitly demonstrated by a 1-loop computation in [36, 66].

Similar expressions are found for the energies of small fluctuations near a given classical solution: as in the BMN case, the fluctuation energies are suppressed by extra factor of J, i.e.

$$\delta E = \tilde{\lambda}(k_1 + \frac{m_1}{J} + ...) + \tilde{\lambda}^2 (k_2 + \frac{m_2}{J} + ...) + ... \tag{1.10}$$

1.4. ADS/CFT DUALITY: NON-BPS STATES

Assuming that the same large J limit is well-defined also on the SYM side, one should then be able to compare the coefficients in (1.9) to the coefficients in the anomalous dimensions of the corresponding SYM operators $\mathrm{tr}(\Phi_1^{J_1} \Phi_2^{J_2} \Phi_3^{J_3}) + ...$ (and also do similar matching for near-by fluctuation modes) [35]. In practice, what is known (at least in principle) is how to compute the dimensions in the different limit: by first expanding in λ and then expanding in $\frac{1}{J}$. One may expect that this expansion of anomalous dimensions may take the form equivalent to (1.9), i.e.

$$\Delta = J + \lambda(\frac{a_1}{J} + \frac{b_1}{J^2} + ...) + \lambda^2 (\frac{a_2}{J^3} + \frac{b_2}{J^4} + ...) + ... , \tag{1.11}$$

and that the respective coefficients in E and Δ may agree with each other. The subsequent work [41, 42, 43, 44, 45, 2, 27, 47, 3, 49] did verify this structure of Δ and, moreover, established the general agreement between the two leading coefficients c_1, c_2 in E (1.9) and the one-loop and two-loop coefficients a_1, a_2 in Δ (1.11) (as usual, by "n-loop" term in Δ we mean the term multiplied by λ^n).

To compute Δ one is to diagonalize the anomalous dimension matrix defined on a set of "long" scalar operators, and this is obviously a non-trivial problem. The important step to this goal was made in [41] where it was observed that the one-loop planar dilatation operator in the scalar sector can be interpreted as a Hamiltonian of an integrable $SO(6)$ spin chain and thus can be diagonalized even for large length $L = J$ by the Bethe ansatz method.[2] In the simplest case of the "$SU(2)$" sector of operators $\mathrm{tr}(\Phi_1^{J_1}\Phi_2^{J_2}) + ...$ built out of two chiral scalars, the dilatation operator can be interpreted as "spin up" and "spin down" states of periodic $\mathrm{XXX}_{1/2}$ spin chain with length $L = J = J_1 + J_2$. Then the 1-loop dilatation operator becomes equivalent to the Hamiltonian of the ferromagnetic Heisenberg model

$$D_1 = \frac{\lambda}{(4\pi)^2} \sum_{l=1}^{J} (I - \vec{\sigma}_l \cdot \vec{\sigma}_{l+1}) \,. \tag{1.12}$$

By considering the thermodynamic limit ($J \to \infty$) of the corresponding Bethe ansatz equations, the proposal of [35] was confirmed at the leading order of expansion in $\tilde{\lambda}$ [42, 43]: for eigen-operators with $J_1 \sim J_2 \gg 1$ it was shown (i) that $\Delta - J = \lambda \frac{a_1}{J} + ...$, and (ii) a remarkable agreement was found between $a_1 = a_1(\frac{J_1}{J_2})$ and the coefficient c_1 in the energies (1.9) of various 2-spin string solutions. It was also possible also to match (as in the BMN case) the energies of fluctuations near the circular $J_1 = J_2$ solution with the corresponding eigenvalues of (1.12) [42].

Similar leading-order agreement between string energies and SYM dimensions was observed also in other sectors of states with large quantum numbers:

(1) in the $SU(3)$ sector: for specific solutions [35, 38, 39] with 3 spins in S^5 which are dual to the operators $\mathrm{tr}(\Phi_1^{J_1}\Phi_2^{J_2}\Phi_3^{J_3}) + ...$ [45, 50];

(2) in the $SL(2)$ [51] sector: for a folded string state [11] with one spin in AdS_5 and one spin in S^5 (with $E = J + S + \frac{\lambda}{J} c_1(\frac{S}{J}) + ...$ [11, 35]) which is dual to the operators $\mathrm{tr}(D^S \Phi^J) + ...$ [43];

(3) in a "subsector" of $SO(6)$ states containing pulsating (and rotating) solutions which again have regular [52] expansion of the energy in the limit of large oscillation number L, i.e. $E = L + c_1 \frac{\lambda}{L} + ...$ [42, 45].

This agreement between the leading-order terms in the expansion of energies of certain semiclassical string states and dimensions of the corresponding "long" SYM operators leaves, however, many questions, in particular:

(i) How to understand this agreement beyond specific examples, i.e. in a more universal way?

(ii) Which is the precise relation between profiles of string solutions and the structure of the dual SYM operators?

[2] Relations between 1-loop anomalous dimension matrix for a certain class of composite operators and integrable spin chain Hamiltonians were observed previously in the large N QCD context [63] (for a review and connections to AdS/CFT see [64]).

(iii) How to characterise the set of semiclassical string states and dual SYM operators to which this direct relation should apply?

(iv) Why the agreement holds at all, i.e. why the two limits (first $J \to \infty$, and then $\tilde{\lambda} \to 0$, or vice versa) taken on the string and the SYM sides give equivalent results to the first two orders in expansion in $\tilde{\lambda}$? Why/when it does not work to all orders in expansion in $\tilde{\lambda}$ (and $\frac{1}{J}$)?

The questions (i),(ii) were addressed in [2, 3, 55, 56, 57] using the low-energy effective action approach for coherent states; an alternative approach based on matching the general solution (and the integrable structure) of the string sigma model with that of the thermodynamic limit of the Bethe ansatz in the $SU(2)$ sector was developed in [47]. The question (iii) was addressed in [15, 16, 17, 57], and the question (iv) – in [27, 28, 29]. Still, our understanding of why there is a direct agreement with gauge theory at the first two $\tilde{\lambda}$ and $\tilde{\lambda}^2$ orders of expansion and why it does not [43, 28] continue to the $\tilde{\lambda}^3$ order is still rather rudimentory.[3]

Below we shall review the effective action approach as developed in [2, 3, 56, 57], concentrating mostly on the leading ("1-loop") order in expansion in $\tilde{\lambda}$.

2. Effective actions for coherent states

The suggestion of how to understand the agreement between leading-order terms in the multispin string energies and the corresponding one-loop anomalous dimensions in a universal way was made in [2] and was clarified and elaborated further in [3, 57]. The key idea was that instead of comparing particular solutions one should try to match effective sigma models which appear on the string side and the SYM side. Another related idea of [2, 3, 57] was that since the "semiclassical" string states carrying large quantum numbers are represented in the quantum theory by coherent states, one should be comparing coherent string states to coherent SYM states (and thus to coherent states of the spin chain). In view of the ferromagnetic nature of the dilatation operator (1.12), in the thermodynamic limit $J = J_1 + J_2 \to \infty$ with fixed large number of impurities (i.e. with fixed $\frac{J_1}{J_2}$) it is favorable to form large clusters of spins. Then a "low-energy" approximation and continuum limit should apply, leading to an effective "non-relativistic" sigma model for a coherent-state expectation value of the spin operator.

At the same time, on the string side, taking the "large space-time energy" (or large J) limit directly in the classical string action produces a reduced "non-relativistic" sigma model that describes in a universal way the leading-order $O(\tilde{\lambda})$ corrections to energies of all string solutions in the two-spin sector. The resulting sigma model action turns out to agree exactly [2] with the semiclassical coherent state action found from the $SU(2)$ sector of the spin chain in the $J \to \infty$, $\tilde{\lambda}$ =fixed limit. This demonstrates how a string action can directly "emerge" from a gauge theory in the large-N limit and provides a direct map between the "coherent" SYM states (or the corresponding operators built out of two holomorphic scalars) and all two-spin classical string states. Furthermore, the correspondence established at the level of the action implies also (i) the matching of the integrable structures and (ii) the matching of the fluctuations around particular solutions and thus it goes beyond the special examples of rigidly rotating strings.

[3]Similar (dis)agreements were found for the $1/J$ corrections to the BMN states [14].

2.1. COHERENT STATES

Let us briefly review the definition of coherent states (see, e.g., [53]). For a harmonic oscillator ($[a, a^\dagger] = 1$) one can define the coherent state $|u\rangle$ as $a|u\rangle = u|u\rangle$, where u is a complex number. Equivalently, $|u\rangle = R(u)|0\rangle$, where $R = e^{ua^\dagger - u^* a}$ so that acting on the vacuum $|0\rangle$ the operator R is simply proportional to $e^{ua^\dagger}$. Note that $|u\rangle$ can be written as a superposition of the eigenstates $|n\rangle$ of the harmonic oscillator Hamiltonian, $|u\rangle \sim \sum_{n=0}^{\infty} \frac{u^n}{\sqrt{n!}} |n\rangle$. An alternative definition of a coherent state is that it is a state with minimal uncertainty for both the coordinate $\hat{q} = \frac{1}{\sqrt{2}}(a + a^\dagger)$ and the momentum $\hat{p} = -\frac{i}{\sqrt{2}}(a - a^\dagger)$ operators, $\Delta hatp^2 = \Delta \frac{1}{2} tq^2 = \frac{1}{2}$, $\Delta\hat{p}^2 \equiv \langle u|\hat{p}^2|u\rangle - (\langle u|\hat{p}|u\rangle)^2$. For that reason it is the "best" approximation to a classical state. If one defines a time-dependent state $|u(t)\rangle = e^{-iHt}|u\rangle$ then the expectation values of $\hat{q}$ and $\hat{p}$, i.e. $\langle u|\hat{q}|u\rangle = \frac{1}{\sqrt{2}}(u+u^*)$, $\langle u|\hat{p}|u\rangle = -\frac{i}{\sqrt{2}}(u - u^*)$ follow the classical trajectories.

Starting instead of the Heisenberg algebra with the $SU(2)$ algebra $[S_3, S_\pm] = \pm S_\pm$, $[S_+, S_-] = 2S_3$ and considering the $s = 1/2$ representation where $\vec{S} = \frac{1}{2}\vec{\sigma}$ one can define a spin coherent state as a linear superposition of spin up and spin down states: $|u\rangle = R(u)|0\rangle$. Here $R = e^{uS_+ - u^* S_-}$, $|0\rangle = |\frac{1}{2}, \frac{1}{2}\rangle$ and u is a complex number. An equivalent way to label the coherent state is by a unit 3-vector $\vec{n}$ defining a point of S^2. Then $|\vec{n}\rangle = R(\vec{n})|0\rangle$ where $|0\rangle$ corresponds to a 3-vector $(0, 0, 1)$ along the 3rd axis. One can write $\vec{n} = U^\dagger \vec{\sigma} U$, $U = (u_1, u_2)$, and then $R(\vec{n})$ is an $SO(3)$ rotation from a north pole to a generic point of S^2 defined by $\vec{n}$. The key property of the coherent state is that $\vec{n}$ determines the coherent state expectation value of the spin operator:

$$\langle \vec{n}|\vec{S}|\vec{n}\rangle = \frac{1}{2}\vec{n} \,. \tag{2.1}$$

Similar definition of coherent states can be given in the case when $SU(2)$ is replaced by a generic group G. Given a semisimple group G with the Cartan basis of its algebra $(\mathrm{H}_i, \mathrm{E}_\alpha, \mathrm{E}_{-\alpha})$ ($[\mathrm{H}_i, \mathrm{H}_j] = 0$, $[\mathrm{H}_i, \mathrm{E}_\alpha] = \alpha_i \mathrm{E}_\alpha$, $[\mathrm{E}_\alpha, \mathrm{E}_{-\alpha}] = \alpha^i \mathrm{H}_i$, $[\mathrm{E}_\alpha, \mathrm{E}_\beta] = N_{\alpha\beta}\mathrm{E}_{\alpha+\beta}$) whose interpretation will be a symmetry group of a quantum Hamiltonian (acting in a unitary irreducible representation Λ on the Hilbert space V_Λ) one may define a set of coherent states by choosing a particular state $|0\rangle$ (with $\langle 0||0\rangle = 1$) in V_Λ and acting on it by the elements of G. A subroup H of G that leaves $|0\rangle$ invariant up to a phase ($\Lambda(h)|0\rangle = e^{i\phi(h)}|0\rangle$) is called maximum stability subgroup. One may then define the coset space G/H, the elements of which ($g = \omega h$, $h \in H$, $\omega \in G/H$, $\Lambda(g) = \Lambda(\omega)\Lambda(h)$) will parametrize the coherent states, $|\omega, \Lambda\rangle = \Lambda(\omega)|0\rangle$.

This definition depends on a choice of group G, its representation Λ and the vector $|0\rangle$. It is natural to assume also that $|0\rangle$ is an eigenstate of the Hamiltonian H, e.g., a ground state. For a unitary representation Λ we may choose $\mathrm{H}_i^\dagger = \mathrm{H}_i$, $\mathrm{E}_\alpha^\dagger = \mathrm{E}_{-\alpha}$ and select $|0\rangle$ to be the highest-weight vector of the representation Λ, i.e. demand that it is annihilated by "raising" generators and is an eigen-state of the Cartan generators: (i) $\mathrm{E}_\alpha|0\rangle = 0$ for all positive roots α; (ii) $\mathrm{H}_i|0\rangle = h_i|0\rangle$. In addition, we may demand that $|0\rangle$ is annihilated also by some "lowering" generators, i.e. (iii) $\mathrm{E}_{-\beta}|0\rangle = 0$ for *some* negative roots β; the remaining negative roots will be denoted by γ. Then the coherent states are given by

$$|\omega, \Lambda\rangle = \exp\left[\sum_\gamma (w_\gamma E_{-\gamma} - w_\gamma^* E_\gamma)\right] |0\rangle \,, \tag{2.2}$$

where γ are the negative roots for which $\mathrm{E}_\gamma|0\rangle \neq 0$. w_γ may be interpreted as coordinates on G/H where H is generated by $(\mathrm{H}_i, \mathrm{E}_\alpha, \mathrm{E}_{-\beta})$.

For example, in the case of $G = SU(3)$ with the Cartan basis $(\mathrm{H}_1, \mathrm{H}_2, \mathrm{E}_\alpha, \mathrm{E}_\beta, \mathrm{E}_{\alpha+\beta}, \mathrm{E}_{-\alpha}, \mathrm{E}_{-\beta}, \mathrm{E}_{-\alpha-\beta})$ and with $|0\rangle$ being the highest-weight of the fundamental representation, i.e. $\mathrm{E}_{-\beta}|0\rangle = 0$, $\mathrm{E}_{-\alpha}|0\rangle \neq 0$, $\mathrm{E}_{-\alpha-\beta}|0\rangle \neq 0$, the subgroup H is generated by $(\mathrm{H}_1, \mathrm{H}_2, \mathrm{E}_\beta, \mathrm{E}_{-\beta})$, i.e. is $SU(2)\times U(1)$ and $G/H = SU(3)/(SU(2)\times U(1)) = CP^2$ (see also [56]). We shall apply this general definition of coherent states in section 3.

2.2. LANDAU-LIFSHITZ MODEL FROM SPIN CHAIN

In general, one can rewrite the usual phase space path integral as an integral over the overcomplete set of coherent states (for the harmonic oscillator this is simply a change of variables from q, p to $u = \frac{1}{\sqrt{2}}(q + ip)$):

$$Z = \int [du]\; e^{i\mathrm{S}[u]} , \qquad \mathrm{S} = \int dt \Big(\langle u|i\frac{d}{dt}|u\rangle - \langle u|H|u\rangle \Big) . \tag{2.3}$$

The first ("Wess-Zumino" or "Berry phase") term in the action $\sim iu^* \frac{d}{dt} u$ is the analog of the usual $p\dot{q}$ term in the phase-space action. Applying this to the case of the Heisenberg spin chain Hamiltonian (1.12) one ends up with with the following action for the coherent state variables $\vec{n}_l(t)$ at sites $l = 1, ..., J$ (see, e.g., [54]):

$$\mathrm{S} = \int dt \sum_{l=1}^{J} \Big[\vec{C}(n_l) \cdot \dot{\vec{n}}_l - \frac{\lambda}{2(4\pi)^2} (\vec{n}_{l+1} - \vec{n}_l)^2 \Big] . \tag{2.4}$$

Here $dC = \epsilon^{ijk} n_i dn_j \wedge dn_k$ (i.e. $\vec{C}$ is a monopole potential on S^2). In local coordinates (at each site l) one has $\vec{n} = (\sin\theta\, \cos\phi,\ \sin\theta\, \sin\phi,\ \cos\theta)$, $\vec{C} \cdot d\vec{n} = \frac{1}{2} \cos\theta d\phi$. In the limit $J \to \infty$ with fixed $\tilde{\lambda} = \frac{\lambda}{J^2}$ (which we are interested in) we can take a continuum limit by introducing the 2-d field $\vec{n}(t, \sigma) = \{\vec{n}(t, \frac{2\pi}{J} l)\}$, $l = 1, ..., J$. Then

$$\mathrm{S} = J \int dt \int_0^{2\pi} \frac{d\sigma}{2\pi} \Big[\vec{C} \cdot \partial_t \vec{n} - \frac{1}{8} \tilde{\lambda} (\partial_\sigma \vec{n})^2 + ... \Big] , \tag{2.5}$$

where dots stand for higher derivative terms suppressed by $\frac{1}{J}$. Since J appears as a factor in front of the action, in the limit $J \to \infty$ all quantum corrections should be also suppressed by $\frac{1}{J}$, and thus the above action can be treated classically. The corresponding equations of motion

$$\partial_t n_i = \frac{1}{2} \tilde{\lambda} \epsilon_{ijk} n_j \partial_\sigma^2 n_k \tag{2.6}$$

are the Landau-Lifshitz equations for a classical ferromagnet. An alternative derivation of them is based on first writing down the Heisenberg equation for the time evolution of the spin operator directly from the spin chain Hamiltonian, then considering the coherent state expectation value and finally taking the continuum limit.

2.3. LANDAU-LIFSHITZ MODEL FROM STRING ACTION

The action (2.5) should be describing the coherent states of the Heisenberg spin chain in the above thermodynamic limit. One may wonder how a similar "non-relativistic" action may appear on the string side where one starts with the usual "relativistic" sigma model (1.1). To obtain such an effective action one is to perform the following sequence of steps [2, 3, 57]:

(i) isolate a "fast" coordinate α whose momentum p_α is large for a class of string configurations we consider;

(ii) gauge-fix $t \sim \tau$ and $p_\alpha \sim J$ (or $\tilde{\alpha} \sim \sigma$ where $\tilde{\alpha}$ is "T-dual" to α);

(iii) expand the action in derivatives of "slow" or "transverse" coordinates (to be identified with $\vec{n}$).

To illustrate this procedure let us consider the $SU(2)$ sector of string states carrying two large spins in S^5, with string motions restricted to S^3 part of S^5. The relevant part of the $AdS_5 \times S^5$ metric is then $ds^2 = -dt^2 + d\mathrm{X}_i d\mathrm{X}_i^*$, with $\mathrm{X}_i\mathrm{X}_i^* = 1$. Let us set

$$\mathrm{X}_1 = X_1 + iX_2 = u_1 e^{i\alpha}\,, \qquad \mathrm{X}_2 = X_3 + iX_4 = u_2 e^{i\alpha}\,, \qquad u_i u_i^* = 1\,.$$

Here α will be a coordinate associated to the total spin in the two planes (which in general will be the sum of orbital and internal spin). u_i (defined modulo $U(1)$ gauge transformations) will be the "slow" coordinates determining the "transverse" string profile. Then

$$d\mathrm{X}_i d\mathrm{X}_i^* = (d\alpha + C)^2 + Du_i Du_i^*, \qquad C = -iu_i^* du_i, \qquad Du_i = du_i - iCu_i,$$

where the second $|Du_i|^2$ term represent the metric of CP^1 (this parametrisation corresponds to the Hopf fibration $S^3 \sim S^1 \times S^2$). Introducing $\vec{n} = U^\dagger \vec{\sigma} U$, $U = (u_1, u_2)$ we get

$$d\mathrm{X}_i d\mathrm{X}_i^* = (D\alpha)^2 + \frac{1}{4}(d\vec{n})^2\,, \qquad D\alpha = d\alpha + C(n)\,. \tag{2.7}$$

Writing the resulting sigma model action in phase space form and imposing the (non-conformal) gauge $t = \tau$, $p_\alpha =$const$= J$, one gets [3] the same action (2.5) with the WZ term $\vec{C} \cdot \partial_t \vec{n}$ originating from the $p_\alpha D\alpha$ term in the phase-space Lagrangian (cf. its origin on the spin chain side as an analog of the $p\dot{q}$ in the coherent state path integral action).

An equivalent approach [57] leading to the same action (2.5) is based on first applying a 2-d duality (or "T-duality") transformation $\alpha \to \tilde{\alpha}$ and then choosing the "static" gauge $t = \tau$, $\tilde{\alpha} = \frac{1}{\sqrt{\tilde{\lambda}}}\sigma$ with $\frac{1}{\sqrt{\tilde{\lambda}}} = \frac{J}{\sqrt{\lambda}}$. Indeed, starting with

$$\mathcal{L} = -\frac{1}{2}\sqrt{-g}\, g^{pq}\big(- \partial_p t \partial_q t + D_p\alpha D_q\alpha + D_p u_i^* D_q u_i\big) \tag{2.8}$$

and applying the 2-d duality in α we get

$$\mathcal{L} = -\frac{1}{2}\sqrt{-g} g^{pq}\big(- \partial_p t \partial_q t + \partial_p\tilde{\alpha}\partial_q\tilde{\alpha} + D_p u_i^* D_q u_i\big) + \epsilon^{pq} C_p \partial_q \tilde{\alpha}\,. \tag{2.9}$$

Thus the "T-dual" background has no off-diagonal metric component but has a non-trivial NS-NS 2-form coupling in the $(\tilde{\alpha}, u_i)$ sector. It is useful not to use conformal gauge here. Eliminating the 2-d metric g^{pq} we then get the Nambu-type action

$$\mathcal{L} = \epsilon^{pq} C_p \partial_q \tilde{\alpha} \;-\; \sqrt{h}\,, \tag{2.10}$$

where $h = |\det\, h_{pq}|$ and $h_{pq} = -\partial_p t \partial_q t + \partial_p \tilde{\alpha} \partial_q \tilde{\alpha} + D_{(p} u_i^* D_{q)} u_i$. If we now fix the static gauge, $t = \tau$, $\tilde{\alpha} = \frac{1}{\sqrt{\tilde{\lambda}}}\sigma$, we finish with the action $I = J \int dt \int_0^{2\pi} \frac{d\sigma}{2\pi}\, \mathcal{L}$, where

$$\mathcal{L} = C_t - \sqrt{(1 + \tilde{\lambda}|D_\sigma u_i|^2)(1 - |D_t u_i|^2) + \frac{1}{4}\tilde{\lambda}(D_t u_i^* D_\sigma u_i + c.c.)^2}\ . \tag{2.11}$$

Making the key assumption that the evolution of u_i in t is slow, i.e. the time derivatives are suppressed (which can be implemented by rescaling t by $\tilde{\lambda}$ and expanding in powers of $\tilde{\lambda}$), we find, to the leading order in $\tilde{\lambda}$,

$$\mathcal{L} = -i u_i^* \partial_t u_i - \frac{1}{2}\tilde{\lambda}|D_\sigma u_i|^2\ . \tag{2.12}$$

This is the same as the CP^1 Landau-Lifshitz action (2.5) when written in terms of $\vec{n}$. Thus the string-theory counterpart of the WZ term in the spin-chain coherent state effective action originates from the 2-d NS-NS WZ term in the action for the "T-dual" coordinate $\tilde{\alpha}$ upon the static gauge fixing of the latter [57].

To summarize:

(i) $(t, \tilde{\alpha})$ are the "longitudinal" coordinates that are gauge-fixed (with $\tilde{\alpha}$ playing the role of the string direction or the spin chain direction on the SYM side);

(ii) $U = (u_1, u_2)$ or $\vec{n} = U^\dagger \vec{\sigma} U$ are the "transverse" coordinates that determine the semiclassical string profile and also the structure of the coherent operator on the SYM side, $\mathrm{tr}(\Pi_\sigma u_i(\sigma)\Phi_i)$ (see [2, 57] and below).

The agreement between the low-energy effective actions on the spin chain and the string side explains not only the matching of energies of coherent states representing configurations with two large spins (and also the matching of near-by fluctuations) but also the equivalence of the integrable structures (which was observed on specific examples in [44, 45]).

2.4. HIGHER ORDERS IN λ

The above leading-order agreement in $SU(2)$ sector has several generalizations.

First, we may include higher-order terms on the string side. Expanding (2.11) in $\tilde{\lambda}$ and eliminating higher powers of time derivatives by field redefinitions (which can be done since the leading-order equation of motion is first order in time derivative) we end up with [3] ($n' \equiv \partial_\sigma n$)

$$\mathcal{L} = \vec{C} \cdot \dot{\vec{n}} - \frac{\tilde{\lambda}}{8}\vec{n}'^2 + \frac{\tilde{\lambda}^2}{32}(\vec{n}''^2 - \frac{3}{4}\vec{n}'^4)$$
$$-\frac{\tilde{\lambda}^3}{64}\left[\vec{n}'''^2 - \frac{7}{4}\vec{n}'^2\vec{n}''^2 - \frac{25}{2}(\vec{n}'\vec{n}'')^2 + \frac{13}{16}\vec{n}'^6\right] + O(\tilde{\lambda}^4)\ . \tag{2.13}$$

The same $\tilde{\lambda}^2$ term is obtained [3] in the coherent state action on the spin chain side by starting with the sum of the 1-loop dilatation operator (1.12) and the 2-loop term found in [20]

$$D_2 = \frac{2\lambda^2}{(4\pi)^4}\sum_{l=1}^{J}(Q_{l,l+2} - 4Q_{l,l+1})\ , \qquad Q_{k,l} \equiv \frac{1}{2}(I - \vec{\sigma}_k \cdot \vec{\sigma}_l)\ . \tag{2.14}$$

The $\vec{n}'^4$ term originates from a non-trivial quantum correction on the spin chain side. This explains the matching of energies and dimensions to the first two orders in $\tilde{\lambda}$, observed on specific examples (using the generalized Bethe ansatz on the spin chain side) in [27]. The equivalent general conclusion about 2-loop matching was obtained in the integrability-based approach in [47].

The order-by-order agreement seems to break down at $\tilde{\lambda}^3$ term. A natural explanation [27, 28] is that the string limit (first $J \to \infty$, then $\tilde{\lambda} \to 0$) and the SYM limit (first $\lambda \to 0$, then $J \to \infty$) need not produce the same result when applied to the 2-parameter functions $E = \Delta(\lambda, J)$. A proposal of how to "complete" the perturbative gauge-theory expression to make the agreement with string theory manifest appeared in [28]; ref. [29] also suggested a possible form of generalized Bethe ansatz on the string side that would naturally interpolate between the string and gauge theory results (see also [62]). These proposals also suggest a resolution of the order $\tilde{\lambda}^3$ disagreement [14] between the string theory and gauge theory predictions for $\frac{1}{J}$ corrections to the BMN spectrum.

A possible explanation of why we found the agreement of $\tilde{\lambda}$ and $\tilde{\lambda}^2$ terms is that the structure of the dilatation operator at one and two loop orders is, in a sense, fixed by the BMN limit, which thus essentially determines the low energy effective action in a unique way. This is no longer so starting with the 3-loop order, where the dilatation operator already contains [20] the 4-spin QQ interactions (cf. (2.14)) which do not contribute to anomalous dimensions in the strict BMN limit.

By analogy with non-renormalization (due to underlaying supersymmetry) of few leading terms in low-energy effective actions, one may suggest that here the 3-loop problem may be related to the appearance of non-trivial interpolating functions as coefficients of the Q^n $(n \geq 2)$ terms in the dilatation operator. This would, in particular, explain why different structures of the QQ terms are found in the "gauge theory" [20] and "string theory" [62] limits.

As was suggested in [27, 28], the disagreement between the string and gauge theory results at 3-loop order in λ and the leading order in J can be repaired by adding "wrapping" contributions to the dilatation operator (and thus to the Bethe ansatz relations) on the gauge theory side. To illustrate this possibility, let us consider the circular solution case [35], and use the function $\frac{\lambda^J}{(1+\lambda)^J}$ [28], which is equal to 1 in the string theory limit ($J \to \infty$ with fixed $\frac{\lambda}{J^2} \equiv \tilde{\lambda}$) but zero in the perturbative gauge theory limit, in order to interpolate between the different $\frac{\lambda^3}{J^5}$ coefficients as follows:

$$\Delta = J + \frac{\lambda}{2J} - \frac{\lambda^2}{8J^3} + \frac{\lambda^3}{16J^5} \frac{\lambda^{J-3}}{(1+\lambda)^{J-3}} + \dots \,. \tag{2.15}$$

This expression agrees with both the string result ($E = \sqrt{J^2 + \lambda}$ in (1.8) with $n = 1$) and the perturbative gauge theory result ($\Delta_{\text{pert}} = J + \frac{\lambda}{2J} - \frac{\lambda^2}{8J^3} + 0 \times \lambda^3 + \dots$ [27]).

Let us add few details about the coherent-state expectation value of higher-loop SYM dilatation operator. This expectation value appears in the action in the coherent state path integral (2.3) of the quantum spin chain. Written in terms of independent permutations or $Q_{l,k} \equiv I - P_{l,k} = \frac{1}{2}(1 - \sigma_l \cdot \sigma_k)$ the "r-loop" term in the planar dilatation operator is expected [20, 61] to contain Q in maximal power $[\frac{r+1}{2}]$, i.e. $D_1, D_2 \sim \sum Q,\ D_3, D_4 \sim$

$\sum Q + \sum QQ$, etc. Explicitly,

$$D = \sum_{r=0}^{\infty} \frac{\lambda^r}{(4\pi)^{2r}} D_r \ , \qquad D_r = \sum_{l=1}^{J} \mathcal{D}_r(l) \ , \tag{2.16}$$

where as in (1.12) [41] and (2.14) [20]

$$\mathcal{D}_0 = I \ , \quad \mathcal{D}_1 = 2Q_{l,l+1} \ , \quad \mathcal{D}_2 = -2(4Q_{l,l+1} - Q_{l,l+2}) \ . \tag{2.17}$$

The 3-loop term [20, 61] is a special case of a 2-parameter family [21, 62]

$$\mathcal{D}_3(\alpha, c) = 4(15Q_{l,l+1} - 6Q_{l,l+2} + Q_{l,l+3})$$

$$+ b_1 Q_{l,l+2} Q_{l+1,l+3} + b_2 Q_{l,l+3} Q_{l+1,l+2} + b_3 Q_{l,l+1} Q_{l+2,l+3} \ , \tag{2.18}$$

$$b_1 = 4 + 2c - \alpha \ , \quad b_2 = -4 - 4c + \alpha \ , \quad b_3 = -2c + 5\alpha \ . \tag{2.19}$$

The choice of $c = 0, \ \alpha = 0$ corresponds to the 3-loop gauge theory operator of [20] whose form is fixed [61] by the superconformal algebra and the structure of Feynman diagrams (and the BMN scaling). This choice is also consistent with integrability of the spin chain. To preserve integrability [21] one should set $\alpha = 0$ (this is the parameter α_2 of [21]), while to have consistency with the gauge-theory perturbative expansion [62] one should set $c = 0$ ($c \equiv c_4$ in [62]). The case of $\alpha = 2, \ c = 0$ corresponds to the operator mentioned in [21] which seemed to agree with some string-theory results. The case of $\alpha = 0, \ c = 1$ is the "string" operator [62] which should correspond to the "string" modification of the Bethe ansatz equations in [29].

Starting with an $SU(2)$ coherent state satisfying (2.1) for which

$$\langle n|Q_{l,k}|n\rangle = \frac{1}{2}(1 - n_l \cdot n_k) = \frac{1}{4}(n_l - n_k)^2 \ , \tag{2.20}$$

computing $\langle n|D|n\rangle$ and then taking the continuum limit (by introducing a spatial coordinate $0 < \sigma \leq 2\pi$, and $n(\sigma_l) = n(\frac{2\pi l}{J})$, so that $n_{l+1} - n_l = \frac{2\pi}{J}\partial_\sigma n + \ldots$, etc.) we find, using Taylor expansion and dropping a total derivative over σ, see [3],

$$\frac{\lambda}{(4\pi)^2}\langle n|\mathcal{D}_1|n\rangle \ \to \ \frac{\tilde\lambda}{8}\Big[\vec n'^2 + O(\frac{1}{J^2}\partial^4 \vec n)\Big], \qquad \tilde\lambda = \frac{\lambda}{J^2} \ , \tag{2.21}$$

$$\frac{\lambda^2}{(4\pi)^4}\langle n|\mathcal{D}_2|n\rangle \ \to \ -\frac{\tilde\lambda^2}{32}\Big[\vec n''^2 + O(\frac{1}{J^2}\partial^6 \vec n)\Big] \ , \tag{2.22}$$

$$\frac{\lambda^3}{(4\pi)^6}\langle n|\mathcal{D}_3|n\rangle \ \to \ \frac{\tilde\lambda^3}{64}\Big[\ k_1 \frac{J^2}{(2\pi)^2}\vec n'^4$$

$$+ \vec n'''^2 + \ k_2 \vec n'^2 \vec n''^2 + \ k_3 (\vec n' \vec n'')^2 + O(\frac{1}{J^2}\partial^8 \vec n)\Big] \ . \tag{2.23}$$

Here

$$k_1 = \frac{1}{16}(16b_1 + 9b_2 + b_3) = \frac{1}{8}(14 - 3c - \alpha)$$

$$k_2 = -\frac{1}{96}(64b_1 + 45b_2 + b_3) = -\frac{1}{48}(38 - 27c - 7\alpha) ,$$

$$k_3 = -\frac{5}{48}(32b_1 + 9b_2 + 5b_3) = -\frac{5}{24}(46 + 9c + \alpha) . \tag{2.24}$$

Note a relation: $k_1 + k_2 + \frac{1}{10}k_3 = 0$. The problematic scaling-violating term $J^2(\partial_1^2 n)^4$ in (2.23) does not cancel automatically; it should cancel after one takes into account quantum corrections (which survive the continuum limit beyond the order $\tilde{\lambda}$ approximation [3]). Quantum corrections are expected also to be important in order to demonstrate the equivalence of the spin-chain result (for the "string" choice of $c = 1,\ \alpha = 0$) with the string-theory result in (2.13). Verifying this equivalence beyond the quadratic $\vec{n}^2$-terms (which obviously agree) remains an open problem.

Let us also mention that one can sum up all terms in the string effective Hamiltonian that are of second order in $\vec{n}$ but to arbitrary order in σ-derivatives [3]

$$\mathcal{L} = \vec{C} \cdot \dot{\vec{n}} - \frac{1}{4}\vec{n}\left(\sqrt{1 - \tilde{\lambda}\,\partial^2} - 1\right)\vec{n} + O(\vec{n}^3) . \tag{2.25}$$

This expression is in agreement with the leading-order results (2.13) and with the exact BMN spectrum (see also [1]). The coherent analogs of BMN states correspond to small fluctuations near the vacuum state $\vec{n}_0 = (0, 0, 1)$. On the spin chain side these correspond (in discrete version) to the microscopic spin wave excitations or magnons.

2.5. OTHER SECTORS

It is possible to extend the approach of [2, 3] to other sectors of rotating string states [55, 56, 57]. First, one is to identify subsectors of operators of the SYM theory which are closed under renormalization at least to one-loop. Such bosonic subsectors are:

(i) the three-spin "$SU(3)$" sector of string configurations with all three S^5 angular momenta (J_1, J_2, J_3) being non-zero. They are dual to chiral operators $\mathrm{tr}(\Phi_1^{J_1}\Phi_2^{J_2}\Phi_3^{J_3})+$ .... These form a set closed only under one-loop renormalization [20], but in the limit when $L = J_1 + J_2 + J_3$ is large one can treat this sector as closed even beyond one loop (mixings with fermionic operators are suppressed by $1/L$ [59]).

(ii) the "$SO(6)$" sector of generic (pulsating and rotating) string motions in S^5 which are dual to operators built out of 6 real scalars. Examples of pulsating string states were considered in [52, 45] and more generally in [57]; this sector will be discussed in the next section. As was pointed out in [59], one can also consider $SO(3)$ and $SO(4)$ subsectors of the $SO(6)$ sector which are again closed modulo $1/L$ corrections.

(iii) the two-spin "$SL(2)$" sector of string configurations with one AdS_5 spin $S = S_1$ and one S^5 angular momentum $J = J_3$, which are dual to operators $\mathrm{tr}(D_{1+i2}^S\Phi^J) + ...$ (forming a set closed under renormalization to all orders [51, 64]).

(iv) the three-spin "$SU(1,2)$" sector of string configurations with two AdS_5 angular momenta S_1, S_2 and one S^5 spin $J = J_3$, dual to operators $\mathrm{tr}(D_{1+i2}^{S_1}D_{3+i4}^{S_2}\Phi^J)$ $+ ...$ which form a set closed under one-loop renormalization.

Operators carrying more general combinations of non-zero spins from the list $(S_1, S_2, J_1, J_2, J_3)$ mix with fermionic and field-strenth operators already at one loop and would require to consider the full superspin chains [51, 65]; on the string side one would then need to include fermions of the GS action [18]. It may happen that one can again isolate some more general subsectors closed modulo $1/L$ corrections, but it appears that in order to apply a derivation of the reduced sigma model action on the string side one would need to impose additional constraints on the form of string configurations [56].

It is indeed straightforward to generalize [55, 56] the leading-order agreement observed in the $SU(2)$ sector to the $SU(3)$ sector of states with three large S^5 spins J_i, $i = 1, 2, 3$, finding the CP^2 analog of the CP^1 Landau-Lifshitz Lagrangian in (2.5),(2.12) [56] $\mathcal{L} = -iu_i^* \partial_0 u_i - \frac{1}{2}\tilde{\lambda}|D_1 u_i|^2$ on both the string and the spin chain sides.

Similar conclusion is reached [56] (see also [69]) in the $SL(2)$ sector of (S, J) states dual to operators like $\mathrm{tr}(D^S\Phi^J) + \ldots$. Like in the $SU(2)$ sector here the one-loop dilatation operator D_1 may be interpreted as the XXX$_{-1/2}$ Heisenberg spin chain Hamiltonian [51]. The corresponding coherent states (related to the $SU(2)$ ones by an "analytic continuation" from the 2-sphere to the 2-hyperboloid, $\vec{n} \to \vec{\ell}$, $\eta^{ij}\ell_i\ell_j = -\ell_1^2 + \ell_2^2 + \ell_3^2 = -1$) are defined so that for the $SL(2)$ generators S_i one has $\langle \ell | S_i | \ell \rangle = -\frac{1}{2}\ell_i$ and then [56]

$$\langle \ell | (D_1)_{SL(2)} | \ell \rangle = \frac{2\lambda}{(4\pi)^2} \sum_{l=1}^{J} \ln[\frac{1}{2}(1 - \eta_{ij}\ell_l^i \ell_{l+1}^j)]$$

$$= \frac{2\lambda}{(4\pi)^2} \sum_{l=1}^{J} \ln\left[1 + \frac{1}{4}\eta_{ij}(\ell_l^i - \ell_{l+1}^i)(\ell_l^j - \ell_{l+1}^j)\right] . \tag{2.26}$$

Note that this is the direct $(- + +)$ signature analog on the classically integrable lattice Hamiltonian for the Heisenberg magnetic [58]. Since we are interested in comparing to the semiclassical string case, S as well as J should be large, and, in view of the ferromagnetic nature of the spin chain, this effectively amounts to a low-energy semiclassical limit of the chain. Considering the limit $J \to \infty$ with fixed $\tilde{\lambda} = \frac{\lambda}{J^2}$ we get, as in the $SU(2)$ case, $\ell(\sigma_l) = \ell(\frac{2\pi l}{J}) \equiv \ell_l$ and $\ell_{l+1} - \ell_l = \frac{2\pi}{J}\partial_\sigma \ell + \mathcal{O}(\frac{1}{J^2}\partial_\sigma^2 \ell)$. To the one loop order, i.e. with only one power of λ in (2.26), in expanding the logarithm we need to keep only the order $\frac{1}{J^2}$ term, i.e. the term quadratic in first derivatives. This leads to

$$\langle \ell | (D_1)_{SL(2)} | \ell \rangle \to J \int_0^{2\pi} \frac{d\sigma}{2\pi} \left[\frac{\tilde{\lambda}}{8}\eta_{ij}\ell'^i \ell'^j + \mathcal{O}(\frac{1}{J^2}\partial^4 \ell) \right] . \tag{2.27}$$

As in the $SU(2)$ case, the same term in the action (containing also the WZ term) is found on the string theory side. This implies the general agreement between the string and SYM theory predictions for the energies/dimensions at leading order in $\tilde{\lambda}$ in the $SL(2)$ sector and thus generalizes the previous results [43] found for particular solutions.

3. General fast motion in S^5 and $SO(6)$ scalar operators

One would like to try to understand the general conditions on string states and SYM operators for which the above correspondence works, and incorporate also states with large oscillation numbers. Here we will follow [56, 57] (a closely related approach was developed in [16, 17, 18]). For strings moving in S^5 with large oscillation number the energy is $E = L + c_1 \frac{\lambda}{L} + \ldots$, i.e. it is again regular in the limit $L \to \infty$, $\tilde{\lambda} = \frac{\lambda}{L^2} \to 0$ [52].

The leading-order duality relation between string energies and anomalous dimensions in this case was checked in [42, 45, 59]. The general condition on string solutions for which $E/L = f(\tilde{\lambda})$ has a regular expansion in $\tilde{\lambda}$ appears to be that the world sheet metric should degenerate [16] in the $\tilde{\lambda} \to 0$ limit, i.e. the string motion should be ultra-relativistic in the small effective string tension limit (in the strict tensionless limit the corresponding states become BPS [15]).

For example, in the conformal gauge the 2-d induced metric in general scales as $g_{00} = -\kappa^2 + ...$ (assuming $t = \kappa\tau$, etc.), or, after a rescaling of the 2-d time coordinate, $g_{00} = -1 + O(\tilde{\lambda}) + ...$, where we used that $\kappa = \frac{1}{\sqrt{\tilde{\lambda}}}$. For the fast-moving strings, the leading $O(1)$ term in the metric gets cancelled out, and thus the metric degenerates in the $\tilde{\lambda} \to 0$ limit.

In the strict tensionless $\tilde{\lambda} \to 0$ limit each string piece is following a geodesic (big circle) in S^5, while switching on tension leads to a slight deviation from geodesic flow, i.e. to a nearly-null world surface [16]. The dual coherent SYM operators are then "locally BPS", i.e. each string bit corresponds to a BPS linear combination of 6 scalars (see below). In general, the scalar operators can be written as

$$\mathcal{O} = C_{m_1...m_L} \mathrm{tr}(\phi_{m_1}...\phi_{m_L}) \,. \tag{3.1}$$

The planar 1-loop dilatation operator D_1 acting on $C_{m_1...m_L}$ was found in [41] to be equivalent to an integrable $SO(6)$ spin chain Hamiltonian

$$\mathrm{H}^{n_1...n_L}_{m_1\cdots m_L} = \frac{\lambda}{(4\pi)^2} \sum_{l=1}^{L} \left(\delta_{m_l m_{l+1}} \delta^{n_l n_{l+1}} + 4\delta^{n_l}_{[m_l} \delta^{n_{l+1}}_{m_{l+1}]} \right) \,. \tag{3.2}$$

To find the analog of the coherent-state action (2.12) we choose a natural set of coherent states $\Pi_l|v_l\rangle$, where at each site $|v\rangle = R(v)|0\rangle$. Here R is an $G = SO(6)$ rotation and $|0\rangle$ is the BPS vacuum state corresponding to $\mathrm{tr}(\phi_1 + i\phi_2)^L$ or $v_{(0)} = (1, i, 0, 0, 0, 0)$, which is invariant under the subgroup $H = SO(2) \times SO(4)$. Then the rotation $R(v)$ and thus the coherent state is parametrized by a point in

$$G/H = SO(6)/[SO(4) \times SO(2)] \,,$$

i.e. v belongs to the Grassmanian $G_{2,6}$ [56]. $G_{2,6}$ is thus the coherent state target space for the spin chain sigma model since it parametrizes the orbits of the half-BPS operator $\phi_1 + i\phi_2$ under the $SO(6)$ rotations. This is the space of 2-planes passing through zero in R^6, or the space of big circles in S^5, i.e. the moduli space of geodesics in S^5 [17]. It can be represented also as an 8-dimensional quadric in CP^5: a complex 6-vector v_m should be subject, in addition to $v_m v_m^* = 1$ (and gauging away the common phase) also to $v_m v_m = 0$ condition. Taking the limit $L \to \infty$ with fixed $\tilde{\lambda} = \frac{\lambda}{L^2}$ and the continuum limit $v_{lm}(t) \to v_m(t, \sigma)$ we then get the $G_{2,6}$ analog of the CP^1 action (2.5),(2.12)

$$\mathrm{S} = L \int dt \int_0^{2\pi} \frac{d\sigma}{2\pi} \left(-i v_m^* \partial_t v_m - \frac{1}{2} \tilde{\lambda} |D_\sigma v_m|^2 \right) \,, \tag{3.3}$$

$$v_m v_m^* = 1 \,, \quad v_m v_m = 0 \,, \quad D_\sigma v_m = \partial_\sigma v_m - (v^* \partial_\sigma v) v_m \,.$$

This is also a generalization of the CP^2 action found in the $SU(3)$ sector [56].

One may wonder how this 8-dimensional sigma model may be related to the string sigma model on $R \times S^5$ where the coordinate space of transverse motions is only 4-dimensional. The crucial point is that the coherent state action is defined on the phase space (cf. the harmonic oscillator case in sect. 2.1), and $8 = (1+5) \times 2 - 2 \times 2$ is indeed the phase space dimension of a string moving in $R_t \times S^5$.

On the string side, the need to use the phase space description is related to the fact that to isolate a "fast" coordinate α for a generic string motion we need to specify both the position and velocity of each string piece. Given $\mathcal{L} = -(\partial t)^2 + (\partial X_m)^2$ in conformal gauge ($\dot{X}X' = 0,\ \dot{X}^2 + X'^2 = \kappa^2,\ X_m^2 = 1$) we find that the point-like trajectories (geodesics) are described by

$$X_m = a_m \cos\alpha + b_m \sin\alpha \ , \qquad \alpha = \kappa\tau, \ a_m^2 = 1, \ b_m^2 = 1, \ a_m b_m = 0 \ .$$

Equivalently,

$$X_m = \frac{1}{\sqrt{2}}(e^{i\alpha} v_m + e^{-i\alpha} v_m^*) \ , \qquad v_m = \frac{1}{\sqrt{2}}(a_m - ib_m), \ |v|^2 = 1, \ v^2 = 0 \ ,$$

where the constant vector v_m thus belongs to $G_{2,6}$. In general, for near-relativistic string motions v_m should change slowly with τ and σ. Then starting with the phase space Lagrangian for (X_m, p_m)

$$\mathcal{L} = p_m \dot{X}_m - \frac{1}{2} p_m p_m - \frac{1}{2} X'_m X'_m \ , \tag{3.4}$$

we may change the variables according to [57] (cf. again the harmonic oscillator case)

$$X_m = \frac{1}{\sqrt{2}}(e^{i\alpha} v_m + e^{-i\alpha} v_m^*) \ , \qquad p_m = \frac{i}{\sqrt{2}} p_\alpha (e^{i\alpha} v_m - e^{-i\alpha} v_m^*) \ , \tag{3.5}$$

where α and v_m now depend on τ and σ and v_m again belongs to $G_{2,6}$. There is an obvious $U(1)$ gauge invariance, $\alpha \to \alpha - \beta$, $v_m \to e^{i\beta} v_m$. Gauge-fixing the 2-d reparametrizations by $t \sim \tau$, $p_\alpha \sim L$ (or, doing an approximate T-duality $\alpha \to \tilde{\alpha}$ and setting $\tilde{\alpha} \sim \sigma$ as in sect. 2.3) one finds, after an additional rescaling of the time coordinate, that the phase-space Lagrangian becomes [57]:

$$\mathcal{L} = p_\alpha D_t \alpha - \frac{1}{2}\tilde{\lambda}|D_\sigma v|^2 - \frac{1}{4}\tilde{\lambda}\left[e^{2i\alpha}(D_\sigma v)^2 + c.c.\right] . \tag{3.6}$$

The first term here produces the WZ term $-iv_m^* \partial_t v_m$ and the last one averages to zero since $\alpha \approx \kappa\tau + ...$, where $\kappa = (\sqrt{\tilde{\lambda}})^{-1} \to \infty$.

Equivalently, the α-dependent terms in the action (that were absent in the $SU(2)$ or $SU(3)$ sectors) can be eliminated by canonical transformations [57]. We then end up with the following 8-dimensional phase-space Lagrangian for the "transverse" string motions, $\mathcal{L} = -iv_m^* \partial_t v_m - \frac{1}{2}\tilde{\lambda}|D_\sigma v_m|^2$, which is the same as found on the spin chain side (3.3). The 3-spin $SU(3)$ case is the special case when $v_m = (u_1, iu_1, u_2, iu_2, u_3, iu_3)$, where u_i belongs to the CP^2 subspace of $G_{2,6}$. The agreement between the spin chain and the string sides in this general $G_{2,6} = SO(6)/[SO(4) \times SO(2)]$ case explains not only the matching for pulsating solutions [52, 45] but also for near-by fluctuations.

Let us now discuss the reason for the restriction $v^2 = 0$ on the spin chain side and also clarify the structure of the coherent operators corresponding to semiclassical string states. Given the scalar operator $\mathcal{O} = C_{m_1...m_L} \mathrm{tr}\,(\phi_{m_1} \cdots \phi_{m_L})$ one may obtain the Schrödinger equation for the wave function $C_{m_1...m_L}(t)$ from[4]

$$S = -\int dt \Big(iC^*_{m_1...m_L} \partial_t C_{m_1...m_L} + C^*_{m_1...m_L} \mathrm{H}^{n_1...n_L}_{m_1...m_L} C_{n_1...n_L} \Big) . \qquad (3.7)$$

In the limit $L \to \infty$ we may consider the coherent state description and assume the factorized ansatz [57]

$$C_{m_1...m_L} = v_{m_1} ... v_{m_L} \,, \qquad (3.8)$$

where each $v_l = \{v_{m_l}\}$ $(l = 1, ..., L)$ is a complex unit-norm 6-vector. The BPS case corresponding to the totally symmetric traceless $C_{m_1...m_L}$ is represented by $v_l = v_{(0)}$, $v^2_{(0)} = 0$. Using (3.2) and substituting the ansatz for $C_{m_1...m_L}$ into the above action one finds

$$S = -\int dt \sum_{l=1}^{L} \Big(iv^*{}_l \partial_t v_l$$

$$+ \frac{\lambda}{(4\pi)^2} \Big[(v^*{}_l v^*{}_{l+1})(v_l v_{l+1}) + 2 - 2(v^*{}_l v_{l+1})(v_l v^*{}_{l+1}) \Big] \Big) , \qquad (3.9)$$

where we suppressed the 6-vector indices in the scalar products. As expected [41], the coherent state expectation value of the Hamiltonian (i.e. order λ term in (3.9)) vanishes for the BPS case when v_l does not depend on l and $v^2 = 0$. More generally, if we assume that v_l is changing slowly with l (i.e. $v_l \simeq v_{l+1}$), then we find that (3.9) contains a potential term $(v^*{}_l v^*{}_l)(v_l v_l)$ coming from the first "trace" structure in (3.2). This term will lead to large (order λL [41]) shifts of anomalous dimensions, invalidating low-energy expansion, i.e. prohibiting one from taking the continuum limit $L \to \infty$, $\tilde{\lambda} = \frac{\lambda}{L^2}$ = fixed, and thus from establishing direct correspondence with string theory along the lines of [2, 3, 56].[5]

To get solutions with small variations of v_l from site to site we are thus to impose

$$v_l^2 = 0 \,, \qquad l = 1, \ldots, L \qquad (3.10)$$

which minimizes the potential energy coming from the first term in (3.2). This condition implies that the operator at each site is invariant under half of supersymmetries: if $v^2 = 0$ the matrix $v_m \Gamma^m$ appearing in the variation of the operator $v_m \phi_m$, i.e. $\delta_\epsilon(v_m \phi_m) = \frac{i}{2} \bar{\epsilon}(v_m \Gamma^m)\psi$, satisfies $(v_m \Gamma^m)^2 = 0$. This means that $v_m \phi_m$ is invariant under the variations associated with the null eigenvalues [57]. One may thus call $v^2 = 0$ a "*locally BPS*" condition since the preserved combinations of supercharges in general are different for each v_l, i.e. the operator corresponding to $C = v_1 ... v_L$ is not BPS. Here

[4]For the coherent states we consider the corresponding equation of motion may be interpreted as a (non-trivial) RG equation for the coupling constant associated to the operator $\mathcal{O}$.

[5]Even $SU(2)$ sector has in general other higher-energy states with dimensions $\sim \lambda L$ (in addition to magnons with energies $\sim \frac{\lambda}{L^2}$ and macroscopic spin waves with $\Delta \sim \frac{\lambda}{L}$ there are also spinons with $\Delta \sim \lambda L$) but these do not correspond to fast strings – they are not captured by the low-energy continuum limit of the coherent state path integral.

"local" should be understood in the sense of the spin chain, or, equivalently, the spatial world-sheet direction.[6]

In the case when v_l are slowly changing we can take the continuum limit as in [2, 3, 56] by introducing the 2-d field $v_m(t,\sigma)$ with $v_{ml}(t) = v_m(t, \frac{2\pi l}{L})$. Then (3.9) reduces to (3.3) (all higher derivative terms are suppressed by powers of $\frac{1}{L}$ and the potential term is absent due to the condition $v^2 = 0$). Then (3.9) becomes equivalent to the $G_{2,6}$ Landau-Lifshitz sigma model (3.3) which was derived from the phase space action on the string side.

To summarize, considering ultra-relativistic strings in S^5 one can isolate a fast variable α (a "polar angle" in the string phase space) whose momentum p_α is large. One may gauge-fix p_α to be constant $\sim L$ or set $\tilde{\alpha} \sim L\sigma$, so that σ or the "operator direction" on the SYM side gets interpretation of "T-dual to fast coordinate" direction. As a result, one finds a local phase-space action with 8-dimensional target space (where one can not eliminate 4 momenta without spoiling the locality). This action is equivalent to the Grassmanian $G_{2,6}$ Landau-Lifshitz sigma model action appearing on the spin chain side.

As a by-product, we thus get a precise mapping between string solutions and SYM operators representing coherent spin chain states [2, 57]. Explicit examples corresponding to pulsating and rotating solutions were given in [57]. In the continuum limit we may write the operator corresponding to the solution $v(t,\sigma)$ as $\mathcal{O} = \mathrm{tr}[\prod_\sigma \mathrm{v}(t,\sigma)]$, $\mathrm{v} \equiv v_m(t,\sigma)\phi_m$. This locally BPS coherent operator is the SYM operator naturally associated to a ultra-relativistic string solution. The t-dependence of the string solution thus translates into the RG scale dependence of $\mathcal{O}$, while the σ-dependence describes the ordering of scalar field factors under the trace.

In general, semiclassical string states represented by classical string solutions should be dual to coherent spin chain states or coherent operators, which are different from the exact eigenstates of the dilatation operator but which should lead to the same energy or anomalous dimension expressions. At the same time, the Bethe ansatz approach [41, 42, 43, 47] is determining the exact eigenvalues of the dilatation operator. The reason why the two approaches happen to be in agreement is that in the limit we consider the problem is essentially semiclassical, and because of the integrability of the spin chain, its exact eigenvalues are not just well-approximated by the classical solutions but are actually exactly reproduced by them for $L \to \infty$, i.e. (just as in the harmonic oscillator or flat space string theory case, cf. also [67]) the semiclassical coherent state sigma model approach happens to be exact.

4. Concluding remarks

As discussed above, there exists a remarkable generalization of the near-BPS (BMN) limit to non-BPS but "locally-BPS" sector of string/SYM states (for related reviews see [12, 40, 22]). It remains to understand better when and why this direct relation works or fails, but the hope is to use it as a guide towards finding the string/SYM spectrum exactly in λ, at least in a subsector of states. The relation between phase-space action

[6]This generalizes the argument implicit in [2]; an equivalent proposal was made in [17]. This is related to but different from the "nearly BPS" operators discussed in [15] (which, by definition, were those which become BPS in the limit $\lambda \to 0$).

for "slow" variables on the string side and the coherent-state action on the SYM (spin chain) side gives a very explicit picture of how string action "emerges" on the conformal gauge theory side (with the central role played by the dilatation operator). This implies not only an equivalence between string energies and SYM dimensions (established to first two orders in expansion in the effective coupling $\tilde{\lambda}$) but also a direct relation between the string profiles and the structure of coherent SYM operators.

One may try also to use the duality as a tool to uncover the structure of planar SYM theory to all orders in λ by assuming the exact correspondence between particular SYM and string states. For example, demanding the consistency with the BMN scaling limit (along with superconformal algebra) determines the structure of the full 3-loop SYM dilatation operator in the $SU(2)$ sector [20, 21]. One can also use the BMN limit to fix only part of the dilatation operator but to all orders in λ [1]. Generalizing (1.12),(2.14) and the 3- and 4-loop expressions in [20, 21] one can organize [27, 3, 1] the dilatation operator as an expansion in powers of $Q_{k,l} = \frac{1}{2}(I - \vec{\sigma}_k \cdot \vec{\sigma}_l)$ which reflect interactions between spin chain sites,

$$D = \sum Q + \sum QQ + \sum QQQ + \ldots .$$

Here the products $Q...Q$ are "irreducible", i.e. index of each site appears only once. The Q^2-terms first appear at 3 loops, Q^3-terms – at 5 loops, etc. [20, 21]. Concentrating on the order-Q part $D^{(1)}$ of D one can write:

$$D^{(1)} = \sum_{r=0}^{\infty} \frac{\lambda^r}{(4\pi)^r} \sum_{l=1}^{L} \mathcal{D}_r(l) , \qquad \mathcal{D}_r(l) = 2 \sum_{k=1}^{r} a_{r,k} Q_{l,l+k} , \tag{4.1}$$

or $D^{(1)} = \sum_{l=1}^{L} \sum_{k=1}^{L-1} h_k(L,\lambda)\, Q_{l,l+k}$. Demanding the agreement with the BMN limit one can then determine the coefficients $a_{r,k}$ and thus the function h_k explicitly to all orders in λ [1]. In particular, for large L, i.e. when D acts on "long" operators, one finds

$$D^{(1)} = 2 \sum_{l=1}^{L} \sum_{k=1}^{\infty} f_k(\lambda)\, Q_{l,l+k} , \qquad f_k(\lambda) = \sum_{r=k}^{\infty} \frac{\lambda^r}{(4\pi)^{2r}}\, a_{r,l} , \tag{4.2}$$

where the coefficients $f_k(\lambda)$ can be summed up in terms of the standard Gauss hypergeometric function [1]

$$f_k(\lambda) = \left(\frac{\lambda}{4\pi^2}\right)^k \frac{\Gamma(k-\frac{1}{2})}{4\sqrt{\pi}\,\Gamma(k+1)}\ {}_2F_1(k-\frac{1}{2}, k+\frac{1}{2}; 2k+1; -\frac{\lambda}{\pi^2}) , \tag{4.3}$$

or, equivalently,

$$f_k(\lambda) = \frac{1}{4\pi(2k-1)} \left(\frac{\lambda}{\pi^2}\right)^k \int_0^1 du \left[\frac{u(1-u)}{1+\frac{\lambda}{\pi^2}u}\right]^{k-1/2} . \tag{4.4}$$

f_k goes rapidly to zero at large k, so we get a spin chain with short-range interactions.

One may hope that imposing additional constraints coming from correspondence with other string solutions (and using recent insights in [28, 29]) may help to determine the structure of the dilatation operator further.

The function $f_k(\lambda)$ in (4.3) smoothly interpolates between the usual perturbative expansion at small λ and $f_k(\lambda) \sim \sqrt{\lambda}$ behaviour at large λ. The latter is the expected behaviour of anomalous dimensions of "long" operators dual to "semiclassical" string states.

Similar interpolating functions should appear also in anomalous dimensions of other SYM operators, though for "short" operators the strong-coupling asymptotics of the dimensions should be $\lambda^{1/4}$. Let us consider, for example, the following dimension 4 supersymmetry descendant of the Konishi operator, $\mathrm{tr}([\Phi_1, \Phi_2]^2)$ (which belongs to the $SU(2)$ sector and should have the same anomalous dimension as the Konishi operator). The first few terms in the perturbative λ-expansion of its anomalous dimension are known to be [68, 20, 33, 61]

$$\Delta_{\lambda \ll 1} = 4 + 3\bar{\lambda} - 3\bar{\lambda}^2 + \frac{21}{4}\bar{\lambda}^3 - \frac{705}{64}\bar{\lambda}^4 + O(\bar{\lambda}^5) , \tag{4.5}$$

$$\bar{\lambda} \equiv \frac{\lambda}{4\pi^2} .$$

If one would to ignore all non-linear in Q terms in the dilatation operator (4.1), then the resummed expression for the anomalous dimension would be [1]

$$\Delta^{(1)} = 4 + \frac{3}{2}(\sqrt{1 + 4\bar{\lambda}} - 1) , \tag{4.6}$$

which does not, however, have the expected large λ asymptotics,

$$\Delta_{\lambda \gg 1} = 2\lambda^{1/4} + ... = 2\sqrt{2\pi}(\bar{\lambda})^{1/4} + ... \tag{4.7}$$

Note that one cannot reproduce such asymptotics with an interpolating expression for Δ built out of rational functions of the square of the effective string $\bar{\lambda} = T^2$: while the expansion of a rational function (like $\sqrt{\sqrt{a + b\bar{\lambda}} + d}$) at small λ would have the same form as (4.5), the factors of π would not match in the strong coupling limit (4.7).

The reason for the above extra factor of $\sqrt{\pi}$ in $\Delta_{\lambda \gg 1}$ in (4.7) expressed in terms of $\bar{\lambda}$ can be understood following ref. [2]. The Konishi operator should correspond to the lowest-level scalar string mode. The masses of the $AdS_5 \times S^5$ string modes are, in general, non-trivial functions of the string tension (the corresponding wave equations receive α'-corrections), but, in the large-tension limit, they should simply be the same as in flat space, i.e. (in units where $R = 1$)

$$m^2 = \frac{4n}{\alpha'} = 4n\sqrt{\lambda} = 8n\pi T , \quad T = \frac{\sqrt{\lambda}}{2\pi} = \sqrt{\bar{\lambda}} . \tag{4.8}$$

Then the standard AdS_5 formula $\Delta(\Delta - 4) = m^2$ for the dimension of a scalar field with mass m (again, expected to be valid in the large tension limit) predicts that for $n = 1$

$$\Delta - 2 = \sqrt{4 + m^2} \to m + ... = 2\sqrt{2\pi\sqrt{\bar{\lambda}}} + ... \tag{4.9}$$

It thus appears that instead of being a rational function of $\bar{\lambda}$, the dimension Δ of the Konishi operator should be a transcendental function. In fact, the hypergeometric function like the one appearing in (4.3), i.e. ${}_2F_1(a, b; c; -k\bar{\lambda})$, seems to be a natural candidate: one can choose its arguments (and an overall coefficient) so that to match the powers of λ and π in both the strong and the weak coupling limits (one example with required strong-coupling asymptotics has $a = -1/4,\ b = 3/4,\ c = 3/2,\ k = 4$). With $\Delta \sim {}_2F_1$ (or Δ being a rational function of ${}_2F_1$) it is possible also to satisfy the string-theory requirement that the strong-coupling expansion should be organized as an expansion in powers of the inverse string tension, i.e. $\Delta = \lambda^{1/4}(c_1 + \frac{c_2}{\sqrt{\lambda}} + \frac{c_3}{(\sqrt{\lambda})^2} + ...)$ (cf. (4.9)).

At the same time, for "long" operators with large canonical dimensions (like BMN operators or non-BPS operators discussed in the previous sections) the interpolating functions appearing in Δ may be simple rational functions like square roots: here both the weak (gauge theory) and the strong (string theory) expansions are organized in terms of $\bar{\lambda} = \frac{\lambda}{4\pi^2}$ with the leading coefficients which do not contain extra powers of π (see also the discussion below eq. (1.6)).

Acknowledgements

I would like to thank the organizers of the Cargèse 2004 Summer School for the invitation to speak. I am grateful to G. Arutyunov, N. Beisert, S. Frolov, M. Kruczenski, J. Russo, A. Ryzhov and B. Stefanski for useful discussions and collaboration on the work described above. This work was supported in part by the DOE grant DE-FG02-91ER40690, the INTAS contract 03-51-6346 and the RS Wolfson award.

References

1. J. M. Maldacena, "The large N limit of superconformal field theories and supergravity," Adv. Theor. Math. Phys. **2**, 231 (1998) [Int. J. Theor. Phys. **38**, 1113 (1999)] [hep-th/9711200].
2. S. S. Gubser, I. R. Klebanov and A. M. Polyakov, "Gauge theory correlators from non-critical string theory," Phys. Lett. B **428**, 105 (1998) [hep-th/9802109].
3. E. Witten, "Anti-de Sitter space and holography," Adv. Theor. Math. Phys. **2**, 253 (1998) [hep-th/9802150].
4. O. Aharony, S. S. Gubser, J. M. Maldacena, H. Ooguri and Y. Oz, "Large N field theories, string theory and gravity," Phys. Rept. **323**, 183 (2000) [hep-th/9905111]. I. R. Klebanov, "TASI lectures: Introduction to the AdS/CFT correspondence," hep-th/0009139.
5. A. M. Polyakov, "Gauge fields and space-time," Int. J. Mod. Phys. A **17S1**, 119 (2002) [hep-th/0110196].
6. D. Berenstein, J. M. Maldacena and H. Nastase, "Strings in flat space and pp waves from N =4 super Yang Mills," JHEP **0204**, 013 (2002) [hep-th/0202021].
7. R. R. Metsaev and A. A. Tseytlin, "Type IIB superstring action in $AdS_5 \times S^5$ background," Nucl. Phys. B **533**, 109 (1998) [hep-th/9805028].
8. S. S. Gubser, I. R. Klebanov and A. M. Polyakov, "A semi-classical limit of the gauge/string correspondence," Nucl. Phys. B **636**, 99 (2002) [hep-th/0204051].
9. M. Blau, J. Figueroa-O'Farrill, C. Hull and G. Papadopoulos, "A new maximally supersymmetric background of IIB superstring theory," JHEP **0201**, 047 (2002) [hep-th/0110242].
10. R. R. Metsaev, "Type IIB Green-Schwarz superstring in plane wave Ramond-Ramond background," Nucl. Phys. B **625**, 70 (2002) [hep-th/0112044]. R. R. Metsaev and A. A. Tseytlin, "Exactly solvable model of superstring in plane wave Ramond-Ramond background," Phys. Rev. D **65**, 126004 (2002) [hep-th/0202109].
11. S. Frolov and A. A. Tseytlin, "Semiclassical quantization of rotating superstring in $AdS_5 \times S^5$," JHEP **0206**, 007 (2002) [hep-th/0204226].

12. A. A. Tseytlin, "Semiclassical quantization of superstrings: $AdS_5 \times S^5$ and beyond," Int. J. Mod. Phys. A **18**, 981 (2003) [hep-th/0209116].
13. A. Parnachev and A. V. Ryzhov, "Strings in the near plane wave background and AdS/CFT," JHEP **0210**, 066 (2002) [hep-th/0208010].
14. C. G. Callan, H. K. Lee, T. McLoughlin, J. H. Schwarz, I. Swanson and X. Wu, "Quantizing string theory in $AdS_5 \times S^5$: Beyond the pp-wave," Nucl. Phys. B **673**, 3 (2003) [hep-th/0307032]. C. G. Callan, T. McLoughlin and I. Swanson, "Holography beyond the Penrose limit," hep-th/0404007; "Higher impurity AdS/CFT correspondence in the near-BMN limit," hep-th/0405153.
15. D. Mateos, T. Mateos and P. K. Townsend, "Supersymmetry of tensionless rotating strings in $AdS_5 \times S^5$, and nearly-BPS operators," hep-th/0309114. "More on supersymmetric tensionless rotating strings in $AdS_5 \times S^5$," hep-th/0401058.
16. A. Mikhailov, "Speeding strings," JHEP **0312**, 058 (2003) [hep-th/0311019].
17. A. Mikhailov, "Slow evolution of nearly-degenerate extremal surfaces," hep-th/0402067. "Supersymmetric null-surfaces," hep-th/0404173.
18. A. Mikhailov, "Supersymmetric null-surfaces," hep-th/0404173.
19. D. J. Gross, A. Mikhailov and R. Roiban, "Operators with large R charge in N = 4 Yang-Mills theory," Annals Phys. **301**, 31 (2002) [hep-th/0205066].
20. N. Beisert, C. Kristjansen and M. Staudacher, "The dilatation operator of N = 4 super Yang-Mills theory," Nucl. Phys. B **664**, 131 (2003) [hep-th/0303060].
21. N. Beisert, "The su(2|3) dynamic spin chain," hep-th/0310252.
22. N. Beisert, "The dilatation operator of N = 4 super Yang-Mills theory and integrability," hep-th/0407277. "Higher-Loop Integrability in N=4 Gauge Theory," hep-th/0409147.
23. B. Eden, C. Jarczak and E. Sokatchev, "A three-loop test of the dilatation operator in N = 4 SYM," hep-th/0409009.
24. A. Santambrogio and D. Zanon, "Exact anomalous dimensions of N = 4 Yang-Mills operators with large R charge," Phys. Lett. B **545**, 425 (2002) [hep-th/0206079].
25. J. C. Plefka, "Lectures on the plane-wave string / gauge theory duality," Fortsch. Phys. **52**, 264 (2004) [hep-th/0307101]. A. Pankiewicz, "Strings in plane wave backgrounds," Fortsch. Phys. **51**, 1139 (2003) [hep-th/0307027]. D. Sadri and M. M. Sheikh-Jabbari, "The plane-wave / super Yang-Mills duality," hep-th/0310119.
26. I. R. Klebanov, M. Spradlin and A. Volovich, "New effects in gauge theory from pp-wave superstrings," Phys. Lett. B **548**, 111 (2002) [hep-th/0206221].
27. D. Serban and M. Staudacher, "Planar N = 4 gauge theory and the Inozemtsev long range spin chain," JHEP **0406**, 001 (2004) [hep-th/0401057].
28. N. Beisert, V. Dippel and M. Staudacher, "A novel long range spin chain and planar N = 4 super Yang-Mills," hep-th/0405001.
29. G. Arutyunov, S. Frolov and M. Staudacher, "Bethe ansatz for quantum strings," hep-th/0406256.
30. K. Pohlmeyer, "Integrable Hamiltonian Systems And Interactions Through Quadratic Constraints," Commun. Math. Phys. **46**, 207 (1976).
31. D. J. Gross and F. Wilczek, "Asymptotically Free Gauge Theories. I," Phys. Rev. D **8**, 3633 (1973).
32. A. V. Kotikov, L. N. Lipatov and V. N. Velizhanin, "Anomalous dimensions of Wilson operators in N = 4 SYM theory," Phys. Lett. B **557**, 114 (2003) [hep-ph/0301021].
33. A. V. Kotikov, L. N. Lipatov, A. I. Onishchenko and V. N. Velizhanin, "Three-loop universal anomalous dimension of the Wilson operators in N = 4 SUSY Yang-Mills model," hep-th/0404092.
34. S. S. Gubser, I. R. Klebanov and A. A. Tseytlin, "Coupling constant dependence in the thermodynamics of N = 4 supersymmetric Yang-Mills theory," Nucl. Phys. B **534**, 202 (1998) [hep-th/9805156].
35. S. Frolov and A. A. Tseytlin, "Multi-spin string solutions in $AdS_5 \times S^5$," Nucl. Phys. B **668**, 77 (2003) [hep-th/0304255].
36. S. Frolov and A. A. Tseytlin, "Quantizing three-spin string solution in $AdS_5 \times S^5$," JHEP **0307**, 016 (2003) [hep-th/0306130].

37. S. Frolov and A. A. Tseytlin, "Rotating string solutions: AdS/CFT duality in non-supersymmetric sectors," Phys. Lett. B **570**, 96 (2003) [hep-th/0306143].
38. G. Arutyunov, S. Frolov, J. Russo and A. A. Tseytlin, "Spinning strings in $AdS_5 \times S^5$ and integrable systems," Nucl. Phys. B **671**, 3 (2003) [hep-th/0307191].
39. G. Arutyunov, J. Russo and A. A. Tseytlin, "Spinning strings in $AdS_5 \times S^5$: New integrable system relations," Phys. Rev. D **69**, 086009 (2004) [hep-th/0311004].
40. A. A. Tseytlin, "Spinning strings and AdS/CFT duality," hep-th/0311139.
41. J. A. Minahan and K. Zarembo, "The Bethe-ansatz for N = 4 super Yang-Mills," JHEP **0303**, 013 (2003) [hep-th/0212208].
42. N. Beisert, J. A. Minahan, M. Staudacher and K. Zarembo, "Stringing spins and spinning strings," JHEP **0309**, 010 (2003) [hep-th/0306139].
43. N. Beisert, S. Frolov, M. Staudacher and A. A. Tseytlin, "Precision spectroscopy of AdS/CFT," JHEP **0310**, 037 (2003) [hep-th/0308117].
44. G. Arutyunov and M. Staudacher, "Matching higher conserved charges for strings and spins," JHEP **0403**, 004 (2004) [hep-th/0310182]. "Two-loop commuting charges and the string / gauge duality," hep-th/0403077.
45. J. Engquist, J. A. Minahan and K. Zarembo, "Yang-Mills duals for semiclassical strings on $AdS_5 \times S^5$," JHEP **0311**, 063 (2003) [hep-th/0310188].
46. M. Kruczenski, "Spin chains and string theory," hep-th/0311203.
47. V. A. Kazakov, A. Marshakov, J. A. Minahan and K. Zarembo, "Classical/quantum integrability in AdS/CFT," hep-th/0402207.
48. M. Kruczenski, A. V. Ryzhov and A. A. Tseytlin, "Large spin limit of $AdS_5 \times S^5$ string theory and low energy expansion of ferromagnetic spin chains," hep-th/0403120.
49. M. Lubcke and K. Zarembo, "Finite-size corrections to anomalous dimensions in N = 4 SYM theory," JHEP **0405**, 049 (2004) [hep-th/0405055].
50. C. Kristjansen, "Three-spin strings on $AdS_5 \times S^5$ from N = 4 SYM," Phys. Lett. B **586**, 106 (2004) [hep-th/0402033]. L. Freyhult, "Bethe ansatz and fluctuations in SU(3) Yang-Mills operators," hep-th/0405167. C. Kristjansen and T. Mansson, "The Circular, Elliptic Three Spin String from the SU(3) Spin Chain," hep-th/0406176.
51. N. Beisert and M. Staudacher, "The N=4 SYM integrable super spin chain", Nucl. Phys. B **670**, 439 (2003) [hep-th/0307042].
52. J. A. Minahan, "Circular semiclassical string solutions on $AdS_5 \times S^5$," Nucl. Phys. B **648**, 203 (2003) [hep-th/0209047].
53. A. Perelomov, "Generalized Coherent States and Their Applications", Berlin, Germany: Springer (1986) 320 p. W. M. Zhang, D. H. Feng and R. Gilmore, "Coherent States: Theory And Some Applications," Rev. Mod. Phys. **62**, 867 (1990).
54. E. H. Fradkin, "Field Theories Of Condensed Matter Systems," Redwood City, USA: Addison-Wesley (1991) 350 p. (Frontiers in physics, 82). S. Sachdev, "Quantum phase transitions", Cambridge U. Press (1999) 352 p.
55. R. Hernandez and E. Lopez, "The SU(3) spin chain sigma model and string theory," JHEP **0404**, 052 (2004) [hep-th/0403139].
56. B. J. Stefanski, Jr. and A. A. Tseytlin, "Large spin limits of AdS/CFT and generalized Landau-Lifshitz equations," JHEP **0405**, 042 (2004) [hep-th/0404133].
57. M. Kruczenski and A. A. Tseytlin, "Semiclassical relativistic strings in S5 and long coherent operators in N = 4 SYM theory," hep-th/0406189.
58. L. D. Faddeev and L. A. Takhtajan, "Hamiltonian Methods In The Theory Of Solitons," Springer, Berlin (1987) 592 p. L. D. Faddeev, "How Algebraic Bethe Ansatz works for integrable model," hep-th/9605187.
59. J. A. Minahan, "Higher loops beyond the SU(2) sector," hep-th/0405243.
60. A. V. Ryzhov and A. A. Tseytlin, "Towards the exact dilatation operator of N = 4 super Yang-Mills theory," hep-th/0404215.
61. N. Beisert, "Higher loops, integrability and the near BMN limit," JHEP **0309**, 062 (2003) [hep-th/0308074].
62. N. Beisert, "Spin chain for quantum strings," hep-th/0409054.
63. A. V. Belitsky, V. M. Braun, A. S. Gorsky and G. P. Korchemsky, "Integrability in QCD and

beyond," hep-th/0407232.

64. A. V. Belitsky, A. S. Gorsky and G. P. Korchemsky, "Gauge / string duality for QCD conformal operators," Nucl. Phys. B **667**, 3 (2003) [hep-th/0304028].
65. A. V. Belitsky, S. E. Derkachov, G. P. Korchemsky and A. N. Manashov, "Superconformal operators in N = 4 super-Yang-Mills theory," hep-th/0311104. "Quantum integrability in (super) Yang-Mills theory on the light-cone," hep-th/0403085.
66. S. A. Frolov, I. Y. Park and A. A. Tseytlin, "On one-loop correction to energy of spinning strings in S^5," hep-th/0408187.
67. A. Jevicki and N. Papanicolaou, "Semiclassical Spectrum Of The Continuous Heisenberg Spin Chain," Annals Phys. **120**, 107 (1979).
68. D. Anselmi, "The N = 4 quantum conformal algebra," Nucl. Phys. B **541**, 369 (1999) [hep-th/9809192].
69. S. Bellucci, P. Y. Casteill, J. F. Morales and C. Sochichiu, "sl(2) spin chain and spinning strings on $AdS_5 \times S^5$," hep-th/0409086.

Part II

Seminars

BARYOGENESIS WITH LARGE EXTRA DIMENSIONS

KARIM BENAKLI
LPTHE, Université Paris 6,
4 pl Jussieu, 75252 Paris cedex, France

Abstract. Baryogenesis is a challenge, particularly for low string scale models because there is an upper bound on the reheat temperature of the Universe, and because certain baryon number violating operators must be suppressed. We review the model using horizontal family symmetries presented in [1] to emphasis its prediction for the weak angle 0.239 very close to experimental value.

In these proceedings I wish to return to the model proposed in [1]: it shows that it is in principle possible to reconcile explanation of the observed baryonic asymmetry of the universe (BAU),with the possibility of large extra dimensions (references will be found in [1]).

Baryogenesis is a challenge in the class of models, mainly because the reheat temperature T_{reh} is constrained to be so low that most existing scenarios are ruled out. For example, the reheat temperature must be less than ~ 100 GeV for a large range of values of the string scale M_s, so the electroweak $B + L$ violating processes are not available for baryogenesis.

The low T_{reh} creates a generic difficulty. One of the Sakharov conditions for baryogensis is that one needs some out-of-equilibrium dynamics. This can be found at phase transitions, or when some interaction is not fast enough to keep up with the expansion of the Universe. However when the temperature (or energy density) of the Universe is low, the expansion rate is too ($H \sim 10^{-18}T$ at $T \sim$GeV), so interactions have no difficulty keeping up with the expansion. Getting the out-of-equilibrium anywhere but a phase transition is hard. If the reheat temperature is less than ~ 0.1 GeV, then the only phase transition available appears to be the one out of inflation.

Another difficulty for baryogenesis models is the bounds on baryon number violation. For instance, to avoid fast proton decay through $|\Delta B| = |\Delta L| = 1$ operators, and neutron-anti-neutron oscillations through $\Delta B = 2$ operators, one may assume that B is conserved mod 3 . This is problematic for scenarios where the BAU is generated in the out-of-equilibrium decay of a particle X. X must have at least two decay modes with different baryon number in the final state, and approximately the same branching ratios. Otherwise the baryon asymmetry generated will be small as a consequence of CPT. If B is conserved mod 3, then X must decay to final states with $B = 1$ and with $B = 2$ (or

L. Baulied et al. (eds.), String Theory: From Gauge Interactions to Cosmology, 293–296.

generation	q	u^c	d^c	ℓ	e^c
1	5	5	2	1	5
2	4	2	1	0	4
3	1	0	1	0	1

TABLE 1. Possible charges for the fermions and the Higgs under the horizontal $U(1)$, for three generations. The first generation is u, d, e, and so on. These charges generate approximately the right Yukawa couplings.

$B = 0$ and $B = 3$), so that X exchange generates a vertex that conserves B mod 3. But $B = 2$ operators are of higher dimension that $B = 1$ operators, so the branching ratio of X to the $B = 2$ final state will be very small.

Suppose that $M_s \leq 10^5$ GeV, so symmetries are required to forbid the $n - \bar{n}$ operator $udsuds$, and the fast proton decay vertices. Then, one can consider that the BAU is generated in an out-of-equilibrium decay of a scalar particle X. It should decay out of equilibrium to final states with different baryon numbers, with enough CP violation in the decay rates to generate a baryon to photon ratio $\eta \sim 3 \times 10^{-10}$. X can be the inflaton and it would decay before nucleosynthesis, at about the right time to reheat the Universe. Or,X is a particle generated in the reheating process, with a number density $n_X = \delta n_\gamma$. The condition such that X annihilations will be out of equilibrium at the reheat temperature and thereafter, so all the Xs will decay is that it is very weakly coupled. Large suppression factor in the couplings could originate from the volume of internal dimensions.

We assume that the SM yukawa couplings are generated by some horizontal $U(1)$ gauge symmetry, which is spontaneously broken below M_s. The quarks (q, u^c, d^c) and leptons (ℓ, e^c) carry positive charges under this symmetry, and the charges are higher for the lighter fermions. The Higgs that breaks the horizontal $U(1)$ with vev θ carries negative charge. By choosing the horizontal charges of the fermions Q^H_f with care, one can generate approximately the right structure for the Yukawa matrices, because the interaction $u^c u H \sim m_u \bar{u} u$ appears multiplied by $(\theta/\Lambda)^{Q^H_{u^c}+Q^H_u}$ and $t^c t H \sim m_t \bar{t} t$ appears multiplied by $(\theta/\Lambda)^{Q^H_{t^c}+Q^H_t}$. Such a mechanism is probably required in models with a low M_s to avoid FCNC. It will also suppress the problematic $u^c d^c s^c u^c d^c s^c$ operator: at M_s where θ is zero, it is forbidden by the horizontal symmetry (if all the fermions are positively charged), and once the horizontal symmetry is broken, $u^c d^c s^c u^c d^c s^c$ can appear suppressed by $(\theta/\Lambda)^{2(Q^H_{u^c}+Q^H_{d^c}+Q^H_{s^c})}$. For $\theta/\Lambda \equiv \epsilon \sim .2$ and the charges in table 2, the operator $u^c d^c s^c u^c d^c s^c$ will be multiplied by ϵ^{16}, which is compatible with the experimental limit for $\Lambda \leq$ few TeV. The proton is stable enough provided that L is conserved mod 2.

Suppose that X is a light ($\sim$ 10 GeV) gauge singlet scalar with $L = 1$. It can decay to SM particles via the dimension 7 operators $Xqqq\ell$ and $Xu^c u^c d^c e^c$. These violate B respectively by 1 and -1 units, so a baryon asymmetry could be generated. We suppose that the fermions have the charges under the horizontal $U(1)$ that are listed in table 2. In this case the principle decay rates will be

$$\Gamma_{\bar{p}} \sim \epsilon^{18}\frac{m_X^7}{\Lambda^6} \qquad X \to c^c\, u^c\, b^c\, \tau^c\ (\bar{D}\,\bar{B}\,\bar{p}\,\tau^+) \tag{0.1}$$

$$\Gamma_p \sim \epsilon^{18}\frac{m_X^7}{\Lambda^6} \qquad X \to c\, s\, b\, \nu_\tau\ (D\,B\,K\,p\,\nu_\tau) \tag{0.2}$$

$$\Gamma_{p2} \sim \epsilon^{20}\frac{m_X^7}{\Lambda^6} \qquad X \to c\, d\, b\, \nu_\tau\ (D\,B\,p\,\nu_\tau) \tag{0.3}$$

where $\epsilon = \theta/\Lambda \sim .2$. We neglect kinematics, factors of 4π, and so on, so these are very approximate estimates. However, for $\Lambda \sim 3$ TeV, and $T_{reh} \sim 3$ MeV, we get $m_X \sim 25$ GeV. This is heavy enough to decay to $B-$ and D–mesons, but light enough to (possibly) be produced in the reheating process, or to be the inflaton. For larger Λ, we would need a larger m_X.

We now need to consider whether a sufficient baryon asymmetry can be generated in the decays. We assume that $\Gamma_p \gg \Gamma_{p2}$ so we neglect Γ_{p2} and all the other smaller decay modes. The net number of baryons produced per X particle will be

$$\frac{n_b}{n_X} \simeq \frac{\Gamma_p - \Gamma_{\bar{p}} + \bar{\Gamma}_{\bar{p}} - \bar{\Gamma}_p}{\Gamma_p + \Gamma_{\bar{p}}} \equiv \theta_{CP} \tag{0.4}$$

where $\bar{\Gamma}$ is the CP conjugate decay. The baryon-to-photon ratio $n_b/n_\gamma \equiv \eta \simeq 3 \times 10^{-10}$ will be

$$\eta \simeq \frac{n_X}{n_\gamma}\theta_{CP} \quad . \tag{0.5}$$

If X is the inflaton, then $n_X/n_\gamma \sim T_{reh}/m_X \sim 10^{-3}$. If X is produced in the reheating process, then $n_X/n_\gamma = \delta$ is a model dependent parameter. One would not expect to make more than one or two Xs in the decay of each inflaton, so in this case $\delta \leq 10^{-3}$. This means that we need $\theta_{CP} \geq 10^{-7}$. If we assume that the CP violation arises through loop corrections involving new particles at the scale M_s, then $\theta_{CP} \sim (m_X/\Lambda)^2 \sim 10^{-6}$, which is approximately right.

The family symmetry presented here obviously suffers from anomalies. These might be cancelled in two different ways. The first is to assume that massive particles in a hidden sector are charged under this $U(1)$, standard model symmetries and some hidden gauge group. The hidden symmetry might suppress any undesirable non-renormalisable operator.

A more interesting possibility is to appeal to a Green-Schwarz mechanism to cancel the anomalies. If the standard model gauge couplings are all given by the vacuum expectation value of a single modulus (dilaton) then anomaly cancellation implies particular tree level relations between the couplings. For the model at hand, the strong, weak and hypercharge $U(1)$ couplings are in the ratio $1 : 1 : 105/33$ at $M_s \sim$ TeV instead of the usual relation $1 : 1 : 5/3$ at 10^{16} GeV. To compare the tree level prediction with experimental measurements we see that the value obtained for $\sin 0.239$ is extremely close to the measured one.! Here I would like to stress this point which we did not in [1].

In conclusion, if the large extra dimension models deal quite easily with particle experiments, explaining the cosmological observation, as to generate in a natural way the measured small or big ratio in the universe, remains a challenge. If one accepts a certain amount of fine-tuning, as it is fashionable today, then models can be easily build as shown above.

References

1. K. Benakli, S. Davidson, Phys.Rev. **D60** (1999) 025004, hep-ph/9810280.

BRANEWORLD COSMOLOGY ALMOST WITHOUT BRANES

GIANLUCA CALCAGNI

Dipartimento di Fisica, Università di Parma
INFN – Gruppo collegato di Parma
Parco Area delle Scienze 7/A, I-43100 Parma, Italy

Abstract. We review some general aspects of braneworld cosmologies in which an inflationary period driven by a scalar field confined on the brane is described by a nonstandard effective Friedmann equation. The perturbation spectra, consistency equations and observational consequences of these models are considered.

1. Introduction

Motivated by recent developments in string, superstring and M theory, several models for a multidimensional target spacetime have been proposed. Among them, particular attention has been devoted to brane-world scenarios, according to which the visible universe is a (3+1)-dimensional variety (a 3-brane) embedded in a bulk with some either non-compact or compactified extra dimensions. Typically, the background metric on the brane is assumed to be the Friedmann-Robertson-Walker (FRW) metric and the Einstein equations are modified in accordance with the gravity model describing spacetime. Their projection on the brane results in the basic FRW equations for the cosmological evolution. For an introduction to the subject and some lists of references, see [1].

In this paper we review how to look for cosmic signatures of high-energy, higher-derivative gravity models. In particular, the construction of a nontrivial set of consistency equations permits us to compare theoretical predictions with the perturbation spectra of the cosmic microwave background (CMB). It turns out that CMB experiments of this and next generation might be able to discriminate between the standard four-dimensional lore and the braneworld paradigm. In Secs. 2 and 3 we introduce the basic ingredients of the patch and slow-roll formalisms, taking as examples the five-dimensional Randall-Sundrum (RS) scenarios and their Gauss-Bonnet (GB) generalization. In Sec. 4 we outline some results on cosmological perturbations in the presence of an extra dimension and find their observational consequences. Conclusions are in Sec. 5

L. Baulied et al. (eds.), String Theory: From Gauge Interactions to Cosmology, 297–303.

2. Setup

One of the first problems one has to deal with when constructing braneworld models is how to stabilize the extra dimension. This can be achieved in a number of ways; in the RS example, Goldberger and Wise have provided a mechanism according to which a 5D massive scalar is put into the bulk with a potential of the same order of the brane tension λ [2]. If the energy density ρ on the brane is smaller than the characteristic energy of the scalar potential, $\rho/V \sim \rho/\lambda \ll 1$, then the radion is stabilized and one gets the standard Friedmann equation $H^2 \propto \rho$ on the brane. On the contrary, if the brane energy density is comparable with the stabilization potential, $\rho/\lambda \gtrsim 1$, the bulk backreacts because it feels the presence of the brane matter, the minimum of the potential is shifted and the well-known quadratic corrections to the Friedmann equation arise:

$$H^2 = \frac{\kappa_4^2}{6\lambda}\rho(2\lambda + \rho), \tag{2.1}$$

where H is the effective Hubble rate experienced by an observer on the brane and κ_4 is the 4D gravitational coupling.

The RS model can be viewed as a particular energy limit of a Gauss-Bonnet braneworld, characterized by the 5D Planck mass $M_5 \equiv \kappa_5^{-2/3}$ and a Friedmann equation

$$H^2 = \frac{c_+ + c_- - 2}{8\alpha}, \tag{2.2}$$

where $\alpha = 1/(8g_s^2)$ is the Gauss-Bonnet coupling (g_s is the string coupling) and, defining $\sqrt{\alpha/2}\,\kappa_5^2 \equiv \sigma_0^{-1}$,

$$c_\pm = \left\{\left[(1 + 4\alpha\Lambda_5/3)^{3/2} + (\sigma/\sigma_0)^2\right]^{1/2} \pm \sigma/\sigma_0\right\}^{2/3}. \tag{2.3}$$

$\Lambda_5 < 0$ is the bulk cosmological constant and σ is the matter energy density which is decomposed into a matter contribution plus the brane tension λ: $\sigma = \rho + \lambda$. When $\sigma/\sigma_0 \gg 1$, one gets the "pure Gauss-Bonnet" high-energy regime,

$$H^2 = \left(\frac{\kappa_5^2}{16\alpha}\right)^{2/3} \rho^{2/3}. \tag{2.4}$$

When the energy density is far below the 5D scale but $\rho \gg \lambda$, one recovers the Friedmann equation of the RS scenario with vanishing 4D cosmological constant, provided that some relations among the parameters of the action are satisfied.

Here we shall consider nonstandard cosmological evolutions on the brane and extend the RS and GB discussion to arbitrary scenarios we dubbed "patch cosmologies" [3], with

$$H^2 = \beta_q^2 \rho^q. \tag{2.5}$$

β_q is a constant and the exponent q is equal to 1 in the pure 4D (radion-stabilized) regime, $q = 2$ in the high-energy limit of the RS braneworld and $q = 2/3$ in the high-energy limit of the GB scenario. In order to simplify the framework, we make the following assumptions:

1. There is a confinement mechanism such that matter lives on the brane only, while gravitons are free to propagate in the bulk. This is guaranteed as long as $\rho < M_5^4$.
2. The contribution of the Weyl tensor is neglected.
3. The contribution of the anisotropic stress is neglected.
4. We concentrate on the large-scale limit of the cosmological perturbations.

Assumption 2 closes the system of equations on the brane and sets aside the nonlocal contributions from the bulk, while assumptions 3 and 4 reduce the number of degrees of freedom of gauge invariant scalar perturbations.

This list might seem too restrictive and to spoil almost all the interesting features of the model. However, assumptions 4 and 2 nicely fit in the inflationary regime, since the long wavelength region of the spectrum, corresponding to the Sachs-Wolfe plateau, encodes the main physics of the inflationary era. Moreover, the dark radiation term, which is the simplest contribution of the Weyl tensor, scales as a^{-4} and is exponentially damped during the accelerated expansion. Finally, bulk physics mainly affects the small-scale/late-time cosmological structure and can be consistently neglected during inflation. This is a highly nontrivial result which has been confirmed with several methods both analytically and numerically [4].

Imposing a perfect fluid on the brane with equation of state $p = w\rho$, the continuity equation governing the cosmological dynamics is the same as in four dimensions, thanks to assumption 2:

$$\dot{\rho} + 3H(\rho + p) = 0\,. \tag{2.6}$$

There are two candidates for the role of inflation. The first one is an ordinary scalar field ϕ with energy density and pressure

$$\rho = \dot{\phi}^2/2 + V(\phi) = p + 2V(\phi)\,. \tag{2.7}$$

The second one is a Dirac-Born-Infeld (DBI) tachyon T such that

$$\rho = V(T)/c_S = -V(T)^2/p\,, \tag{2.8}$$

$$c_S \equiv \sqrt{1 - \dot{T}^2}\,. \tag{2.9}$$

From a string-theoretical point of view, the evolution of the tachyon proceeds up to the condensation $\dot{T} \to 1$ into the closed string vacuum, where no signal of open excitations propagates (Carrollian limit). Also, near the minimum the strong coupling regime emerges, $g_s = O(1)$, and the perturbative description implicit in the DBI action may fail down. However, from a cosmological perspective Eq. (2.8) is a toy model and, like in standard inflation, an additional reheating mechanism around the condensation is required for gracefully exiting the inflationary period. Here we will not consider this and other (indeed solvable) problems concerning the tachyon and just implement the DBI action in the cosmological dynamics as an alternative model of inflation.

3. Slow-roll parameters

Let ψ denote the inflaton field irrespectively of its action. Expressions involving ψ will be valid for both the ordinary scalar and the tachyon. The first-order slow-roll (SR) parameters are defined as

$$\epsilon \equiv -\frac{\dot{H}}{H^2}, \qquad \eta \equiv -\frac{\ddot{\psi}}{H\dot{\psi}}, \tag{3.1}$$

together with their evolution equations with respect to synchronous time

$$\dot{\epsilon} = H\epsilon\left[\left(2-\widetilde{\theta}\right)\epsilon - 2\eta\right], \qquad \dot{\eta} = H\left(\epsilon\eta - \xi^2\right), \tag{3.2}$$

where $\xi^2 \equiv (\dddot{\psi}/\dot{\psi})^{\cdot}/H^2$ is a second-order parameter, in the sense that it appears only in expressions which are $O(\epsilon^2, \eta^2, \epsilon\eta)$. Here $\widetilde{\theta} = 2$ for the tachyon field and $\widetilde{\theta} = \theta \equiv 2(1-q^{-1})$ for the ordinary scalar field (4D: $\theta = 0$; RS: $\theta = 1$; GB: $\theta = -1$). Note that each time derivative of the SR parameters increases the order of the SR expressions by one.

One can construct infinite towers of SR parameters encoding the full dynamics of the inflationary model. For instance, Eq. (3.1) provides the first entries of the "Hubble" SR tower; another, sometimes more convenient tower is the potential tower defined as

$$\epsilon_{\phi V,0} \equiv \frac{q}{6\beta_q^2}\frac{V'^2}{V^{1+q}}, \tag{3.3}$$

$$\epsilon_{\phi V,n} \equiv \frac{1}{3\beta_q^2}\left[\frac{V^{(n+1)}(V')^{n-1}}{V^{nq}}\right]^{1/n}, \qquad n \geq 1, \tag{3.4}$$

in the case of the normal scalar field. The potential tower can be related to the Hubble tower by approximated relations [3].

The first SR parameter is actually the time derivative of the Hubble radius $R_H \equiv H^{-1}$. Because of its purely geometrical content, it cannot be implemented in these SR towers recursively. By definition, there is inflation when $\epsilon < 1$:

$$\frac{\ddot{a}}{a} = H^2(1-\epsilon). \tag{3.5}$$

Under the *slow-roll approximation*, if the potential term dominates over the kinetic term, then the inflaton slowly rolls down its potential, $\epsilon, \eta \ll 1$, and the perfect fluid mimics that of a cosmological constant, $p \approx -\rho$. Deviations from the de Sitter behaviour generate large-scale perturbations which explain the anisotropies in the CMB.

4. Cosmological perturbations: theory and observations

Quantum fluctuations of the scalar field governing the accelerated era are inflated to cosmological scales because of the superluminal expansion. They constitute the seeds of both the small anisotropies observed in the microwave sky and the large-scale nonlinear structures around which gravitating matter organizes itself. For an introduction of the subject in the general relativistic case, see [5]. The standard procedure to adopt in order to compute the perturbation spetrum is: (a) Write the linearly perturbed metric in terms of gauge-invariant scalar quantities. (b) Compute the effective action of the scalar field fluctuation and the associated equation of motion. (c) Write the perturbation amplitude as a function of an exact solution of the equation of motion with constant SR parameters. (d) Perturb this solution with small variations of the parameters.

In scenarios with an extra dimension the full computation is very nontrivial due to either the extra degrees of freedom in the 5D metric and the complicated geometrical background on which to solve the Einstein equations coupled with the junction conditions on the brane. However, as explained above things become simpler when going to the large-scale limit. In this case, several arguments show that the resulting spectra are, to lowest SR order,

$$A = \frac{k}{5\pi z}, \tag{4.1}$$

$$z(\phi) = \frac{a\dot{\phi}}{H}, \tag{4.2}$$

$$z(T) = \frac{a\dot{T}}{c_S \beta_q^{1/q} H^{\theta/2}}, \tag{4.3}$$

$$z(h) = \frac{\sqrt{2}a}{\kappa_4 F_q}, \tag{4.4}$$

$$F_q^2 \equiv \frac{3q\beta_q^{2-\theta} H^\theta}{\zeta_q \kappa_4^2}, \tag{4.5}$$

where $A(h) = A_t$ is the tensor spectrum of the gravitational sector and ζ_q is a numerical constant which depends on the concrete gravity model one is considering: it is $\zeta_1 = 1 = \zeta_{2/3}$ for the 4D and GB cases and $\zeta_2 = 2/3$ for RS [6].

To lowest order, the scalar and tensor spectral indices are first order in the SR parameter, $n_t \equiv d\ln A_t^2/d\ln k \sim O(\epsilon) \sim n_s - 1 \equiv d\ln A_s^2/d\ln k$, while their running $\alpha_{s,t} \equiv dn_{s,t}/d\ln k$ is second order. Here k is the comoving wave number of the perturbation and the subscripts s and t refer to scalar and tensor perturbations, respectively. In the case of exact scale invariance, $n_s = 1$ and $n_t = 0$. The tensor-to-scalar ratio is

$$r \equiv A_t^2/A_s^2 = \epsilon/\zeta_q + O(\epsilon^2). \tag{4.6}$$

Combining the SR expressions of the observables, one gets the consistency equations

$$n_t = -(2+\theta)\zeta_q r + O(\epsilon^2), \tag{4.7}$$

$$\alpha_t = (2+\theta)\zeta_q r[(2+\theta)\zeta_q r + (n_s - 1)], \tag{4.8}$$

$$\alpha_s(\phi) \approx \zeta_q r[4(3+\theta)\zeta_q r + 5(n_s - 1)], \tag{4.9}$$

$$\alpha_s(T) \approx (3+\theta)\zeta_q r[(2+\theta)\zeta_q r + (n_s - 1)]. \tag{4.10}$$

The key point is that the set of consistency relations is not degenerate when considering different patches θ and θ'. The only known (accidental) degeneracy is for Eq. (4.7) in the RS and 4D case, where $n_t = -2r$ at first SR order. However, the second-order version of this equation together with the expressions for the runnings definitely break the degeneracy. This implies that, at least in principle, braneworld scenarios can be discriminated between each other.

To quantify the effect of the extra dimension, we can use the recent CMB data coming from WMAP [7]. With the upper bound $r < 0.06$ for the tensor-to-scalar ratio and the best-fit value $n_s \approx 0.95$ for the scalar spectral index, the relative scalar running in two different patches is

$$\alpha_s^{(\theta,\psi)} - \alpha_s^{(\theta',\psi')} \sim O(10^{-2}), \tag{4.11}$$

which is close to the WMAP estimate of the experimental error. This estimate will be highly improved by either the updated WMAP data set and near-future experiments, including the European Planck satellite, for which the forecast precision should be ameliorated by one order of magnitude, $\Delta\alpha_s \sim O(10^{-3})$.

5. Conclusions

In this paper we have summarized some results on braneworld inflation and their observable consequences. We have not presented a full 5D calculation but we expect that bulk physics would not dramatically improve large-scale results [4]. The study of the microwave background could give the first clues of a wider spacetime in the next years or even months.

In addition to the brane conjecture, one may insert other exotic ingredients, borrowed from string and M theory, that may give rise to characteristic predictions, although at the price of increasing the number and complexity of concurring models. For instance, the introduction of a noncommutative scale can generate a blue-tilted spectrum and explain, at least partially, the low-multipole suppression of the CMB spectrum detected by WMAP (e.g. [8]).

It would be interesting to find new cosmological scenarios with $\theta \neq 0, \pm 1$ and exploit the compact formalism provided by the patch formulation of the cosmological dynamics. Certainly there could be a lot of work for M/string theorists in this direction.

A final important question is in order: Will the CMB be the smoking gun of extra dimensions? In the context of the patch formalism the answer, unfortunately, is no. Some general relativistic models may predict a set of values for the observables $\{n_t, n_s, r, \alpha_s \dots\}$ close to that of a braneworld within the experimental sensitivity. Even noncommutativity may not escape this "cosmic degeneracy" since, for example, a blue-tilted spectrum can be achieved by the 4D hybrid inflation. So we can talk about clues but not proofs about high-energy cosmologies when examining the experimental data. The subject has to be further explored in a more precise way than that provided here in order to find out more compelling and sophisticated predictions, extending the discussion also to the small-scale region of the spectrum.

Acknowledgements

It is a pleasure to thank the organizers for their kind hospitality at Cargèse Summer School 2004.

References

1. V.A. Rubakov, Phys. Usp. **44** (2001) 871, hep-ph/0104152; R. Maartens, Living Rev. Relativity **7** (2004) 1, gr-qc/0312059; P. Brax, C. van de Bruck, and A.-C. Davis, hep-th/0404011.
2. W.D. Goldberger and M.B. Wise, Phys. Rev. Lett. **83** (1999) 4922, hep-ph/9907447; W.D. Goldberger and M.B. Wise, Phys. Lett. B **475** (2000) 275, hep-ph/9911457.

3. G. Calcagni, Phys. Rev. D **69** (2004) 103508, hep-ph/0402126.
4. D.S. Gorbunov, V.A. Rubakov, and S.M. Sibiryakov, J. High Energy Phys. **10** (2001) 015, hep-th/0108017; C. Gordon and R. Maartens, Phys. Rev. D **63** (2001) 044022, hep-th/0009010; K. Ichiki, M. Yahiro, T. Kajino, M. Orito, and G.J. Mathews, Phys. Rev. D **66** (2002) 043521, astro-ph/0203272; B. Leong, A. Challinor, R. Maartens, and A. Lasenby, Phys. Rev. D **66** (2002) 104010, astro-ph/0208015; K. Koyama, Phys. Rev. Lett. **91** (2003) 221301, astro-ph/0303108; K. Koyama, D. Langlois, R. Maartens, and D. Wands, hep-th/0408222.
5. V.F. Mukhanov, H.A. Feldman, and R.H. Brandenberger, Phys. Rep. **215** (1992) 203.
6. D. Langlois, R. Maartens, D. Wands, Phys. Lett. B **489** (2000) 259, hep-th/0006007; J.F. Dufaux, J.E. Lidsey, R. Maartens, M. Sami, hep-th/0404161; G. Calcagni, hep-ph/0406057.
7. C.L. Bennett *et al.*, Astrophys. J., Suppl. Ser. **148** (2003) 1, astro-ph/0302207; D.N. Spergel *et al.*, Astrophys. J., Suppl. Ser. **148** (2003) 175, astro-ph/0302209; H.V. Peiris *et al.*, Astrophys. J., Suppl. Ser. **148** (2003) 213, astro-ph/0302225; S.L. Bridle, A.M. Lewis, J. Weller, and G. Efstathiou, Mon. Not. R. Astron. Soc. **342** (2003) L72, astro-ph/0302306.
8. G. Calcagni, Phys. Rev. D **70** (2004) 103525, hep-th/0406006; G. Calcagni and S. Tsujikawa, Phys. Rev. D **70** (2004) 103514, astro-ph/0407543.

CHAOTIC CASCADES FOR D-BRANES ON SINGULARITIES

SEBASTIÁN FRANCO
Center for Theoretical Physics, Massachusetts Institute of Technology, Cambridge, MA 02139, USA

YANG-HUI HE
Dept. of Physics and Math/Physics RG, Univ. of Pennsylvania, Philadelphia, PA 19104, USA

CHRISTOPHER HERZOG
Kavli Institute for Theoretical Physics, University of California, Santa Barbara, CA 93106, USA

AND

JOHANNES WALCHER
Institute for Advanced Study, Princeton, NJ 08540, USA

Abstract. We briefly review our work on the cascading renormalization group flows for gauge theories on D-branes probing Calabi-Yau singularities. Such RG flows are sometimes chaotic and exhibit duality walls. We construct supergravity solutions dual to logarithmic flows for these theories. We make new observations about a surface of conformal theories and more complicated supergravity solutions.

1. Introduction

Extending the revolutionary AdS/CFT correspondence [1] beyond the original relation between $\mathcal{N} = 4$ SYM on N D3-branes and Type IIB supergravity (sugra) in $AdS_5 \times S^5$ with N units of RR 5-form flux on the S^5 is important to understanding realistic strongly coupled field theories such as QCD.

Two standard extensions have been (1) *reducing the SUSY to* $\mathcal{N} = 1$ by placing the D3-branes transverse to a Calabi-Yau singularity (the dual sugra background becomes $AdS_5 \times X^5$, where X^5 is some non-spherical horizon); and (2) *breaking conformal invariance and inducing an RG flow*, by introducing fractional branes, i.e., D5-branes wrapped over collapsing 2-cycles of the singularity (in the sugra dual, 3-form fluxes are turned on). A fascinating type of RG flow is the *duality cascade*: Seiberg duality is used to switch to an alternative description whenever infinite coupling is reached. This idea was introduced in [2] for the gauge theory on D-branes probing the conifold.

L. Baulied et al. (eds.), String Theory: From Gauge Interactions to Cosmology, 305–309.

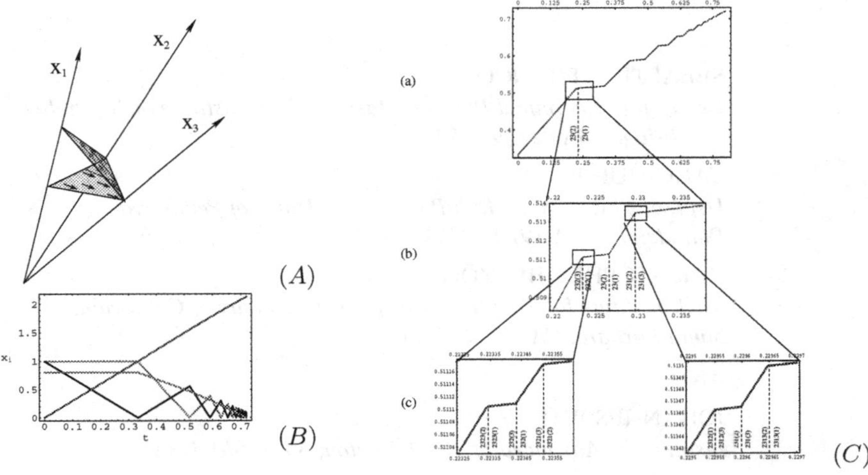

Figure 1.

2. Cascades in coupling space

There is an interesting way to look at cascading RG flows. In a gauge theory described by a quiver with k gauge groups, the inverse squared couplings $x_i \equiv 1/g_i^2$ are positive and define a k-dimensional cone $(\mathbb{R}_+)^k$. Inside this cone, the RG flow generates a trajectory dictated by the beta functions and satisfying $\sum_i^k r^i/g_i^2 \;=\; \text{constant}$. Each step between dualizations then corresponds to a straight line in the simplex defined by the intersection between this hyperplane and the $(\mathbb{R}_+)^k$ cone. We show such a trajectory in Figure 1(A). Now, each wall of the cone corresponds to one of the gauge couplings going to infinity. Therefore, whenever one of them is reached, we switch to a Seiberg dual theory at weak coupling. There will be then a different simplex associated to the dual theory. The entire cascade corresponds to a flow in the space of glued simplices. From this perspective, which resembles a billiard bouncing in coupling space, one foresees that cascading RG flows will exhibit chaotic behavior.

3. Duality Walls and Fractals

After introducing the notion of a duality cascade, it is natural to wonder whether some supersymmetric extension of the Standard Model, such as the MSSM, can sit at the IR endpoint of a cascade. This question was posed by Matthew Strassler [3]. Generically, while trying to reconstruct such a RG flow, one encounters a UV accumulation point beyond which Seiberg duality cannot proceed. This phenomenon is dubbed a *duality wall* and has been constructed for gauge theories engineered with D-branes on singularities [4]. Figure 1(B) shows the behavior of couplings for a cascade with a duality wall for the theory on D-branes over a complex cone over the Zeroth-Hirzebruch surface F_0 [1].

Postponing the question of a possible UV completion of duality walls, we can study the dependence of its position on initial couplings. Illustrating with F_0, the result is remarkable and is presented in Figure 1(C). The curve is a fractal, with concave and convex cusps. Whenever we zoom in on a convex cusp, an infinite, self-similar structure of more cusps emerges.

One subtle point which was not emphasized in [5] involves the existence in coupling space of a codimension two surface of conformal theories for F_0 and the other del Pezzo quiver gauge theories. If the number of gauge couplings is $n+2$, then a naive counting of the linearly independent β-functions constrains only two combinations of gauge couplings when the theory is conformal, leaving an n-dimensional surface of conformal theories. This n-dimensional surface is parametrized on the gravity side by the dilaton and the integral of the NSNS B_2 form through $n-1$ independent 2-cycles.

The existence of this codimension two surface may well affect the existence and behavior of the duality wall for F_0. In [4], it was assumed that a generic choice of initial couplings would lie on the conformal surface. However, if the initial conditions do not lie on the conformal surface, one expects large coupling constant corrections to the anomalous dimensions, which will in turn affect the strengths of the β-functions.

4. Supergravity Duals

The main support for the idea of a cascading RG flow in the original case of the conifold comes from a supergravity dual construction. This dual reproduces the logarithmic decrease in the effective number of colors towards the IR and also matches the beta functions for the gauge couplings.

In [5], analog supergravity solutions were constructed describing logarithmic cascades for the gauge theories on D-branes probing complex cones over del Pezzo surfaces. The fact that this was possible is remarkable, since they were obtained without knowing the explicit metric. These supergravity solutions are of the general type studied by Graña and Polchinski [6].

The general form of the metric is a warped product of flat four-dimensional Minkowski space and a Calabi-Yau $\mathbf{X}$

$$ds^2 = Z^{-1/2}\eta_{\mu\nu}dx^\mu dx^\nu + Z^{1/2}ds^2_{\mathbf{X}} \, , \tag{4.1}$$

The solution also carries 3-form flux $G_3 = F_3 - \frac{i}{g_s}H_3$. In order to preserve $\mathcal{N} = 1$ supersymmetry, G_3 must be supported only on X, imaginary self-dual, a $(2,1)$ form and harmonic. Indeed, it is possible to construct a G_3 satisfying all these condition. It has the form

$$G_3 = \sum_{I=1}^{n} a^I(\eta + i\frac{dr}{r}) \wedge \phi_I \tag{4.2}$$

where the ϕ_I, $I = 1\ldots n$, are a basis of $(1,1)$ forms orthogonal to the Kähler class of the del Pezzo and $\eta = \left(\frac{1}{3}d\psi + \sigma\right)$. The one-form σ satisfies $d\sigma = 2\omega$, with ω the Kähler form on dP_n, and $0 \le \psi < 2\pi$ is the angular coordinate on the circle bundle over dP_n.

[1] We refer the reader to [4, 5] for a detailed description of the associated quiver theory.

The intersection product between the ϕ_I is $\int_{dP_n} \phi_I \wedge \phi_J = -A_{IJ}$, where A_{IJ} is the Cartan matrix for the exceptional Lie algebra $\mathcal{E}_n$. There is a different type of fractional brane associated to each ϕ_I, given by D5-branes wrapping the 2-cycle in the del Pezzo Poincaré dual to ϕ_I.

Let us now study the number of D5-branes and D3-branes associated to these solutions

D5-Branes: The number of D5-branes is given by the Dirac quantization of the RR 3-form F_3: $a^J = 6\pi\alpha' M^J$. Hence, this family of solutions are dual to cascades in which the number of fractional branes of each type remains constant.

D3-Branes: Similarly, the effective number of D3-branes is computed from $F_5 = \mathcal{F}_5 + *\mathcal{F}_5$ where $\mathcal{F}_5 = d^4x \wedge d(Z^{-1})$ and Z is the warp factor in (3.1). The factor Z satisfies the equation

$$\nabla^2_{\mathbf{X}} Z = -\frac{1}{6}|H_3|^2 \,. \tag{4.3}$$

In [5], $|F_3|^2$ was assumed to be a function only of the radius, in which case

$$Z(r) = \frac{2 \cdot 3^4}{9-n}\alpha'^2 g_s^2 \left(\frac{\ln(r/r_0)}{r^4} + \frac{1}{4r^4}\right) \sum_{i,j} M^I A_{IJ} M^J \tag{4.4}$$

and from Dirac quantization, the number of D3-branes will grow logarithmically: $N = \frac{3}{2\pi} g_s \ln(r/r_0) \sum_{I,J} M^I A_{IJ} M^J$. However, generically, Z may depend on other coordinates on the Calabi-Yau cone $\mathbf{X}$.[2] The function Z averaged over the other coordinates may still be logarithmic in r [7].

5. Recent developments

Recently, there has been further progress in the study of quiver theories and their sugra duals. In [8], a-maximization [9] was used to compute the volume of the 5d horizon of the dual of the dP_1 gauge theory, yielding an irrational value. This result corrected previous computations in the literature and was obtained by carefully taking into account the global symmetries that are actually preserved by the superpotential. In [5], the duality cascade for dP_1 was analyzed using naive R-charges that did not take into account these global symmetries. A stable elliptical region in coupling space was found with a self-similar logarithmic cascade. Redoing the analysis with the new R-charges, we find the same elliptical region albeit with a slightly different shape and center.

The 5d horizon for the complex cone over dP_1 is called $Y^{2,1}$ and is a member of an infinite family of Sasaki-Einstein geometries denoted $Y^{p,q}$. They have $S^2 \times S^3$ topology and their metrics have been explicitly constructed [10]. Furthermore, the gauge theory duals to the entire $Y^{p,q}$ family have been constructed [11]. These developments change profoundly the status of the AdS/CFT, providing an infinite number of field theories with explicit sugra duals.

[2]We would like to thank Q. J. Ejaz for telling us about this possibility.

Acknowledgements

We would like to thank Qudsia Jabeen Ejaz, Ami Hanany, Pavlos Kazakopoulos Igor Klebanov and Joe Polchinski for useful discussions. This research is funded in part by the CTP and the LNS of MIT and by the department of Physics at UPenn. The research of S. F. was supported in part by U.S. DOE Grant #DE-FC02-94ER40818. The research of Y.-H. H. was supported in part by U.S. DOE Grant #DE-FG02-95ER40893 as well as an NSF Focused Research Grant DMS0139799 for "The Geometry of Superstrings". C. H. was supported in part by the NSF under Grant No. PHY99-07949. The research of J. W. was supported by U.S. DOE Grant #DE-FG02-90ER40542. S. F. would like to thank the organizers of Cargese Summer School, where this material was presented.

References

1. J. M. Maldacena, "The large N limit of superconformal field theories and supergravity," Adv. Theor. Math. Phys. **2**, 231 (1998) [Int. J. Theor. Phys. **38**, 1113 (1999)] [arXiv:hep-th/9711200].
2. I. R. Klebanov and M. J. Strassler, "Supergravity and a confining gauge theory: Duality cascades and chiSB-resolution of naked singularities," JHEP **0008**, 052 (2000) [arXiv:hep-th/0007191].
3. M. J. Strassler, "Duality in Supersymmetric Field Theory and an Application to Real Particle Physics," Talk given at International Workshop on Perspectives of Strong Coupling Gauge Theories (SCGT 96), Nagoya, Japan. Available at http://www.eken.phys.nagoya-u.ac.jp/Scgt/proc/
4. S. Franco, A. Hanany, Y. H. He and P. Kazakopoulos, "Duality walls, duality trees and fractional branes," arXiv:hep-th/0306092.
5. S. Franco, Y. H. He, C. Herzog and J. Walcher, Phys. Rev. D **70**, 046006 (2004) [arXiv:hep-th/0402120].
6. M. Grana and J. Polchinski, "Supersymmetric three-form flux perturbations on AdS(5)," Phys. Rev. D **63**, 026001 (2001) [arXiv:hep-th/0009211].
7. C. Herzog, Q. J. Ejaz,and I. Klebanov, "Cascading RG Flows for New Sasaki-Einstein Manifolds," arXiv:hep-th/0412193.
8. M. Bertolini, F. Bigazzi and A. L. Cotrone, "New checks and subtleties for AdS/CFT and a-maximization," arXiv:hep-th/0411249.
9. K. Intriligator and B. Wecht, "The exact superconformal R-symmetry maximizes a," Nucl. Phys. B **667**, 183 (2003) [arXiv:hep-th/0304128].
10. D. Martelli and J. Sparks, "Toric geometry, Sasaki-Einstein manifolds and a new infinite class of AdS/CFT duals," arXiv:hep-th/0411238, and references therein.
11. S. Benvenuti, S. Franco, A. Hanany, D. Martelli and J. Sparks, "An infinite family of superconformal quiver gauge theories with Sasaki-Einstein duals," arXiv:hep-th/0411264.

$\mathcal{N} = 1$ GAUGE THEORY EFFECTIVE SUPERPOTENTIALS

BEN M. GRIPAIOS
Rudolf Peierls Centre for Theoretical Physics, University of Oxford, 1, Keble Road, Oxford, OX1 3NP, United Kingdom

Abstract. Recent advances in computing effective superpotentials of four dimensional $\mathcal{N} = 1$ SUSY gauge theories coupled to matter are described. The correspondence with matrix models and two proofs of this are discussed. By considering the case with matter transforming in the fundamental representation, a Wilsonian derivation of the Veneziano-Yankielowicz effective superpotential for a pure gauge theory is given using the generalized Konishi anomaly.

1. Introduction

This work provides a brief introduction to recent advances in the study of four-dimensional gauge theories with the minimal amount (that is $\mathcal{N} = 1$) of supersymmetry. We consider general theories of this type, with arbitrary gauge group coupled to matter in an arbitrary representation, and with an arbitrary superpotential for the matter. The aim is to compute the effective superpotential for the light gauge superfields obtained (*à la* Wilson) by integrating out the matter[1]. Minimising this effective superpotential then yields the quantum vacua of the theory. In doing so, one discovers two remarkable things [1]. The first remarkable thing is that the 'integrating out' part, which involves evaluating Feynman diagrams in superspace, reduces to the computation of planar diagrams in a matrix model. That is, the entire dependence on (super)space 'disappears'![2] The second remarkable thing is that, having done the perturbative computation in this way to n loops, one finds (upon minimising the effective superpotential) that one has calculated the n-instanton correction to the vacuum structure. Thus, a *perturbative* computation leads to *non-perturbative* information about the physics!

We stress that although the methods apply to general minimally supersymmetric four-dimensional gauge theories, they do not tell us *everything* about such theories. A complete specification of the theory is given by the effective action; here one is only able to calculate

[1]The matter must be massive, so that we can sensibly integrate it out.

[2]The disappearance is not a straightforward dimensional reduction however. If it were, the arguments presented here would hold in arbitrary dimensions; in fact they are specifically four-dimensional.

L. Baulied et al. (eds.), String Theory: From Gauge Interactions to Cosmology, 311–317.

the so-called F-terms (the effective superpotential) in the effective action. The Kähler, or D-terms (which lack the strong constraint of holomorphy) are not determined.

In the next section, we *sketch* two proofs [2, 3] of the gauge theory/matrix model correspondence (conjectured in [1]) and show how the superpotentials are calculated in each case. We then show that there are matter-independent (i.e. pure gauge) contributions to the effective superpotential which are undetermined. These contributions turn out to be non-perturbative and provide the bridge between the perturbative computation and the non-perturbative physics mentioned above. In section 3 we show that these contributions can in fact be determined in this framework [4], and we do this. These pure gauge theory contributions correspond to the Veneziano-Yankielowicz superpotential [5, 6]. In the framework presented here, the effective superpotential is derived by integrating out massive matter and the Veneziano-Yankielowicz superpotential is thus a *bona-fide* Wilsonian effective superpotential in this context.

2. The Gauge Theory/Matrix Model Correspondence

We employ $\mathcal{N} = 1$ superspace notations (see e.g. [7]). The gauge fields (and their superpartners) are written as components of a real vector superfield V; by acting with superspace covariant derivatives, one can form the analogue of the gauge field strength, $W_\alpha \sim \bar{D}^2 e^{-V} D_\alpha e^V$ and the gauge-invariant glueball chiral superfield $S \sim \mathrm{tr} W^\alpha W_\alpha$. The matter is represented by chiral superfields Φ with a tree-level matter superpotential in the action which is polynomial in the matter superfields[3]

$$\int d^4x d^2\theta W_{\text{tree}} = \int d^4x d^2\theta g_k \Phi^k. \tag{2.1}$$

Here, the coefficients g_k are called the tree-level matter couplings. We consider integrating out the matter Φ in some *background* glueball field S to obtain an effective superpotential W_{eff}, which depends on S, the g_k, and the gauge coupling (which we write in terms of the dimensionally-transmuted scale Λ).

It was claimed in the introduction that the perturbative computation of W_{eff} reduces to the evaluation of planar diagrams in a bosonic matrix model, and one might well ask how this can be demonstrated. Two proofs have appeared. The first [2] simply considers the contributing Feynman diagrams in superspace and shows that the momentum dependence of the bosons and their fermion superpartners cancels in all such diagrams. The only things left to consider are insertions of S, factors of g_k coming from the vertices, and numerical symmetry factors. One can show that these can be obtained from planar diagrams of the matrix model

$$\exp \frac{F(S)}{g_s^2} = \int d\phi \exp \frac{W_{\text{tree}}(\phi)}{g_s}, \tag{2.2}$$

where ϕ are $N' \times N'$ bosonic matrices and $S = g_s N'$. The restriction to planar diagrams is enforced by taking the 't Hooft limit: $N' \gg 1$ and $g_s \ll 1$, with S fixed. The action

[3] Only polynomials of degree three or less are renormalizable. However, since we claim that the computation of the effective superpotential reduces to a matrix model, the results must be independent of the momenta and any momentum cutoff. The results are thus independent of the UV completion of the theory and one is free to consider 'non-renormalizable' tree-level superpotentials.

of the matrix model is given by the tree-level matter superpotential W_{tree} with the matter superfields Φ replaced by bosonic matrices ϕ.

To compute the perturbative computation to W_{eff} obtained by integrating out the matter (e.g. for gauge group $SU(N)$), one evaluates

$$N\frac{\partial F(S)}{\partial S}, \tag{2.3}$$

where $F(S)$ is the perturbative free energy of the matrix model in the planar limit.

The second proof [3] is rather different. One considers the effect of general chiral changes of variables $\delta\Phi = \epsilon f(\Phi, W_\alpha)$ in the path integral. These lead to anomalous Ward identities generalizing the Konishi anomaly [8, 9]. For example, the variation $\delta\Phi = \epsilon\Phi'(\Phi)$ yields

$$\left\langle \Phi'\frac{\partial W_{\text{tree}}}{\partial \Phi} - S\frac{\partial \Phi'}{\partial \Phi}\right\rangle = 0. \tag{2.4}$$

From the general chiral change of variables specified by f, one obtains a complete set of anomalous Ward identities for the chiral matter fields, and one can show that these are in one-to-one correspondence with the complete set of Ward identities in the matrix model (which, since the matrix model partition function (2.2) is just an integral, correspond to integration by parts identities). This establishes the correspondence between the SUSY gauge theory and the bosonic matrix model.

Having established the correspondence, one can go on and calculate the effective superpotential for any given theory. To do this, one needs to solve the complete set of Ward identities to obtain the expectation values $\langle\Phi^k\rangle$ appearing in the tree-level matter superpotential in terms of the background glueball superfield S and the couplings g_k. The effective superpotential can then be determined from the partial differential equations

$$\frac{\partial W_{\text{eff}}}{\partial g_k} = \langle\Phi^k\rangle, \tag{2.5}$$

which follow from standard supersymmetry and holomorphy arguments.

We note that these partial differential equations only specify the effective superpotential up to a term which is independent of the matter couplings g_k, but which may depend on both S and the gauge coupling scale Λ. This term must contain any contribution to the effective superpotential coming from the *pure* gauge theory *without* matter. So let us ask the question: is there a pure gauge theory contribution? It turns out that there is, as was shown many years ago by Veneziano and Yankielowicz [5, 6] using the $U(1)_R$ symmetry of the pure gauge theory. For the gauge group $SU(N)$, for example, the pure gauge theory superpotential is

$$W_{\text{eff}}(S, \Lambda) = N\left(-S\log\frac{S}{\Lambda^3} + S\right). \tag{2.6}$$

Such terms are non-perturbative. One way to see this is to minimise W_{eff} with respect to S. This reproduces the vacuum condensate $S^N = \Lambda^{3N}$, due to (non-perturbative) instantons [10]. In the next section, we show how such terms can be derived using the generalized Konishi anomaly in the presence of matter discussed above [4]. This then renders the above approach self-contained, as well as providing an independent derivation of the Veneziano-Yankielowicz terms.

Before doing so, one might ask where the missing terms are hidden in the perturbative approach using Feynman diagrams. One might assume that they correspond to diagrams with gauge superfields in the loops; this is not really correct, since the missing terms are non-perturbative and thus cannot show up in any diagram. Intriguingly, these terms can be generated by the measure of the matrix model (i.e. the volume of the gauge group) [11, 1], though it is not at all clear why.

3. Pure Gauge Terms

In order to derive the pure gauge theory contributions, we determine the effective superpotential in the case where the matter sector consists of F flavours of 'quarks' transforming in the fundamental representation of the gauge group, which we take to be $SU(N)$ (though the argument can be applied to any classical Lie group). Furthermore, we choose a tree-level superpotential in which the quarks can have either zero or non-zero classical expectation values at the minima. If a quark has a non-zero *vev*, then since the quarks transform non-trivially under the gauge group, the gauge group must be spontaneously broken via the Higgs mechanism. By putting each of the F quarks at zero or non-zero minima, we can *engineer* the gauge symmetry breaking such that the unbroken gauge group is anything from $SU(N)$ down to $SU(N-F)$. We then solve the Konishi anomaly Ward identities (2.4) and the resulting partial differential equations (2.5), determining the effective superpotential in each vacuum, up to a constant term (by 'constant' we mean 'independent of the tree-level matter couplings').

The tree-level matter couplings are free parameters in the theory. We vary them such that both the quark masses and the Higgs *vevs* (which determine the masses of the massive gauge bosons) become large. In that limit, the massive matter decouples from the unbroken low energy gauge group, and the effective superpotential contains a sum of contributions from the decoupled matter and the low energy gauge group. Once we have identified the contribution of the massive matter and discarded it, we are left with the superpotential of the low energy gauge group. This includes the constant term.

Now any two distinct vacua have different unbroken gauge groups, but the same constant term. If we subtract the two superpotentials (with the massive matter discarded), the constant cancels and we are left with a difference equation for the pure gauge theory superpotential. The solution to this difference equation yields precisely the Veneziano-Yankielowicz terms (2.6). To determine the constant term in any theory, one then demands that $W_{\text{eff}}(S, g_k, \Lambda)$ reproduces the correct decoupled contributions of the unbroken gauge group and massive matter in any vacuum in the massive limit [12]. Incidentally, the fact that the matching in one vacuum correctly reproduces the superpotential in all vacua justifies *a posteriori* the assumption that the constant term is the same for each vacuum branch.

Having explained the argument, let us now carry it out. Since quarks are Dirac fermions and chiral supermultiplets contain Weyl fermions, we represent F flavours of quarks by F chiral superfields Q_i transforming in the fundamental representation of $SU(N)$ and a further F chiral superfields $\tilde{Q}^j$ transforming in the anti-fundamental representation.

The tree-level matter superpotential is written in terms of the gauge invariant mesons $M_i^j = Q_i\tilde{Q}^j$ as

$$W_{\rm tree} = m{\rm tr}M - \lambda{\rm tr}M^2. \tag{3.1}$$

The classical vacua are then

$$mM_i^j - 2\lambda M_i^k M_k^j = 0, \tag{3.2}$$

with F_- eigenvalues at $M_i^i = 0$ and $F_+ = F - F_-$ eigenvalues at $M_i^i = m/2\lambda$. If M_i^i has a non-zero vev, then so have Q_i and $\tilde{Q}^i$, and the gauge symmetry is broken. The low energy gauge group is thus broken down to $SU(N - F_+)$. The quantum theory has the Konishi anomaly and the classical vacua are modified to (2.4)

$$m\langle M_i^j\rangle - 2\lambda\langle M_i^k M_k^j\rangle = \delta_i^j S, \tag{3.3}$$

with $F_\pm$ eigenvalues at

$$\langle M_i^i\rangle = \frac{m}{4\lambda}\left(1 \pm \sqrt{1 - \frac{8\lambda S}{m^2}}\right). \tag{3.4}$$

The partial differential equations following from holomorphy and supersymmetry are [3]

$$\begin{aligned}\frac{\partial W_{\rm eff}}{\partial m} &= \langle {\rm tr}M\rangle,\\ \frac{\partial W_{\rm eff}}{\partial \lambda} &= -\langle {\rm tr}M^2\rangle.\end{aligned} \tag{3.5}$$

We shall not write here the expression for the effective superpotential $W_{\rm eff}$ which is obtained by integrating these equations (it is rather cumbersome). Taking the limit of $W_{\rm eff}$ in which the quark mass m and Higgs vev $\sqrt{m/2\lambda}$ become large and subtracting the superpotentials for the vacua in which the number of Higgsed quarks is $F_{1,2}$, one obtains

$$W_{\rm eff,1} - W_{\rm eff,2} \to (F_1 - F_2)\frac{m^2}{4\lambda} + (F_1 - F_2)\left[S\log\frac{S}{m^2/2\lambda} - S\right]. \tag{3.6}$$

The first term represents the decoupled matter: it is given (according to the non-renormalization theorem) by the classical expectation value of $W_{\rm tree}$. The second term must therefore represent the contribution of the low energy pure gauge group $SU(N - F_{1,2})$.[4] It seems peculiar that what we have identified as the superpotential of the low energy gauge group contains the matter couplings m and λ. However, these are precisely the factors needed to convert the $SU(N)$ gauge coupling scale Λ to the $SU(N - F_{1,2})$ scales $\Lambda_{1,2}$ via the scale-matching relation

$$\Lambda_1^{3(N-F_1)}\left(\frac{m^2}{2\lambda}\right)^{F_1} = \Lambda^{3N-F}m^F = \Lambda_2^{3(N-F_2)}\left(\frac{m^2}{2\lambda}\right)^{F_2}. \tag{3.7}$$

[4]There is a subtlety here: the glueball superfield S includes the massive gauge bosons, which should be integrated out by replacing them with their *vevs*. However, in the decoupled limit, the massive gauge bosons have zero *vevs*, so the field S is equivalent to the glueball superfield of the low energy gauge group once the massive gauge bosons have been integrated out.

This relation comes from requiring that the coupling constants of the high energy theory (with dynamic matter) and the low energy theory (with matter integrated out) match at the Higgs and quark mass scales (see *e.g.* [13]). Replacing the matter couplings by the appropriate gauge coupling scales in this way (and discarding the massive matter) leads to the difference equation

$$W_{\text{eff},1} - W_{\text{eff},2} = (N - F_1)\left(-S\log\frac{S}{\Lambda_1^3} + S\right) - (N - F_2)\left(-S\log\frac{S}{\Lambda_2^3} + S\right), \quad (3.8)$$

with solution

$$W_{\text{eff}}(S, \Lambda) = N\left(-S\log\frac{S}{\Lambda^3} + S\right) + f(S). \quad (3.9)$$

Here, $f(S)$ is an arbitrary function of S alone; it is independent of all other parameters. On dimensional grounds, $f(S) \propto S$ and one sees that the ambiguity in f (which can be re-written as a pure number multiplying Λ^{3N}) corresponds to the freedom to choose a renormalisation group scheme [13].

4. Discussion

The methods summarised above provide a very powerful framework in which to study gauge theories with $\mathcal{N} = 1$ SUSY, and it is certainly of interest to go on and study the vacuum structure and phases of specific models.

More general extensions to this work include the question of whether similar results hold in dimensions other than four [14], the extension to supergravity (rather than super-gauge) backgrounds [1, 15, 16, 17, 18] and whether dynamical breaking of supersymmetry may be studied in this framework.

References

1. R. Dijkgraaf and C. Vafa, *A perturbative window into non-perturbative physics*, hep-th/0208048.
2. R. Dijkgraaf, M. T. Grisaru, C. S. Lam, C. Vafa, and D. Zanon, *Perturbative computation of glueball superpotentials*, hep-th/0211017.
3. F. Cachazo, M. R. Douglas, N. Seiberg, and E. Witten, *Chiral rings and anomalies in supersymmetric gauge theory, JHEP* **12** (2002) 071, [hep-th/0211170].
4. B. M. Gripaios and J. F. Wheater, *Veneziano-Yankielowicz superpotential terms in N = 1 SUSY gauge theories, Phys. Lett.* **B587** (2004) 150.
5. G. Veneziano and S. Yankielowicz, *An effective Lagrangian for the pure N=1 supersymmetric Yang-Mills theory, Phys. Lett.* **B113** (1982) 231.
6. T. R. Taylor, G. Veneziano, and S. Yankielowicz, *Supersymmetric QCD and its massless limit: An effective Lagrangian analysis, Nucl. Phys.* **B218** (1983) 493.
7. S. J. Gates, M. T. Grisaru, M. Rocek, and W. Siegel, *Superspace, or one thousand and one lessons in supersymmetry, Front. Phys.* **58** (1983) 1–548, [hep-th/0108200].
8. K. Konishi, *Anomalous supersymmetry transformation of some composite operators in SQCD, Phys. Lett.* **B135** (1984) 439.
9. K.-I. Konishi and K.-I. Shizuya, *Functional integral approach to chiral anomalies in supersymmetric gauge theories, Nuovo Cim.* **A90** (1985) 111.

10. V. A. Novikov, M. A. Shifman, A. I. Vainshtein, and V. I. Zakharov, *Instanton effects in supersymmetric theories*, *Nucl. Phys.* **B229** (1983) 407.
11. H. Ooguri and C. Vafa, *Worldsheet derivation of a large N duality*, *Nucl. Phys.* **B641** (2002) 3–34, [hep-th/0205297].
12. A. Brandhuber, H. Ita, H. Nieder, Y. Oz, and C. Romelsberger, *Chiral rings, superpotentials and the vacuum structure of N = 1 supersymmetric gauge theories*, hep-th/0303001.
13. J. Terning, *Tasi-2002 lectures: Non-perturbative supersymmetry*, hep-th/0306119.
14. R. Dijkgraaf and C. Vafa, *N = 1 supersymmetry, deconstruction, and bosonic gauge theories*, hep-th/0302011.
15. A. Klemm, M. Marino, and S. Theisen, *Gravitational corrections in supersymmetric gauge theory and matrix models*, *JHEP* **03** (2003) 051, [hep-th/0211216].
16. R. Dijkgraaf, A. Sinkovics, and M. Temurhan, *Matrix models and gravitational corrections*, hep-th/0211241.
17. J. R. David, E. Gava, and K. S. Narain, *Konishi anomaly approach to gravitational F-terms*, hep-th/0304227.
18. B. M. Gripaios, *Superpotentials for glueball and conformal supergravity backgrounds*, *Commun. Math. Phys.* in press, hep-th/0311025.

GROMOV-WITTEN THEORY AND AUTOMORPHIC FORMS

DANIEL GRÜNBERG
visiting Racah Institute, Jerusalem

Abstract. Here's a quick introduction to Gromov-Witten invariants, their origin in topological strings, their relation to BPS states counting, and the automorphic properties of their generating function (GW potential). The plan is to interpolate between Morozov's and Dijkgraaf's lectures, i.e. between KdV hierarchies (or their solutions as τ-functions) and topological strings (or correlators of operators on a worldsheet).

1. Recap TFT

Topological Field Theory studies maps f from a Riemann surface Σ_g of genus g (string worldsheet) to a target space X. Examples are obtained by twisting non-linear sigma models with CY threefold as targets. So what is twisting ?

For a NLσM, bosons ϕ are simply coordinates on the target space (like the map above), while fermions squat in bundles:

$$\psi_+ \in K^{1/2} \otimes f^*(TX) \qquad \psi_- \in \bar{K}^{1/2} \otimes f^*(TX)$$

K is the canonical bundle[1] on Σ_g. Since X is a complex manifold (even Kähler), we can split $\psi_\pm$ into holomorphic and anti-holomorphic components, $\psi^i_\pm$ and $\psi^{\bar{i}}_\pm$.

Twisting consists in multiplying the bundles by $K^{\pm 1/2}$ so that now

$$\psi^i_+ \in f^*(TX) \qquad \psi^{\bar{i}}_+ \in K \otimes f^*(\bar{T}X).$$

Idem for ψ_-. There are two kinds of twisting, leading to the A- and B-models, according to whether you twist the holomorphic or anti-holomorphic components of ψ_+ and ψ_- in the same way or opposite. Twisting ensures that the Lagrangian can now be rewritten in very compact form (here for the A-model):

$$\mathcal{L} = t \underbrace{\int_{\Sigma_g} f^*(\gamma)}_{=:d \text{ (degree of } f)} + it \int_{\Sigma_g} \{Q, V\}$$

[1] ie the highest power of the cotangent bundle (here: power 1, hence line bundle Ω of differential forms, $c_1(K) = 2g - 2$).

L. Baulied et al. (eds.), String Theory: From Gauge Interactions to Cosmology, 319–324.

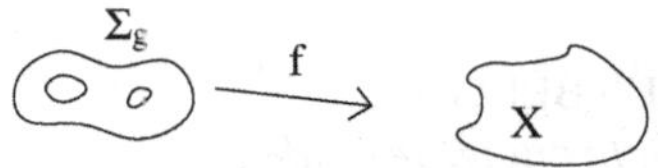

Figure 1. An instanton, ie a map from the worldsheet to the target space.

where t plays the role of coupling constant ($t = 1/g_s$), Q is the BRST operator and $V \sim g_{i\bar{j}}\ \psi_+ \bar{\partial}\phi$. We shall incorporate t into the Kähler form γ on X, so that it becomes a complex Kähler parameter, and $\int_{\Sigma_g} f^*(\gamma) = \int_{f_*[\Sigma_g]} \gamma = \int_d \gamma = d \cdot t$. Here, $d \in H_2(X)$ is called the degree of the map f.

Varying the complex structure of X will vary the metric $g_{i\bar{j}}$ and hence V, which will merely generate vanishing terms of the form $\{Q, \dots\}$; the action will remain unchanged. Thus our model does not depend on the complex structure of X nor on that of the worldsheet ! This is the reason why we call it **topological**. [2]

For the same reason, $e^{-it\{Q,\int V\}}$ is actually independent of t, and so we'll go to the limit Re $t \to \infty$ to evaluate the path integral for the action. What sort of maps (or 'instantons') contribute the most to the path integral for $e^{-\mathcal{L}} = e^{-dt}\, e^{-it\{Q,\int V\}}$? Since t is large, the phase will wobble (and cancel itself out) for any value of V, except if V vanishes, ie if $\bar{\partial}\phi = 0$. Since the boson ϕ is just the map f (I should have used the same letter), this means *only holomorphic maps contribute*, ie holomorphic instantons.

Warning: there are four different wordings in the literature for one and the same limit Re $t \to \infty$: *weak coupling* limit (since $t \sim 1/g_s$), *large volume* limit (since t became a Kähler parameter of X), *holomorphic* limit and *topological* limit (reasons above).

2. Correlation Functions

What operators can we insert on the worldsheet Σ_g to build correlators ? Each operator will stand in correspondence with a complex subvariety of X. Choose n subvarieties $Z_1, \dots, Z_n$; their homology classes have duals $\gamma_1, \dots, \gamma_n \in H^{2*}(X)$. For each choice of Z (or γ), we build the operator

$$\mathcal{O}_\gamma(p) := \gamma_{i_1 \dots i_k}(f(p))\ \psi_+^{i_1} \dots \psi_+^{i_k}$$

of ghost number k, and supported on Z. The point p is a marking on Σ_g such that $f(p) \in Z$, representing the insertion of (the pullback of) $\mathcal{O}_\gamma$. [3]

We are now in a position do define the correlators:

$$\langle \mathcal{O}_{\gamma_1} \dots \mathcal{O}_{\gamma_n} \rangle_{g,d} := \int_{\overline{\mathcal{M}}_{g,n}(X,d)} D\phi D\psi\, e^{-\mathcal{L}}\, \mathcal{O}_{\gamma_1} \dots \mathcal{O}_{\gamma_n},$$

[2]For the B-model, there is no dependence on the Kähler structure of X and on the complex structure of the worldsheet.

[3]As an aside, I can disclose to you that $\{Q, \mathcal{O}_\gamma\} = -\mathcal{O}_{d\gamma}$, that is, the BRST cohomology of the A-model TFT is isomorphic to the de Rham cohomology of X. In other words, the chiral ring is isomorphic to the (quantum) cohomology ring.

where $\overline{\mathcal{M}}_{g,n}(X,d)$ is the moduli space of n-pointed holomorphic maps from Σ_g to X of degree d. We go again to the large t limit to make connection with the correlators en vogue among algebraic geometers:

$$\langle\gamma_1\dots\gamma_n\rangle_{g,d} := \int_{\overline{\mathcal{M}}_{g,n}(X,d)} \mathrm{ev}^*(\gamma_1\wedge\dots\wedge\gamma_n),$$

where the evaluation map at each point p_i is simply $\mathrm{ev}_i : \overline{\mathcal{M}}_{g,n}(X,d) \to X : \{\mathrm{map} f\} \mapsto f(p_i)$. Again, this is to vanish unless $f(p_i) \in Z_i\ \forall i$! Note that from the mathematical perspective, the operators have dropped off of the picture. The enumerative meaning associated with $\langle\gamma_1\dots\gamma_n\rangle_{g,d}$ is the 'number of maps' of degree d such that $f(p_i) \in Z_i$. These are the *Gopakumar-Vafa invariants* (invariants of the complex structure of X).

3. BPS Invariants

One has to be careful in your choice of differential forms γ_i (how many and of what degrees); after all, their total degree has to match the dimension of the moduli space to be integrated over, which is known to be

$$\begin{aligned}\dim_{\mathbb{C}}\overline{\mathcal{M}}_{g,n}(X,d) &= d\cdot c_1(X) + (\dim X - 3)(1-g) + n\\ &= n \quad \text{for CY threefolds}\\ &= 3d-1+n \quad \text{for rational curves (ie. } g=0\text{) in } \mathbf{CP}^2\\ &\overset{!}{=} \tfrac{1}{2}\sum \deg\gamma_i\end{aligned}$$

Thus, when studying plane rational curves (embeddings of the line $\mathbf{CP}^1$), we could choose $3d-1$ four-forms $\gamma_i \simeq [pt.]$ so that lhs and rhs equal $6d-2$. The latter are dual to points Z_i in the complex plane, and so the geometric meaning of $f(p_i) \in Z_i$ is: you need to specify $3d-1$ points to end up with a finite number of rational curves (of degree d) passing through them (eg. degree 1: through 2 points passes 1 line).

That in itself is quite cool to know, but you can do better: compute that number, $\langle[pt.]^{3d-1}\rangle_{0,d}$, via the so-called WDVV equation (asserts associativity of the quantum cohomology ring) and end-up with a recursion formula (by Kontsevich): 1,1,12, 620,... for $d = 1,2,3,4,\dots$.

Similarly, for CY threefolds, we could very well choose $n = 0$, ie no insertion of operators (hence no γ_i), and both the integrand and integration space are 0-dimensional. In that case, the output is defined to be the degree of the 0-cycle $[\overline{\mathcal{M}}_{g,n}(X,d)]$ and denoted by $\langle 1\rangle_{g,d}$ or anew by $N_{g,d}$. There's no constraint $f(p_i) \in Z_i$ needed, ie embedded curves are already isolated, they don't show up in infinite families. Here too, the $N_{0,d}$ were computed (this time using Mirror Symmetry for the quintic, by Candelas et al), and yielded 2875, 4876875/8,... for $d = 1,2,\dots$.

But they are not integers! What went wrong ? In fact, $N_{0,d}$ are not quite the numbers of rational curves in a CY threefold, but rather the 'numbers of maps'. If n_d is the number of rational curves of degree d, then it will also contribute to $N_{0,2d}$ via a double cover; etc. The two quantities are linearly related: $N_{0,d} = \sum_{k|d} k^{-3}\, n_{d/k}$, so that now we do obtain integers: $n_d = 2875, 609250,\dots$ for $d = 1,2,\dots$. For higher genus, the linear relation still holds: $N_{g,d} = \sum_{k|d,g'\geq g}\cdots n_{g',d/k}$ where the dots are universal coefficients (rational numbers, known for CY threefolds).

These new instanton invariants $n_{g,d}$ (believed to be integers by the Gopakumar-Vafa conjecture) count BPS states (or D-branes) wrapped around cycles of X. There are attempts to define them via the moduli space $\mathcal{M}$ of such objects: $n_{g,d} = (-1)^{\dim \mathcal{M}}\chi(\mathcal{M})$ (can be negative!). Their mathematical meaning is still shadowy.

4. GW Potentials

The genus-g topological string amplitude is obtained by summing over all degrees of the instanton:

$$F_g := \sum_{d,n\geq 0} \frac{1}{n!}\langle \mathcal{O}_{\gamma_1} \ldots \mathcal{O}_{\gamma_n}\rangle_{g,d}$$

The algebro-geometric version of this is the *GW potential* at genus g:

$$F_g(t_i) := \sum_{d,n\geq 0} \frac{1}{n!} \sum_{i_1\ldots i_n} t_{i_1}\ldots t_{i_n}\langle \gamma_1 \ldots \gamma_n\rangle_{g,d}$$

The complex parameters t_i are gathered together into one class $\gamma = \sum t_i\gamma_i$ (γ_i form a basis of $H^{2*}(X)$), and the last sum is simply $\langle\gamma^n\rangle_{g,d}$.

For CY threefolds, the correlators $\langle\gamma^n\rangle_{g,d}$ vanish except when $\gamma \in H^2(X)$, ie a Kähler class, in which case they give $(-d\cdot t)^n\langle 1\rangle_{g,d}$ (by the divisor axiom). So the GW potential can be more suggestively written as

$$F_g(t_i) = \sum_{d\geq 0}\langle 1\rangle_{g,d}\sum_{n\geq 0}\frac{(-d\cdot t)^n}{n!} = \sum_{d\geq 0} N_{g,d}\, e^{-d\cdot t} = \sum_{\text{holo maps } f} \exp\int_{\Sigma_g} f^*(\gamma)$$

which is the starting point of topological string theory: summing over all holomorphic maps, whereby $N_{g,d}$ is the size of the moduli space of maps of a given degree.

Let us collect all amplitudes into a single generating function, the full GW potential or 'free energy':

$$F(t_i,\lambda) = \sum_{g\geq 0} F_g\lambda^{2g-2}$$

where λ is the string coupling. The *partition function* is then $Z = e^F$, which can always be rewritten as an infinite product – at least formally:

$$Z = e^F = \prod_{l,d}(1-y^l q^d)^{c_{l,d}}, \qquad y := e^{i\lambda},\ q := e^{-t} \tag{4.1}$$

Compare this with the vacuum τ-function in Morozov's lecture in this volume: $\log\tau = \sum_{g\geq 0} F_g\lambda^{2g-2}$, which is related to the solution u of the KdV hierarchy by $\partial_x^2\log\tau = u$.

5. KdV Hierarchy

Indeed, the *Virasoro conjecture* claims that $u := \partial_{t_0}^2 F$ is a solution of the 'KdV hierarchy', a single equation which can be unwound into an infinite set (or hierarchy) – denoting

t_0 by x:

$$\begin{array}{lll} k=1: & \lambda^{-2}\partial_{t_1}u & = \frac{1}{2}uu_x + \frac{1}{4}u_{xxx} \\ k=2: & \lambda^{-4}\partial_{t_2}u & = \frac{1}{2}u^2u_x + \frac{1}{6}u_xu_{xx} + \frac{1}{12}u_{xxx} + \frac{1}{240}\partial_x^5 u \\ \text{etc,} & & \end{array}$$

of which the first line is the traditional KdV equation (hence the name).

This KdV hierarchy for $\partial_{t_0}^2 F$ is equivalent to another hierarchy for Z, namely $L_m Z = 0$ for $m \geq -1$, where the L_m form the Virasoro algebra $[L_k, L_m] = (k-m)L_{k+m}$ (hence the name for the conjecture). Explicitly:

$$\begin{array}{lll} L_{-1} & = & \partial_{t_0} + t_0^2 + \sum_{i\geq 0} t_{i+1}\partial_{t_i} \\ L_0 & = & \partial_{t_1} + \sum_{i\geq 0} t_i \partial_{t_i} \\ L_1 & = & \ldots \end{array}$$

(see again Morozov's lectures.)

If we rather work with the simpler moduli space of curves, $\overline{\mathcal{M}}_{g,n}$, instead of $\overline{\mathcal{M}}_{g,n}(X,d)$, yet keep all the formalism above, then the KdV hierarchy is known to hold (proof by Kontsevich, conjecture by Witten). In this case, the correlators are simply $\langle \gamma_{i_1} \ldots \gamma_{i_n} \rangle_{g,d} := \int_{\overline{\mathcal{M}}_{g,n}} c_1(\mathcal{L}_1)^{i_1} \ldots c_1(\mathcal{L}_n)^{i_1}$ for the tautological line bundles[4] $\mathcal{L}_i$ (no mention of X nor of any map f anymore).

6. Automorphic Products

For an example of (4.1), the product form is explicitly known for the contribution from constant maps ($d = 0$) in case of CY threefolds with Euler characteristic χ (see 'melting crystals'):

$$Z_{\text{const}} = \left[\prod (1-y^l)^l\right]^{-\chi/2}.$$

The reason we would like to know the explicit product form of the full partition function Z (ie know the powers $c_{l,d}$) is that such products can have enticing automorphic properties, ie the function transforms covariantly under some group (eg: modular forms have automorphism group $SL(2, \mathbf{Z})$). Borcherds has proved that the product (4.1) is automorphic under the Lorentz group $O(2, s+2; \mathbf{Z})$ if the powers $c_{l,d}$ are the coefficients in the Taylor expansion of a modular form. This can be generalised to the lifting of Jacobi forms.

Eg: Set $j(\tau) - 744 =: \sum c(n)q^n$, a modular function of weight 0. Then $p^{-1}\prod_{m>0, n\in\mathbf{Z}} (1-p^m q^n)^{c(mn)}$ is an automorphic form of weight 0 for the group $O(2,2;\mathbf{Z})$. Since this group is isomorphic to $SL(2,\mathbf{Z}) \times SL(2,\mathbf{Z}) \cdot \mathbf{Z}_2$, we expect a function modular in each variable p, q and symmetric under swap of p and q. Indeed, it's easy to show the product boils down to $j(p) - j(q)$. This is the case $s = 0$ from Borcherds' theorem.

Eg: More easily, since $12\,\theta = 12(1 + 2q + 2q^2 + \ldots)$ is a modular form of weight 1/2, the product $q\prod(1-q^n)^{24}$ is an automorphic form for $SL(2,\mathbf{Z})$. Indeed, this is just the discriminant Δ. Similarly, you could write all Eisenstein series in product form

[4]The fibre of $\mathcal{L}_i$ at each point of $\overline{\mathcal{M}}_{g,n}$ (ie above each n-pointed genus-g curve $\Sigma_{g,n}$) is simply the cotangent line at the ith marked point.

with powers given in a closed formula: $E_4 = (1-q)^{240}(1-q^2)^{\cdots} \ldots$. Also: $j(\tau) = q^{-1}(1-q)^{1728} \ldots$.

Now what can we learn from this ? Lots. Automorphic forms – let alone modular forms – tie the knot with virtually all corners of maths. For instance, in number theory, their coefficients (in Taylor expansion of q) often have interesting properties. Ramanujan studied the generating function for partitions, $q^{1/24}/\eta =: \sum c(n)q^n$, and discovered smashing congruences: $c(7n+5) \equiv 0 \bmod 7$, or $c(11n+6) \equiv 0 \bmod 11$, etc.

If GW invariants $N_{g,d}$ or BPS invariants $n_{g,d}$ are contained in powers of GW potentials that have automorphic properties, its natural to expect that they have similar congruences. Wild conjecture! Tiny evidence yet! Only in few examples of heterotic string compactifications on certain CY threefold are there known results for the prepotential F_0 (genus-0) with a suitable product formula for $\exp(F_0)$.

If true (or even half-true), these speculations turn would shed light on the enumerative meaning of these invariants, and on the geometry of the curves or of the moduli space they pertain to.

Acknowledgements

I am particularly grateful for the support from the Racah Institute that enabled my visit to Israel's vibrant string theory community. Ditto for the organisers of the Cargèse 2004 school.

References

1. D. Grünberg, *Gromov-Witten Theory from a Stringy Perspective*, unofficial thesis, unpublished,
`http://www.physik.hu-berlin.de/~grunberg/thesis.ps (or .pdf)`

BLACK HOLE INSTABILITIES AND PHASE TRANSITIONS

SEAN HARTNOLL
DAMTP, Centre for Mathematical Sciences, Cambridge University
Wilberforce Road, Cambridge CB3 OWA, UK

Abstract. Generalised black holes have a horizon given by an arbitrary Einstein manifold. I review a criterion for the classical stability of these black holes. Roughly, spherical horizons are stable but lemon-shaped horizons can be unstable. In Anti-de Sitter space these black holes are dual to gauge theory on a curved background given by the same Einstein manifold. The dual thermal field theory effect is a novel phase transition induced by inhomogeneous Casimir pressures and characterised by a "condensation of pressure".

1. Generalised AdS black holes

The AdS/CFT correspondence relates the physics of black holes in AdS_{d+2} to a thermal field theory living on the 'boundary' $S^1 \times S^d$. Here we discuss an instability and dual phase transition that arises when the black hole has boundary $S^1 \times \mathcal{M}$, with $\mathcal{M}$ a general Einstein manifold.

Let us review the stability of generalised black holes [1, 2] with a negative cosmological constant. Generalised black holes take the form

$$ds^2 = -f(r)dt^2 + \frac{dr^2}{f(r)} + r^2 ds_d^2 \,, \tag{1.1}$$

where $f(r) = 1-(1+r_+^2/L^2)(r_+/r)^{d-1} + r^2/L^2$. The event horizon is at $r = r_+$. If ds_d^2 were the round metric on S^d, then the spacetime would be the usual Schwarzschild black hole in AdS space. However, the $d+2$ dimensional vacuum Einstein equations allow the horizon metric ds_d^2 to be any d dimensional Einstein metric on a Riemannian manifold $\mathcal{M}$

$$R_{\alpha\beta} = (d-1)g_{\alpha\beta} \,, \tag{1.2}$$

where α, β are indices for $\mathcal{M}$. The resulting spacetime is called a generalised black hole. The asymptotic geometry of the background has now changed and the dual thermal field theory lives on $S^1 \times \mathcal{M}$ rather than $S^1 \times S^d$.

The linearised stability of generalised black holes under perturbations $g_{ab} \to g_{ab} + h_{ab}$ has recently been investigated [1, 2, 3]. For large, $r_+/L \gg 1$, generalised AdS black holes

L. Baulied et al. (eds.), String Theory: From Gauge Interactions to Cosmology, 325–328.

one finds

$$\lambda_L < -A^2 \frac{r_+^2}{L^2} \quad \Leftrightarrow \quad \text{instability}, \tag{1.3}$$

where λ_L is the minimum eigenvalue of the Lichnerowicz operator on $\mathcal{M}$ and A^2 is a positive $\mathcal{O}(1)$ number that may be determined numerically [2]. Recall that the Lichnerowicz operator acts on symmetric rank two tensors as

$$(\Delta_L h)_{ab} = 2R^c{}_{abd}h^d{}_c + R_{ca}h^c{}_b + R_{cb}h^c{}_a - \nabla^c\nabla_c h_{ab} . \tag{1.4}$$

Note from (1.3) that the minimum Lichnerowicz eigenvalue needs to be very negative in order for instability to occur. Examples of Einstein manifolds with large negative Lichnerowicz eigenvalues are the Böhm metrics on $S^5 \ldots S^9$ and on products of spheres [3].

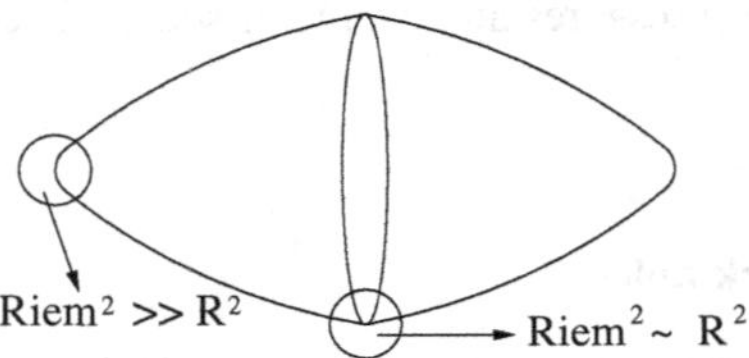

Figure 1. The Böhm metrics have regions of large curvature near $\theta = 0$ and $\theta = \theta_f$. The Ricci scalar is constant.

The Böhm metrics on S^5 have the form

$$ds_5^2 = d\theta^2 + a(\theta)^2 d\Omega_2^2 + b(\theta)^2 d\widetilde{\Omega}_2^2 , \tag{1.5}$$

where $d\Omega_2^2$ and $d\widetilde{\Omega}_2^2$ are round metrics on spheres S^2 and $\widetilde{S}^2$. The coordinate θ has range 0 to θ_f. The Einstein equations imply some nonlinear differential equations for $a(\theta)$ and $b(\theta)$. The boundary conditions for topology S^5 are that $a(0) = 0, b(0) = b_0$ and $a(\theta_f) = a_0, b(\theta_f) = 0$. In fact, the solutions to the equations on S^5 have $a(\theta) = b(\theta_f - \theta)$. There is a discrete infinity of solutions, at specific values of b_0. Solutions exist for arbitrarily small b_0. As $b_0 \to 0$ the metric develops conical singularities at $\theta = 0, \theta = \theta_f$. Thus b_0 gives the resolution of the singularity. Figure 1.1 illustrates a Böhm metric. Also as $b_0 \to 0$ one finds that $\lambda_L \to -\infty$ [3]. Therefore if we use Böhm metrics as the horizon of a generalised black hole then the black hole will be unstable if the curvature at the tips of the manifold is sufficiently large.

The connection between large curvatures and the Lichnerowicz spectrum which appears for the Böhm metrics is more general. On Einstein manifolds with positive curvature the minimum Lichnerowicz eigenvalue is related to the Weyl curvature. Roughly, one expects [3]

$$\lambda_L^2 \sim C_{\alpha\beta\gamma\delta}C^{\alpha\beta\gamma\delta}|_{\text{max.}} , \tag{1.6}$$

where $C_{\alpha\beta\gamma\delta}C^{\alpha\beta\gamma\delta}|_{\text{max.}}$ is the maximum value taken by $C_{\alpha\beta\gamma\delta}C^{\alpha\beta\gamma\delta}$ on the manifold $\mathcal{M}$. One finds that the unstable mode tends to relax the curvatures of the spacetime near the horizon.

2. Dual phase transition

The criterion for instability (1.3) may be translated into a dual field theory language using the standard AdS/CFT dictionary to give [2]

$$T^2 < -\lambda_L \times \mathcal{O}(1) \quad \Leftrightarrow \quad \text{instability}\,, \tag{2.1}$$

where T is the temperature of the field theory. Thus the duality predicts a critical temperature in the dual field theory on $S^1 \times \mathcal{M}$:

$$T_C^2 \sim |\lambda_L|\,. \tag{2.2}$$

The unstable mode of the generalised black hole is a tensor mode on $\mathcal{M}$. That is, the mode has no legs in the t or r directions and satisfies

$$\nabla^\alpha h_{\alpha\beta} = 0\,, \qquad h^\alpha{}_\alpha = 0\,. \tag{2.3}$$

Furthermore the unstable mode has finite energy and is thus the "normalisable mode" in the AdS/CFT language. Therefore the growth in $h_{\alpha\beta}$ due to the instability is mapped via the AdS/CFT correspondence to a growth in the expectation value of the spatial components of the energy momentum tensor

$$h_{\alpha\beta} \sim e^{\omega t} \quad \Rightarrow \quad < T_{\alpha\beta} > \sim e^{\omega t}\,. \tag{2.4}$$

Combining the two facts (2.2) and (2.4) suggests that the dual thermal field theory undergoes a phase transition when cooled below the temperature T_C. The phase transition is characterised by a "condensation of pressure", that is, by the spatial components of the energy momentum tensor gaining a vacuum expectation value.

The appearance of the pressure suggests a thermodynamic characterisation of the instability. Recent work on black brane instabilities has supported the Gubser-Mitra conjecture which states that translationally invariant horizons are classically stable if and only if they are locally thermodynamically stable. Local thermodynamic stability requires the Helmholtz free energy of the system to be a concave function of temperature and a convex function of volume. The known instabilities of black branes are due to a negative heat capacity

$$C_V = -T \left.\frac{\partial^2 F}{\partial T^2}\right|_V < 0\,. \tag{2.5}$$

The large generalised AdS black holes which have been considered here always have positive heat capacity. The alternative thermodynamic instability is to have a negative isothermal compressibility [4]

$$K_T^{-1} = V \left.\frac{\partial^2 F}{\partial V^2}\right|_T = -V \left.\frac{\partial P}{\partial V}\right|_T < 0\,. \tag{2.6}$$

If this were indeed the thermodynamic instability of the dual field theory that triggers the phase transition, then generalised black holes would provide an example of a generalisation of the Gubser-Mitra correspondence to horizons without translational invariance.

A negative compressibility would most likely arise in the regions of the manifold with large curvature, as this is where the Casimir pressures are greatest. The compressibility would want to relax the curvatures to reduce the Casimir effect. This mimics the behaviour of the bulk unstable mode mentioned above. The field theory background is not dynamical, so the condensation of pressure provides an alternative means of cancelling the wrong-sign Casimir pressure.

Acknowledgements

It is my please to thank Gary Gibbons and Chris Pope for the initial collaborations which stimulated my interest in generalised black holes. I would like to thank also the organisers of the Cargèse 2004 ASI.

References

1. G. Gibbons and S. A. Hartnoll, Phys. Rev. D **66** (2002) 064024.
2. S. A. Hartnoll, JHEP **0308** (2003) 019.
3. G. W. Gibbons, S. A. Hartnoll and C. N. Pope, Phys. Rev. D **67** (2003) 084024.
4. S. A. Hartnoll, Phys. Rev. D **70** (2004) 044015.

PROBING SINGULARITIES

VERONIKA E. HUBENY
Department of Physics & LBNL
University of California, Berkeley, CA 94720, USA

Abstract. We review the method of holographically extracting information about the singularity of a Schwarzschild-AdS black hole and contrast this with probing cosmological singularities.

1. Introduction

Curvature singularities are a rather generic feature in general relativity. As demonstrated by the singularity theorems [1], even relatively mild initial data can evolve to a singularity. Indeed, in nature, we expect singularities inside black holes as well as cosmological ones. Of course, the laws of general relativity break down at a singularity, thereby precluding us from evolving the classical spacetime into its causal future. But the real world is not classical, and it should not suffer from singularities; in fact we expect quantum effects to become important long before the actual singularity is reached. Unfortunately, to answer the intriguing and fundamental question of what actually happens requires understanding of quantum gravity, which is beyond the scope of our present knowledge.

Nevertheless, probing singularities within the quantum framework which is available—namely string theory—should provide us with further clues to the full theory of quantum gravity. String theory has already proved very successful in resolving many timelike singularities. In simple examples like orbifold singularities, perturbative string dynamics suffices to explain the resolution, while more complicated examples such as the conifold singularity require stringy non-perturbative effects. Spacelike singularities are even more subtle, and we expect that a fully non-perturbative formulation of string theory will be essential for their resolution.

While there already exist several remarkable nonperturbative formulations of string theory, the one best suited for our purposes is the well-known AdS/CFT correspondence [2]. In particular, we will use the duality between 10-dimensional IIB string theory on asymptotically $AdS_5 \times S^5$ spacetimes and 4-dimensional $\mathcal{N} = 4$ Super Yang Mills gauge theory, living on the boundary of AdS. Since the bulk has gravity, it can contain black holes, as well as cosmological regions. Once we discover how these are represented by

L. Baulied et al. (eds.), String Theory: From Gauge Interactions to Cosmology, 329–335.

the dual (non-gravitational) gauge theory, we will be able to learn how the respective singularities are resolved.

In the following section, we review the method proposed by FHKS [3] of probing the spacelike singularity inside a Schwarzschild-AdS black hole. Although one can extract specific information about the singularity from certain correlators in the dual CFT, this extraction is rather subtle and in particular requires analytic continuation through a branch cut. This reflects the fact that the singularity is hidden behind an event horizon, and therefore is not directly "visible" from the boundary where the CFT resides.

On the other hand, if the singularity could be directly visible from the boundary, one might expect the CFT to encode it more directly and hence provide a more explicit mechanism for singularity resolution. HHM [3] had presented scenarios which seemed to guarantee the requisite naked singularities; however after more detailed study [5], we believe it unlikely that one will be able to find a clean example with naked singularities (see M. Rangamani's talk in this volume).

The set-up of [3], however, included a cosmological region in the bulk, with Big Bang and Big Crunch singularities, cast within the AdS/CFT framework. One may therefore ask whether one can revert to the methods of FHKS to probe these cosmological singularities. In section 3, we show that the answer is *no*; the black hole and cosmological singularities in fact behave very differently from each other.

2. Black hole singularity

To study a singularity and an event horizon in the bulk gravity theory, we will consider the Schwarzschild-AdS geometry, describing a large black hole in AdS, whose gauge dual corresponds to an approximately thermal state. By analyzing the properties of this state, such as the expectation values of various operators, we can hope to extract information about the black hole. Indeed, due to the powerful nature of the AdS/CFT duality, we expect to be able to decode the full quantum near-singularity bulk physics from the gauge dual.

However, since such decoding requires that the gauge theory encodes behind-the-horizon physics in the first place, we should first ask, *how much of the bulk physics does the gauge dual encode?* Naive considerations might lead one to expect that the CFT encodes only the region of the bulk which is causally connected to the boundary, *i.e.* outside the event horizon. On the other hand, event horizon is a global object (defined as the boundary of the past of future infinity), which means that we cannot determine the presence or position of the event horizon without knowing the entire future evolution of the spacetime. Hence if the horizon were to bound what the CFT can encode, then the AdS/CFT correspondence would have to be very nonlocal in time.

We illustrate this point by a simple gedanken-experiment [6]: The holographic nature of the AdS/CFT correspondence suggests that as a boundary observer, one can obtain instantaneous information about the event in the bulk, for instance by measuring appropriately decorated Wilson loop [7]. Once this information is obtained, *i.e.* entirely to the future of this measurement, the boundary observer may collapse in a shell which forms a black hole whose horizon will have encompassed the measured event.

While the above argument indicates that the CFT should encode at least some physics inside a horizon, it does not lend itself to detailed computational analysis. Instead, it is more fruitful to turn to a simpler set-up, namely that of the eternal black hole, which is static outside the horizon and has both future and past singularities, as well as *two* boundaries. This geometry was analyzed previously in 3 dimensions by a number of authors [8, 9, 10]; here we study the higher-dimensional analog [3]. The real-time or thermofield formalism for thermal field theory [11] is especially well-suited for this analysis. One copy of the CFT resides on each of the two asymptotic regions; the CFTs are noninteracting but entangled through the Hartle-Hawking state [9]. In this approach, a single boundary thermal description is recovered by tracing over the Hilbert space of the other boundary CFT.

Moreover, as demonstrated previously for 3 dimensions [10], one can probe physics behind the horizon by studying the correlator of two operators, one on each asymptotic boundary, each creating a large mass bulk particle. As the mass $m \to \infty$, the correlator can be evaluated in the semi-classical geodesic approximation and is given by $\exp(-m\mathcal{L})$, where $\mathcal{L}$ is the (regularized) proper length of the spacelike geodesic joining the boundary points. Because the geodesic passes through spacetime regions inside the horizon, this boundary correlator reveals information about the geometry behind the horizon.

Unfortunately, the three dimensional case, which is simple enough to study analytically, is rather special. The geometry is locally that of pure AdS_3, and the black hole singularity is merely a result of the orbifold nature of this geometry [12]. Consequently, the geodesics are not sensitive to the position of the singularity, and correspondingly the correlation function is relatively structureless.

On the other hand, this situation changes drastically in higher dimensions. In particular, the singularity in $d > 3$ dimensions approaches that of the d-dimensional Schwarzschild black hole singularity, which has a dramatic effect on the spacelike geodesics: pictorially, spacelike radial geodesics "bounce off" from the singularity. For simplicity, we focus on $d = 5$ where the boundary CFT is four dimensional $\mathcal{N} = 4$ SYM; however, our results are qualitatively similar to those in other higher dimensions as well.

The first surprise, underlining the marked difference from the previously-studied 3-dimensional case, is that the Penrose diagram of the Schw-AdS$_5$ spacetime is different. While the Schw-AdS$_3$ geometry has a square-shaped Penrose diagram, this can no longer hold in higher dimensions [3]. One reason, apparent from Fig.1.1, is that outgoing radial null geodesics starting at the past singularity do not reach the boundary at a time-symmetric point. Apart from the differences in the Penrose diagrams, there are two important features in the symmetric radial spacelike geodesics in higher dimensions which are absent in the 3-D case: There exists a particular time t_c beyond which there are no geodesics connecting the two boundaries, and the geodesics cross each other.

This has an important consequence for our present goal of finding a signature of the singularity in the dual field theory. Consider the boundary to boundary correlator $\langle\phi\phi\rangle(t)$ between two high-dimension operators inserted in a symmetric fashion on the two boundaries at time t. Since this correlation function is related to the proper length of a spacelike geodesic connecting the two points, we expect that it reflects the special behaviour of the geodesics as $t \to t_c$. In fact, direct evaluation suggests that we should see a pole in the correlation function:

$$\langle\phi\phi\rangle(t) \sim e^{-m\mathcal{L}(t)} \sim \frac{1}{(t-t_c)^{2m}} \quad \text{as} \quad t \to t_c\,. \tag{2.1}$$

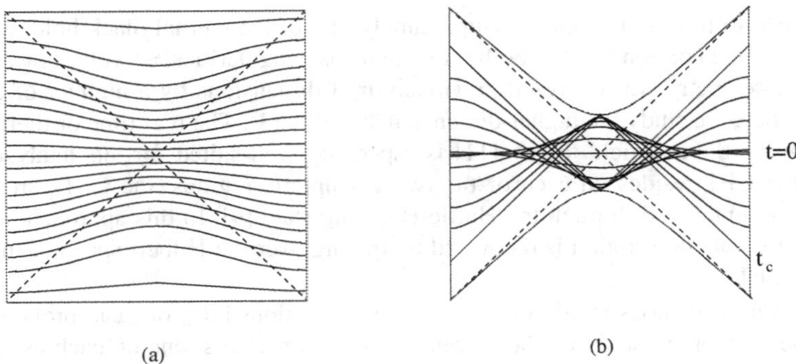

Figure 1. Penrose diagrams of Schw-AdS in (a) 3 and (b) 5 dimensions. The top and bottom curves are the future and past singularities, the vertical lines are the boundaries, the dashed diagonal lines are the event horizons, and the remaining curves are spacelike radial geodesics.

This corresponds to a light-cone singularity in the field theory, since the geodesics are becoming almost null.

Had this been the full story, we would have a striking signature of the black hole singularity in the gauge theory dual. However, general considerations of the boundary field theory rule this out; one can easily show that $|\langle\phi\phi\rangle(t)| \leq |\langle\phi\phi\rangle(0)| < \infty$. What went wrong? In evaluating Eq.(2.1) we assumed that the correlator is dominated by a single geodesic, namely the real, "bounce" geodesic shown in Fig.1.1b. But, in fact, this geodesic does *not* dominate the correlator, for there are in general multiple geodesics that connect the two boundary points. One indicator of this fact is the intersection of nearby geodesics around $t = 0$; in a cut-off field theory there would be *three* geodesics connecting the two points at (cut-off) boundaries. Alternately, in a Euclidean picture, one real and two complex geodesics contribute. At $t = 0$ their proper distances coincide, creating a branch point in the correlator which behaves as $\mathcal{L} \sim t^{4/3}$ for small t. (The 3 in the denominator of the exponent, implying a 3-sheeted Riemann surface, corresponds to the 3 geodesics.)

By studying various resolutions of this branch point, one can show that as t increases from 0, the correlator defined by the boundary CFT is given by a symmetric sum of the two complex branches of this expression, each attributed to a complex geodesic in the complexified spacetime. But the correlator is an analytic function of t and can be continued onto the real sheet. This is the essential feature of the CFT which enables us to extract information which is not directly measurable. On the real sheet, the "light cone singularity" does appear. So the boundary correlator *does* contain information about the singularity, albeit in a subtle way.

3. Cosmological singularity

In the preceding section we have seen that despite the singularity being spacelike, we can study it in the CFT by analyzing analytically continued correlators which correspond to spacelike geodesics which bounce off the singularity. Let us now contrast this story with a different type of spacelike singularity, namely a cosmological (Big Crunch) singularity. Since one can embed cosmological singularities into an AdS/CFT set-up [3], one might expect that CFT correlators, related to bulk geodesics as above, would similarly yield information about the singularity. Surprisingly, this is not the case, due to a classical GR distinction between these two scenarios. Namely, spacelike geodesics can bounce off the black hole singularity, but they cannot bounce off the cosmological one.

To see this, let us for simplicity consider spherically symmetric spacetimes and focus on the radial geodesics. Writing the radial part of the metric suggestively as

$$ds^2 = -dt^2 + a^2(t)\, dr^2 \,, \tag{3.1}$$

we obtain the t-component of the geodesic equation:

$$\ddot{t} + a(t)\, a'(t)\, \dot{r}^2 = 0 \tag{3.2}$$

where $\dot{} \equiv \frac{d}{d\lambda}$ (λ being the affine parameter along the geodesic), and $' \equiv \frac{d}{dt}$. In order for the geodesic to bounce off from a (future) spacelike singularity, there must be a point along the geodesic at which $\dot{t} = 0$ while $\ddot{t} < 0$. Evidently, the first condition can be satisfied for spacelike geodesics; whether the second condition can be satisfied as well depends on the sign of $a'(t)$ (since $a(t) > 0$), as dictated by Eq.(3.2).

Consider now a homogeneous FRW cosmology with a Big Crunch. Then $a(t)$ is simply the scale factor. Near the future singularity the space is imploding, so that $a'(t) < 0$, which implies that $\ddot{t} > 0$ everywhere along the geodesic. Hence, spacelike geodesics cannot bounce off the singularity in these cosmologies.[1] Since spacelike geodesics cannot bounce, they will hit the singularity, which makes these FRW cosmologies spacelike geodesically incomplete. As we have seen, this is very different from the behaviour in the black hole geometry.[2]

As emphasized in the previous section, the classical feature that spacelike geodesics bounce off from the black hole singularity was crucial for extracting information about the Schwarzschild-AdS black hole singularity via the CFT [3]; in fact, studying the CFT correlators may provide further insight into the singularity resolution in quantum gravity. On the other hand, this feature of the singularity being absent in the cosmological case, the corresponding singularity resolution should be very different from that of the black hole[3].

[1]Note that to bounce off from a past Big Bang singularity, we would need $\ddot{t} > 0$ when $\dot{t} = 0$; but this cannot be satisfied either, because in that region $a'(t) > 0$.

[2]In order to write the metric in the form of Eq.(3.1), we need to perform a change of coordinates from the usual Schwarzschild form; near the singularity, this yields $a(t) \sim -1/t$, so that $a'(t) > 0$, which allows $\ddot{t} < 0$ when $\dot{t} = 0$. In other words, the Schwarzschild "time" direction, which is spacelike inside the horizon, is actually expanding as one approaches the future singularity.

[3]This argument assumes that the difference in the behaviour of certain specific *probes* of the singularity are manifestations of the difference in the actual singularity itself; this is by no means guaranteed, merely suggestive.

4. Discussion

In summary, we have seen that the CFT encodes physics behind the horizon, including the near-singularity region. Unlike the previously-studied three dimensional case, in higher dimensions there is a genuine curvature singularity which is bowed in on the Penrose diagram, leading to a special behaviour of the geodesics as a critical time t_c is approached. Despite subtleties related to the nontrivial analytic structure of $\mathcal{L}(t)$, the CFT correlators reveal distinct signals of the black hole singularity. In fact, given[4] the CFT data $\langle\phi\phi\rangle$, the properties of the singularity are computationally accessible [3]. Since the "t_c" singularity persists to all orders in $\frac{1}{m}$ and g_s, as well as for small l_s, we can also extract the stringy and quantum behaviour near the black hole singularity. Thus, using analyticity, we have demonstrated that a significant amount of information from behind the horizon, and in particular from near the singularity, is encoded in boundary theory correlators. Since analyticity was essential for our arguments, it would be worthwhile to understand its implications in the context of AdS/CFT at a much deeper level.

We have given an example of how we can learn about (and presumably resolve) spacelike singularities in the bulk, using AdS/CFT. Hitherto, it has been commonly assumed that understanding the black hole singularity will help us understand the cosmological Big Bang or Big Crunch singularities as well. We have seen that this may not be a well-founded expectation; the black hole singularities should be encoded very differently from the cosmological ones. The fact that, even at the classical level, the near-singularity geodesic properties differ may be viewed as a hint to the singularity resolutions being different. As already evidenced in our experience with timelike singularities, there are many different channels of resolution depending on the details of the singularity in question; it is not surprising that this situation has an analog in the context of spacelike singularities.

Acknowledgements

I wish to thank my collaborators, L. Fidkowski, M. Kleban, and especially M. Rangamani and S. Shenker. I would also like to thank also the organizers of the Cargèse 2004 ASI, and the Kavli Institute for Theoretical Physics for hospitality. This work is supported in part by the funds from the Berkeley Center for Theoretical Physics, and also by the DOE grant DE-AC03-76SF00098 and the NSF grants PHY-0098840 and PHY99-0794.

References

1. S. W. Hawking and G. F. R. Ellis, *The Large Scale Structure of Space-Time*, Cambridge University Press, (1973).
2. O. Aharony, S.S. Gubser, J. Maldacena, H. Ooguri, Y. Oz, Phys. Rept. 323 (2000) 183, hep-th/9905111.
3. L. Fidkowski, V. Hubeny, M. Kleban and S. Shenker, JHEP **0402**, 014 (2004) hep-th/0306170.
4. T. Hertog, G. T. Horowitz and K. Maeda, Phys. Rev. D **69**, 105001 (2004) hep-th/0310054.
5. V. E. Hubeny, X. Liu, M. Rangamani and S. Shenker, hep-th/0403198.

[4] Of course, it is not presently feasible to obtain the CFT "data" $\langle\phi\phi\rangle$ in the relevant regime by a direct computation: the CFT is strongly coupled, and $m \to \infty$ implies infinitely large dimension operators. Nevertheless, it is still of use to ask the matter-of-principle question: what does a given CFT information tell us about the bulk?

6. V. E. Hubeny, Int. J. Mod. Phys. D **12**, 1693 (2003) hep-th/0208047.
7. L. Susskind and N. Toumbas, *Phys. Rev. D* **61**, 044001 (2000) hep-th/9909013. B. Freivogel, S. B. Giddings and M. Lippert, *Phys. Rev. D* **66**, 106002 (2002) hep-th/0207083.
8. V. Balasubramanian and S. F. Ross, *Phys. Rev. D* **61**, 044007 (2000) hep-th/9906226. J. Louko, D. Marolf and S. F. Ross, *Phys. Rev. D* **62**, 044041 (2000) hep-th/0002111.
9. J. M. Maldacena, *JHEP* **0304**, 021 (2003) hep-th/0106112.
10. P. Kraus, H. Ooguri and S. Shenker, *Phys. Rev. D* **67**, 124022 (2003) hep-th/0212277.
11. M. Le Bellac, *Thermal Field Theory*, Cambridge University Press (1966).
12. M. Banados, C. Teitelboim and J. Zanelli, *Phys. Rev. Lett.* **69**, 1849 (1992) hep-th/9204099.

INFLATION UNLOADED

MATTHEW KLEBAN
Institute for Advanced Study, Einstein Drive
Princeton, NJ 08540, USA

Abstract. I present a brief review of astro-ph/0406099, which argues that there is a limit on the number of efolds of inflation which are *observable* in a universe which undergoes an eternally accelerated expansion in the future. Such an acceleration can arise from an equation of state $p = w\rho$, with $w < -1/3$, and it implies the existence of event horizons. In some respects the future acceleration acts as a second period of inflation, and "initial perturbations" (including signatures of the first inflationary period) are inflated away or thermalize with the ambient Hawking radiation. Thus the current CMB data may be looking as far back in the history of the universe as will ever be possible even in principle, making our era a most opportune time to study cosmology.

1. Introduction

Quantum fluctuations produced during inflation are imprinted on the curvature, and are subsequently stretched by inflation to super (Hubble) horizon scales[1]. Once there, they "freeze out", i.e. their amplitude approaches a constant set by the horizon crossing condition, and their wavelength scales with the particle horizon, $\lambda(t) = \lambda_0 \, a(t)/a_0$. What happens next depends on the subsequent evolution of $a(t)$. When inflation ends, the Hubble horizon begins to grow linearly in time, but the wavelength stretches more slowly, as $\lambda(t) \sim a(t)$. If the vacuum energy is zero, this situation will persist indefinitely, and the Hubble horizon eventually catches up with the perturbation (see the left panel of Fig. 1), after which it can collapse and form structure. A patient observer in such a universe can see arbitrarily far back in time by measuring these perturbations: the longer she waits, the farther back during inflation the fluctuations she sees were generated.

However, if at some time the post-inflationary universe begins to accelerate (and continues to do so forever), there will be event horizons [2]. In this case a part of the global spacetime is permanently inaccessible to any given observer, and the evolution of perturbations is very different. Depending on when they were produced, inflationary fluctuations may or may not re-enter the Hubble horizon during matter domination (see the right panel

[1]Because space is tightly constrained in these proceedings, I have included only the most immediately relevant references. Please see [1] for more.

L. Baulied et al. (eds.), String Theory: From Gauge Interactions to Cosmology, 337–340.

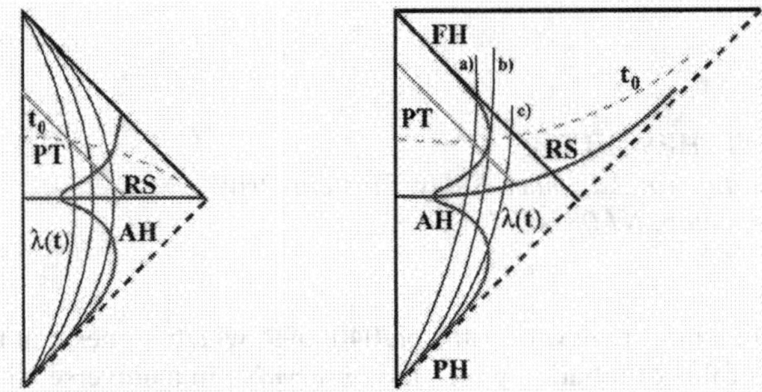

Figure 1. *Evolution of the wavelengths of some typical inflationary perturbations in a universe without (left panel) and with (right panel) event horizons. In the left panel, all fluctuations eventually reenter the Hubble horizon. In the right panel, in the case a), a fluctuation is stretched outside of the Hubble horizon during inflation, remains there for a time, then reenters during a matter dominated era after inflation, and eventually gets expelled out of the horizon once more during the final stage of acceleration. In b), the fluctuation would have reentered about now, but the late acceleration just prevents that. In c), the late acceleration prevents the fluctuation from ever reentering the Hubble horizon. AH = apparent horizon, RS = reheating surface, PT = photon trajectory , F(P) H = future(past) horizon,* $t_0 = now$.

of Fig. 1). If they do not re-enter by the time the universe begins to accelerate, they will never do so and hence will never be directly observable.

The photons which comprise the CMB originate on the slice (i.e. a sphere) of the last scattering surface which is separated from the observer by null geodesics (labelled PT in Fig. 1). In a decelerating universe the radius of this last scattering sphere grows without bound, and new information about inflation continues to become available over time.

In a universe which accelerates, the last-scattering sphere asymptotes to the size of the event horizon at the time of last scattering, which is finite. Therefore, the pattern of temperature anisotropies in the CMB "freezes" after the transition to future acceleration.

Eventually even this remnant will be permanently erased. Spacetimes with event horizons contain Hawking particles, and as the cosmological expansion continues, the CMB redshifts until it is colder than the Hawking radiation. After this time, any remaining information in the CMB will be masked by quantum effects.

2. Quantification

The condition that an initial Hubble-scale perturbation (generated at some time t_i during inflation with scale H_i) has expanded to fill the observable universe today (subscript 0 refers to now) is:

$$a(t_i)H(t_i) = a_0 H_0 \,. \tag{2.1}$$

Using $a(t) = a_e \exp(H_i(t - t_e))$ for times during inflation yields

$$N \equiv H_i(t_e - t_b) = \ln\left(\frac{a_e H_i}{a_0 H_0}\right), \tag{2.2}$$

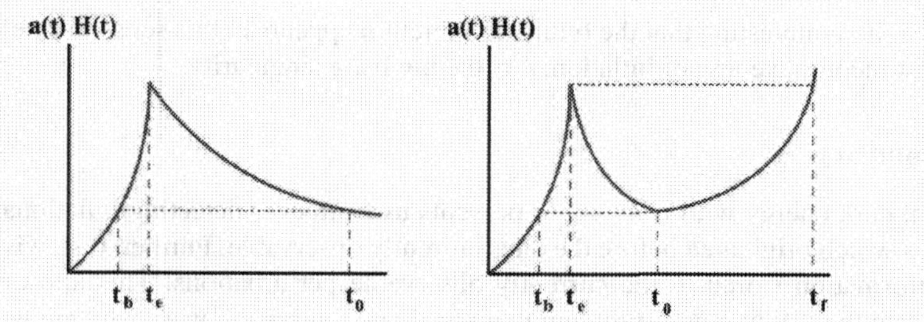

Figure 2. On the left, the evolution of the comoving Hubble scale $a(t)H(t)$ for a universe which inflates followed by radiation and matter domination; on the right, the same quantity for a universe that enters a late-time accelerating phase.

for some time t_b during inflation. After inflation, the universe grew by a factor of about $a_0/a_e \sim T_e/T_0$, where T_e is the reheating temperature and $T_0 \sim 10^{-3}eV$ the current CMB temperature. Taking this ratio to be about $10^{26} - 10^{28}$ and the scale of inflation to be $H_i \sim 10^{14}GeV$, one finds $N \sim 60$. Hence, to use inflation to solve the horizon and flatness problems requires at least 60 efolds (this is somewhat model dependent).

If the universe accelerates in the future, the comoving Hubble scale $a(t)H(t)$ grows at late times. At a time t_f when the comoving Hubble scale equals its value at reheating, the last perturbation generated during inflation will be larger than the horizon, and afterwards no new structure will form from inflationary perturbations. The equality

$$a(t_f)H(t_f) = a(t_e)H_i\,, \tag{2.3}$$

(where t_e is the time at reheating) defines t_f; it is calculated below).

Spacetimes with event horizons contain (approximately) thermal Hawking particles, with a characteristic temperature $T_H = H/2\pi$. Being a quantum effect, this spectrum does not redshift in the usual way. If the universe is accelerating, the CMB temperature T_{CMB} will eventually decrease to a point where it is equal to $T_H \sim H(t)$. This occurs at a time t_T when

$$a(t_T)H(t_T) = a(t_e)T_e\,. \tag{2.4}$$

Taking the ratio $H_i/T_e \sim 1$, $t_f \sim t_T$. Using (2.3), eqs. (2.1) and (2.2), and the scaling $a(t)H(t) \sim a_0H_0(t/t_0)^{-(1+3w)/[3(1+w)]}$ when $-1 < w < -1/3$:

$$t_T \sim t_f \sim 10^{78(1+w)/|1+3w|}t_0\,. \tag{2.5}$$

In the limit $w \to -1/3$ the time diverges. The limit $w \to -1$ yields

$$t_T \sim \frac{60}{H_0}\,. \tag{2.6}$$

Therefore if the cosmic acceleration never ends, only those inflationary fluctuations produced in the interval between the end of inflation and 60 efolds before the end will ever be observable. Further, the information which is accessible now will be lost after the time t_T. It is interesting that the future acceleration appears in this sense to shield us from the past incompleteness of inflation, *e.g.* the big bang singularity.

3. Summary

Eternal dark energy with $w < -1/3$ prevents us from ever detecting inflationary perturbations which originated before the ones currently observable. Further, it slowly degrades the information stored in the currently observable perturbations. This allows us to reformulate the "Why now?" problem in a novel and interesting way: why are we living in the best time to do cosmology; the time at which we can see back the farthest?

References

1. N. Kaloper, M. Kleban and L. Sorbo, Phys. Lett. B **600**, 7 (2004).
2. S. Hellerman, N. Kaloper and L. Susskind, JHEP **0106** (2001) 003; W. Fischler, A. Kashani-Poor, R. McNees and S. Paban, JHEP **0107** (2001) 003.

D-BRANE EFFECTIVE ACTIONS

PAUL KOERBER
UBC, Department of Physics & Astronomy
6224 Agricultural Rd., Vancouver, B.C., V6T 1Z1, Canada

Abstract. We present an alternative way to construct the D-brane effective action in the case where the background gauge field strengths are not slowly varying and/or in the case of multiple coinciding D-branes.

1. Introduction

It has been known for quite some time that the effective action for a single D-brane in the presence of slowly varying background gauge field strengths is the Dirac-Born-Infeld action [1]. A class of derivative corrections was calculated to all orders in α' in [2] using boundary conformal field theory methods. When multiple D-branes coincide the gauge group becomes non-abelian and much less is known about the effective action beyond the leading orders. As became clear from an analysis of the spectrum of strings stretching between intersecting branes, the answer is not just a symmetrized trace version of the abelian Dirac-Born-Infeld action [3]. Furthermore, the corrections containing commutators of field strengths turn out to be intimately related to the derivative corrections. We present an alternative way of constructing these D-brane effective actions based on requiring the existence of a type of BPS equations which are an α'-deformation of the equations introduced in [4]. For more details and further references, I refer to my PhD thesis [5].

2. Derivative corrections

In [4] the (anti-)self-duality equations of euclidian Yang-Mills theory in 4 dimensions were generalized to dimensions greater than 4. One can show that solutions to these generalized equations satisfy a Bogomolny bound and preserve part of the supersymmetry, so we will call these equations BPS equations. Their symmetry is only a subgroup G of the rotation group $SO(d)$. We will focus on the case were $G = U(k) \subset SO(2k)$. Introducing complex coordinates the BPS equations read

$$F_{\alpha\beta} = F_{\bar{\alpha}\bar{\beta}} = 0, \tag{2.1}$$

$$\sum_{\alpha} F_{\alpha\bar{\alpha}} = 0. \tag{2.2}$$

L. Baulied et al. (eds.), String Theory: From Gauge Interactions to Cosmology, 341–343.

The first equation is the condition for a holomorphic vector bundle while the second one is called the DUY condition [6]. It turns out that the abelian Born-Infeld action is (without derivative corrections) the unique deformation of the Maxwell action allowing for an α' deformation of these BPS equations to automatically solve the equations of motion. The holomorphicity condition stays uncorrected while the DUY condition becomes

$$\sum_{\alpha} \operatorname{arctanh} 2\pi\alpha' F_{\alpha\bar{\alpha}} = 0 \,. \tag{2.3}$$

It is still possible to establish a Bogomolny bound and show that part of the supersymmetry is unbroken by solutions to this deformed equation.

Interestingly, if we study the derivative corrections introduced in [2], which can be written as

$$S_{\mathrm{Wyl}} = -\tau_p \int d^{p+1}x \; (1+P) \sqrt{-\det\left(\eta + 2\pi\alpha' F\right)} \,, \tag{2.4}$$

with

$$\begin{aligned} P &= \frac{1}{48} S^{\rho_1}{}_{\rho_2\mu_1\mu_2} S^{\rho_2}{}_{\rho_1\mu_3\mu_4} \frac{\delta}{\delta(2\pi\alpha' F_{\mu_1\mu_2})} \frac{\delta}{\delta(2\pi\alpha' F_{\mu_3\mu_4})} \,, \\ S^{\rho_1}{}_{\rho_2\mu_1\mu_2} &= h^{\rho_1\rho_3} \Big(\left(2\pi\alpha'\right)^2 \partial_{\rho_1}\partial_{\rho_3} F_{\mu_1\mu_2} \\ &\quad + 2 \, (2\pi\alpha')^3 h^{\nu_1\nu_2} \partial_{\rho_1} F_{[\mu_1|\nu_1} \partial_{\rho_3|} F_{\mu_2]\nu_2} \Big) \,. \end{aligned} \tag{2.5}$$

one can still find a deformation of the DUY condition which is now given by

$$\sum_{\alpha} (1+P) \left[\operatorname{arctanh} 2\pi\alpha' F\right]_{\alpha\bar{\alpha}} = 0 \,. \tag{2.6}$$

Assuming that these BPS equations should still exist in the non-abelian case, we were successful in using them as a constraint to construct the non-abelian D-brane effective action up to order α'^4 [7]. Our results passed the spectrum test which the symmetrized trace Born-Infeld action failed [8].

3. Outlook

Another interesting aspect is that the transverse scalars describing the positions of the D-branes become matrices if the D-branes coincide [9]. One can study general coordinate invariance in this context [10]. The most interesting transformations seem to be ones that mix the transversal and the longitudinal coordinates. Invariance under these transformations seems again to impose strong constraints on the form of the non-abelian D-brane effective action [11].

Acknowledgements

It is a pleasure to thank A. Sevrin for the stimulating collaboration and encouragement during the work on my PhD thesis. I would also like to thank the organizers of the Cargèse 2004 summer school for the chance to present this work at the Gong Show.

References

1. E. Fradkin and A. Tseytlin, Phys.Lett. **B163** (1985) 123; R. Leigh, Mod.Phys.Lett. **A4** (1989) 2767.
2. N. Wyllard, Nucl.Phys. **B598** (2001) 247-275, hep-th/0008125; N. Wyllard, JHEP **0108** (2001) 027, hep-th/0107185.
3. A. Hashimoto and W. Taylor, Nucl.Phys. **B503** (1997) 193, hep-th/9703217.
4. E. Corrigan, C. Devchand, D. Fairlie and J. Nuyts, Nucl.Phys. **B214** (1983) 452.
5. P. Koerber, Fortsch.Phys. **52** (2004) 871, hep-th/0405227.
6. K. Uhlenbeck and S.T. Yau, Comm. Pure Appl. Math. **39** (1986) 257 and Comm. Pure Appl. Math. **42** (1989) 703; S.K. Donaldson Duke Math. J. **54** (1987) 231.
7. P. Koerber and A. Sevrin, JHEP **0210** (2002) 046, hep-th/0208044.
8. A. Sevrin and A. Wijns, JHEP **0308** (2003) 059, hep-th/0306260.
9. E. Witten, Nucl.Phys. **B460** (1996) 335, hep-th/9510135.
10. J. de Boer and K. Schalm, JHEP **0302** (2003) 041, hep-th/0108161; D. Brecher, K. Furuuchi, H. Ling and M. Van Raamsdonk, JHEP **0406** (2004) 020, hep-th/0403289.
11. D. Brecher, H. Ling, P. Koerber and M. Van Raamsdonk, work in progress.

$ADS_2 \times S^2$ AS AN EXACT HETEROTIC STRING BACKGROUND

DOMENICO ORLANDO

CPHT - École Polytechnique 91128 Palaiseau, France

Abstract. An exact heterotic string theory on an $\mathrm{AdS}_2 \times S^2$ background is found as deformation of an $SL(2,\mathbb{R}) \times SU(2)$ WZW model. Based on [1].

1. Intro

Anti de Sitter in three dimensions and S^3 are among the most simple and yet interesting string backgrounds. They are exact solutions to the string equations beyond the supergravity approximation and, at the same time, are simple to deal with although non-trivial thanks to the presence of non-vanishing curvatures. For this reason they constitute an unique setting in which to analyze AdS/CFT correspondence, black-hole physics, little-string theory.

String propagation in these backgrounds is described in terms of WZW models for the $SL(2,\mathbb{R})$ and $SU(2)$ groups, hence marginal deformations of such models allow to study moduli space of the string vacua. In particular well-known class of marginal deformations for wzw models are those driven by left–right current bilinears [2, 3]. On the other hand S^3 and AdS_3 are embedded in larger structures so one can consider marginal deformations where just one of the currents belongs to the $SU(2)$ or AdS_3 algebra, the other belonging to some other $U(1)$ corresponding to an internal magnetic or electric field.

This kind of deformation generates a continuous line of exact CFT's. In this note we will show how with an appropriate choice for the deforming current we obtain a boundary in moduli space and that this boundary can be given a simple geometric interpretation in terms of the $\mathrm{AdS}_2 \times S^2$ near-horizon geometry of the Bertotti-Robinson black hole [4, 5].

2. $SU(2)$ asymmetric deformation

In the $SU(2)$ case, there exists just one possible choice for the deforming current the two other being related by inner automorphisms, since the group has rank one, is compact and

L. Baulied et al. (eds.), String Theory: From Gauge Interactions to Cosmology, 345–349.

its Lie algebra simple. Take the WZW model for $SU(2)$:

$$S_{SU(2)_k} = \frac{1}{2\pi} \int \mathrm{d}^2 z \left\{ \frac{k}{4} \left(\partial\alpha\bar{\partial}\alpha + \partial\beta\bar{\partial}\beta + \partial\gamma\bar{\partial}\gamma + 2\cos\beta\, \partial\alpha\bar{\partial}\gamma \right) + \sum_{a=1}^{3} \psi^a \bar{\partial} \psi^a \right\} \tag{2.1}$$

where ψ^a are the left-moving free fermions, superpartners of the bosonic $SU(2)_k$ currents, and (α, β, γ) are the usual Euler angles parameterizing the $SU(2)$ group manifold. The left-moving fermions transform in the adjoint of $SU(2)$; there are no right-moving superpartners but a right-moving current algebra of total charge $c = 16$ can be realized in terms of right-moving free fermions. This means that we can build a $\mathcal{N} = (1, 0)$ worldsheet supersymmetry-compatible deformation given by:

$$\delta S_{\text{magnetic}} = \frac{\sqrt{k k_G} H}{2\pi} \int \mathrm{d}^2 z \left(J^3 + \imath \psi^1 \psi^2 \right) \bar{J}_G; \tag{2.2}$$

where J^3 belongs to the $SU(2)$ algebra and $\bar{J}_g$ is the current of the algebra at level k_g realized by the right-moving free fermions. An exact CFT is obtained for any value of the deformation parameter H.

2.1. GEOMETRY

These new backgrounds all present a constant dilaton, a magnetic field, a NS-NS field proportional to the unperturbed one and a metric retaining a residual $SU(2) \times U(1)$ isometry [6]. The most remarkable property is that the deformation line in moduli space has a boundary corresponding to a critical value of the deformation parameter $H^2 = 1/2$. At this point the $U(1)$ subgroup decompactifies and the resulting geometry is the left coset $SU(2)/U(1) \sim S^2$ which is thus found to be an exact CFT background only supported by a magnetic field (the dilaton remains constant and NS field vanishes). A geometrical interpretation for this process can be given as follows: the initial S^3 sphere is a Hopf fibration of an S^1 fiber generated by the J^3 current over an S^2 base; the deformation only acts on the fiber, changing its radius up to the point where this seems to vanish, actually marking the trivialization of the fibration:

$$S^3 \xrightarrow[H^2 \to H^2_{\max}]{} \mathbb{R} \times S^2, \tag{2.3}$$

If we turn our attention to the gauge field one can show that a quantization of the magnetic charge is only compatible with levels of the affine algebras such that $\frac{k}{k_G} = p^2, \; p \in \mathbb{Z}$. We will find the same condition in terms of the partition function for the boundary deformation.

Although this construction has been implicitly carried on for first order in α' background fields, it is important to stress that the resulting metric is nevertheless exact at all orders since the renormalization boils down to the redefinition of the level k that is simply shifted by the dual Coxeter number (just as in the WZW case).

2.2. PARTITION FUNCTION

Consider the case of $k_g = 2$ (one right-moving $\mathbb{C}$ fermion). The relevant components of the initial partition funciton are given by a $SU(2)_{k-2}$-modular-invariance-compatible combination of $SU(2)_{k-2}$ supersymmetric characters and fermions from the gauge sector. For our pourposes it is useful to further decompose the supersymmetric $SU(2)_k$ characters in terms of those of the $\mathcal{N} = 2$ minimal models:

$$\chi^j(\tau)\,\vartheta\begin{bmatrix}a\\b\end{bmatrix}(\tau,\nu) = \sum_{m\in\mathbb{Z}_{2k}} \mathcal{C}^j_m\begin{bmatrix}a\\b\end{bmatrix}\Theta_{m,k}\left(\tau, -\frac{2\nu}{k}\right). \tag{2.4}$$

The deformation acts as a boost on the left-lattice contribution of the Cartan current of the supersymmetric $SU(2)_k$ and on the right current from the gauge sector:

$$\begin{aligned}\Theta_{m,k}\,\bar\vartheta\begin{bmatrix}h\\g\end{bmatrix} &= \sum_{n,\bar n} \mathrm{e}^{-\imath\pi g\left(\bar n+\frac{h}{2}\right)} q^{\frac{1}{2}\left(\sqrt{2k}n+\frac{m}{\sqrt{2k}}\right)^2} \bar q^{\frac{1}{2}\left(\bar n+\frac{h}{2}\right)^2} \\ &\longrightarrow \sum_{n,\bar n} \mathrm{e}^{-\imath\pi g\left(\bar n+\frac{h}{2}\right)} q^{\frac{1}{2}\left[\left(\sqrt{2k}n+\frac{m}{\sqrt{2k}}\right)\cosh x+\left(\bar n+\frac{h}{2}\right)\sinh x\right]^2} \\ &\qquad\times \bar q^{\frac{1}{2}\left[\left(\bar n+\frac{h}{2}\right)\cosh x+\left(\sqrt{2k}n+\frac{m}{\sqrt{2k}}\right)\sinh x\right]^2}, \end{aligned} \tag{2.5}$$

where the boost parameter x is given by $\cosh x = \frac{1}{1-2H^2}$.

Although an exact CFT is obtained for any value of the deformation parameter H we will concentrate, as before, on the boundary value $H^2 = 1/2$. In this case the boost parameter diverges thus giving the following constraints: $4(k+2)n + 2m + 2\sqrt{2k}\bar n + \sqrt{2k}h = 0$. Therefore, the limit is well-defined only if the level of the supersymmetric $SU(2)_k$ satisfies the quantization condition $k = 2p^2\ ,\ p \in \mathbb{Z}$ *i.e.* the charge quantization for the flux of the gauge field. Under these constraints the $U(1)$ corresponding to the combination of charges orthogonal our condition decouples and can be removed. In this way we end up with the expression for the S^2 partition function:

$$Z_{S^2}\begin{bmatrix}a;h\\b;g\end{bmatrix} = \sum_{j,\bar\jmath} M^{j\bar\jmath} \sum_{N\in\mathbb{Z}_{2p}} \mathrm{e}^{\imath\pi g\left(N+\frac{h}{2}\right)}\, \mathcal{C}^j_{p(2N-h)}\begin{bmatrix}a\\b\end{bmatrix} \bar\chi^{\bar\jmath} \tag{2.6}$$

in agreement with the result found in [8] by using the coset construction. The remaining charge N labels the magnetic charge of the state under consideration.

3. $SL(2,R)$ deformation

The same construction as above can be repeated for the $SL(2,\mathbb{R})$ WZW model. In this case the moduli space is somewhat richer for it is possible to realize three different asymmetric deformations using the three generators of the group. These are not equivalent ($SL(2,\mathbb{R})$ is not compact) and in fact they lead to three physically different backgrounds. The elliptic deformation line, in example, contains the Gödel universe [7], the parabolic deformation gives the superposition of AdS_3 and a gravitational plane wave. Two of these deformation lines present the same boundary effect as the $SU(2)$ deformation. In particular the elliptic deformation leads to the hyperbolic space $H_2 = SL(2,\mathbb{R})/U(1)$ supported by an immaginary magnetic field, *ie* an exact but non-unitary CFT. The hyperbolic deformation, on the other hand, leads to $\mathrm{AdS}_2 = SL(2,\mathbb{R})/U(1)$ supported by an electric field. No charge quantization is present in this case, because of the non-compact nature of the background.

In this latter case it is not yet possible to give the same construction for the partition function as for the $SU(2)$ case since this would require the decomposition of the initial partition function in a basis of hyperbolic characters which is not a simple exercize. Nevertheless by following the same procedure as before it is possible to evaluate the effect of the deformation on the spectrum of primaries and hence give the resulting AdS_2 background spectrum.

4. $\mathrm{AdS}_2 \times S^2$

The S^2 and AdS_2 backgrounds can be combined so to give an exact CFT corresponding to the $\mathrm{AdS}_2 \times S^2$ near-horizon geometry of the BR black-hole.

Let us now consider the complete heterotic string background which consists of the $\mathrm{AdS}_2 \times S^2$ space–time times an $\mathcal{N} = 2$ internal conformal field theory $\mathcal{M}$, that we will assume to be of central charge $\hat{c} = 6$ and with integral R-charges. The levels k of $SU(2)$ and $\hat{k}$ of $SL(2,\mathbb{R})$ are such that the string background is critical:

$$\hat{c} = \frac{2(k-2)}{k} + \frac{2(\hat{k}+2)}{\hat{k}} = 4 \Longrightarrow k = \hat{k}. \tag{4.1}$$

This translates into the equality of the radii of the corresponding S^2 and AdS_2 factors, which is in turn necessary for supersymmetry. Furthermore, the charge quantization condition for the two-sphere restricts further the level to $k = 2p^2$, $p \in \mathbb{N}$.

The combined $\mathrm{AdS}_2 \times S^2$ background can give new insights about the physics of the BR black hole in particular by analizing the Schwinger-pair production in such background, or the study of the stability and propagation of D-branes.

Acknowledgements

Research partially supported by the EEC under the contracts HPRN-CT-2000-00131, HPRN-CT-2000-00148, MEXT-CT-2003-509661, MRTN-CT-2004-005104 and MRTN-CT-2004-503369.

References

1. D. Israel, C. Kounnas, D. Orlando, and P. M. Petropoulos, *Electric / magnetic deformations of s**3 and ads(3), and geometric cosets,* [hep-th/0405213].
2. S. Chaudhuri and J. A. Schwartz, *A criterion for integrably marginal operators, Phys. Lett.* **B219** (1989) 291.
3. S. Forste and D. Roggenkamp, *Current current deformations of conformal field theories, and wzw models, JHEP* **05** (2003) 071, [hep-th/0304234].
4. B. Bertotti, *Uniform Electromagnetic Field In The Theory Of General Relativity, Phys. Rev.* **116** (1959) 1331.
5. I. Robinson, *Bull. Acad. Polon. Sci.* **7** (1959) 351.

6. E. Kiritsis and C. Kounnas, *Infrared behavior of closed superstrings in strong magnetic and gravitational fields*, *Nucl. Phys.* **B456** (1995) 699–731, [hep-th/9508078].
7. D. Israël, *Quantization of heterotic strings in a Gödel/anti de Sitter spacetime and chronology protection*, *JHEP* **01** (2004) 042, [hep-th/0310158].
8. P. Berglund, C. V. Johnson, S. Kachru and P. Zaugg, *Heterotic Coset Models and (0,2) String Vacua*, *Nucl. Phys. B* **460** (1996) 252, [hep-th/9509170].

STRINGS AND D-BRANES IN HOLOGRAPHIC BACKGROUNDS

DAN ISRAËL AND ARI PAKMAN
The Racah Institute of Physics,
The Hebrew University of Jerusalem, 91904 Jerusalem, Israel

Abstract. We review recent progress in the study of non-rational (boundary) CFTs and their applications to holographic backgrounds in superstring theory. We focus on the supersymmetric coset $SL(2,\mathbb{R})/U(1)$ and its dual N=2 Liouville. We discuss the modular properties of their characters, their partition function and boundary states for their D-branes. Then these results are used to study the CFT of the NS5-brane background, with applications to Little String Theories.

1. Introduction

Conformal field theories (CFT) are the natural building blocks for the exact perturbative description of superstring theory in non-trivial backgrounds without RR fluxes. When the curved space is non-compact, the CFTs needed are *non-rational* (NRCFTs). They are more difficult to handle that the rational ones, and the simplest examples have been fairly understood only in the last years.

This progress has been largely motivated by their relevance to describe holographic backgrounds. Among the latter, the most famous are AdS(3) and strings in $d \leq 2$. The holographic nature of five-branes solutions is less clear, mainly due to the non-triviality of the dual non-gravitational theory, called *Little String Theory*.

These constructions can typically be recast as non-rational analogues of Gepner models, with a non-compact factor given by the supersymmetric coset $SL(2,\mathbb{R})/U(1)$. It is this last theory that we shall study in detail in this note.

The non-rational CFTs are defined in opposition to rational ones. The techniques used to solve the latter do not work for the former, or need to be adapted. The complications arise because the Hilbert space has a continuous spectrum, due to the non-compactness of the target space. In general the spectrum of these theories splits into two categories of states. On the one hand the spectrum contains *continuous representations*, corresponding to asymptotic states propagating in the non-compact "radial" direction(s). On the other hand the states of the *discrete representations* correspond to (a finite set of) localized bound states. As we may infer from general considerations of scattering theory, these two kinds of states mix. For example, in the coset $SL(2,\mathbb{R})/U(1)$, the reflection amplitude

L. Baulied et al. (eds.), String Theory: From Gauge Interactions to Cosmology, 351–354.

for continuous representations has poles whenever it is analytically continued to a discrete representation.

The modular properties of the characters exhibit a similar pattern. While the continuous representations modular transform into themselves, the discrete representations yield both discrete and continuous ones. These results hold both in the bosonic and susy cosets [1], and are shown using a lemma proven in [1]. These new features complicate the construction of partition functions and boundary states by a large amount.

A last important aspect is that for specific values of the level, these theories acquire some kind of rational behavior. Remember that a free a boson on a circle at rational radius, has an extended chiral symmetry, and summing over the orbits of this symmetry one gets a finite set of *extended characters*. The same holds for the (super) coset $SL(2,\mathbb{R})/U(1)$ at rational level, since the target space asymptotes a cylinder of radius $\sqrt{2k}$. In particular, in the supersymmetric case one obtain a finite set of rational N=2 R-charges, which is desirable to construct space-time supersymmetric vacua.

2. SL(2,R)/U(1) and N=2 Liouville

The theory is obtained applying the rules of an elliptic coset construction to the supersymmetric WZW model $SL(2,\mathbb{R})$ at level k. For the *axial* coset the sigma model is well defined because the action of the gauge field has no fixed point, and corresponds to an Euclidean two-dimensional black hole. The spectrum of primaries is obtained by descent from AdS_3.

The worldsheet-supersymmetric partition function has been computed using the powerful techniques of marginal deformations of WZW models[3]. This amplitude splits naturally into (non-minimal) characters of the N=2 superconformal algebra. An exact decomposition of the partition function has been carried out in[2]. We have found first a contribution of discrete representations, with the correct multiplicities for all the descendants. For the continuous representations, their contribution is divergent, due to the infinite volume available for them. An infrared regularization of the partition function is possible, leading to a finite non-trivial density of continuous representations, compatible with N=2 supersymmetry. However this regulator breaks (super)conformal symmetry, and there is a price to pay: the partition function contains an extra non-universal contribution which is not related to the N=2 algebra.

The super-coset $SL(2,\mathbb{R})/U(1)$ has been conjectured to be dual to the N=2 Liouville theory. Evidence for this equivalence comes from a sigma-model mirror symmetry and from the agreement between perturbative computations of correlators. However, we argue that these results come from a more fundamental structure of these theories. Indeed, both theories possess the same chiral algebra, which is the (non-minimal) N=2 SCA. This algebra can be decomposed into the bosonic coset $SL(2,\mathbb{R})/U(1)$ and a free boson[1]. Since conformal bootstrap results for the Euclidean AdS_3 are obtained using *only the chiral symmetries* of the model (without an explicit reference to a specific action), they can be applied to the supersymmetric $SL(2,\mathbb{R})/U(1)$ coset and to the N=2 Liouville theory as well, because they both lift to the same current algebra[4]. Therefore the theories can differ only by the way the left and right representations of the algebra are glued in the closed string spectrum.

The vector coset $SL(2,\mathbb{R})/U(1)$ has a singular sigma-model, therefore it receives substantial corrections; its single cover is a $\mathbb{Z}_k$ orbifold of the cigar[3]. It is likely that the N=2 Liouville is the corrected form of this vector coset.

3. D-branes from boundary AdS$_3$

The study of D-branes in these exact superstrings backgrounds is essential, in order to understand the non-perturbative dynamics in these non-compact manifolds, and it may give indications about its holographic degrees of freedom. The construction of the exact boundary states, which contain all the information about the couplings of the D-branes to the closed string states, follows the same logic as before. Conformal bootstrap methods were employed successfully to study D-branes in the bosonic Liouville theory in Euclidean AdS$_3$. The results have been used later to study the bosonic coset.

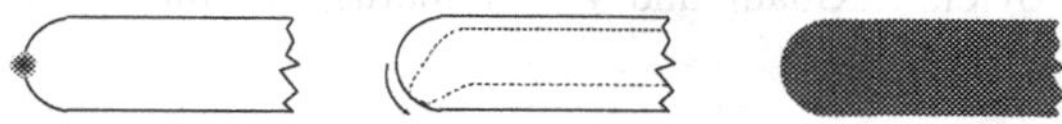

Figure 1. The consistent D-branes in the cigar: D0-, D1- and D2-branes (left to right)

In[4] we constructed the D-branes in the super-coset $SL(2,\mathbb{R})/U(1)$ using similar methods. An important aspect of this analysis is that the boundary conformal bootstrap uses only chiral symmetries. Therefore the arguments given above about the duality super-coset $SL(2,\mathbb{R})/U(1)$ / N=2 Liouville extend straightforwardly in the presence of a boundary. However, one should be careful about the way left and right representations are glued, in order to construct the basis of Ishibashi states. Then the various boundary states of the theory can be constructed by descent from D-branes of Euclidean AdS$_3$. There is a direct connection between the boundary conditions for the currents of the $\widehat{\mathfrak{sl}}(2,\mathbb{R})$algebra and the gluing conditions of type A or B defined in. The D-branes that we obtained satisfy by construction the factorization constraints, since they descend from consistent D-branes in Euclidean AdS$_3$. We also checked the Cardy condition, i.e. the open-closed channel consistency of the annulus diagram. The various D-branes of the cigar are depicted in fig. 1.

4. strings near NS5 branes

The results given above can be lifted to the background created by NS5-branes distributed on a topologically trivial circle, known to be T-dual to an exact CFT. When all the NS5-branes are separated from each other, the background is perturbative and one can take a *double scaling limit* where gravity decouples and this perturbative nature holds. We have shown in[2] that the complicated solution for the ring of NS5 in this limit can be obtained as a *null gauging* of the super-WZW model $SL(2,\mathbb{R}) \times SU(2)$. This CFT has worldsheet N=4 SCA, so there are no perturbative corrections metric. However, instantons corrections show up, and they are indeed captured by the supergravity solution. The null coset can be recast as a $\mathbb{Z}_k$ orbifold of $SL(2,\mathbb{R})/U(1) \times SU(2)/U(1)$. The $\mathbb{Z}_k$ orbifold can be thought as coming from the GSO projection generalizing Gepner models. However, this orbifold changes deeply the background of the effective theory, because in the semi-classical limit its twisted sectors become very light; hence the correct geometry is not given by the sum of the two coset factors.

Once this identification is understood it is possible to write the one-loop amplitude for this NS5-brane backgroud[2]. Various BPS D-branes in this background are now under study. In particular there are new non-factorizable D-branes that can be constructed out of the coset D-branes[5]. In type IIA we have also D4-branes stretched between the NS5-branes on which a D=4, N=2 SYM theory lives. Quite remarkably, the one-point function for these D4-branes can be related to the beta-function of the gauge theory.

Acknowledgements

It is a pleasure to thank C. Kounnas and J. Troost for the collaborations leading to the present results. We also thank C. Bachas, S. Elitzur, A. Giveon, E. Kiritsis, D. Kutasov, B. Pioline, E. Rabinovici, S. Ribault and V. Schomerus for stimulating discussions.

References

1. D. Israel, A. Pakman and J. Troost, JHEP **0404**, 045 (2004)
2. D. Israel, C. Kounnas, A. Pakman and J. Troost, JHEP **0406**, 033 (2004)
3. D. Israel, C. Kounnas and M. P. Petropoulos, JHEP **0310**, 028 (2003)
4. D. Israel, A. Pakman and J. Troost, arXiv: hep-th/0405259.
5. D. Israel, A. Pakman and J. Troost, to appear

COSMIC CENSORSHIP IN ADS/CFT

MUKUND RANGAMANI
Department of Physics & LBNL, Univesity of California
Berkeley, CA 94720, USA

Abstract. We review recent attempts at violations of cosmic censorship in asymptotically Anti-deSitter spacetimes. The essential logic behind the potential counter-examples was that there exist smooth initial data that are guaranteed to evolve in the future to a singularity, which nonetheless are not going to be cloaked by a black hole horizon, owing to the configuration possessing insufficient energy to produce the desired black hole. We demonstrate that a refined analysis indicates that the kinematics always allows for black hole formation, thereby upholding cosmic censorship in this context.

1. Introduction

Singularities are a ubiquitous feature in classical general relativity. As attested to by the singularity theorems, mild initial data can often evolve to singular spacetimes. Within the classical framework, formation of singularities implies lack of predictability. Moreover, it implies that formation of singularities in some local region would lead to drastic effects arbitrarily far away. To avert this consequence it was proposed by Penrose [1] that the is a universal *cosmic censor* that prevents these singularities from being causally accessible to asymptotic observers. In a more colloquial phrasing the cosmic censorship conjecture may be stated as follows: whenever initial data under evolution form singularities, the singularities are hidden behind event horizons of black holes, thereby preventing them from influencing the physics at large distances.

While the spirit of the cosmic censorship conjecture is appropriate within the limited sense of classical general relativity, it is not clear that this is a desirable state of affairs in the quantum gravity context. As such one would expect that accessing the large curvature effects should provide insight into the workings of quantum gravity (see V. Hubeny's talk in this volume for a discussion of these issues). To this end it is useful and instructive to analyze the domain of validity of the cosmic censorship conjecture. At the same time given its recalcitance to being proven, one wonders whether there might be some important piece of physics that we are missing.

Ideally one would like to talk about cosmic censorship violation in contexts where we have some level of confidence about what we expect in the full quantum gravity context.

L. Baulied et al. (eds.), String Theory: From Gauge Interactions to Cosmology, 355–362.

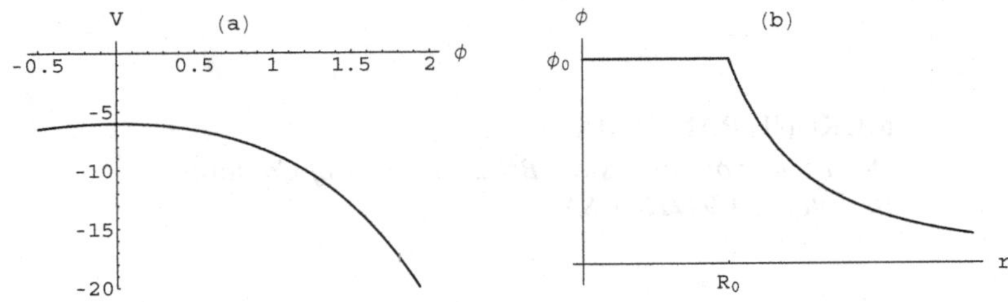

Figure 1. The scalar potential and the field profile.

As of date the framework wherein this criterion is met is the AdS/CFT correspondence. Here we know that there is an in principle formulation of quantum gravity in asymptotically Anti-deSitter spacetimes in terms of a quantum field theory. This makes it ideal to explore consequences of cosmic censorship in asymptotically AdS spacetimes.

The first steps in the direction of potential cosmic censorship violation were taken in [2] and more recently this was extended to a context with greater control from the AdS/CFT point of view in [3]. The essence of their argument is to produce a simple configuration that reliably evolves into a singularity, and then show that there is insufficient energy available to make a black hole whose horizon is big enough to cloak the singularity. A careful study of this scenario with the purpose of identifying a region in parameter space where we can reliably argue that naked singularities will form and moreover the dual CFT will be well-behaved was carried out in [5]. It was argued that there aren't any initial conditions that satisfy the aforementioned criteria. In particular, we carry out a refined estimate of the proper size of the singularity and the energitics in the system to show that there is always a potential for black hole formation. This is in agreement with the numerical results of [4], [6], and [7], wherein full evolution of the initial data was carried out.

We begin by reviewing the framework in which we will consider cosmic censorship violation and then proceed to proceed to detail how we can make reliable estimates of the size of the singularity and the energetics involved and end with a short discussion.

2. The Framework

The simplest context in which we can discuss the formation of naked singularities in asymptotically AdS is to consider a scalar gravity system with the scalar potential being picked arbitrarily, with the only constraint that it admits AdS vacua. In [2], such a set-up was chosen, where it was argued that with a suitable choice of scalar potential one can ensure that it is possible to find configurations with very little energy, which nonetheless make black holes. Choosing an arbitrary scalar potential is not viable within the AdS/CFT context. Nonetheless, [3] argue that it is possible to create naked singularities in this case as well. The idea is to use the well known fact that in AdS, scalar fields can have negative mass squared, but nevertheless be stable. In fact such fields are available in the context of $\mathcal{N} = 8$ gauged supergravity on AdS_5. Consider then the following action, which is a

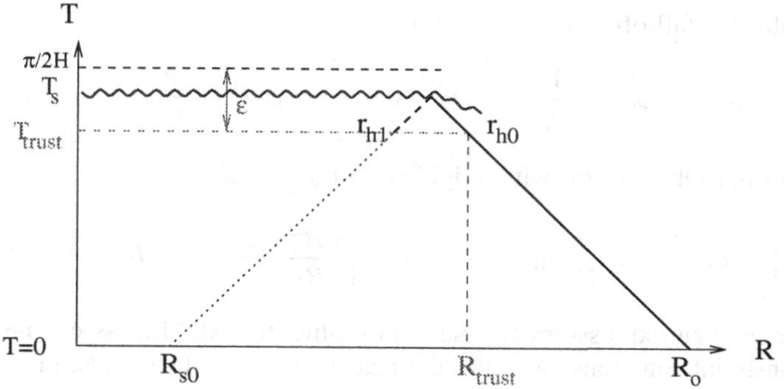

Figure 2. A sketch of the causal diagram for the evolution of the scalar field profile presented in Fig 1.1b.

consistent truncation of the supergravity theory:

$$S = \int \sqrt{-g} \left[\frac{1}{2} R - \frac{1}{2} (\nabla \phi)^2 - V(\phi) \right] , \tag{2.1}$$

where the potential is plotted in Fig. 1.1.a, and is given by

$$V(\phi) = -2\, e^{2\phi/\sqrt{3}} - 4\, e^{-\phi/\sqrt{3}} . \tag{2.2}$$

The initial data is prescribed in the form of the field profile $\phi(r)$ as indicated in Fig. 1.1b. We are assuming that we have time symmetric initial data and will be parametrising the metric on the inital data surface as

$$ds_\Sigma^2 = \left(1 - \frac{m(r)}{3\,\pi^2\, r^2} + r^2 \right)^{-1} dr^2 + r^2\, d\Omega_3^2 , \tag{2.3}$$

so that r denotes the proper size of the $\mathbf{S}^3$ on the initial data surface and $m(r)$ is related to the ADM mass of the configuration and is determined in terms of $\phi(r)$ by virtue of the constraint equations.

As is clear from Fig. 1.1.b, there are two parameters at our disposal: R_0 which is size of the homogeneous region of the scalar field and ϕ_0 the value of the scalar field in the homogeneous region. The choice of this homogeneous profile has been made to ensure that the future evolution of the initial data results in a singularity. Thinking in terms of FRW cosmologies, a homogenous scalar field tends to roll down the potential and collapse to a singularity in open universes with a negative cosmological constant. The asymptotic fall-off of the scalar field is dictated by the requirement that the geometry be asymptotically AdS. In Fig. 1.2 we sketch the conformal diagram for the evolution of the scalar field in the domain of dependence of the region $r \leq R_0$.

For the sake of concreteness there are two distinct scalar profiles which we can consider:

– Normalizable fall-off

$$\phi(r) = \phi_0 \equiv \frac{A}{R_0^2} \qquad r \le R_0 \,, \qquad \phi(r) = \frac{A}{r^2} \qquad r > R_0 \,. \tag{2.4}$$

– Non-normalizable fall-off with a rigid cut-off at $r = R_1$

$$\phi(r) = \phi_0 \,, \qquad r \le R_0 \qquad \phi(r) = \frac{\phi_1 \, R_1^2}{\ln R_1} \, \frac{\ln r}{r^2} \,, \qquad R_0 < r \le R_1 \,. \tag{2.5}$$

With the explicit expressions for the field profile, the ADM mass can be calculated from the constraint equations. We refer the reader to [5] for the details. In what follows we assume that we denote $M_{\text{config}}(\phi_0, R_0)$ as the energy contained in the initial data.

Given the kinematical information about how much energy is contained in the initial data, the strategy is to ask whether we have enough energy to form a black hole which can enclose the singularity which is guaranteed to form upon evolution. To this end we need to be aware of the size of the singularity, in order to ascertain the size of the event horizon of the black hole. This we turn to next.

3. The Size of the singularity

The first step is to model the evolution of a homogeneous scalar field in hyperbolic slices of AdS. This is an accurate description of the dynamics in the domain of dependence of the homogeneous region $r \le R_0$ for the configurations considered in the body of the paper. The metric in d dimensions can be written as

$$ds^2 = -dT^2 + a(T)^2 \left(\frac{dR^2}{1+R^2} + R^2 \, d\Omega_{d-2}^2 \right) \,. \tag{3.1}$$

where the scale factor $a(T)$ and the scalar field $\phi(T)$ are to be determined from the Einstein's equations, which read:

$$\begin{aligned} \dot{a}^2 - \frac{2}{(d-1)(d-2)} \, a^2 \left(\frac{1}{2} \, \dot{\phi}^2 + V \right) &= 1 \\ \frac{\ddot{a}}{a} - \frac{2}{(d-1)(d-2)} \left(V - \frac{(d-2)}{2} \, \dot{\phi}^2 \right) &= 0 \\ \ddot{\phi} + (d-1) \, \frac{\dot{a}}{a} \, \dot{\phi} + V_{,\phi} = 0 \,. & \end{aligned} \tag{3.2}$$

Using (3.2) it can be shown that we get a curvature singularity as $a \longrightarrow 0$, unless $a(T) \sim \cos T$ as for pure AdS, where $a = 0$ corresponds merely to a coordinate singularity.

Evolving these equations is difficult to do analytically, but it is relatively straightforward numerically. There are two cases where we can make analytic statements, The first is where we assume that the field is small and study only the Gaussian fluctuations in the potential, and the other is the near-singularity regime, where the potential is negligible. Armed with this information it is possible to show that a curvature singularity will occur in $T < \frac{\pi}{2}$.

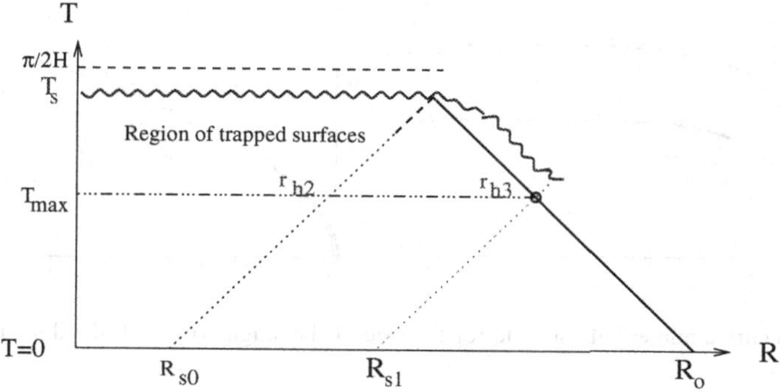

Figure 3. Refined causal digram for the FRW region where the homogeneous scalar field collapses.

Let us now turn to the next step, namely estimating the mass of the black hole required to cloak the singularity. This involves a bit more detailed analysis, since it depends on the behaviour of the null geodesics in the dynamical background. What we can do is to estimate when the pure AdS metric cease to approximate the actual geometry. We will then use that as the cutoff for the ingoing radial null geodesic starting at $T = 0$ at radius R_0; namely, we approximate the black hole size r_{h0} by the proper radius reached by this null geodesic at the cut-off time $T_{\rm trust}$, *cf.*, Fig. 1.2. Carrying out this exercise we find:

$$r_{h0} = 0.27\, R_0\, \phi_0^{2/3} \; . \tag{3.3}$$

We must emphasise here that the above serves as an estimate for the lower bound on the size of the black hole that needs to form.

To get a more reliable estimate on the size of the black hole necessary to cloak the singularity, we solve FRW equations (3.2) numerically. With an explicit numerical solution to the scale factor $a(T)$, we can study the property of null geodesics in the resulting big crunch geometry. Note that this is still a much easier problem than having to solve for the full evolution.

In particular we can obtain the behavior of $r_{h0}(\phi_0)$. As indicated in Fig. 1.3, we start from a point on the initial data slice where the spheres have proper size R_0 and consider a radially ingoing null geodesic. At some point very close to the singularity we cut-off the geometry and from the end-point of the geodesic from R_0 onto this cut-off surface we take a radially ingoing null geodesic. This would intersect the initial data slice at R_{s0}. However, the proper size of the spheres will peak at some intermediate time T_{max}. Call this radius r_{h2}; this will be the lower bound on the size of the black hole. We use the numerical solution for $a(T)$ to estimate r_{h2}. In Fig. 1.4 we show the behavior of $r_{h2}(\phi_0)/R_0$. We see that for large ϕ_0, r_{h2} is independent of ϕ_0. This follows from the fact that the potential is exponential in ϕ. Moreover, we find that for small ϕ_0,

$$r_{h2}(\phi_0) = 0.37\, R_0\, \phi_0^{2/3} \tag{3.4}$$

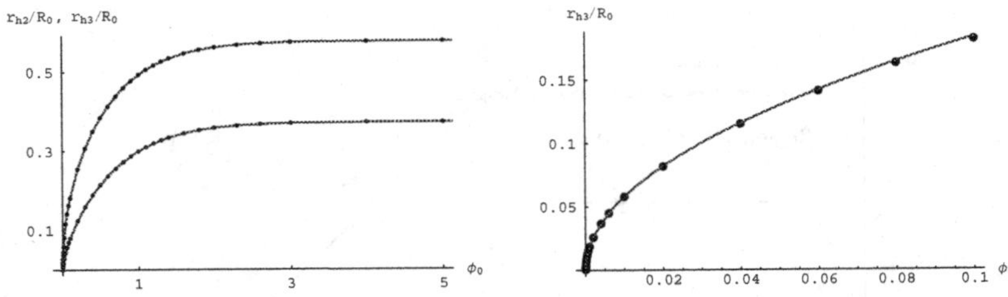

Figure 4. Accurate numerical estimate for the size of the singularity with the domain of dependence.

The numerical evaluation of r_{h2} differs from our weak field estimate r_{h0} only in the numerical factor, which is attributable to a more accurate determination of the point where the geometry deviates significantly from pure AdS.

The estimate for the black hole size (3.4), does not constitute the tightest lower bound. In deriving it we assumed that the singularity stretched out to the boundary of the domain of dependence at time T_s. However, we can argue that it necessarily has to be bigger and so obtain more stringent bounds. By following an outgoing congruence of null geodesics from the initial data surface, we can show that the best estimate of the size of the singularity is given by r_{h3} in Fig. 1.3. The basic idea is to exploit the fact that we are able to move all the way to the boundary of the domain of dependence (see [5] for details). In fact, with a bit of numerical work we can show that for small ϕ_0

$$r_{h3}(\phi_0) = 0.58\, R_0\, \phi_0^{1/2} \,, \qquad \text{for} \quad \phi_0 \ll 1 \,, \tag{3.5}$$

is the best estimate that can be obtained within the domain of dependence. Moreover we can show that for

$$r_{h3} \longrightarrow \sigma_s\, R_0 = 0.577\, R_0 \,, \qquad \text{for} \quad \phi_0 \gg 1 \,. \tag{3.6}$$

Plots of r_{h3} as function of ϕ_0 are shown in Fig. 1.4.

With the numerical information about the singularity in hand, we can ask what is the energy required to form the black hole of the requisite size. For the normalizable scalar mode, the black hole in question is the Schwarzschild AdS black hole, while for the non-normalizable configuration it is the scalar hair black hole. Referring the interested reader to [5] for details; we find at the end of the day:

$$M_{\rm config} - M_{\rm BH} > 0 \tag{3.7}$$

indicating that there is always enough energy in the system to form the black hole of the appropriate size. One might worry that since we only estimate the lower bound on the black size by the ray tracing arguments it might be possible that the singularity will be a bit bigger and there would be a potential for a naked singularity. However, numerical evolution of the data [7] that our estimate for the size of the singularity is quite robust and hence there is no violation of cosmic censorship.

4. Discussion

In this talk we have discussed the potential violation of cosmic censorship within the AdS/CFT context. At the face value one is tempted to think that naked singularity formation should be easy in AdS spacetimes, since it is easy to form singularities (as attested to by the FRW collapse scenario) and moreover black holes are quite heavy in AdS spacetimes. Nonetheless in the context of AdS/CFT correspondence we see that there is an interesting obtacle to violation of cosmic censorship. In fact, the arguments can be extended to the original context of cosmic censorship violation in AdS spacetimes as suggested in [2]; see [5] and [8]. The essential reason for the upholding of the cosmic censorship conjecture is that the singularity that forms from the initial data is rather small, and there is always sufficient energy in the system to cloak it with a black hole of the desired size.

At the same time, one begins to wonder whether there is an interesting lesson to be learnt from the CFT. Although we have not explicitly discussed the issue, the estimates of the energy in the initial configuration are most reliably carried out from the CFT side owing to the fact that the bulk ADM energy has finite boundary contributions. In any event it is the CFT which lives on the boundary that has a well defined Hamiltonian evolution and it is the eigenvalues of this conserved Hamiltonian that are of interest to us.

There are other physical effects from the string theoretic viewpoint that we have not discussed here. It is shown in [5] that in the situations where we turn on the non-normalizable modes we have to content with a D-brane instability of the system. In these contexts one can imagine that there are stringy effects at work which serve to cloak the singularity. At this stage one might speculate whether there is an underlying mechanism in the field theory that serves to prevent cosmic censorship violation. After all one might argue that smooth initial data in the bulk maps to a well defined state in the CFT and since the Hamiltonian evolution is well behaved, there is no room for singularities. Fleshing out this statement in a quantitative fashion could serve to prove the cosmic censorship conjecture.

Acknowledgements

I wish to thank my collaborators, Veronika Hubeny, Xiao Liu and Stephen Shenker for a wonderful collaboration. I would also like to thank also the organizers of the Cargèse 2004 ASI for an excellent summer school and the Kavli Institute for Theoretical Physics for hospitality. This work was supported by the funds from the Berkeley Center for Theoretical Physics, and also by the DOE grant DE-AC03-76SF00098 and the NSF grant PHY-0098840.

References

1. R. Penrose, Riv. Nuovo Cim. **1**, 252 (1969) [Gen. Rel. Grav. **34**, 1141 (2002)].
2. T. Hertog, G. T. Horowitz and K. Maeda, Phys. Rev. Lett. **92**, 131101 (2004) [arXiv:gr-qc/0307102].
3. T. Hertog, G. T. Horowitz and K. Maeda, Phys. Rev. D **69**, 105001 (2004) [arXiv:hep-th/0310054].
4. M. Gutperle and P. Kraus, JHEP **0404**, 024 (2004) [arXiv:hep-th/0402109].
5. V. E. Hubeny, X. Liu, M. Rangamani and S. Shenker, arXiv:hep-th/0403198.

6. D. Garfinkle, Phys. Rev. D **69**, 124017 (2004) [arXiv:gr-qc/0403078].
7. A. V. Frolov, arXiv:hep-th/0409117.
8. T. Hertog, G. T. Horowitz and K. Maeda, arXiv:gr-qc/0405050.

TOPICS IN BLACK HOLE PRODUCTION

VYACHESLAV S. RYCHKOV
Institute for Theoretical Physics, University of Amsterdam, 1018XE Amsterdam, The Netherlands

Abstract. We revisit Voloshin's model of multiple black hole production in trans-Planckian elementary particle collisions in $D = 4$. Our revised computation shows that the cross section to produce n additional black holes is suppressed by s^{-1}, rather than being enhanced as was originally found. We also review the semiclassical gravity picture of black hole production from hep-th/0409131, making additional comments about the meaning of wavepacket subdivision.

1. Introduction

Black hole (BH) production in trans-Planckian elementary particle collisions ($E \gg E_{Planck}$) has long been considered a theoretical possibility. If TeV-scale gravity scenarios based on large extra dimensions or warped compactifications are realized in nature, this possibility may be realized in practice at future accelerators (see [1] for a recent review). The key question is the cross section of this process, which is usually assumed to be set by the horizon radius of the produced BH (the so called "geometric cross section" $\sigma \sim \pi r_h^2$).

In this note I would like to discuss two aspects of the BH production problem relevant for justifying the geometric cross section estimate. First I will explain how to derive this estimate from a controlled semiclassical gravity approximation to the BH production process, adding some comments to the original discussion of [2] (also recently reviewed in [3]). Then I will estimate the cross section of multiple BH production due to collisions between virtual gravitons emitted by the primary particles, in a model first proposed by Voloshin [4]. The conclusion (contrary to [4]) is that multiple BH production gives a subdominant contribution to the total cross section.

The size of produced BHs in large extra dimension scenarios is typically much smaller than the size of extra dimensions, and their production may be considered as happening in flat D dimensional spacetime. Since we are focussing on theoretical issues, to keep the discussion clear we will work in $D = 4$. We will use Planck units, setting $E_{Planck} = 1$.

L. Baulied et al. (eds.), String Theory: From Gauge Interactions to Cosmology, 363–369.

2. Wavepackets and Semiclassics

In this note we will adhere to the standard believe that in quantum theory of gravity classical BHs with mass $M \gg 1$ will be realized as long-lived resonance states, decaying via Hawking radiation, with lifetime $\sim M^3$. It is then energetically allowed to produce such BHs in trans-Planckian ($E \gg 1$) elementary particle collisions.

To estimate the cross section of this process, Eardley and Giddings [5] looked at the grazing collision of two ultrarelativistic point particles in classical general relativity, using formation of a closed trapped surface (CTS) as a sufficient condition for BH formation. In this totally classical description, lower bound for the cross section is given by πb_{max}^2, where b_{max} is the maximal impact parameter for which we are able to find a CTS in the spacetime formed by two colliding Aichelburg-Sexl shock waves. [5] found $b_{max} \sim E \sim r_h$, and this implies $\sigma \sim \pi r_h^2$. Recourse to such an indirect method is necessary, because explicit solutions of Einstein's equations exhibiting the final BH state are out of reach.

How do we justify this approach from quantum gravity point of view? A point of immediate concern is that particles in a collider experiment are described by wide wavepacket states of macroscopic size (set essentially by the beam radius, which in turn is determined by the focussing ability of accelerator magnets). These wavepackets are vastly larger than BHs whose production we are trying to describe. If we use the energy momentum tensor of these wavepacket states $|\psi\rangle$ in the RHS of the semiclassical Einstein's equations

$$R_{\mu\nu} - \tfrac{1}{2} g_{\mu\nu} R = \langle\psi|T_{\mu\nu}|\psi\rangle, \tag{2.1}$$

we won't see any BH production whatsoever, since the energy is spread out over a huge volume, and the energy density is insufficient to cause collapse. Informally speaking, particles "do not fit" inside a BH.

However, this does not mean that BHs do not form. The correct interpretation is that the part of the gravitational field wavefunction corresponding to BH production was erased—averaged away—by eq. (2.1). Some averaging is always inevitable when using the semiclassical field equations, since we substitute $T_{\mu\nu}$ by its expectation value. Unfortunately, in this case it destroys precisely the part of the wavefunction we are interested in. To see BH production, one should instead proceed as follows. First of all, we have to subdivide the initial wavepackets into much smaller wavepackets of size $w \ll r_h$:

$$|\psi\rangle = N^{-1/2} \sum\nolimits_{i=1}^{N} |\psi_i\rangle. \tag{2.2}$$

This subdivision is carried out so that the small wavepackets $|\psi_i\rangle$ in the RHS are almost orthogonal. This orthogonality is quite obvious in the position representation (see Fig. 1). Because of the orthogonality, collisions between different pairs of small wavepackets are mutually excluded possibilities, and probabilities of BH production in each such elementary collision should be added. Now, it is the collisions of the small wavepackets that we are going to analyze using eq. (2.1). Condition $w \ll r_h$ ensures that the small wavepackets produce a collision spacetime which is a small perturbation of the one corresponding to point particles of the same energy. Thus the Eardley-Giddings analysis applies, and adding probabilities results in the geometric cross section.

Using finite-size wavepackets instead of point particles has an additional bonus in that it puts the conditions of applicability of the semiclassical approximation under control. For example, curvature blows up when the shock fronts of Aichelburg-Sexl waves corresponding to point particles collide [6]. However, taking wavepacket size into account regulates the curvature and brings it below the Planck value, so that we can trust the semiclassical gravity approximation [2].

3. Multiple Black Hole Production

According to the above discussion, the geometric cross section formula provides a lower bound for a single large BH production cross section in a trans-Planckian collision. However, as the energy of the particles grows, multiple BH production also becomes energetically allowed. It is important to understand which process is dominant at asymptotically high energies. If multiple BHs dominate, it will be much harder to observe dipole patterns of emitted particles expected in the Hawking evaporation of a single large BH.

We are going to discuss a model proposed by Voloshin [4], in which multiple BHs are produced due to collisions between virtual gravitons emitted by the trans-Planckian projectiles. Such virtual graviton emission is a quantum effect: classically, any radiation happens after the particles collide. The process is studied diagrammatically, with a typical diagram shown in Fig. 2. Only the case of small peripheral BHs ($1 \ll m_i \ll E$, $i = 1 \ldots n$) is considered, so that the gravitons are "soft".

The production amplitude is computed from the diagrams using the standard QFT propagators and particular vertices for soft graviton emission and for BH production (Fig. 3). The amplitude to emit a positive helicity graviton of energy $\omega \ll E$ and small transverse momentum $\mathbf{k} = (k_2, k_3)$, $|\mathbf{k}| \ll \omega$ is given by (see Appendix A)

$$A \propto (E/\omega)^2 (k_2 + ik_3)^2 \,. \tag{3.1}$$

(We will not pay attention to constant numerical factors. Thus our final result (3.3) is valid up to a factor c^n.)

Although we are unable to compute the elementary BH production vertex $f(q^2)$, the geometric cross section allows to fix the combination

$$|f(q^2)|^2 \rho(q^2) \sim (q^2)^2, \tag{3.2}$$

where $\rho(q^2)$ is the density of BH states at mass $\sqrt{q^2}$.

A crucial final element of the model is a condition which ensures that the emitted gravitons do not subsequently fall into a common large BH. Such an infall may happen

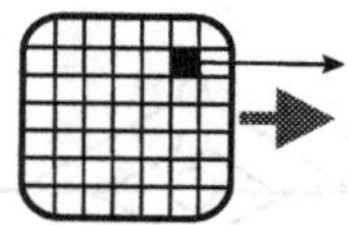

Figure 1. Subdivision of the right-moving particle wavepacket. Similar subdivision has to be done for the left-moving particle.

due to graviton rescattering diagrams, which we are not going to compute. Thus, without a "fall safe" condition we would be in danger of greatly overestimating the multiple BH amplitude. Voloshin's "fall safe" condition limits the transverse momenta of emitted gravitons by $|\mathbf{k}| \lesssim 1/E$. To derive this condition, we note that a typical emitted graviton will be off-shell by $\Delta E \sim \mathbf{k}^2/\omega$. It will exist for a time interval $\Delta t \sim 1/\Delta E$, during which it will reach transverse separation $\Delta z \sim (\mathbf{k}/\omega)\Delta t$ from the projectile. Voloshin's condition arises if we require that this transverse separation is larger than the horizon radius of the big BH formed in the collision of the primary particles: $\Delta z \gtrsim E$.

In the described model, our computation (see Appendix B) gives the following amplitude to produce n additional BHs with 4-momenta q_i:

$$f^{(n)}(s, q_i^2) \sim f(s) \prod\nolimits_{i=1}^{n} (q_i^2)^{-2} f(q_i^2) \,. \tag{3.3}$$

The original computation of [4], Eq. (5), gave an amplitude larger than (3.3) by a factor of $(s q_i^2)^n$. We believe that our result is correct; see Appendix B for an explanation.

Using (3.2), we can compute from (3.3) the contribution of the diagrams from Fig. 2 into the total cross section σ_n to produce one large ($m^2 \sim s$) and n small BHs. This contribution will behave like s^{1-n}, the suppression being due to the phase space restriction $|\mathbf{k}| \lesssim 1/E$ satisfied by the small BHs as a consequence of the "fall safe" condition.

However, for $n \geq 2$ there are diagrams which give a larger contribution, so that $\sigma_n \sim const$ is likely for any $n \geq 1$. Consider, e.g., Fig. 4, where the primary particles emit "fall safe" gravitons of energy E_1, $1 \ll E_1 \ll E$, and it is these gravitons that form n smaller BHs according to the previous model. The allowed phase space for this diagram will be much bigger, since the individual small BHs can now have much larger transverse momenta $|\mathbf{k}| \lesssim 1/E_1$, only their sum being $\lesssim 1/E$. Choosing E_1 above the threshold of n BH production, we will get an s-independent contribution to σ_n.[1] In any case, we see that σ_n is suppressed compared to the geometric cross section value $\sigma_0 \sim s$.

4. Conclusions

In this note we discussed two aspects of the BH production problem. In Section 2 we discussed how wavepacket arguments can be used to justify the use of semiclassical gravity in this problem. In Section 3, we revisited Voloshin's model of multiple BH production, and presented a revised computation which shows that this process is suppressed compared to the production of a single large BH. Our main conclusion is that the geometric

[1]It is also easy to see that the *inclusive* cross section of multiple BH production cannot decay with s. This is because the particles may first reduce their energy by emitting one or more "fall safe" gravitons (which costs no s-dependent factor, see [4], Eq. (3)), and then collide to form BHs. I am grateful to M. Voloshin for this remark.

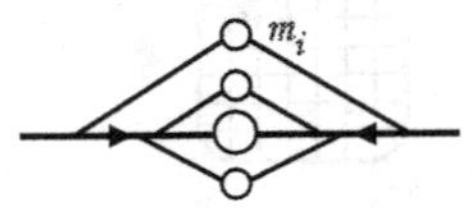

Figure 2. A typical diagram for multiple BH production in Voloshin's model

cross section estimate seems to be in rather good health, surviving all checks and resisting any disproving attempts.

Acknowledgements

I would like to thank Steve Giddings for the opportunity to collaborate on [2]. A part of Section 2 arose as an answer to Marco Cavaglià's interesting questions. I am especially grateful to Misha Voloshin for the e-mail correspondence which helped me to understand his model, and for the critical remarks about the early versions of the computation. It is also a pleasure to thank the organizers of the Cargèse 2004 String Theory school. This work was supported by Stichting FOM.

Appendix

A. Soft graviton emission vertex

Consider emission of a positive helicity collinear graviton with energy ω and small transverse momentum $\mathbf{k} = (k_2, 0)$, $k_2 \ll \omega$. Its polarization tensor, satisfying the constraints $h_{ij}k_j = 0$, $h_{ii} = 0$, is given by

$$h_{ij} \approx \begin{pmatrix} k_2^2/\omega^2 & -k_2/\omega & -ik_2/\omega \\ -k_2/\omega & 1 & i \\ -ik_2/\omega & i & -1 - k_2^2/\omega^2 \end{pmatrix} . \tag{4.1}$$

The emission amplitude is $A \propto \langle p - k|T_{ij}|p\rangle h_{ij}$, where the energy momentum tensor matrix element is

$$\langle P|T_{\mu\nu}|p\rangle = \langle P|\phi_{,\mu}\phi_{,\nu} - \frac{1}{2}\eta_{\mu\nu}(\partial\phi)^2|p\rangle = p_\mu P_\nu + p_\nu P_\mu - \eta_{\mu\nu}(pP) . \tag{4.2}$$

From this we find $A \propto E^2 k_2^2/\omega^2$.

B. Multiple BH production amplitude

We will compute the $n = 2$ amplitude in the case when both small BHs are produced at rest in the c.m. frame: $q_i = (m_i, 0, 0, 0)$ in Fig. 5. The calculation naturally splits into 3 steps: 1) compute the loop integral over the longitudinal momenta; 2) multiply by the emission vertices and integrate over the transverse momenta; 3) multiply by the BH production vertices. It is convenient to write the graviton momenta as

$$k_i = (\frac{m_i}{2} + \frac{x_i - y_i}{2}, \frac{m_i}{2} - \frac{x_i + y_i}{2}, \mathbf{k}_i), \quad k_i' = q_i - k_i \tag{4.3}$$

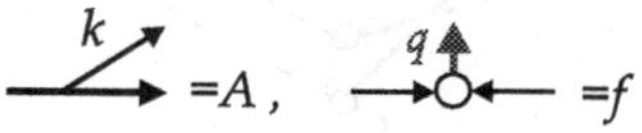

$$\xrightarrow{\;k\;} = A\,, \qquad \xrightarrow{}\!\overset{q\uparrow}{\circ}\!\xleftarrow{} = f$$

Figure 3. Graviton emission and BH production vertices

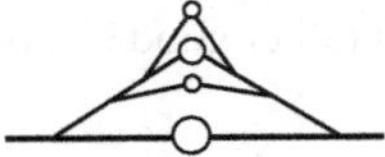

Figure 4. A diagram giving an s-independent contribution to σ_n

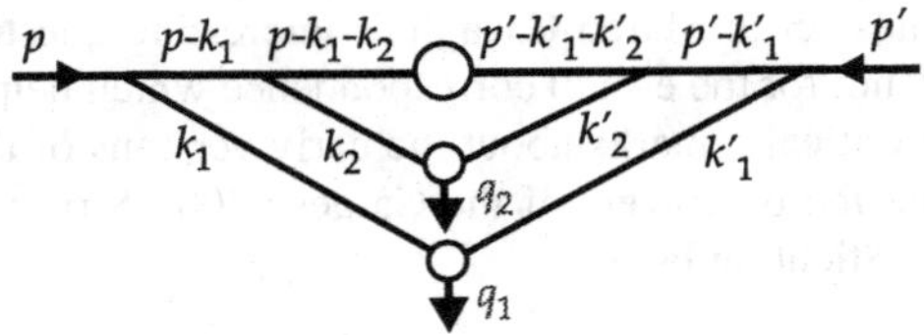

Figure 5. A diagram for one large and two small BHs in the final state

$(i = 1, 2)$. The longitudinal loop integral now separates into the integrals over x_i and y_i. The part depending on x_i is

$$I = \int dx_1\, dx_2 \frac{1}{k_1^2 k_2^2 (p - k_1)^2 (p - k_1 - k_2)^2} \approx \int \frac{dx_1\, dx_2}{(m_1 x_1 - \mathbf{k}_1^2)(m_2 x_2 - \mathbf{k}_2^2)(-E x_1 - \mathbf{k}_1^2)(-E(x_1 + x_2) - \mathbf{k}_{12}^2)} . \tag{4.4}$$

The $+i\epsilon$ is implicit in each denominator. This integral is easy to compute by closing the contour in the lower half-plane. Omitting corrections of the order $m/E \ll 1$, we have

$$I \propto [E^2 \mathbf{k}_1^2 (\mathbf{k}_1^2 + \mathbf{k}_2^2)]^{-1} . \tag{4.5}$$

Since we used complex analysis, it is important to check that the integral is dominated by soft, almost real gravitons. This is indeed true, the important region being $|x_i| \lesssim \mathbf{k}_i^2/m_i$. The tail corresponding to $|x_i| \gg \mathbf{k}_i^2/m_i$ can be estimated directly as:

$$\frac{1}{m_1 m_2 E^2} \int \frac{dx_1}{x_1^2} \frac{dx_2}{x_2^2} \ll I . \tag{4.6}$$

This check also justifies *post factum* neglecting the dependence of the graviton emission vertices on x_i, as well as the omission of x^2 terms in the denominators of (4.4).

Before proceeding to the next step, we have to add diagrams differing by the order of graviton emission (Fig. 6). This summation has the effect $I \to [E^2 \mathbf{k}_1^2 \mathbf{k}_2^2]^{-1}$. Finally, we

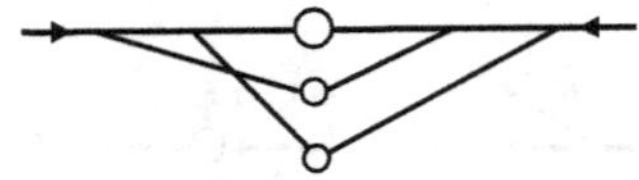

Figure 6. One of the 3 permuted diagrams

multiply by the same factor arising from the y-integration, and get:

$$[E^2\mathbf{k}_1^2\mathbf{k}_2^2]^{-2} \quad \text{(longitudinal part)} . \tag{4.7}$$

We will assume that the small BHs are produced in the spin 0 state, so that the colliding gravitons have opposite helicities. The product of the corresponding emission vertices is

$$\propto (E/m)^4(k_2 + ik_3)^2(k_2 - ik_3)^2 = (E/m)^4(\mathbf{k}^2)^2 . \tag{4.8}$$

Multiplying (4.7) by such factors for both BHs and integrating over $|\mathbf{k}_i| \lesssim 1/E$ (the "fall safe" condition), we get a number $\sim (m_1 m_2)^{-4}$.

Finally, we multiply by the BH production vertices and arrive at

$$f^{(2)} \sim f(s)\, f(m_1^2)\, f(m_2^2)\, (m_1 m_2)^{-4} . \tag{4.9}$$

This formula agrees with the general result (3.3) in the considered case.

Extension of the above computation to the general case is quite straightforward. A point worth mentioning is the use of the standard identity

$$\sum\nolimits_{perm} [a_1(a_1 + a_2) \cdots (a_1 + \ldots + a_n)]^{-1} = [a_1 \cdots a_n]^{-1} \tag{4.10}$$

when summing over the order of graviton emission in the $n > 2$ case.

In [4], the estimate $\int d^4k_i/(k_i^2 k_i'^2) \sim O(1)$ was used in computing the amplitude. However, our analysis shows that the correct estimate is:

$$\int \frac{d^4k}{k^2 k'^2} \propto \int \frac{dx\, dy\, d^2\mathbf{k}}{(mx - \mathbf{k}^2 + i\epsilon)(my - \mathbf{k}^2 + i\epsilon)} \sim \frac{1}{m^2 E^2} . \tag{4.11}$$

This extra factor, for each of n BHs, explains the difference between (3.3) and the result of [4].

References

1. P. Kanti, hep-ph/0402168.
2. S. B. Giddings and V. S. Rychkov, Phys. Rev. D **70**, 104026, hep-th/0409131.
3. V. S. Rychkov, hep-th/0410041.
4. M. B. Voloshin, Phys. Lett. B **524**, 376 (2002), hep-ph/0111099.
5. D. M. Eardley and S. B. Giddings, Phys. Rev. D **66**, 044011 (2002), gr-qc/0201034.
6. V. S. Rychkov, Phys. Rev. D **70**, 044003 (2004), hep-ph/0401116; hep-ph/0405104.

TOWARDS THE EXACT DILATATION OPERATOR OF $\mathcal{N} = 4$ SUPER YANG-MILLS THEORY

ANTON V. RYZHOV
Department of Physics, Brandeis University
Waltham, MA 02454, USA

Abstract. I present a summary of hep-th/0404215, which suggested a novel way of organizing the dilatation operator D of planar $\mathcal{N} = 4$ SYM in the $SU(2)$ sector. Instead of the usual perturbative expansion in powers of λ, we split D into parts $D^{(n)}$ according to the number n of independent pairwise interactions between spins at different sites. The BMN limit fixes $D^{(1)}$ completely, and it has regular expansions at both small and large values of λ. Anomalous dimensions of "long" operators in the two-scalar sector then generically scale as $\sqrt{\lambda}$ at large λ, i.e. in the same way as energies of semiclassical states in the dual $AdS_5 \times S^5$ string theory.

1. Introduction

The $\mathcal{N} = 4$ supersymmetric $SU(N)$ Yang-Mills theory is a family of CFT's parametrized by N and the 't Hooft coupling $\lambda = g_{\rm YM}^2 N$. The problem of computing dimensions of local gauge invariant operators should simplify in the planar limit of $N \to \infty$, λ fixed. In this limit the AdS/CFT duality conjecture suggests that conformal dimensions should be smooth functions of λ, and have regular expansions at both large and small λ.

The main obstacle to verifying the correspondence explicitly is our lack of tools for obtaining exact all-order results on either gauge theory ($\sum c_n \lambda^n$) or string theory ($\sum \frac{b_n}{(\sqrt{\lambda})^n}$) side. One potentially fruitful idea of how to go beyond the first few orders in SYM perturbation theory is to try to determine the exact structure of the dilatation operator D by imposing additional conditions (like superconformal symmetry, BMN limit, integrability, etc.), implied by the expected correspondence with $AdS_5 \times S^5$ string theory. One may then be able to see if they admit a regular expansion not only at small but also at large λ.

This is the approach explored in [1]. Using regularity of a BMN-type scaling limit [2, ?] as an input, we concentrate on the planar $SU(2)$ sector of single trace SYM operators built out of chiral combinations X and Z of two the 6 SYM adjoint scalars, i.e. $\mathrm{tr}(X...XZ...ZX...)$ with canonical dimension L. This sector is closed under renormalization. The eigen-operators of D with J_1 Z's and J_2 X's (so $L = J_1 + J_2$) should be

L. Baulied et al. (eds.), String Theory: From Gauge Interactions to Cosmology, 371–377.

dual to string states with two components of the $SO(6)$ spin. On general grounds, the SYM dilatation operator computed in the planar limit should be a series in λ

$$D=\sum_{r=0}^{\infty}\frac{\lambda^r}{(4\pi)^{2r}}D_{2r}. \tag{1.1}$$

Restricting D to planar graphs suggests that D_{2r} should be given by local sums over sites $a=1,...,L$ with Z and X interpreted as a spin "up" and spin "down" state of a periodic ($a+L\equiv a$) spin chain for which D is the Hamiltonian. The one-loop term D_2 turns out to be equivalent to the Hamiltonian of the ferromagnetic $\mathrm{XXX}_{\frac{1}{2}}$ Heisenberg spin chain, and r-loop contributions involve interactions of $r+1$ nearest neighbors. Generic r-loop contributions to 1.1 contain terms linear in the projectors $Q_{a,b}=\mathbf{1}-P_{a,b}$ ($P_{a,b}$ permutes spins at sites a and b), a term quadratic in $Q_{a,b}$, and so on:

$$D_{2r} = D_{2r}^{(1)}+D_{2r}^{(2)}+..., \tag{1.2}$$

$$D_{2r}^{(1)}=2\sum_{c=1}^{r}\mathfrak{a}_{r,c}Q_{a,a+c}, \quad D_{2r}^{(n)}\sim\sum Q^n. \tag{1.3}$$

Q^n in $D_{2r}^{(n)}$ stands for products of independent projectors, i.e. with all indices corresponding to different sites. Then

$$D=D_0+D^{(1)}+D^{(2)}+..., \quad D^{(1)}=2\sum_{r=1}^{\infty}\frac{\lambda^r}{(4\pi)^{2r}}\sum_{a=1}^{L}\sum_{c=1}^{r}\mathfrak{a}_{r,c}Q_{a,a+c}. \tag{1.4}$$

Using the periodicity of the chain ($Q_{a,b+L}=Q_{a,b}$, etc.) $D^{(1)}$ can be rewritten as

$$D^{(1)}=\sum_{a=1}^{L}\sum_{c=1}^{L-1}h_c(L,\lambda)Q_{a,a+c} \tag{1.5}$$

In [1] we determined the general expression for the coefficients $\mathfrak{a}_{r,c}$ and thus the functions $h_c(L,\lambda)$, i.e. found the spin-spin (linear in Q) part of the exact dilatation operator D. The key point was to demand that the BMN-type scaling limit

$$L\to\infty, \quad \tilde{\lambda}\equiv\frac{\lambda}{L^2}=\text{fixed} \tag{1.6}$$

of the coherent-state expectation value of $D^{(1)}$ is well defined [2, 3]. This turns out to be (nearly) equivalent to the consistency with the BMN expression for the anomalous dimensions of the 2-impurity operators. Imposing the condition of agreement with the BMN square root formula fixes one remaining free coefficient at each order in r.

2. Constraints on $D^{(1)}$

Following [2, 3], we consider the coherent state path integral representation for the quantum mechanics of D as a generalized spin chain Hamiltonian. The corresponding action for a collection of unit 3-vectors $\vec{n}_a(t)$ at each of L sites of the chain ($\langle n| \sigma_a^i |n\rangle = n_a^i$, $(n_a^i)^2 = 1, i = 1, 2, 3$) is then given by

$$S = \int dt \left[\sum_{a=1}^{L} \mathcal{L}_{\rm WZ}(n_a) - \langle D \rangle \right], \quad \langle D \rangle = \langle n| D |n\rangle \ , \tag{2.1}$$

where $\mathcal{L}_{\rm WZ}(\vec{n}_a) = C_i(\vec{n}_a)\dot{n}_a^i$ ensures the proper $SU(2)$ commutation relations if one considers $\vec{n}_a$ back as spin operators. The spin chain in question is ferromagnetic,[1] and in the long spin chain limit $L \to \infty$ one expects that the low energy excitations of the spin chain will be captured by the semiclassical dynamics of 2.1. The correspondence with string theory then suggests that the low energy effective action for the system governed by (2.1) should have a well defined continuum limit. To take the continuum limit one may introduce a field $\vec{n}(\sigma, t)$, $0 < \sigma \leq 2\pi$, with $\vec{n}_a(t) = \vec{n}(\frac{2\pi a}{L}, t)$, so that 2.1 becomes

$$S \to L \int dt \int_0^{2\pi} \frac{d\sigma}{2\pi} \left[C_i(\vec{n})\dot{n}_a^i - \mathcal{H}(\partial\vec{n}, \partial^2\vec{n}, ...; \tilde{\lambda}) \right] . \tag{2.2}$$

$\mathcal{H}$ which originated from $\langle D \rangle$ should be a regular function of the effective coupling $\tilde{\lambda}$ and σ-derivatives of $\vec{n}(t, \sigma)$ in the limit $L \to \infty$, $\tilde{\lambda}$ fixed (with subleading $\frac{1}{L^n}$ terms omitted). Quantum corrections are then suppressed because of the large prefactor L in front of the action.

Since the $Q_{a,b}$ in (1.4) satisfy

$$\langle n|Q_{a,b}|\rangle n = \frac{1}{2}(1 - \vec{n}_a \cdot \vec{n}_b) = \frac{1}{4}(\vec{n}_a - \vec{n}_b)^2 \ , \tag{2.3}$$

$\langle n| D^{(1)} |n\rangle$ contains terms *quadratic* in $\vec{n}$ (but all orders in derivatives); $\langle n| D^{(2)} |n\rangle$, terms *quartic* in n, etc. The approximation that distinguishes $D^{(1)}$ from all higher $D^{(k)}$ in 1.2 is the one in which one keeps only small fluctuations of $\vec{n}(t, \sigma)$ near its ("all-spins-up") ground-state value $\vec{n}_0 = (0, 0, 1)$. Then $\vec{n} = \vec{n}_0 + \vec{\delta n}$, where $|\vec{\delta n}| \ll 1$, so that higher powers of the fluctuating field $\vec{\delta n}$ are suppressed, regardless the number of spatial derivatives acting on them. Such configurations correspond to semiclassical spinning string states with $J_1 \gg J_2$, and are close to a single-spin BPS state. They should indeed represent semiclassical or coherent-state analogs of few-impurity BMN states, having the same BMN energy-spin relation which is indeed reproduced in the limit $J_1 \gg J_2$ by the classical two-spin string solutions. In this BMN-type approximation, demanding that the continuum version of $\langle n| D^{(1)} |n\rangle$ have a regular scaling limit 1.6 implies (after integrating by parts)

$$\begin{aligned} \frac{\lambda^r}{(4\pi)^{2r}} \langle D_{2r}^{(1)} \rangle &= \frac{\lambda^r}{2(4\pi)^{2r}} \sum_{c=1}^{r} \mathrm{a}_{r,c}(\vec{n}_a - \vec{n}_{a+c})^2 \\ &\to d_r \tilde{\lambda}^r \left[(\partial^r \vec{n})^2 + \mathcal{O}\left(\frac{\partial^{2r+2}}{L^2} \right) \right], \end{aligned} \tag{2.4}$$

[1]Then the state with all spins "up" (represented by the operator $\mathrm{tr} Z^L$) is a true vacuum.

with the coefficients $a_{r,c}$ satisfying

$$a_{r,c} = \frac{(-1)^{r-c}(2r)!}{(r-c)!(r+c)!} a_{r,r}\,, \quad c = 1, ..., r-1\,. \tag{2.5}$$

The first non-vanishing coefficient becomes

$$d_r = \frac{(-1)^{r-1}}{2^{2r}(2r)!} \sum_{c=1}^{r} c^{2r} a_{r,c} = \frac{(-1)^{r-1}}{2^{2r+1}} a_{r,r}\,, \tag{2.6}$$

and the r-loop contribution to the expectation value of $D^{(1)}$ takes the form (cf. 2.4)

$$\frac{\lambda^r}{(4\pi)^{2r}} \langle D^{(1)}_{2r} \rangle \quad \rightarrow \quad L\left[d_r \tilde{\lambda}^r \int_0^{2\pi} \frac{d\sigma}{2\pi} (\partial^r \vec{n})^2 + \mathcal{O}(\frac{1}{L})\right] \tag{2.7}$$

What remains is to find the values of $a_{r,r}$, which can be done by analyzing the spectrum of two-impurity BMN operators

$$\mathrm{O}_n^{\mathrm{BMN}} \quad = \quad \frac{1}{\sqrt{J+1}} \sum_{p=0}^{J} \cos\left[\frac{\pi n(2p+1)}{J+1}\right] \mathrm{tr}(XZ^pXZ^{J-p})\,. \tag{2.8}$$

Their anomalous dimensions can be computed in both string theory and gauge theory in the large J, fixed $\frac{\lambda}{J^2}$ limit, and one finds

$$\Delta_{\mathrm{BMN}} = J + 2\sqrt{1 + \frac{\lambda}{J^2} n^2} + \mathcal{O}(\frac{1}{J})\,. \tag{2.9}$$

To reproduce 2.9 by acting with the dilatation operator (1.1) on $\mathrm{O}_n^{\mathrm{BMN}}$, we should have

$$a_{r,c} = \frac{(-1)^{r-c}\,\Gamma(2r+1)\,\Gamma(2r-1)}{\Gamma(r-c+1)\,\Gamma(r+c+1)\,\Gamma(r)\,\Gamma(r+1)}\,. \tag{2.10}$$

3. Summing $D^{(1)}$ to all orders

The expression for $D^{(1)}$ can actually be summed up explicitly. First, summing 2.7 over r with d_r in 2.6 gives a very simple formula for the quadratic in $\vec{n}$ ("small fluctuation" or "BMN") part of the coherent state effective Hamiltonian in 2.2:

$$\begin{aligned} \langle D^{(1)} \rangle \quad &= \quad \sum_{r=1}^{\infty} \frac{\lambda^r}{(4\pi)^{2r}} \langle D^{(1)}_{2r} \rangle \\ &\rightarrow \quad L \int_0^{2\pi} \frac{d\sigma}{2\pi} \left[\frac{1}{4} \vec{n} \left(\sqrt{1 - \tilde{\lambda}\, \partial^2} - 1 \right) \vec{n} + \mathcal{O}(\frac{1}{L}) \right]\,. \end{aligned} \tag{3.1}$$

Remarkably, this is the same expression that follows from the classical $AdS_5 \times S^5$ string sigma model action expanded in the limit 1.6. Next, let us substitute the values 2.10 for the coefficients $\mathfrak{a}_{r,c}$ we have found above into $D^{(1)}$ in 1.4 and try to formally perform the summation over r first, independently for each $Q_{a,a+c}$ term. We get

$$D^{(1)} = 2\sum_{a=1}^{L}\sum_{c=1}^{\infty} f_c(\lambda)\, Q_{a,a+c}\,, \quad f_c(\lambda) = \sum_{r=c}^{\infty} \frac{\lambda^r}{(4\pi)^{2r}}\, \mathfrak{a}_{r,c}\,. \tag{3.2}$$

Remarkably, the series representation for the coefficients $f_c(\lambda)$ can then be summed up in terms of the standard hypergeometric functions,

$$f_c(\lambda) = \left(\frac{\lambda}{4\pi^2}\right)^c \frac{\Gamma(c-\frac{1}{2})}{4\sqrt{\pi}\,\Gamma(c+1)}\; {}_2F_1(c-\frac{1}{2}, c+\frac{1}{2}; 2c+1; -\frac{\lambda}{\pi^2}) \tag{3.3}$$

The coefficient in front of ${}_2F_1$ is equal to $\frac{2\lambda^c}{(4\pi)^{2c}}\mathfrak{a}_{c,c}$. The f_c go to 0 rapidly at large c, so we effectively have a spin chain with short range interactions. The resulting coefficients $f_c(\lambda)$ are smooth positive functions of λ having regular expansion at both small λ and large λ, in which case we find

$$f_c(\lambda)_{\lambda\to\infty} = \frac{\sqrt{\lambda}}{\pi^2}\left[\frac{1}{4c^2-1} + \mathcal{O}(\frac{1}{\lambda})\,,\right] \tag{3.4}$$

The square root $f_c \to \sqrt{\lambda}$ asymptotics of (3.4) is related to the cut structure of ${}_2F_1$.

So far, we have treated all $Q_{a,a+c}$ terms as independent but for finite L the terms with c and $c+mL$ are the same because of the periodicity of the chain (implied by cyclicity of the trace in the operators). Also, under the sum over a one has $Q_{a,a+c} = Q_{a,a+L-c}$. Therefore, for finite L the sum over c should, in fact, be restricted to run from $c = 1$ to $c = L-1$,

$$D^{(1)} = 2\sum_{a=1}^{L}\sum_{c=1}^{\infty} f_c(\lambda)\, Q_{a,a+c} = \sum_{a=1}^{L}\sum_{c=1}^{L-1} h_c(L,\lambda)\, Q_{a,a+c}\,. \tag{3.5}$$

The new coefficients h_c depend on both the the length of the chain and 't Hooft coupling

$$\begin{aligned} h_c(L,\lambda) = \sum_{m=0}^{\infty} \Bigg[&\left(\frac{\lambda}{4\pi^2}\right)^{c+mL} \frac{\Gamma(c+mL-\frac{1}{2})}{4\sqrt{\pi}\,\Gamma(c+mL+1)} \\ &\times {}_2F_1(c+mL-\frac{1}{2}, c+mL+\frac{1}{2}; 2c+2mL+1; -\frac{\lambda}{\pi^2})\Bigg] \\ &+ (c \to L-c)\,. \end{aligned} \tag{3.6}$$

For finite L, we may expand the hypergeometric functions in 3.6 at large λ as in (3.4) and then do the sum over m. Ignoring the issue of convergence of the resulting strong-coupling expansion, that leads to the following simple result for the leading-order term

$$h_c(L,\lambda)_{\lambda\to\infty} = \frac{\sqrt{\lambda}}{2\pi L}\left[\frac{\sin\frac{\pi}{L}}{\cos\frac{\pi}{L} - \cos\frac{2\pi c}{L}} + \mathcal{O}(\lambda^{-1})\right] \tag{3.7}$$

We can apply the above relations to the first non-trivial case of small length. The non-BPS operator with $L = 4$ is the level four descendant K of the Konishi scalar operator

$$K = \mathrm{tr}[X, Z]^2 = 2\,\mathrm{tr}(XZXZ - XXZZ)\,. \tag{3.8}$$

The action of $D^{(1)}$ 3.5 on K is determined by noting that

$$\Pi_1 K = \Pi_3 K = 6K, \quad \Pi_2 K = 0\,, \quad D^{(1)} K = \gamma^{(1)} K\,, \tag{3.9}$$

and one finds the following surprisingly simple result

$$\gamma^{(1)} = \frac{3}{2}\left(\sqrt{1 + \frac{\lambda}{\pi^2}} - 1\right)\,. \tag{3.10}$$

4. Concluding remarks

The resulting $D^{(1)}$ may be interpreted as a Hamiltonian of a periodic spin chain with long-range interactions. One could conjecture that this spin chain may be integrable; and, furthermore, the higher-order terms $D^{(2)}, D^{(3)}, \ldots$ in 1.4 may be effectively determined by $D^{(1)}$, e.g., expressed in terms of higher conserved charges of the chain. This would then determine the full D.

Inspired by recent work in spin chains, in [1] we suggested to organize the dilatation operator as an expansion 1.4, 1.2 in powers of independent projection operators $Q_{a,b}$ [3] at L sites of spin chain

$$D = D_0 + \sum_{n=1}^{\infty} D^{(n)} \text{ with } D_0 = L, \quad D^{(1)} = \sum_{a=1}^{L}\sum_{c=1}^{L-1} h_c(L,\lambda)\, Q_{a,a+c}, \; \ldots \tag{4.1}$$

where $D^{(n)}$ are given by sums of products of n Q's at independent sites of the spin chain. We determined the coefficients in $D^{(1)}$ by demanding that its BMN-type scaling limit [2, 3] be regular, and found that it admits a very simple representation 3.2, applicable at least in the large L limit. This representation includes all orders in λ and suggests that the corresponding anomalous dimensions should grow as $\sqrt{\lambda}$ for large λ.

A natural extension of this work would be to try to find the next term in the expansion (4.1), namely

$$D^{(2)} = \sum_{a=1}^{L} \sum_{c_1,c_2,c_3=1}^{L-1}{}' \; h_{c_1;c_2,c_3}(L,\lambda)\, Q_{a,a+c_1} Q_{a+c_2,a+c_2+c_3} \tag{4.2}$$

The prime on the sum here means that certain terms should be omitted: since $Q_{a,b}Q_{b,c} + Q_{b,c}Q_{a,b} = Q_{a,b} + Q_{b,c} - Q_{a,c}$ the terms with $c_2 = c_1$ and $c_2 = c_3 - c_1$ have already been included in $D^{(1)}$. The contributions of higher $D^{(n)}$ terms should be crucial at finite L, resolving, in particular, the above-mentioned contradiction between the $\sqrt{\lambda}$ asymptotics of the coefficients in $D^{(1)}$ and the expected $\sqrt[4]{\lambda}$ scaling of dimensions of operators corresponding to string modes.

The AdS/CFT duality suggests that D should correspond to an integrable spin chain. The simplest possibility could be that, by analogy with the Inozemtsev chain , the operator $D^{(1)}$ (with interaction coefficients given by 3.6) represents a Hamiltonian of an integrable spin 1/2 chain, while all higher order terms $D^{(n)}$ are effectively determined by $D^{(1)}$ through integrability. Satisfying the requirements of integrability, BMN scaling and consistency with gauge theory should be essentially unique.

References

1. A. V. Ryzhov and A. A. Tseytlin, "Towards the exact dilatation operator of N=4 super Yang-Mills theory," hep-th/0404215.
2. M. Kruczenski, "Spin chains and string theory," hep-th/0311203.
3. M. Kruczenski, A. V. Ryzhov and A. A. Tseytlin, "Large spin limit of $AdS_5 \times S^5$ string theory and low energy expansion of ferromagnetic spin chains," hep-th/0403120.

MODULI TRAPPING AND STRING GAS COSMOLOGY

SCOTT WATSON
Department of Physics, Brown University, Box 1843,
Providence, RI 02912, USA

Abstract. In this talk I will discuss the role of finite temperature quantum corrections in string cosmology and show that they can lead to a stabilization mechanism for the volume moduli. I will show that from the higher dimensional perspective this results from the effect of states of enhanced symmetry on the one-loop free energy. These states lead not only to stabilization, but also suggest an alternative model for ΛCDM. At late times, when the low energy effective field theory gives the appropriate description of the dynamics, the moduli will begin to slow-roll and stabilization will generically fail. However, stabilization can be recovered by considering cosmological particle production near the points of enhanced symmetry leading to the process known as moduli trapping.

1. Initial Conditions

One problem in string cosmology is the issue of initial conditions. Not only must models of string cosmology address the standard initial condition problems in cosmology, but string theory also predicts the existence of extra dimensions. The usual prescription for dealing with the extra dimensions is to take them small, stable, and unobservable. However a complete model of string cosmology should explain how this came about and why the explicit breaking of Lorentz invariance should be allowed. A step in this direction was first proposed by Brandenberger and Vafa[1]. They argued that, by considering the dynamics of a string gas in nine, compact spatial dimensions initially taken at the string scale, one could explain why three dimensions grow large while six stay compact. The crux of their argument is based on the fact that in addition to the usual Kaluza Klein modes of a particle on a compact space, strings also possess winding modes. These extra degrees of freedom will generically halt cosmological expansion; however, if these modes could annihilate with their anti-partners through the process,

$$W^{+} + W^{-} \rightleftarrows \text{unwound string}, \tag{1.1}$$

this would allow the dimension they occupy to expand. Then, the fact that strings generically intersect (interact) in at most three spatial dimensions means that in the remaining six dimensions thermal equilibrium cannot be maintained. Thus, the winding modes will

L. Baulied et al. (eds.), String Theory: From Gauge Interactions to Cosmology, 379–383.

drop out of equilibrium and the six spatial dimensions will be frozen near the string scale. Furthermore, once all the winding modes in the three large dimensions annihilate, the universe emerges filled with a gas of momentum modes, which evolves as a radiation dominated universe.

2. Strings at Finite Temperature

The usual starting point of string cosmology is the action,

$$S = \frac{1}{2\kappa^2} \int d^{d+1}x \sqrt{-g} e^{-2\varphi} \left(R + 4(\nabla\varphi)^2 - \frac{1}{12} H^2 + \mathcal{O}(\alpha') \right) + \mathcal{O}(g_s). \qquad (2.1)$$

For simplicity we will ignore the Ramond sector and set $H = 0$. Here we have in mind the heterotic string on a toroidal background. Motivated by the Brandenberger-Vafa scenario we will take the background to be $\mathbb{R}^4 \times T^6$, where we assume that the three spatial dimensions have grown large enough to be approximated by an FRW universe and the six small dimensions are toroidal and near the string scale. To include time dependence we make use of the adiabatic approximation, which implies that we can replace static quantities by slow varying functions of time[1].

The action (2.1) represents a double expansion in both the string coupling $g_s \sim e^{\langle\varphi\rangle}$ and the string tension, $T = \frac{1}{4\pi\alpha'}$. We now want to include terms coming from g_s corrections at finite temperature[2, 3]. Let us consider the 1-loop free energy

$$F = TZ_1 \sim \beta^{-1} \sum e^{-\beta M(n,\omega,N,\bar{N})}, \qquad (2.2)$$

where Z_1 is the one loop partition function, $M(n, \omega, N, \bar{N})$ is the string mass, and $\beta = 1/T$ is the inverse temperature. In the early universe we are interested in temperatures near or below the string scale ($\beta \gtrsim \sqrt{\alpha'}$) where the major contribution to the one-loop free energy can be seen to come from the massless modes of the string.

In the case of the heterotic string there are additional massless states that occur at the special radius $R = \sqrt{\alpha'}$. These extra states include winding and momentum modes of the string. To understand the dynamics that result by including these states we can find the energy density and pressure, which follow from the free energy as

$$\rho = \frac{1}{V} \frac{\partial}{\partial\beta}(\beta F)$$
$$p_i = -\frac{1}{9V} \frac{\partial F}{\partial(lnR_i)}, \qquad (2.3)$$

where V is the spatial volume and R_i is the scale factor in the ith direction. As discussed above, we take initial conditions where three dimensions have grown large and six remain near the string scale. For such initial data, the above pressure in three dimensions corresponds to the equation of state of a radiation dominated universe $p_3 = \frac{1}{3}\rho$, whereas the

[1]This is known to be a good approximation in the early universe, and is equivalent to saying that the time scale for string interactions is greater than the rate of expansion. However, one should keep in mind that such an approximation will mostly likely break down near a curvature singularity.

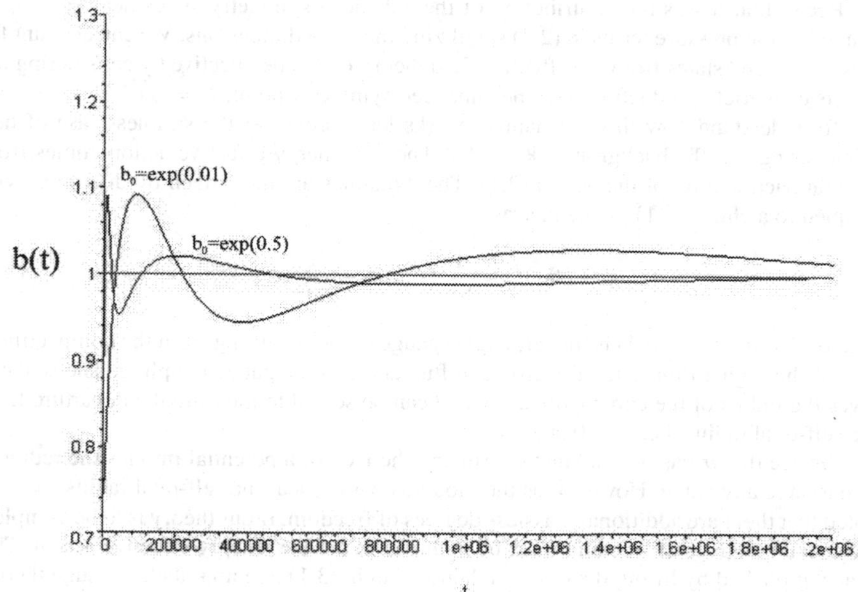

Figure 1. The evolution of the volume modulus of the extra dimensions for different initial values near the self-dual radius. We see that the moduli are eventually stabilized at the self-dual radius where the pressure is found to vanish.

pressure in the small dimensions gives the behavior,

$$p_6 < 0 \text{ for } R_6 > \sqrt{\alpha'} \quad \text{and} \quad p_6 > 0 \text{ for } R_6 < \sqrt{\alpha'}. \tag{2.4}$$

In *dilaton gravity* negative pressure implies a contracting universe, whereas positive pressure leads to expansion. Thus, as can be seen in Figure 2, pressure leads to a stabilizing effect for the scale factor of the extra dimensions driving the radius toward the enhanced symmetry point $R_{(6)} = \sqrt{\alpha'}$ where the pressure vanishes. At this location the gauge symmetry of the heterotic string is enhanced, $E_8 \times E_8 + U(1)^6 \longrightarrow E_8 \times E_8 + SU(2)^6$.

In addition to stabilizing the volume moduli, it has been shown that the remnant string modes, if taken in the dark sector, can also lead to an interesting cold dark matter candidate[4, 5] (see also[6]).

3. Moduli Trapping

If one attempts to extend the arguments above to the $4D$ effective field theory, one finds that the stabilization mechanism no longer holds. This is not surprising since the pressure in the extra dimensions has no analog from the $4D$ perspective. However, one thing that should remain is the idea of enhanced symmetry.

Recall that it was the contribution of the enhanced symmetry states near $R = \sqrt{\alpha'}$ that led to the pressure terms in (2.4) stabilizing the extra dimensions. We can account for these enhanced states from the effective field theory (EFT) perspective by considering the effects of particle production near the enhanced symmetry point, $R = \sqrt{\alpha'}$.

To understand how this mechanism works let us consider the simplest case of heterotic strings on the background $\mathbb{R}^4 \times S^1$. The low energy effective action comes from the compactification of the action (2.1). The dynamics are then given by dilaton gravity coupled to a chiral $U(1)$ gauge theory,

$$\mathcal{L}\mathrm{m} = (\partial\sigma)^2 - \frac{1}{4g^2}(F_{\mu\nu})^2 - \frac{1}{4g^2}(\bar{F}_{\mu\nu})^2, \tag{3.1}$$

where $F = dA$ ($\bar{F} = d\bar{A}$) is the left (right) gauge theory resulting from the compactification of the higher dimensional metric and flux and g is the gauge coupling. The scalar σ gives the radius of the compactification and can be scaled to measure the departure from the self-dual radius, i.e. $\sigma = 0$ at $R = \sqrt{\alpha'}$.

We see that σ has only a kinetic term and the lack of a potential implies the radius is free to take any value. However, as the modulus passes near the self-dual radius we have noted that there are additional massless degrees of freedom. If our theory is to be complete these extra degrees of freedom must be included in the low energy effective action. This is accomplished by lifting the effective lagrangian in (3.1) to a non-abelian gauge theory, in this case chiral $SU(2)$. We introduce the covariant derivative,

$$D_\mu\sigma = \partial_\mu\sigma + gA_\mu\sigma. \tag{3.2}$$

This leads to a time dependent mass for the new vector A_μ

$$\frac{1}{2}g^2\sigma(t)^2A_\mu^2. \tag{3.3}$$

This time dependent mass implies particle production in our cosmological space-time. The stabilization of the radius σ can now be realized as follows; initially σ is dominated strictly by the kinetic term, however once it passes near the enhanced symmetry point $\sigma = 0$, A_μ particles will be produced. Then, as σ continues its trajectory the mass of the A_μ's will increase and this leads to backreaction on σ. This force, along with friction from the cosmological expansion, will eventually stabilize σ at the enhanced symmetry point $\sigma = 0$.

This is a simple example of moduli trapping[7, 8, 9]. Although we have considered here a simple toy model of a string on a circle, points of enhanced symmetry are present in nearly all string and M-theory compactifications. Moreover, it is worth mentioning that this mechanism need not apply only to volume moduli. In fact, in Kofman, et. al.[7] the modulus of interest was the distance between two branes which is of course related by T-duality to the case we have considered here.

4. Conclusions

We have seen that from the $10D$ perspective it is possible to stabilize the volume modulus of a heterotic string compactification on a T^6. The stabilization was found to be the result of the pressure exerted on the compact space due to the presence of enhanced symmetry states contributing to the one-loop free energy of the strings at finite temperature. In the $4D$ effective field theory such effects can be understood by considering particle production near the enhanced symmetry point. Near this point additional massless states are allowed which can be particle produced in the cosmological space-time. These states then get masses via the string Higgs Effect and backreact on the modulus stabilizing the extra dimensions at the self-dual radius.

Acknowledgements

I would like to thank Robert Brandenberger, Sera Cremonini, Steve Gubser, Laur Jarv, Liam McAllister, and Thomas Mohaupt for useful comments and discussions during this work. I would also like to thank the organizers of the Cargèse 2004 ASI. This work was supported by NASA GSRP.

References

1. R. H. Brandenberger and C. Vafa, "Superstrings In The Early Universe," Nucl. Phys. B **316**, 391 (1989).
2. J. Kripfganz and H. Perlt, "Cosmological Impact Of Winding Strings," Class. Quant. Grav. **5**, 453 (1988).
3. A. A. Tseytlin and C. Vafa, "Elements of string cosmology," Nucl. Phys. B **372**, 443 (1992) [arXiv:hep-th/9109048].
4. S. S. Gubser and P. J. E. Peebles, "Cosmology with a dynamically screened scalar interaction in the dark sector," arXiv:hep-th/0407097.
5. S. S. Gubser and P. J. E. Peebles, "Structure formation in a string-inspired modification of the cold dark matter model," arXiv:hep-th/0402225.
6. T. Battefeld and S. Watson, "Effective field theory approach to string gas cosmology," JCAP **0406**, 001 (2004) [arXiv:hep-th/0403075].
7. L. Kofman, A. Linde, X. Liu, A. Maloney, L. McAllister and E. Silverstein, "Beauty is attractive: Moduli trapping at enhanced symmetry points," JHEP **0405**, 030 (2004) [arXiv:hep-th/0403001].
8. S. Watson, "Moduli stabilization with the string Higgs effect," arXiv:hep-th/0404177.
9. L. Jarv, T. Mohaupt and F. Saueressig, "Singular compactifications and cosmology," arXiv:hep-th/0311016.

STRINGY GEOMETRY: GENERALIZED COMPLEX STRUCTURE

MAXIM ZABZINE
LPTHE, Université Paris 6,
4 pl Jussieu, 75252 Paris cedex, France

Abstract. We consider a worldsheet realization of generalized complex geometry, a notion introduced recently by Hitchin which interpolates between complex and symplectic manifolds.

The recently developed notion of generalized complex geometry naturally extends and unifies complex and symplectic geometries, in general interpolating between the two [1, 2]. There have been many hints that this geometry should be relevant to string theory. In this note, we present toy 2D models which provide the "physical" derivation of the generalized complex geometry. This note is based on the joint with U. Lindström, R. Minasian and A. Tomasiello [3].

1. Mathematical preliminaries

In this Section we review some basic notions and fix notations. Namely we collect general facts concerning the generalized complex structure, see [1] and [2] for further details. Also we work out the coordinate form of the integrability conditions for the generalized complex structure.

Let us start by recalling the definition of the standard complex structure on a manifold $\mathcal{M}$ ($\dim \mathcal{M} = d$). An almost complex structure is defined as a linear map on the tangent bundle[1] $J : T \to T$ such that $J^2 = -1_d$. This allows the definition of projectors on T,

$$\pi_{\pm} = \frac{1}{2}(1_d \pm iJ). \tag{1.1}$$

An almost complex structure is called integrable if the projectors $\pi_{\pm}$ define integrable distributions on T, namely if

$$\pi_{\mp}[\pi_{\pm}X, \pi_{\pm}Y] = 0 \tag{1.2}$$

for any $X, Y \in T$ where $[\ ,\ ]$ is a standard Lie bracket on T.

[1] In this note we consider the complexified tangent and cotangent bundles.

L. Baulied et al. (eds.), String Theory: From Gauge Interactions to Cosmology, 385–391.

A generalization of the notion of complex structure has been proposed by Hitchin [1]. In Hitchin's construction T is replaced by $T \oplus T^*$ and the Lie bracket is replaced by the appropriate bracket on $T \oplus T^*$, the so called Courant bracket. Thus a generalized complex structure is an almost complex structure $\mathcal{J}$ on $T \oplus T^*$ whose $\pm i$-eigenbundles are Courant involutive. In other words, a generalized complex structure is splitting of $T \oplus T^*$ into a sum of two complementary Dirac subbundles. A detailed study of generalized complex geometry can be found in Gualtieri's thesis [2].

Now let us give detailed definitions. On $T \oplus T^*$ there is a natural indefinite metric defined by $(X + \xi, X + \xi) = -i_X \xi$. In the coordinate basis (∂_μ, dx^μ) we can write this metric as follows

$$\mathcal{I} = \begin{pmatrix} 0 & 1_d \\ 1_d & 0 \end{pmatrix}. \tag{1.3}$$

A generalized almost complex structure is a map $\mathcal{J} : T \oplus T^* \to T \oplus T^*$ such that $\mathcal{J}^2 = -1_{2d}$ and that $\mathcal{I}$ is hermitian with respect to $\mathcal{J}$, $\mathcal{J}^t \mathcal{I} \mathcal{J} = \mathcal{I}$. On $T \oplus T^*$ there is a Courant bracket which is defined as follows

$$[X + \xi, Y + \eta]_c = [X, Y] + \mathcal{L}_X \eta - \mathcal{L}_Y \xi - \frac{1}{2} d(i_X \eta - i_Y \xi). \tag{1.4}$$

This bracket is skew-symmetric but in general does not satisfy the Jacobi identity. However if there is a subbundle $L \subset T \oplus T^*$ which is involutive (closed under the Courant bracket) and isotropic with respect to $\mathcal{I}$ then the Courant bracket on the sections of L does satisfy the Jacobi identity. This is a reason for imposing hermiticity of $\mathcal{I}$ with respect to $\mathcal{J}$. One important feature of the Courant bracket is that, unlike the Lie bracket, this bracket has a nontrivial automorphism defined by a closed two-form b,

$$e^b(X + \xi) = X + \xi + i_X b. \tag{1.5}$$

such that

$$[e^b(X + \xi), e^b(Y + \eta)]_c = e^b[X + \xi, Y + \eta]_c. \tag{1.6}$$

We can construct the projectors on $T \oplus T^*$

$$\Pi_\pm = \frac{1}{2}(I \pm i\mathcal{J})\,; \tag{1.7}$$

the almost generalized complex structure $\mathcal{J}$ is integrable if

$$\Pi_\mp [\Pi_\pm (X + \xi), \Pi_\pm (Y + \eta)]_c = 0, \tag{1.8}$$

for any $(X + \xi), (Y + \eta) \in T \oplus T^*$. This is equivalent to the single statement

$$[X + \xi, Y + \eta]_c - [\mathcal{J}(X + \xi), \mathcal{J}(Y + \eta)]_c + \mathcal{J}[\mathcal{J}(X + \xi), Y + \eta]_c + \\ + \mathcal{J}[X + \xi, \mathcal{J}(Y + \eta)]_c = 0 \tag{1.9}$$

which resembles the definition of the Nijenhuis tensor.

To relate this construction to the physical models we have to reexpress the above definitions in coordinate form. The map $\mathcal{J}$ can be written in the form

$$\mathcal{J} = \begin{pmatrix} J & P \\ L & K \end{pmatrix}, \tag{1.10}$$

where $J : T\mathcal{M} \to T\mathcal{M}$, $P : T^*\mathcal{M} \to T\mathcal{M}$, $L : T\mathcal{M} \to T^*\mathcal{M}$ and $K : T^*\mathcal{M} \to T^*\mathcal{M}$ and hence they correspond to the tensor fields, $J^\mu{}_\nu$, $L_{\mu\nu}$, $P^{\mu\nu}$ and $K_\mu{}^\nu$. Then the condition $\mathcal{J}^2 = -1_{2d}$ becomes

$$J^\mu{}_\nu J^\nu{}_\lambda + P^{\mu\nu} L_{\nu\lambda} = -\delta^\mu{}_\lambda, \tag{1.11}$$

$$J^\mu{}_\nu P^{\nu\lambda} + P^{\mu\nu} K_\nu{}^\lambda = 0, \tag{1.12}$$

$$K_\mu{}^\nu K_\nu{}^\lambda + L_{\mu\nu} P^{\nu\lambda} = -\delta^\mu{}_\lambda, \tag{1.13}$$

$$K_\mu{}^\nu L_{\nu\lambda} + L_{\mu\nu} J^\nu{}_\lambda = 0. \tag{1.14}$$

The hermiticity of $\mathcal{I}$ with respect to $\mathcal{J}$ translates into the following conditions

$$J^\mu{}_\nu + K_\mu{}^\nu = 0, \qquad P^{\mu\nu} = -P^{\nu\mu}, \qquad L_{\mu\nu} = -L_{\nu\mu}. \tag{1.15}$$

In local coordinates the integrability condition (1.9) is equivalent to the following four conditions

$$J^\nu{}_{[\lambda} J^\mu{}_{\rho],\nu} + J^\mu{}_\nu J^\nu{}_{[\lambda,\rho]} + P^{\mu\nu} L_{[\lambda\rho,\nu]} = 0 \tag{1.16}$$

$$P^{[\mu|\nu} P^{|\lambda\rho]}{}_{,\nu} = 0 \tag{1.17}$$

$$J^\mu{}_{\nu,\rho} P^{\rho\lambda} + P^{\rho\lambda}{}_{,\nu} J^\mu{}_\rho - J^\lambda{}_{\rho,\nu} P^{\mu\rho} + J^\lambda{}_{\nu,\rho} P^{\mu\rho} - P^{\mu\lambda}{}_{,\rho} J^\rho{}_\nu = 0 \tag{1.18}$$

$$J^\lambda{}_\nu L_{[\lambda\rho,\gamma]} + L_{\nu\lambda} J^\lambda{}_{[\gamma,\rho]} + J^\lambda{}_{[\rho} L_{\gamma]\nu,\lambda} + L_{\lambda\rho} J^\lambda{}_{\gamma,\nu} + J^\lambda{}_\rho L_{\lambda\gamma,\nu} = 0 \tag{1.19}$$

To summarize, the generalized complex structure $\mathcal{J}$ is defined by three tensor fields $J^\mu{}_\nu$, $L_{\mu\nu}$ and $P^{\mu\nu}$ which satisfy the algebraic conditions (1.11)-(1.15) and the differential conditions (1.16)-(1.19).

The usual complex structure J is embedded in the notion of generalized complex structure

$$\mathcal{J} = \begin{pmatrix} J & 0 \\ 0 & -J^t \end{pmatrix}. \tag{1.20}$$

One can check that all properties (1.11)-(1.19) are satisfied provided that J is a complex structure. Also, a symplectic structure is an example of a generalized complex structure

$$\mathcal{J} = \begin{pmatrix} 0 & -\omega^{-1} \\ \omega & 0 \end{pmatrix} \tag{1.21}$$

where ω is an ordinary symplectic structure ($d\omega = 0$). More exotic examples exist and are given by manifolds, that do not admit any known complex or symplectic structure, but do admit a generalized complex structure [2].

Consider a generalized complex structure $\mathcal{J}$; a new generalized complex structure can be generated by

$$\mathcal{J}_b = \begin{pmatrix} 1 & 0 \\ b & 1 \end{pmatrix} \mathcal{J} \begin{pmatrix} 1 & 0 \\ -b & 1 \end{pmatrix} \tag{1.22}$$

if $b \in \Omega^2_{closed}(\mathcal{M})$. The structure $\mathcal{J}_b$ is integrable due to the fact that the transformation (1.5) is an automorphism of the Courant bracket. The transformation (1.5) is called a b-transform.

The key feature of a complex manifold is that is locally equivalent to $\mathbb{C}^k$ via a diffeomorphism. For symplectic manifolds the Darboux theorem states that a symplectic structure is locally equivalent, via diffeomorphism, to the standard symplectic structure $(\mathbf{R}^{2k}, \omega)$, where

$$\omega = dx_1 \wedge dx_2 + ... + dx_{2k-1} \wedge dx_{2k}. \tag{1.23}$$

For generalized complex manifolds there exists a generalized Darboux theorem [2], which states that in a neighborhood of a regular point[2] a generalized complex structure on a manifold $\mathcal{M}$ is locally equivalent via a diffeomorphism and a b-transform (see (1.22)), to the product of an open set in $\mathbb{C}^k$ and an open set in the standard symplectic space $(\mathbf{R}^{d-2k}, \omega)$.

The Courant bracket on $T \oplus T^*$ can be twisted by a closed three form H. Namely given a closed three form H one can define another bracket on $T \oplus T^*$ by

$$[X+\xi, Y+\eta]_H = [X+\xi, Y+\eta]_c + i_X i_Y H. \tag{1.24}$$

This bracket has similar properties to the Courant bracket. Again if a subbundle $L \subset T \oplus T^*$ is closed under the twisted Courant bracket and isotropic with respect to $\mathcal{I}$, then the Courant bracket on the sections of L does satisfy the Jacobi identity. Thus in the integrability condition (1.9) the Courant bracket $[\ ,\ ]_c$ can be replaced by the new twisted Courant bracket $[\ ,\ ]_H$. In local coordinates the new integrability condition is equivalent to four expressions

$$J^\nu{}_{[\lambda} J^\mu{}_{\rho],\nu} + J^\mu{}_\nu J^\nu{}_{[\lambda,\rho]} + P^{\mu\nu}(L_{[\lambda\rho,\nu]} + J^\sigma{}_{[\lambda} H_{\rho]\sigma\nu}) = 0 \tag{1.25}$$

$$P^{[\mu|\nu} P^{|\lambda\rho]}{}_{,\nu} = 0 \tag{1.26}$$

$$J^\mu{}_{\nu,\rho} P^{\rho\lambda} + P^{\rho\lambda}{}_{,\nu} J^\mu{}_\rho + J^\lambda{}_{[\nu,\rho]} P^{\mu\rho} - P^{\mu\lambda}{}_{,\rho} J^\rho{}_\nu - P^{\lambda\sigma} P^{\mu\rho} H_{\sigma\rho\nu} = 0 \tag{1.27}$$

$$J^\lambda{}_\nu L_{[\lambda\rho,\gamma]} + L_{\nu\lambda} J^\lambda{}_{[\gamma,\rho]} + J^\lambda{}_\rho L_{\gamma\nu,\lambda} + J^\lambda{}_\gamma L_{\nu\rho,\lambda} + L_{\lambda\rho} J^\lambda{}_{\gamma,\nu} + \\ + J^\lambda{}_\rho L_{\lambda\gamma,\nu} + H_{\rho\gamma\nu} - J^\lambda{}_{[\rho} J^\sigma{}_\gamma H_{\nu]\lambda\sigma} = 0 \tag{1.28}$$

2. Topological model

In this Section we consider a toy topological model which will provide a "physical" derivation of generalized complex geometry. The model has the following action

$$S_{top} = \int d^2\sigma\, d\theta\ S_{+\mu} \partial_{=} \Phi^\mu \tag{2.1}$$

written in $N=(1,0)$ superfields where $(++,=)$ are worldsheet indices and $(+,-)$ are two-dimensional spinor indices. This is a topological system which describes the holomorphic maps $\Phi : \Sigma \to \mathcal{M}$. The model is manifestly $N=(1,0)$ supersymmetric and can be defined over any differential manifold $\mathcal{M}$. We would like to find the restrictions on $\mathcal{M}$ arising from the requirement that the model admits $(2,0)$ supersymmetry.

[2] P is a Poisson structure and it will define a symplectic foliation. The point is called regular if P has constant rank in a neighborhood.

We have to look for additional (non manifest) supersymmetry transformations. The general transformations of S_+ and Φ are given by the following expressions

$$\delta(\epsilon)\Phi^\mu = \epsilon^+ D_+\Phi^\nu J^\mu{}_\nu - \epsilon^+ S_{+\nu} P^{\mu\nu}, \tag{2.2}$$

$$\begin{aligned}\delta(\epsilon)S_{+\mu} = i\epsilon^+\partial_{++}\Phi^\nu L_{\mu\nu} - \epsilon^+ D_+ S_{+\nu} K_\mu{}^\nu + \epsilon^+ S_{+\nu} S_{+\rho} N_\mu{}^{\nu\rho} + \\ +\epsilon^+ D_+\Phi^\nu D_+\Phi^\rho M_{\mu\nu\rho} + \epsilon^+ D_+\Phi^\rho S_{+\nu} Q_{\mu\rho}{}^\nu\end{aligned} \tag{2.3}$$

Classically the Ansatz (2.2) and (2.3) is unique on dimensional grounds and by Lorentz covariance [4]. This Ansatz involves seven different tensors on $\mathcal{M}$. We have to require the standard $N = (2, 0)$ supersymmetry algebra, i.e. the manifest and non-manifest supersymmetry transformations commute and the nonmanifest supersymmetry transformations satisfy the following conditions

$$[\delta(\epsilon_2), \delta(\epsilon_1)]\Phi^\mu = 2i\epsilon_1^+\epsilon_2^+\partial_{++}\Phi^\mu, \qquad [\delta(\epsilon_2), \delta(\epsilon_1)]S_{+\mu} = 2i\epsilon_1^+\epsilon_2^+\partial_{++}S_{+\mu}. \tag{2.4}$$

Since the nonmanifest transformations are written in $(1, 0)$ superfield then the first requirement is automatically satisfied. Next we have to calculate the commutator of two nonmanifest supersymmetry transformations. Imposing the condition (2.4) implies four algebraic and eleven differential conditions on the seven tensors introduced in (2.2) and (2.3).

Before analyzing the algebra in detail it is useful to look at the invariance of the action. The action (2.1) is invariant under (2.2) and (2.3) if the following algebraic conditions

$$J^\mu{}_\nu + K_\nu{}^\mu = 0, \qquad L_{\mu\nu} = -L_{\nu\mu}, \qquad P^{\mu\nu} = -P^{\nu\mu} \tag{2.5}$$

as well as the differential conditions

$$\frac{1}{2}P^{\mu\nu}{}_{,\rho} = -N_\rho{}^{\mu\nu}, \qquad J^\mu{}_{[\nu,\rho]} = Q_{\nu\rho}{}^\mu, \qquad \frac{1}{2}L_{[\mu\nu,\rho]} = M_{\rho\nu\mu} \tag{2.6}$$

are satisfied. The differential conditions (2.6) allow us to express all three index tensors in terms of appropriate derivatives of two index tensors J, P, L and K. These two index tensors can be combined as a single object

$$\mathcal{J} = \begin{pmatrix} J & P \\ L & K \end{pmatrix}, \tag{2.7}$$

where $\mathcal{J} : T \oplus T^* \to T \oplus T^*$. It is easy to see that the algebraic part of the supersymmetry algebra (2.4) can be written as a single equation, namely that $\mathcal{J}^2 = -1_{2d}$. Passing then to the action, the algebraic condition (2.5) is equivalent to a hermiticity of $\mathcal{I}$ with respect to $\mathcal{J}$ (i.e., the natural pairing on $T \oplus T^*$, see previous Section). Therefore $\mathcal{J}$ is an almost generalized complex structure. Finally we have to analyze the eleven differential conditions coming from the algebra using (2.6). Using the results from the previous Section, we can show that all differential conditions arising from (2.4) are the same as the conditions (1.16)-(1.19). Therefore we have proved that the differential conditions that come from the supersymmetry algebra are equivalent to integrability of $\mathcal{J}$ with respect to the Courant bracket.

To summarize the topological model (2.1) admits $(2, 0)$ supersymmetry if and only if manifold $\mathcal{M}$ is generalized complex manifold.

As we briefly mentioned in the previous Section, a generalized complex manifold is equivalent locally, via diffeomorphism and b-transform, to a product of a symplectic and a complex manifolds. If we choose the Darboux coordinates (label n) along the symplectic leaf and the standard complex coordinates (label i, $\bar{i}$) transverse to the leaf the supersymmetry transformations (2.2) and (2.3) is simplified drastically and have the following form

$$\delta\Phi^i = i\epsilon^+ D_+\Phi^i, \qquad \delta\Phi^{\bar{i}} = -i\epsilon^+ D_+\Phi^{\bar{i}} \tag{2.8}$$

$$\delta S_{+i} = i\epsilon^+ D_- S_{+i}, \qquad \delta S_{+\bar{i}} = -i\epsilon^+ D_+ S_{+\bar{i}} \tag{2.9}$$

$$\delta\Phi^n = -\epsilon^+ S_{+(n+1)}, \qquad \delta S_{+(n+1)} = -i\epsilon^+\partial_{++}\Phi^n \tag{2.10}$$

$$\delta\Phi^{n+1} = \epsilon^+ S_{+n}, \qquad \delta S_{+n} = i\epsilon^+\partial_{++}\Phi^{n+1} \tag{2.11}$$

3. Topological model with WZ term

In the previous Section we presented the topological model for which the extended supersymmetry is related to the generalized complex structure with integrability defined with the respect to the Courant bracket. The natural question is now the following: if in the integrability condition the Courant bracket is replaced by the twisted Courant bracket, can we then construct a model which incorporates twisted integrability? This is in fact possible and the solution is related to the WZ term.

We consider the topological model with an additional term

$$S_{top} = \int d^2\sigma\, d\theta\; S_{+\mu}\partial_= \Phi^\mu - \frac{1}{2}\int d^2\sigma\, d\theta\; D_+\Phi^\mu \partial_=\Phi^\nu B_{\mu\nu} \tag{3.1}$$

The last term is a WZ term and it depends only on a closed three-form H

$$H_{\mu\nu\lambda} = \frac{1}{2}(B_{\mu\nu,\lambda} + B_{\lambda\mu,\nu} + B_{\nu\lambda,\mu}) \tag{3.2}$$

if the world-sheet does not have a boundary. The model (3.1) has $N = (1,0)$ supersymmetry and can be defined over any differential manifold $\mathcal{M}$ equipped with a closed three-form H. The Ansatz for the nonmanifest supersymmetry transformations is given by the same expressions as before, (2.2) and (2.3). The off-shell supersymmetry algebra is exactly the same, (2.4).

The main difference comes from the action. Namely invariance of the new action (3.1) under the transformations (2.2) and (2.3) leads to new relations between the three and two index tensors in the supersymmetry transformations. The action (3.1) is invariant under (2.2) and (2.3) if the following algebraic conditions are satisfied

$$J^\mu{}_\nu + K_\nu{}^\mu = 0, \qquad L_{\mu\nu} = -L_{\nu\mu}, \qquad P^{\mu\nu} = -P^{\nu\mu} \tag{3.3}$$

as well as the differential conditions

$$\frac{1}{2}P^{\mu\nu}{}_{,\rho} = -N_\rho{}^{\mu\nu}, \qquad J^\mu{}_{[\nu,\rho]} + P^{\mu\lambda}H_{\lambda\nu\rho} = Q_{\nu\rho}{}^\mu,$$

$$\frac{1}{2}L_{[\mu\nu,\rho]} + \frac{1}{2}J^\lambda{}_{[\mu}H_{\nu]\lambda\rho} = M_{\rho\nu\mu}. \tag{3.4}$$

The algebraic part of all conditions remains the same as in the previous Section and therefore the two-index tensors can be combined in a single object $\mathcal{J}$ which is an almost generalized complex structure. However the differential conditions will change. Using (3.4) we have to study the supersymmetry algebra (2.4). We can show that the three differential conditions arising from (2.4) are the same as conditions (1.25)-(1.28). Therefore we have proved that the differential conditions coming from the supersymmetry algebra are equivalent to integrability of $\mathcal{J}$ with respect to the *twisted* Courant bracket.

Finally let us remark that both actions (2.1) and (3.1) are invariant under following transformation with $b \in \Omega^2_{closed}(\mathcal{M})$

$$\Phi \rightarrow \Phi, \quad S_{+\mu} \rightarrow S_{+\mu} + \frac{1}{2} b_{\mu\nu} D_+ \Phi^\nu \tag{3.5}$$

which correspond to a b-transform in $T \oplus T^*$ (since $\partial_= \Phi \rightarrow \partial_= \Phi$).

Acknowledgements

It is a pleasure to thank Ulf Lindström, Ruben Minasian and Alessandro Tomasiello for the stimulating collaboration. I would like to thank also the organizers of the Cargèse 2004 ASI.

References

1. N. Hitchin, "Generalized Calabi-Yau manifolds," Q. J. Math. **54** (2003), no. 3, 281–308, arXiv:math.DG/0209099.
2. M. Gualtieri, "Generalized complex geometry," Oxford University DPhil thesis, arXiv:math.DG/0401221.
3. U. Lindström, R. Minasian, A. Tomasiello and M. Zabzine, "Generalized complex manifolds and supersymmetry," arXiv:hep-th/0405085, to appear in Comm.Math.Phys.
4. U. Lindström, "Generalized N = (2,2) supersymmetric non-linear sigma models," Phys. Lett. B **587** (2004) 216 [arXiv:hep-th/0401100].

INTEGRABILITY OF SUPERCONFORMAL FIELD THEORY AND SUSY N=1 KDV

ANTON M. ZEITLIN
Department of High Energy Physics, Physics Faculty,
St. Petersburg State University, Ul'yanovskaja 1, Petrodvoretz,
St.Petersburg, 198904, Russia

Abstract. The quantum SUSY N=1 hierarchy based on $sl(2|1)^{(2)}$ twisted affine superalgebra is considered. The construction of the corresponding Baxter's Q-operators and fusion relations is outlined. The relation with the superconformal field theory is discussed.

One of the most famous integrable systems (IS) is the Korteweg-de Vries hierarchy. It is related with the superconformal field theory because its Poisson brackets give the Virasoro algebra and the involutive family of integrals of motion (IM) providing the integrability of the conformal field theory (CFT). Since the late 1980-s the supersymmetric and fermionic extensions of the KdV system have been known (see e.g. [1], [2], [3] and references therein), which in turn are related with superconformal field theory (SCFT). During the following years it was extensively studied both on the classical and the quantum level.

However, up to the present nobody has applied the most successful method in the theory of integrable systems, the so-called quantum inverse scattering method (QISM) to these IS. In this short paper we demonstrate some algebraic tools giving possibility to study SUSY N=1 KdV via QISM.

1. RTT-relation

The SUSY N=1 KdV model is related to the following L-operator:

$$\mathcal{L}_F = D_{u,\theta} - D_{u,\theta}\Phi h_\alpha - (e_{\delta-\alpha} + e_\alpha),$$

where h_α, $e_{\delta-\alpha} \equiv e_{\alpha_0}$, e_α are the Chevalley generators of twisted affine Lie superalgebra $sl(2|1)^{(2)} \cong osp(2|2)^{(2)} \cong C(2)^{(2)}$, $D_{u,\theta} = \partial_\theta + \theta\partial_u$ is a superderivative, the variable u lies on a cylinder of circumference 2π, θ is a Grassmann variable, $\Phi(u,\theta) = \phi(u) - \frac{i}{\sqrt{2}}\theta\xi(u)$ is a bosonic superfield with the following Poisson brackets: $\{D_{u,\theta}\Phi(u,\theta),$ $D_{u',\theta'}\Phi(u',\theta')\} = D_{u,\theta}(\delta(u-u')(\theta-\theta'))$. Making a gauge transformation of the L-operator we obtain a new superfield $\mathcal{U}(u,\theta) \equiv D_{u,\theta}\Phi(u,\theta)\partial_u\Phi(u,\theta) - D^3_{u,\theta}\Phi(u,\theta)$ $=-\theta U(u) - i\alpha(u)/\sqrt{2}$, where U and α generate the superconformal algebra under the

L. Baulied et al. (eds.), String Theory: From Gauge Interactions to Cosmology, 393–396.

Poisson brackets:

$$\begin{aligned}\{U(u),U(v)\} &= \delta'''(u-v)+2U'(u)\delta(u-v)+4U(u)\delta'(u-v),\\ \{U(u),\alpha(v)\} &= 3\alpha(u)\delta'(u-v)+\alpha'(u)\delta(u-v),\\ \{\alpha(u),\alpha(v)\} &= 2\delta''(u-v)+2U(u)\delta(u-v).\end{aligned}$$

The SUSY N=1 KdV system has an infinite number of conservation laws and the first nontrivial one gives the SUSY N=1 KdV equation: $\mathcal{U}_t = -\mathcal{U}_{uuu} + 3(\mathcal{U}D_{u,\theta}\mathcal{U})_u$. The integrals of motion are generated by the logarithm of the supertrace of the corresponding monodromy matrix, which has the following form:

$$\mathbf{M}^{(cl)} = e^{2\pi i p h_{\alpha_1}} P\exp\int_0^{2\pi} du\Big(\frac{i}{\sqrt{2}}\xi(u)e^{-\phi(u)}e_{\alpha_1}$$

$$-\frac{i}{\sqrt{2}}\xi(u)e^{\phi(u)}e_{\alpha_0} - e^2_{\alpha_1}e^{-2\phi(u)} - e^2_{\alpha_0}e^{2\phi(u)} - [e_{\alpha_1}, e_{\alpha_0}]\Big).$$

Its quantum generalization can be represented in the quantum P-exponential form (for the explanation of this notion see below and [3] for details):

$$\mathbf{M}^{(q)} = e^{2\pi i P h_{\alpha_1}} Pexp^{(q)}\int_0^{2\pi} du(W_-(u)e_{\alpha_1} + W_+(u)e_{\alpha_0}).$$

Vertex operators $W_\pm$ are defined in the following way $W_\pm(u) = \int d\theta : e^{\pm\Phi(u,\theta)} := \mp\frac{i}{\sqrt{2}}\xi(u) : e^{\pm\phi(u)} :$. The universal R-matrix with the lower Borel subalgebra represented by $(q^{-1}-q)^{-1}\int_0^{2\pi} duW_\pm(u)$ is equal to $\mathbf{L} = e^{-\pi i P h_{\alpha_1}}\mathbf{M}^{(q)}$. Due to this fact $\mathbf{L}$ satisfies the RTT-relation:

$$\mathbf{R}_{ss'}\Big(\mathbf{L}_s\otimes\mathbf{I}\Big)\Big(\mathbf{I}\otimes\mathbf{L}_{s'}\Big) = (\mathbf{I}\otimes\mathbf{L}_{s'})\Big(\mathbf{L}_s\otimes\mathbf{I}\Big)\mathbf{R}_{ss'},$$

where s, s' mean that the corresponding object is considered in some representation of $C_q(2)^{(2)}$. Thus the supertraces of the monodromy matrix ("transfer matrices") $\mathbf{t}_s = str\mathbf{M}_s$ commute, providing the quantum integrability. It is very useful to consider the evaluation representations of $C_q(2)^{(2)}$, $\rho_s(\lambda)$, where now the symbol s means integer and half-integer numbers. Denoting $\rho_s(\lambda)(\mathbf{M})$ as $\mathbf{M}_s(\lambda)$ we find that $\mathbf{t}_s(\lambda) = str\mathbf{M}_s(\lambda)$ commute: $[\mathbf{t}_s(\lambda), \mathbf{t}_{s'}(\mu)] = 0$. The expansion of $\log(\mathbf{t}_{\frac{1}{2}}(\lambda))$ in λ (the transfer matrix in the fundamental 3-dimensional representation) is believed to give us as coefficients the local IM, the quantum counterparts of the mentioned IM of SUSY $N = 1$ KdV.

2. The Q-operator

Using the super q-oscillator representations of the upper Borel subalgebra of the quantum affine superalgebra $C_q(2)^{(2)}$ we define the $\mathbf{Q}_\pm$ operators (see [4] for details). The transfer-matrices in different evaluation representations can be expressed in such a way:

$$2cos(\pi P)\mathbf{t}_s(\lambda) = \mathbf{Q}_+(q^{s+\frac{1}{4}}\lambda)\mathbf{Q}_-(q^{-s-\frac{1}{4}}\lambda) + \mathbf{Q}_+(q^{-s-\frac{1}{4}}\lambda)\mathbf{Q}_-(q^{s+\frac{1}{4}}\lambda),$$

where s runs over integer and half-integer nonnegative numbers. $\mathbf{Q}_{\pm}$ operators satisfy quantum super-Wronskian relation:

$$2cos(\pi P) = \mathbf{Q}_{+}(q^{\frac{1}{4}}\lambda)\mathbf{Q}_{-}(q^{-\frac{1}{4}}\lambda) + \mathbf{Q}_{+}(q^{-\frac{1}{4}}\lambda)\mathbf{Q}_{-}(q^{\frac{1}{4}}\lambda).$$

One should note, that we use only $4s + 1$-dimensional "$osp(1|2)$-induced" representations (sometimes called atypical) of $C(2)^{(2)}$. It allows, however, to construct the fusion relations, see below. To construct the relations like Baxter's ones we introduce additional "quarter"-operators, constructed "by hands" from the $\mathbf{Q}$-operators:

$$2cos(\pi P)\mathbf{t}_{\frac{k}{4}}(\lambda) = \mathbf{Q}_{+}(q^{\frac{k}{4}+\frac{1}{4}}\lambda)\mathbf{Q}_{-}(q^{-\frac{k}{4}-\frac{1}{4}}\lambda) - \mathbf{Q}_{+}(q^{-\frac{k}{4}-\frac{1}{4}}\lambda)\mathbf{Q}_{-}(q^{\frac{k}{4}+\frac{1}{4}}\lambda)$$

for odd integer k. The Baxter's relations are:

$$\mathbf{t}_{\frac{1}{4}}(\lambda)\mathbf{Q}_{\pm}(\lambda) = \pm\mathbf{Q}_{\pm}(q^{\frac{1}{2}}\lambda) \mp \mathbf{Q}_{\pm}(q^{-\frac{1}{2}}\lambda),$$

$$\mathbf{t}_{\frac{1}{2}}(q^{\frac{1}{4}}\lambda)\mathbf{Q}_{\pm}(\lambda) = \mathbf{t}_{\frac{1}{4}}(q^{\frac{1}{2}}\lambda)\mathbf{Q}_{\pm}(q^{-\frac{1}{2}}\lambda) + \mathbf{Q}_{\pm}(q\lambda).$$

The fusion relations have the following form very similar to the $A_1^{(1)}$ case:

$$\mathbf{t}_j(q^{\frac{1}{4}}\lambda)\mathbf{t}_j(q^{-\frac{1}{4}}\lambda) = \mathbf{t}_{j+\frac{1}{4}}(\lambda)\mathbf{t}_{j-\frac{1}{4}}(\lambda) + (-1)^{4j}.$$

But they are only "fusion-like" because the "quarter"-operators do not seem to correspond to any representation of $C(2)^{(2)}$. The truncation of these relations for different values of q, being the root of unity: $q^N = \pm 1$, $N \in \mathbb{Z}$, $N > 0$ has the following form:

$$\mathbf{t}_{\frac{N}{2}}(\lambda) + \mathbf{t}_{\frac{N}{2}-\frac{1}{2}}(\lambda) = 2cos(\pi NP).$$

In the case when $p = \frac{l+1}{N}$, where $l \geq 0$, $l \in \mathbb{Z}$ there exists an additional number of truncations:

$$\mathbf{t}_{\frac{N}{2}-\frac{1}{4}}(\lambda) = 0, \quad \mathbf{t}_{\frac{N}{2}}(\lambda) = \mathbf{t}_{\frac{N}{2}-\frac{1}{2}}(\lambda) = (-1)^{l+1},$$

$$\mathbf{t}_{\frac{N}{2}-\frac{1}{2}-s}(\lambda q^{\frac{N}{2}}) = (-1)^{4s}\mathbf{t}_s(\lambda)(-1)^{l+1}.$$

These relations allow us to rewrite the fusion relation system in the Thermodynamic Bethe Ansatz Equations of D_{2N} type.

3. Conclusions

In this paper we studied algebraic relations arising from the integrable structure of CFT provided by the SUSY N=1 KdV hierarchy. The construction of the $\mathbf{Q}$-operator as a "transfer"-matrix corresponding to the infinite-dimensional q-oscillator representation could be also applied to the lattice models. The relations like Baxter's and fusion ones will be also valid in the lattice case because they depend only on the decomposition properties of the representations.

In the following we also plan to study the quantization of $N > 1$ SUSY KdV hierarchies, related with super-W conformal/topological integrable field theories.

Acknowledgements

I am very grateful to my supervisor Prof. P. Kulish. It is a pleasure to thank Prof. M. Semenov-Tian-Shansky and F. Smirnov for useful discussions and Prof. L. Baulieu, B. Pioline and LPTHE, Univ. Paris 6 for support and hospitality. This work was supported by Dynasty Foundation.

References

1. P.P. Kulish, A.M. Zeitlin, Zapiski Nauchn. Seminarov POMI, 291 (2002) 185 (in Russian), English translation in: Journal of Mathematical Sciences, Kluwer Academic Publishers, hep-th/0312158.
2. P.P. Kulish, A.M. Zeitlin, Phys. Lett. **B 581** (2004) 125, hep-th/0312159; P.P. Kulish, A.M. Zeitlin, Theor. Math. Phys. 142 (2005), in press.
3. P.P. Kulish, A.M. Zeitlin, Phys. Lett. **B 597** (2004) 229, hep-th/0407154.
4. P.P. Kulish, A. M. Zeitlin, Nucl. Phys. **B**, 2005, in press.

Appendix

LIST OF SPEAKERS

Arkani-Hamed Nima, arkani@carnot.harvard.edu,
Dept. of Physics, Harvard University, Cambridge, MA 02138, USA
Baulieu Laurent, baulieu@lpthe.jussieu.fr,
L.P.T.H.E., Universités Paris VI et Paris VII, Boîte 126,
4 Place Jussieu, 75252 Paris Cedex 05, France
Bernard Denis, dbernard@spht.saclay.cea.fr,
SPhT-Saclay, CEA/Saclay, 91191 Gif-sur-Yvette Cedex, France
Brandenberger Robert, rhb@het. brown. edu,
Box 1843, Brown University, Providence, RI 02912-1843, USA
De Boer Jan, jdeboer@science.uva.nl,
Institute for Theoretical Physics
Valckenierstraat 65, 1018 XE Amsterdam, The Netherlands
Dijkgraaf Robbert, rhd@science.uva.nl,
Institute for Theoretical Physics
Valckenierstraat 65, 1018 XE Amsterdam, The Netherlands
Elitzur Shmuel, elitzur@vms.huji.ac.il,
Racah Institute of Physics, Hebrew University, Jerusalem, 91904 Israel
Morozov Alexei, morozov@itep.ru,
ITEP, Bol. Cheremushkinskaya, 25 117218, Moscow, Russia
Okun Lev, okun@itep.ru,
ITEP, Bol. Cheremushkinskaya, 25 117218, Moscow, Russia
Ooguri Hirosi, ooguri@theory.caltech.edu,
Caltech, Pasadena, CA 91125, USA
Pioline Boris, pioline@lpthe.jussieu.fr,
L.P.T.H.E., Universités Paris VI et Paris VII, Boîte 126,
4 Place Jussieu, 75252 Paris Cedex 05, France
Pokorski Stefan, pokorski@fuw.edu.pl,
IFT UW, Hoza 69, PL-00-681 Warszawa, Poland
Polchinski Joseph, joep@kitp.ucsb.edu,
KITP, University of California, Santa Barbara, CA 93106, USA
Rabinovici Eliezer, eliezer@vms.huji.ac.il,
Racah Institute of Physics, Hebrew University, Jerusalem, 91904 Israel
Tseytlin Arkady, tseytlin@mps.ohio-state.edu,
Physics Department, The Ohio State University,
Columbus OH43210-1106, USA

LIST OF PARTICIPANTS

Alkalaev Konstantin, alkalaev@lpi.ru,
P.N.Lebedev Physical Institute, 53 Leninsky prospect,
Moscow, 119991 Russia

Angelantonj Carlo, carlo.angelantonj@cern.ch,
Humboldt Univ., Unter den Linden 6, 10099 Berlin, Germany

Anisimov Alexey, anisimov@mppmu.mpg.de,
Max-Planck-Institute, Föhringer Ring 6 80805 München, Germany

Arcioni Giovanni, arcionig@phys.huji.ac.il,
Racah Institute of Physics, Hebrew University, Jerusalem, 91904 Israel

Arvidsson Par, par@fy.chalmers.se,
Institute of Theoretical Physics, Chalmers Univerity of Technology
S-412 96 Gothenburg, Sweden

Ashok Sujay, ashok@physics.rutgers.edu,
Department of Physics k Astronomy, Rutgers University
136 Frelinghuysen Road, Piscataway, New Jersey 08854, USA

Banks Thomas, banks@scipp.ucsc.edu,
University of California, Santa Cruz, 1156 High Street Santa Cruz, CA
95064, USA / Department of Physics & Astronomy, Rutgers University
136 Frelinghuysen Road, Piscataway, New Jersey 08854, USA

Belov Dmitriy, belov@physics.rutgers.edu,
Department of Physics & Astronomy, Rutgers University
136 Frelinghuysen Road, Piscataway, New Jersey 08854, USA

Benakli Karim, kbenakli@lpthe.jussieu.fr,
L.P.T.H.E., Universités Paris VI et Paris VII, Boîte 126,
4 Place Jussieu, 75252 Paris Cedex 05, France

Boels Rutger, rhboels@science.uva.nl,
University of Amsterdam, Postbus 19268, 1000 GG Amsterdam,
The Netherlands

Calcagni Gianluca, calcagni@fis.unipr.it,
Parma University, Via Universit, 12 - 1-43100 Parma, Italy

Ceresole Anna, anna.ceresole@to.infn.it,
Department of Theoretical Physics, University of Turin
Via P. Giuria 1, 10125 Torino, Italy and INFN

Dell'aquila Eleonora, dellaqui@physics.rutgers.edu,
Department of Physics and Astronomy, Rutgers University,
136 Frelinghuysen Road, NJ 08854-8019 Piscataway, USA

L. Baulied et al. (eds.), String Theory: From Gauge Interactions to Cosmology, 401–404.

Durin Bruno, bdurin@lpthe.jussieu.fr,
LPTHE, 4 place Jussieu, tour 24, 5eme étage, boîte 126, 75252 Paris cedex 05, France

Escoda Cristina, ce225@cam.ac.uk,
DAMTP, Centre for Mathematical Sciences, Wilberforce Road, CB3 0WA Cambridge, United Kingdom

Evslin Jarah, jarah@df.unipi.it,
INFN Sezione di Pisa, via Buonarroti, 2, 56127 Pisa, Italy

Falkowski Adam, afalkows@fuw.edu.pl,
Institute of Theoretical Physics, Warsaw University, ul. Hoza 69, 00-681 Warsaw, Poland

Flink Erik, erik.flink@fy.chalmers.se,
Department of Theoretical Physics, Chalmers University of Technology, SE-41296 Göteborg, Sweden

Flournoy Alex, flournoy@physics.technion.ac.il,
Department of Physics, Technion, 32000 Haifa, Israel

Franco Sebastian, sfranco@mit.edu,
Center for Theoretical Physics, Massachusetts Institute of Technology, 77 Massachusetts Avenue, MA 02139 Cambridge, USA

Giryavets Alexander, giryav@stanford.edu,
Department of Physics, Stanford University, Stanford CA 94305, USA

Grana Mariana, mariana@lpt.ens.fr,
LPTENS, Département de Physique de l'ENS, 24, Rue Lhomond, 75005 Paris, France

Gripaios Ben, b.gripaios1@physics.ox.ac.uk,
Theoretical Physics, University of Oxford, 1 Keble Road, OX1 3NP Oxford, United Kingdom

Grunberg Daniel, grunberg@mccme.ru,
KdV Institute, Plantage Muidergr 24, 1018 TV Amsterdam, The Netherlands

Gursoy Umut, umut@mit.edu,
Massachusetts Institute of Technology, 77 Massachussetts Ave, Cambridge MA 02139, USA

Halmagyi Nick, halmagyi@physics.usc.edu,
Department of Physics and Astronomy, University of Southern California, Los Angeles CA 90089, USA

Hartnoll Sean, s.a.hartnoll@damtp.cam.ac.uk,
DAMTP, Centre for Mathematical Sciences, Wilberforce road, CB3 0WA Cambridge, United Kingdom

Heise Rainer, rahe@fy.chalmers.se,
Institute for Theoretical Physics, Chalmers University of Technology and Goteborg University, 41296 Göteborg, Sweden

Herbst Manfred, manfred.herbst@cern.ch,
CERN, Theory Division, 1211 Geneva 23, Switzerland

Hubeny Veronika, veronika@itp.stanford.edu,
Department of Physics,382 Via Pueblo Mall,Stanford CA 94305, USA

Jarv Laur, l.jaerv@tpi.uni-jena.de,
Friedrich-Schiller-Universität Jena, D - 07740 Jena, Germany
Kirsch Ingo, ik@physik.hu-berlin.de,
Humboldt-Univ. Berlin, Unter den Linden 6, 10099 Berlin, Germany
Kleban Matthew, matthew@ias.edu,
Institute for Advanced Study, Einstein Drive, Princeton, NJ 08540, USA
Koerber Paul, koerber@tena4.vub.ac.be,
Vrije Univ. Brussel, Plainlaan 2, 1050 Brussel, Belgium
Kohlprath Emmanuel, emmanuel.kohlprath@cpht.polytechnique.fr,
CPHT, École Polytechnique, F91128 Palaiseau Cedex, France
Kuperstein Stanislav, kupers@post.tau.ac.il,
Tel Aviv University, P.O. Box 39040, Tel Aviv 69978, Israel
Lee Sangmin, sangmin.lee@cern.ch,
CERN, Theory Division, 1211 Geneva 23, Switzerland
Lindstrom Ulf, ulf.lindstrom@teorfys.uu.se,
Department of Theoretical Physics, Box 803, SE-751 08 Uppsala, Sweden
Maillard Tristan, mtristan@itp.phys.ethz.ch,
Eidgenssische Technische Hochschule Zürich,
ETH Zentrum, Rämistrasse 101, CH-8092 Zürich, Switzerland
Maoz Liat, lmaoz@science.uva.nl,
Institute for Theoretical Physics, Amsterdam Valckenierstraat 65,
1018XE Amsterdam, the Netherlands
Naqvi Asad, anaqvi@science.uva.nl,
Institute for Theoretical Physics, Amsterdam Valckenierstraat 65,
1018XE Amsterdam, the Netherlands
Olsson Martin, martin.olsson@teorfys.uu.se,
Department of Theoretical Physics, Box 803, SE-751 08 Uppsala, Sweden
Orlando Domenico, domenico.orlando@cpht.polytechnique.fr,
CPHT, École Polytechnique, F91128 Palaiseau Cedex, France
Ozer Aybike, aybike@maths.tcd.ie,
Trinity College Dublin, College Green, Dublin 2, Ireland
Pakman Ari, pakman@racah.huji.ac.il,
Racah Institute of Physics, Hebrew University, Jerusalem, 91904 Israel
Papadodimas Kyriakos, papadod@fas.harvard.edu,
Faculty of Arts and Sciences, Harvard University,
Cambridge, MA 02138, USA
Park Jongwon, jongwon@theory.uchicago.edu,
Particle Theory Group, Enrico Fermi Institute, University of Chicago,
5640 South Ellis Avenue, Chicago, Illinois 60637, USA
Patir Assaf, assaf.patir@weizmann.ac.il,
Department of Particle Physics, Weizmann Institute of Science,
Rehovot, Israel
Paulot Louis, lpaulot@ulb.ac.be,
Université Libre de Bruxelles, Campus Plaine, boulevard du Triomphe,
Physique Théorique et Mathématique C.P. 231 1050 Bruxelles, Belgium

Prezas Nikolaos, prezas@physik.hu-berlin.de,
Humboldt Univ., Unter den Linden 6, 10099 Berlin, Germany
Rangamani Mukund, mukund@socrates.berkeley.edu,
Physics Dept University of California, Berkeley, CA 94720 and Physics Dept., Harvard University, 17 Oxford St. Cambridge, MA 02138, USA
Roest Diederik, d.roest@phys.rug.nl,
Institute for Theoretical Physics, University of Gröningen,
Nijenborgh 4 9747 AG, Gröningen, The Netherlands
Rotaev Mikhail, mrotaev@mail.ru,
Institute of Theoretical and Experimental Physics (ITEP),
B. Cheremushkinskaya, 25, Moscow, 117259, Russia
Rulik Ksenya, rulik@to.infn.it,
Department of Theoretical Physics, University of Turin
Via P. Giuria 1, 10125 Torino, Italy
Rychkov Vyacheslav, rychkov@science.uva.nl,
University of Amsterdam, Postbus 19268,
1000 GG Amsterdam, The Netherlands
Ryzhov Anton, ryzhovav@brandeis.edu,
Brandeis University, 415 South St. Waltham, MA 02454-9110, USA
Sever Amit, asever@cc.huji.ac.il,
Racah Institute of physics, Hebrew University, Jerusalem, 91904 Israel
Shaynkman Oleg, shayn@lpi.ru,
P.N. Lebedev Physical Institute, 53 Leninsky prospect,
Moscow, 119991 Russia
Sulkowski Piotr, piotr.sulkowski@fuw.edu.pl,
Institute of Theoretical Physics, Warsaw University, ul. Hoa 69,
PL-00-681 Warsaw, Poland
Tanzini Alessandro, tanzini@lpthe.jussieu.fr,
L.P.T.H.E., Universités Paris VI et Paris VII, Boîte 126,
4 Place Jussieu, 75252 Paris Cedex 05, France
Vikman Alexander, vikman@theorie.physik.uni-muenchen.de,
Ludwig-Maximilians-Universitt, Theresienstrafle 37, 80333 München, Germany
Watson Scott, watson@het.brown.edu,
Physics Department, Brown University, Providence RI 02912, USA
Williams Brook, brook@physics.ucsb.edu,
University of California, Santa Barbara, CA 93106, USA
Zabzine Maxim, zabzine@lpthe.jussieu.fr,
L.P.T.H.E., Universités Paris VI et Paris VII, Boîte 126, 4 Place Jussieu, 75252 Paris Cedex 05, France
Zeitlin Anton, zam@math.ipme.ru,
Institute of Problems of Mechanical Engineering, Russian Academy of Sciences, V.O., Bolshoj pr., 61 St. Petersburg, 199178 Russia